Modeling and Optimization in Science and Technologies

Volume 10

About this Series

The book series *Modeling and Optimization in Science and Technologies (MOST)* publishes basic principles as well as novel theories and methods in the fast-evolving field of modeling and optimization. Topics of interest include, but are not limited to: methods for analysis, design and control of complex systems, networks and machines; methods for analysis, visualization and management of large data sets; use of supercomputers for modeling complex systems; digital signal processing; molecular modeling; and tools and software solutions for different scientific and technological purposes. Special emphasis is given to publications discussing novel theories and practical solutions that, by overcoming the limitations of traditional methods, may successfully address modern scientific challenges, thus promoting scientific and technological progress. The series publishes monographs, contributed volumes and conference proceedings, as well as advanced textbooks. The main targets of the series are graduate students, researchers and professionals working at the forefront of their fields.

More information about this series at http://www.springer.com/series/10577

Srikanta Patnaik · Xin-She Yang
Kazumi Nakamatsu
Editors

Nature-Inspired Computing and Optimization

Theory and Applications

 Springer

Editors
Srikanta Patnaik
Department of Computer Science
  and Engineering
SOA University
Bhubaneswar, Odisha
India

Kazumi Nakamatsu
School of Human Science and Environment
University of Hyogo
Himeji
Japan

Xin-She Yang
School of Science and Technology
Middlesex University
London
UK

ISSN 2196-7326   ISSN 2196-7334   (electronic)
Modeling and Optimization in Science and Technologies
ISBN 978-3-319-84522-7   ISBN 978-3-319-50920-4   (eBook)
DOI 10.1007/978-3-319-50920-4

This Springer imprint is published by Springer Nature
The registered company is Springer International Publishing AG
The registered company address is: Gewerbestrasse 11, 6330 Cham, Switzerland

Preface

Nature-inspired computing provides promising and effective approaches for problem solving in optimization, machine intelligence, data mining and resource management. Nature has evolved over millions of years under a variety of challenging environments and can thus provide a rich source of inspiration for designing algorithms and approaches to tackle challenging problems in real-world applications.

The success of these algorithms in applications has increased their popularity in recent years, and active research has also led to the significant increase in the number of algorithms in recent years. It is estimated that about 140 different types of algorithms now exist in the literature, and this number is certainly gradually increasing. Researchers have tried to find inspiration from various sources in nature, such as ants, bees, fish, birds, mammals, plants, physical and chemical systems such as gravity, river systems, waves and pheromone. This leads to a diverse of range of algorithms with different capabilities and different levels of performance.

However, such diversity may also cause confusion and distractions from important research topics. For example, many researchers wonder why such algorithms work and what their mathematical foundations for different search algorithms are. At the moment, it still lacks good theoretical understanding of metaheuristics. In fact, without a good mathematical framework, it is difficult to establish any solid mathematical foundation for analysing such algorithms. Such lack of theoretical analysis, together with different claims of results, it is understandable that misunderstanding and criticism have arisen in the research community concerning some metaheuristic algorithms.

There is a strong need for the whole research community to review carefully the developments concerning metaheuristics and bio-inspired computation so as to identify the key challenges, to inspire further research and to encourage innovative approaches that can help to develop effective tools for tackling hard problems in applications.

This book provides readers a timely snapshot of the state-of-the-art developments in the field of nature-inspired computing and its application in optimization, with the emphasis on both new applications and analysis of algorithms in their implementation context. Therefore, this book is intended as a practice-oriented reference guide for students, researchers and professionals.

Despite the recent developments and the success of these algorithms, there are still some key issues that need further investigation. For example, there still lacks a general mathematical framework for analysing nature-inspired algorithms. In addition, the applications and case studies in the current literature have not focused on the large-scale problems yet. Thus, it is hoped that this book can inspire further research with a special focus on finding solutions for key challenges, including theoretical analysis, benchmarking, performance evaluation and large-scale applications.

We would like to thank the review board members for their constructive comments on the manuscripts of all the chapters during the peer-review process, which have greatly helped to ensure and improve the quality of the book. We also would like to thank the editors and staff at Springer for their help and professionalism. Last but not least, we thank our families for their help and support.

Bhubaneswar, India   Srikanta Patnaik
London, UK   Xin-She Yang
Himeji, Japan   Kazumi Nakamatsu
October 2016

Contents

The Nature of Nature: Why Nature-Inspired Algorithms Work   1
David Green, Aldeida Aleti and Julian Garcia
1   Introduction: How Nature Works   1
2   The Nature of Nature   2
   2.1   Fitness Landscape   3
   2.2   Graphs and Phase Changes   4
3   Nature-Inspired Algorithms   6
   3.1   Genetic Algorithm   6
   3.2   Ant Colony Optimization   7
   3.3   Simulated Annealing   7
   3.4   Convergence   8
4   Dual-Phase Evolution   9
   4.1   Theory   9
   4.2   GA   9
   4.3   Ant Colony Optimization   11
   4.4   Simulated Annealing   12
5   Evolutionary Dynamics   13
   5.1   Markov Chain Models   13
   5.2   The Replicator Equation   15
6   Generalized Local Search Machines   18
   6.1   The Model   18
   6.2   SA   19
   6.3   GA   20
   6.4   ACO   21
   6.5   Discussion   21
7   Conclusion   22
References   24

**Multimodal Function Optimization Using an Improved
Bat Algorithm in Noise-Free and Noisy Environments**.   29
Momin Jamil, Hans-Jürgen Zepernick and Xin-She Yang
1   Introduction   30
2   Improved Bat Algorithm   31
3   IBA for Multimodal Problems   34
     3.1   Parameter Settings   34
     3.2   Test Functions   35
     3.3   Numerical Results.   35
4   Performance Comparison of IBA with Other Algorithms.   41
5   IBA Performance in AWGN   42
     5.1   Numerical Results.   44
6   Conclusions   47
References   47

**Multi-objective Ant Colony Optimisation in Wireless
Sensor Networks**   51
Ansgar Kellner
1   Introduction   51
2   Multi-objective Combinatorial Optimisation Problems   52
     2.1   Combinatorial Optimisation Problems.   52
     2.2   Multi-objective Combinatorial Optimisation Problems   53
     2.3   Pareto Optimality   54
     2.4   Decision-Making.   55
     2.5   Solving Combinatorial Optimisation Problems   59
3   Multi-objective Ant Colony Optimisation.   60
     3.1   Origins   60
     3.2   Multi-objective Ant Colony Optimisation   66
4   Applications of MOACO Algorithms in WSNs   72
5   Conclusion   74
References   74

**Generating the Training Plans Based on Existing Sports Activities
Using Swarm Intelligence**   79
Iztok Fister Jr. and Iztok Fister
1   Introduction   79
2   Artificial Sports Trainer   81
3   Generating the Training Plans   82
     3.1   Preprocessing   84
     3.2   Optimization Process   87
4   Experiments   91
5   Conclusion with Future Ideas.   93
References   93

Limiting Distribution and Mixing Time for Genetic Algorithms 95
S. Alwadani, F. Mendivil and R. Shonkwiler
1   Introduction 95
2   Preliminaries. 96
      2.1   Random Search and Markov Chains. 98
      2.2   Boltzmann Distribution and Simulated Annealing. 99
3   Expected Hitting Time as a Means of Comparison 101
      3.1   "No Free Lunch" Considerations 104
4   The Holland Genetic Algorithm. 105
5   A Simple Genetic Algorithm 109
6   Shuffle-Bit GA. 115
      6.1   Results 117
      6.2   Estimate of Expected Hitting Time. 118
7   Discussion and Future Work 120
References 121

**Permutation Problems, Genetic Algorithms, and Dynamic
Representations**. 123
James Alexander Hughes, Sheridan Houghten and Daniel Ashlock
1   Introduction 123
2   Problem Descriptions 125
      2.1   Bin Packing Problem 125
      2.2   Graph Colouring Problem. 125
      2.3   Travelling Salesman Problem 126
3   Previous Work on Small Travelling Salesman Problem Instances 127
4   Algorithms. 128
      4.1   2-Opt 128
      4.2   Lin–Kernighan 129
      4.3   Genetic Algorithm Variations 129
      4.4   Representation 134
5   Experimental Design 135
      5.1   Bin Packing Problem 135
      5.2   Graph Colouring Problem. 136
      5.3   Travelling Salesman Problem 137
6   Results and Discussion. 138
      6.1   Bin Packing Problem 138
      6.2   Graph Colouring Problem. 142
      6.3   Travelling Salesman Problem 144
7   Conclusions 147
References 148

**Hybridization of the Flower Pollination Algorithm—A Case Study in
the Problem of Generating Healthy Nutritional Meals for Older
Adults** 151
Cristina Bianca Pop, Viorica Rozina Chifu, Ioan Salomie,
Dalma Szonja Racz and Razvan Mircea Bonta
1   Introduction 152
2   Background 153
    2.1   Optimization Problems 153
    2.2   Meta-Heuristic Algorithms 154
3   Literature Review 156
4   Problem Definition 158
    4.1   Search Space and Solution Representation 158
    4.2   Fitness Function 159
    4.3   Constraints 161
5   Hybridizing the Flower Pollination Algorithm for Generating
    Personalized Menu Recommendations 162
    5.1   Hybrid Flower Pollination-Based Model 162
    5.2   Flower Pollination-Based Algorithms for Generating
          Personalized Menu Recommendations 164
    5.3   The Iterative Stage of the Hybrid Flower Pollination-Based
          Algorithm for Generating Healthy Menu Recommendations 166
6   Performance Evaluation 168
    6.1   Experimental Prototype 168
    6.2   Test Scenarios 171
    6.3   Setting the Optimal Values of the Algorithms' Adjustable
          Parameters 172
    6.4   Comparison Between the Classical and Hybrid Flower
          Pollination-Based Algorithms 179
7   Conclusions 181
References 182

**Nature-inspired Algorithm-based Optimization for Beamforming
of Linear Antenna Array System** 185
Gopi Ram, Durbadal Mandal, S.P. Ghoshal and Rajib Kar
1   Introduction 186
2   Problem Formulation 187
3   Flower Pollination Algorithm [55] 189
    3.1   Global Pollination: 190
    3.2   Local Pollination: 190
    3.3   Pseudo-code for FPA: 191
4   Simulation Results 193
    4.1   Optimization of Hyper-Beam by Using FPA 194
    4.2   Comparisons of Accuracies Based on t test 196
5   Convergence Characteristics of Different Algorithms 211

6   Conclusion 212
7   Future Research Topics 212
References 212

**Multi-Agent Optimization of Resource-Constrained Project
Scheduling Problem Using Nature-Inspired Computing** 217
Pragyan Nanda, Sritam Patnaik and Srikanta Patnaik
1   Introduction 217
     1.1   Multi-agent System................................ 218
     1.2   Scheduling 219
     1.3   Nature-Inspired Computing........................ 220
2   Resource-Constrained Project Scheduling Problem................ 220
3   Various Nature-Inspired Computation Techniques for RCPSP 222
     3.1   Particle Swarm Optimization (PSO) 223
     3.2   Particle Swarm Optimization (PSO) for RCPSP 224
     3.3   Ant Colony Optimization (ACO) 227
     3.4   Ant Colony Optimization (ACO) for RCPSP 229
     3.5   Shuffled Frog-Leaping Algorithm (SFLA) 230
     3.6   Shuffled Frog-Leaping Algorithm (SFLA) for RCPSP 232
     3.7   Multi-objective Invasive Weed Optimization 235
     3.8   Multi-objective Invasive Weed Optimization for MRCPSP...... 235
     3.9   Discrete Flower Pollination......................... 236
     3.10   Discrete Flower Pollination for RCPSP 237
     3.11   Discrete Cuckoo Search 237
     3.12   Discrete Cuckoo Search for RCPSP 238
     3.13   Multi-agent Optimization Algorithm (MAOA) 238
4   Proposed Approach 240
     4.1   RCPSP for Retail Industry 240
     4.2   Cooperative Hunting Behaviour of Lion Pride 240
5   A Lion Pride-Inspired Multi-Agent System-Based Approach
     for RCPSP.. 242
6   Conclusion.. 244
References 244

**Application of Learning Classifier Systems to Gene Expression
Analysis in Synthetic Biology** 247
Changhee Han, Kenji Tsuge and Hitoshi Iba
1   Introduction 248
2   Learning Classifier Systems: Creating Rules that Describe
     Systems 249
     2.1   Basic Components 250
     2.2   Michigan- and Pittsburgh-style LCS 250
3   Examples of LCS...................................... 251
     3.1   Minimal Classifier Systems......................... 251

3.2   Zeroth-level Classifier Systems............................ 252
3.3   Extended Classifier Systems.............................. 254
4   Synthetic Biology: Designing Biological Systems 255
4.1   The Synthetic Biology Design Cycle 255
4.2   Basic Biological Parts................................... 256
4.3   DNA Construction 257
4.4   Future Applications..................................... 257
5   Gene Expression Analysis with LCS 259
6   Optimization of Artificial Operon Structure 261
7   Optimization of Artificial Operon Construction
by Machine Learning.. 262
7.1   Introduction 262
7.2   Artificial Operon Model................................ 262
7.3   Experimental Framework 263
7.4   Results 266
7.5   Conclusion 270
8   Summary 272
References 272

**Ant Colony Optimization for Semantic Searching of Distributed
Dynamic Multiclass Resources**................................ 277
Kamil Krynicki and Javier Jaen
1   Introduction 277
2   P2p Search Strategies....................................... 279
3   Nature-Inspired Ant Colony Optimization 284
4   Nature-Inspired Strategies in Dynamic Networks................. 287
4.1   Network Dynamism Inefficiency.......................... 288
4.2   Solution Framework 289
4.3   Experimental Evaluation................................ 290
5   Nature-Inspired Strategies of Semantic Nature 292
5.1   Semantic Query Inefficiency............................. 292
5.2   Solution Framework 293
5.3   Experimental Evaluation................................ 298
6   Conclusions and Future Developments......................... 301
References 302

**Adaptive Virtual Topology Control Based on Attractor
Selection**... 305
Yuki Koizumi, Shin'ichi Arakawa and Masayuki Murata
1   Introduction 306
2   Related Work.. 308
3   Attractor Selection 308
3.1   Concept of Attractor Selection 309
3.2   Cell Model 309

    3.3   Mathematical Model of Attractor Selection.   310
4   Virtual Topology Control Based on Attractor Selection   312
    4.1   Virtual Topology Control   312
    4.2   Overview of Virtual Topology Control Based on Attractor
           Selection   312
    4.3   Dynamics of Virtual Topology Control.   314
    4.4   Attractor Structure.   316
    4.5   Dynamic Reconfiguration of Attractor Structure   317
5   Performance Evaluation   318
    5.1   Simulation Conditions.   318
    5.2   Dynamics of Virtual Topology Control Based on Attractor
           Selection   321
    5.3   Adaptability to Node Failures   323
    5.4   Effects of Noise Strength   324
    5.5   Effects of Activity.   324
    5.6   Effects of Reconfiguration Methods of Attractor Structure   325
6   Conclusion   326
References   327

CBO-Based TDR Approach for Wiring Network Diagnosis   329
Hamza Boudjefdjouf, Francesco de Paulis, Houssem Bouchekara,
Antonio Orlandi and Mostafa K. Smail
1   Introduction   330
2   The Proposed TDR-CBO-Based Approach.   332
    2.1   Problem Formulation   332
    2.2   The Forward Model   333
    2.3   Colliding Bodies Optimization (CBO)   337
3   Applications and Results   339
    3.1   The Y-Shaped Wiring Network   340
    3.2   The YY-shaped Wiring Network   344
4   Conclusion   347
References   348

**Morphological Filters: An Inspiration from Natural Geometrical
Erosion and Dilation**.   349
Mahdi Khosravy, Neeraj Gupta, Ninoslav Marina, Ishwar K. Sethi
and Mohammad Reza Asharif
1   Natural Geometrical Inspired Operators   350
2   Mathematical Morphology   351
    2.1   Morphological Filters   352
3   Morphological Operators and Set Theory.   353
    3.1   Sets and Corresponding Operators   354
    3.2   Basic Properties for Morphological Operators.   356
    3.3   Set Dilation and Erosion.   357

3.4 A Geometrical Interpretation of Dilation
and Erosion Process 359
3.5 Direct Effect of Edges and Borders on the Erosion
and Dilation 360
3.6 Closing and Opening 363
3.7 A Historical Review to Definitions and Notations 367
4 Practical Interpretation of Binary Opening and Closing 369
5 Morphological Operators in Grayscale Domain 370
5.1 Basic Morphological Operators in Multivalued Function
Domain. 370
5.2 Dilation and Erosion of Multivalued Functions. 374
5.3 Two Forms of Presentation for Dilation
and Erosion Formula. 375
6 Opening and Closing of Multivalued Functions 376
7 Interpretation and Intuitive Understanding of Morphological
Filters in Multivalued Function Domain. 377
8 Conclusion 379
References 379

Brain Action Inspired Morphological Image Enhancement 381
Mahdi Khosravy, Neeraj Gupta, Ninoslav Marina, Ishwar K. Sethi
and Mohammad Reza Asharif
1 Introduction 382
2 Human Visual Perception. 383
3 Visual Illusions 384
4 Visual Illusions 386
4.1 Rotating Snakes 388
5 Mach Bands Illusion 393
6 Image Enhancement Inspiration from Human Visual Illusion. 393
7 Morphological Image Enhancement Based on Visual Illusion 394
8 Results and Discussion. 398
9 Summary 403
References 406

**Path Generation for Software Testing: A Hybrid Approach Using
Cuckoo Search and Bat Algorithm** 409
Praveen Ranjan Srivastava
1 Introduction 409
2 Related Work. 410
3 Motivational Algorithm 411
3.1 Cuckoo Search Algorithm. 411
3.2 Bat Algorithm [12] 412
4 Proposed Algorithm 414

5   Path Sequence Generation and Prioritization 416
6   Analysis of Proposed Algorithm............................. 421
7   Conclusions and Future Scope 422
References 423

An Improved Spider Monkey Optimization for Solving a Convex
Economic Dispatch Problem.................................. 425
Ahmed Fouad Ali
1   Introduction 425
2   Related Work... 426
3   Economic Dispatch Problem 427
     3.1   Problem Constraints 427
     3.2   Penalty Function................................. 428
4   Social Behavior and Foraging of Spider Monkeys 429
     4.1   Fission–Fusion Social Behavior 429
     4.2   Social Organization and Behavior...................... 429
     4.3   Communication of Spider Monkeys 430
     4.4   Characteristic of Spider Monkeys 430
     4.5   The Standard Spider Monkey Optimization Algorithm 430
     4.6   Spider Monkey Optimization Algorithm.................. 434
5   Multidirectional Search Algorithm 436
6   The Proposed MDSMO Algorithm........................... 439
7   Numerical Experiments 439
     7.1   Parameter Setting 439
     7.2   Six-Generator Test System with System Losses 440
     7.3   The General Performance of the Proposed MDSMO with
           Economic Dispatch Problem.......................... 441
     7.4   MDSMO and Other Algorithms 441
8   Conclusion and Future Work 443
References 446

Chance-Constrained Fuzzy Goal Programming with Penalty
Functions for Academic Resource Planning in University Management
Using Genetic Algorithm 449
Bijay Baran Pal, R. Sophia Porchelvi and Animesh Biswas
1   Introduction 449
2   FGP Problem Formulation 453
     2.1   Membership Function Characterization................... 453
     2.2   Deterministic Equivalents of Chance Constraints 454
3   Formulation of Priority Based FGP Model..................... 456
     3.1   Euclidean Distance Function for Priority
           Structure Selection 457
4   FGP Model with Penalty Functions 458
     4.1   Penalty Function Description 458

    4.2   Priority Based FGP Model with Penalty Functions............   460
    4.3   GA Scheme for FGP Model   460
5   FGP Formulation of the Problem   461
    5.1   Definitions of Decision Variables and Parameters............   461
    5.2   Descriptions of Fuzzy Goals and Constraints   462
6   A Case Example...   464
    6.1   An Illustration for Performance Comparison.................   469
7   Conclusions   471
References   472

Swarm Intelligence: A Review of Algorithms   475
Amrita Chakraborty and Arpan Kumar Kar
1   Introduction   476
2   Research Methodology.....................................   477
3   Insect-Based Algorithms....................................   478
    3.1   Ant Colony Optimization Algorithm.....................   478
    3.2   Bee-Inspired Algorithms.............................   480
    3.3   Firefly-Based Algorithms   481
    3.4   Glow-Worm-Based Algorithms.........................   483
4   Animal-Based Algorithms   484
    4.1   Bat-Based Algorithm   484
    4.2   Monkey-Based Algorithm............................   485
    4.3   Lion-Based Algorithm   486
    4.4   Wolf-Based Algorithm   486
5   Future Research Directions..................................   487
6   Conclusions   488
References   488

Contributors

Aldeida Aleti Faculty of Information Technology, Monash University, Clayton, Australia

Ahmed Fouad Ali Faculty of Computers and Informatics, Department of Computer Science, Suez Canal University, Ismailia, Egypt

S. Alwadani Acadia University, Wolfville, NS, Canada

Shin'ichi Arakawa Osaka University, Osaka, Japan

Mohammad Reza Asharif Faculty of Engineering, Information Department, University of the Ryukyus, Okinawa, Japan

Daniel Ashlock University of Guelph, Guelph, ON, Canada

Animesh Biswas Department of Mathematics, University of Kalyani, Kalyani, West Bengal, India

Razvan Mircea Bonta Computer Science Department, Technical University of Cluj-Napoca, Cluj-Napoca, Romania

Houssem Bouchekara Constantine Electrical Engineering Laboratory, LEC, Department of Electrical Engineering, University of Constantine 1, Constantine, Algeria

Hamza Boudjefdjouf Constantine Electrical Engineering Laboratory, LEC, Department of Electrical Engineering, University of Constantine 1, Constantine, Algeria

Amrita Chakraborty Department of Electronics and Telecommunication Engineering, Jadavpur University, Kolkata, West Bengal, India

Viorica Rozina Chifu Computer Science Department, Technical University of Cluj-Napoca, Cluj-Napoca, Romania

Iztok Fister Faculty of Electrical Engineering and Computer Science, University of Maribor, Maribor, Slovenia

Iztok Fister Jr. Faculty of Electrical Engineering and Computer Science, University of Maribor, Maribor, Slovenia

Julian Garcia Faculty of Information Technology, Monash University, Clayton, Australia

S.P. Ghoshal Department of ECE, NIT Durgapur, India

David Green Faculty of Information Technology, Monash University, Clayton, Australia

Neeraj Gupta Faculty of Machine Intelligence and Robotics (MIR), University of Information Science and Technology, Ohrid, Republic of Macedonia

Changhee Han Graduate School of Information Science and Technology, The University of Tokyo, Tokyo, Japan

Sheridan Houghten Brock University, St. Catharines, ON, Canada

James Alexander Hughes Brock University, St. Catharines, ON, Canada

Hitoshi Iba Graduate School of Information Science and Technology, The University of Tokyo, Tokyo, Japan

Javier Jaen DSIC, Universitat Politècnica de València, Valencia, Spain

Momin Jamil Harman/Becker Automotive Systems GmbH, Karlsbad, Germany

Arpan Kumar Kar Information Systems, DMS, Indian Institute of Technology, Delhi, New Delhi, India

Rajib Kar Department of ECE, NIT Durgapur, India

Ansgar Kellner Institute of System Security, Technische Universität Braunschweig, Braunschweig, Germany

Mahdi Khosravy Faculty of Computers Networks and Security (CNS), University of Information Science and Technology, Ohrid, Republic of Macedonia

Yuki Koizumi Osaka University, Osaka, Japan

Kamil Krynicki DSIC, Universitat Politècnica de València, Valencia, Spain

Durbadal Mandal Department of ECE, NIT Durgapur, India

Ninoslav Marina University of Information Science and Technology, Ohrid, Republic of Macedonia

F. Mendivil Acadia University, Wolfville, NS, Canada

Masayuki Murata Osaka University, Osaka, Japan

Pragyan Nanda Faculty of Engineering and Technology, Department of Computer Science and Engineering, SOA University, Bhubaneswar, Odisha, India

Antonio Orlandi UAq EMC Laboratory, Department of Industrial and Information Engineering and Economics, University of L'Aquila, L'Aquila, Italy

Bijay Baran Pal Department of Mathematics, University of Kalyani, Kalyani, West Bengal, India

Srikanta Patnaik Faculty of Engineering and Technology, Department of Computer Science and Engineering, SOA University, Bhubaneswar, Odisha, India

Sritam Patnaik Department of Electrical and Computer Engineering, School of Computer Engineering, National University of Singapore, Singapore, Singapore

Francesco de Paulis UAq EMC Laboratory, Department of Industrial and Information Engineering and Economics, University of L'Aquila, L'Aquila, Italy

Cristina Bianca Pop Computer Science Department, Technical University of Cluj-Napoca, Cluj-Napoca, Romania

R. Sophia Porchelvi Department of Mathematics, ADM College for Women, Nagapattinam, India

Dalma Szonja Racz Computer Science Department, Technical University of Cluj-Napoca, Cluj-Napoca, Romania

Gopi Ram Department of ECE, NIT Durgapur, India

Ioan Salomie Computer Science Department, Technical University of Cluj-Napoca, Cluj-Napoca, Romania

Ishwar K. Sethi Department of Computer Science and Engineering, Oakland University, Rochester, Oakland, MI, USA

R. Shonkwiler Georgia Tech, Atlanta, Georgia

Mostafa K. Smail Institut Polytechnique des Sciences Avancées (IPSA), Ivry-sur-Seine, France

Praveen Ranjan Srivastava Information System and System Area, Indian Institute of Management (IIM) Rohtak, Rohtak, India

Kenji Tsuge Institute for Advanced Biosciences, Keio University, Tsuruoka, Yamagata, Japan

Xin-She Yang Middlesex University, School of Science and Technology, London, UK

Hans-Jürgen Zepernick Blekinge Institute of Technology, Karlskrona, Sweden

Members of Review Board

The Nature of Nature: Why Nature-Inspired Algorithms Work

David Green, Aldeida Aleti and Julian Garcia

Abstract Nature has inspired many algorithms for solving complex problems. Understanding how and why these natural models work leads not only to new insights about nature, but also to an understanding of deep relationships between familiar algorithms. Here, we show that network properties underlie and define a whole family of nature-inspired algorithms. In particular, the network defined by neighbourhoods within landscapes (real or virtual) underlies the searches and phase transitions mediate between local and global search. Three paradigms drawn from computer science—dual-phase evolution, evolutionary dynamics and generalized local search machines—provide theoretical foundations for understanding how nature-inspired algorithms function. Several algorithms provide useful examples, especially genetic algorithms, ant colony optimization and simulated annealing.

Keywords Nature-inspired algorithms · Dual-phase evolution · Evolutionary dynamics · Generalized local search machines

1 Introduction: How Nature Works

Borrowing from nature has become a well-established practice in computing. The reasons for this are easy to understand. Computing has to deal with increasingly complex problems where traditional methods often do not work well. Natural systems have evolved ways to solve such problems. Methods borrowed from nature include both ways to represent and model systems, such as cellular automata or neural networks, and procedures to solve complex problems.

D. Green · A. Aleti (✉) · J. Garcia
Faculty of Information Technology, Monash University, Clayton 3800, Australia
e-mail: aldeida.aleti@monash.edu

D. Green
e-mail: david.green@monash.edu

J. Garcia
e-mail: julian.garcia@monash.edu

© Springer International Publishing AG 2017
S. Patnaik et al. (eds.), *Nature-Inspired Computing and Optimization*,
Modeling and Optimization in Science and Technologies 10,
DOI 10.1007/978-3-319-50920-4_1

The motivation for basing algorithms on nature is that the natural processes concerned are known to produce desirable results, such as finding an optimal value of some feature. This observation has inspired many algorithms based on nature. Despite their effectiveness, methods modelled on nature have often been treated with suspicion. Traditional mathematical methods, such as linear programming, are based on well-known theoretical foundations. So their interpretation, and their limitations, can be tested analytically. In contrast, nature-based methods are *ad hoc* heuristics based on phenomena whose properties are not always understood, even by biology.

The above issues raise a need to identify theoretical foundations to underpin nature-based algorithms. To address this need, we set out to do the following in this account. First, we identify features that are common to many nature-inspired algorithms and show how these are described by a formal model that explains why the algorithms work. Secondly, we describe three frameworks for describing nature-inspired algorithms and their operation. Finally, we discuss some deeper issues about the differences between natural processes and methods based on them. This includes both the danger of simplifying nature and further lessons we can derive from the way processes actually work in nature.

2   The Nature of Nature

Many natural processes have inspired optimization algorithms. However, most of them share a common underlying model, which reveals how they work. We begin with a simple example that captures the main features. The great deluge algorithm (GDA) is based on the idea of an agent moving about in a landscape, and the landscape is in the process of being flooded [18]. Initially, the agent can wander freely anywhere in the landscape. However, the rising water level prevents entry into low-lying areas. These areas expand and coalesce, until eventually the agent is trapped on a single peak, after which all it can do is to climb to the top of that peak.

An optimization algorithm based on GDA interprets model parameters as coordinates in an N-dimensional 'landscape'. The object variable, which is to be maximized, is treated as elevation. The flood level then becomes a constraining lower limit on the object function, which is gradually raised. The agent is a model that performs a random walk through the solution space, rejecting any move to a region where the object function falls below the flood level. The algorithm terminates when the agent can no longer find anywhere above flood level. An important feature of the algorithm is a transition from 'global search' in which it can move freely anywhere, to 'local search', in which it is constrained to a local neighbourhood.

In the great deluge algorithm, the agent continues to take a random walk within the local region in which it is confined and is forced to move uphill as rising water level progressively shrinks the size of the region. A more efficient approach to this final search is 'hill climbing'. Based on the analogy with a mountaineer ascending a mountain, in hill climbing the agent does not wander at random, but moves uphill on every step, until it reaches a local peak.

By mimicking natural processes, nature-inspired algorithms (NIAs) provide ways to find a target by navigating a search through a network of possibilities. The manner of these searches sometimes reveals deep similarities between completely different systems. Other algorithms, for example simulated annealing (see Sect. 3.3) and the great deluge algorithm use analogues with natural processes to manage a switch from what is initially a global search to a local search in their final stages.

We can see another example of deep similarity if we compare the gravitational search algorithm (GSA) [67] and the firefly algorithm [81]. Both involve attraction between potential solutions. Initially, these are scattered across a search space, and weaker solutions move towards better ones. GSA represents solutions as objects in an N-dimensional space. Their locations are given by values of N parameters, and their masses are values of the object function. Gravitational attraction draws the objects towards each other, with smaller masses moving closer to larger ones. The firefly algorithm mimics the flashing behaviour of fireflies. As in GSA, the fireflies are scattered over an N-dimensional space. The brightness of a firefly's flashes corresponds to the quality of its solution. Fireflies are attracted to others with brighter flashes, but brightness decreases with distance.

The simple models discussed above capture features that are present in many nature-inspired algorithms (NIAs). The procedures are characterized as a search through a space defined by values of the variables, with a transition from a global search to a local search. For optimization problems, the idea of a search space leads to the concept of a 'fitness landscape'.

2.1 Fitness Landscape

The notion of a 'fitness landscape' is an abstraction developed in evolutionary theory and borrowed by the optimization community to explain the relationship between an optimization problem and the search algorithm. For optimization problems, many NIAs provide ways to search the fitness landscape.

Formally, a fitness landscape has the following components: (i) a set of possible solutions s, also known as the search space, (ii) the distance (neighbourhood) operator, which assigns each solution $s \in s$ to a set of neighbours $n \subset s$ and (iii) the fitness function $F : s \rightarrow \Re$. An example of a fitness landscape is shown in Fig. 1.

As the neighbourhood of a solution depends on the distance operator, a given problem can have any number of fitness landscapes. The neighbourhoods can be very large, such as the ones arising from the crossover operator of a genetic algorithm or as small as the ones created by a 1-flip operator.

The neighbourhood operator influences the topology of the fitness landscape and, as a consequence, the performance of the algorithm. A neighbourhood operator that creates a fitness landscape with one local optimum that is also the global optimum would be easy to search with a local search method. When the fitness landscape has plateaux, the progress of a search algorithm could easily stagnate, because of the

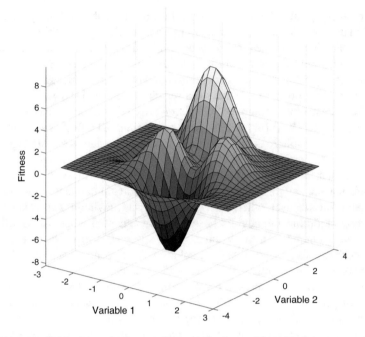

Fig. 1 An example of a fitness landscape of a maximization problem. Note the two *small peaks* (local optima) in front of the *large one* (global optimum)

search iterating between equally fit solutions. Plateaux can be detrimental to most search algorithms which rely on gradients they can follow.

The shape and size of gradients and the distribution of local optima are also features that decide the effectiveness of one algorithm over another. For instance, if the difference in fitness between any two neighbouring solutions is on average small, then the landscape is more likely to be easy to negotiate for a wide range of local search methods. In contrast, if a significant fitness difference is observed in the neighbourhood, the choice of the search algorithm becomes important.

2.2 Graphs and Phase Changes

As we saw earlier with the great deluge algorithm, many optimization algorithms involve a combination of *global search* and *local search*. In terms of a fitness land-scape, a global search aims to find a high mountain to climb, and a local search tries to find the top of a single mountain. The challenge is how to strike a balance between the two. Many natural processes achieve this balance by exploiting a mathematical property of connectivity in graphs.

(a)   **(b)**

Fig. 2 The spread of an epidemic process (e.g. a fire) shows how connectivity in a fitness landscape produces a transition from global to local search. Here, a grid represents space, with *white cells* denoting empty space, *grey* denoting fuel and *black cells* denoting burnt areas. When a fire starts at the centre of the grid, its spread can be viewed as a search for neighbouring cells that contain fuel. **a** If the density of fuel is subcritical, then the fire is restricted in a small, isolated area (local search). **b** If the density of fuel is supercritical, then the fire spreads across the entire landscape

A *graph* is a set of *nodes* together with *edges*, which link pairs of nodes. Points in a fitness landscape form a graph, with neighbourhoods defining the edges. The larger the neighbourhood, the more edges connect the point to others.

If a graph is formed by adding edges at random to a set of nodes, then as the number of edges grows steadily in size, the resulting clumps that form remain small until a critical density is reached, whereupon a *connectivity avalanche* occurs, absorbing almost all the nodes into a single 'giant component' [19, 20]. In effect, the connectivity avalanche means that a graph has two phases: *connected* and *fragmented*.

The connectivity avalanche explains how natural processes are able to mediate between local and global searches. In a fitness landscape, removing areas in the manner of the great deluge creates holes, thus removing edges from the underlying graph until the graph fragments (Fig. 2). Other natural processes, and the algorithms based on them, achieve the same effect by other means.

The flower pollination algorithm uses analogies with the way flowers pollinate to manage the balance between local search and global search [80, 83]. Here, the 'flowers' are potential solutions to a problem, and like several other algorithms, variables determine their location in 'space'. Breeding occurs by pollination. Cross-pollination ('global pollination') occurs when different 'flowers' trade pollen. Self-pollination ('local pollination') occurs when a flower pollinates itself. The crucial step is the transition between the two processes. The bat algorithm mimics the way bats home in on prey [82]. Like the flower pollination algorithm, the organisms (bats) are potential solutions to a problem and parameter values determine their location in 'space'. The bat makes a random walk through the environment, but uses echolocation to detect food (global search). When food is located, the bat switches to a local random walk to close in on its goal.

3 Nature-Inspired Algorithms

Nature-inspired algorithms (NIAs) are sometimes termed *meta-heuristics*. This means either they incorporate other, simpler methods as part of their operation, or they can be set over other methods to overcome problems in behaviour. For instance, local search methods, such as hill climbing, can become trapped on a foothill in the fitness landscape, rather than the mountain. NIAs can prevent searches becoming trapped this way. As we saw earlier, many NIAs achieve this by enabling transitions from global to local search.

Among the large number of NIAs, we have selected three prominent algorithms to analyse in this work. Each method mimics a unique natural process. These are *genetic algorithms, ant colony optimization* and *simulated annealing*. The way they carry out the search is based on quite different processes. Genetic algorithms are iterative methods, as they iteratively evolve (change) a population of solutions that are randomly initialized. Ant colony optimization belongs to the class of constructive methods, since they build a solution one step at a time, based on the knowledge learnt in the previous steps. Simulated annealing, on the other hand, is an iterative search method that adapts the search procedure based on a time-dependent schedule. The following sections describe the natural processes that have inspired the development of these three unique NIAs and the theoretical investigations that have been carried out to prove their convergence.

3.1 Genetic Algorithm

Genetic algorithms [44] are optimization methods that mimic the process of evolution and natural selection. They are part of a larger class of methods known as evolutionary algorithms (EAs).

The main components of natural evolution that are part of a GA are *inheritance, mutation, crossover* and *selection*. They are known as genetic operators. GAs maintain a population of solutions that evolves by means of these genetic operators. The optimization process starts with a set of solutions as initial population that are randomly initialized or created by applying some heuristic. Genetic operators are applied with predefined rates. The new individuals are added to the population, and the replacement procedure selects the solutions that will survive to the next generation to maintain the prescribed population size.

3.2 Ant Colony Optimization

The algorithm of ant colony optimization (ACO) [16] was built from inspiration by the foraging behaviour of ants. When ants search for food, they initially explore the area around their nest in a random way. When a food source is found, the ants carry it back to the nest, at the same time depositing a chemical pheromone trail on the ground. This trail is intended to guide the other ants to the food source, which will also deposit more pheromone in the trail. As time goes by, some of the pheromone is evaporated, and the trail may disappear if it is not updated with more pheromone.

This indirect communication between ants has been shown to enable the ant colony to establish the shortest path to the food source [15]. The natural phenomenon of intensification of short (high-quality) paths, and evaporation (forgetting paths that are not so good), has inspired a set of methods known as ant colony optimization (ACO).

ACO algorithms are based on a parameterized probabilistic model (pheromone model) that builds on high-quality results found in the past. This usage of memory is considered one of the most important elements of a powerful approximate method [7].

In ACO, ants build a solution by performing randomized walks through the solution space. Each move is made stochastically, biased by the pheromone levels. In the initial phase of the algorithm, pheromone levels are equal and the first solutions are created through uniformly randomly chosen values.

The local pheromone update prescribed by ACO specifies that every time an ant makes a move, the amount of pheromone on all paths is decreased according to the evaporation rate in the interval [0, 1]. The local pheromone update enables the diversity of exploration within the same iteration. Ants of the same iteration are less likely to build the same path when pheromone is evaporated immediately after a move has been made. As a result, the diversity of the solutions provided is enhanced. The pheromone information on each move is increased by a quantity measured according to the fitness of solutions.

3.3 Simulated Annealing

Simulated annealing (SA) [48] is inspired from the technique of annealing in metallurgy. This involves the heating and slow cooling of a material in a controlled way to reduce defects that are related to the material's thermodynamic free energy. Heating and cooling the material affect both the thermodynamic free energy and the temperature. In SA, energy is interpreted as fitness, and the slow cooling process is interpreted as the decrease in the probability of accepting worse solutions while exploring the fitness landscape of an optimization problem. The acceptance criterion is known as the Metropolis condition, defined for a maximization problem as follows:

$$p(T, s, s') = \begin{cases} 1 & \text{if } f(s') > f(s) \\ e^{\frac{f(s')-f(s)}{T}} & \text{otherwise,} \end{cases}$$

where s is the current search position and s' is one of its neighbours. The high acceptance of worse solutions at the beginning of the search (when the temperature is
high) allows for a more extensive exploration of the search space. As the temperature decreases, worse solutions become less likely to be accepted, which leads to
convergence of the algorithm.

3.4 Convergence

Theoretical investigations of NIAs have mainly focused on convergence [13, 14, 28,
77]; that is, given enough resource, will NIAs find an optimal solution? In GAs,
Schmitt [69] establishes mutation as the main force in the random generator phase
that assures ergodicity of the Markov chain describing the algorithm. The crossover
operator was found to help mutation in the random generation by adding to the
contraction process over the state space. With no mutation, however, crossover and
selection operators show a convergence effect known as *genetic drift* [69].

A Markov chain describing an algorithm without mutations is usually not ergodic,
and its dynamics depend on the initial population. Schmitt [69] also shows that GAs
do not necessarily converge to populations with only local optima when employed
with the most widely used selection methods.

The first theoretical aspect of ACO considered was convergence of the algorithm.
Gutjahr [38] proved convergence of a graph-based ant system (GBAS), whereas in
[39] the convergence proofs were more general. GBAS, however, is quite different
from any implementation of the original ACO algorithm.

Stützle and Dorigo [72] and Dorigo and Birattari [17] proved convergence in value
and in solution of a class of ACO algorithms known as $ACO_{\tau_{min}}$. These algorithms
apply a positive lower bound τ_{min} to all pheromone values, to prevent any probability
to generate any solution from becoming zero. These methods are among the best
performing in practice.

To represent the decreasing value of the temperature, the SA algorithm has been
described as a sequence of homogeneous Markov chains of finite length. This leads
to convergence proofs by Hajek [40], which show that for asymptotic convergence
to occur, the cooling has to be carried out sufficiently slowly. The analysis of NIAs
using Markov models will be discussed further in Sect. 5.1.

While convergence proofs are helpful in understanding the working of an algorithm, they do not give any insight into how the algorithm interacts with the search
space of the problem it is solving. The behaviour (and success) of an optimization
method depends on the problem; hence, the ability to model this interaction would
reveal why certain methods are more appropriate for a particular class of problems.

4 Dual-Phase Evolution

As we have seen, many natural processes, and the NIAs based on them, involve transitions from a connected phase to a fragmented phase. Such phase transitions are very common in nature and can occur in both directions. In the NIAs we have discussed so far, the phase transitions always involved a global search in a connected phase and a local search in a fragmented phase. In nature, however, the processes that occur are not necessarily global search and local search. More generally, we can describe them as *variation* and *selection*. Systems with these properties give rise to the phenomenon of dual-phase evolution.

4.1 Theory

Dual-phase evolution (DPE) is a process that occurs in many natural systems [34, 35, 61, 62]. It explains how repeated transitions between connected and fragmented phases promote self-organization in nature and how patterns emerge out of interacting elements.

DPE occurs when networks of agents undergo repeated changes between connected and fragmented phases. The theory was first introduced to explain the punctuated nature of vegetation types in long-term forest history [31, 32] and in species evolution over geologic time [36]. Subsequent research has shown that it occurs in a wide variety of physical, social and other systems [35, 62].

The formal requirements of DPE are as follows:

1. There is a set of objects, which form the nodes of a network, and interactions or relationships between the objects define the edges of the network.
2. The connectivity of the network shifts repeatedly, alternating between a connected phase and a fragmented phase.
3. The objects are subject to two processes, broadly described as *selection* and *variation*, and each dominates in one of the above phases (Fig. 3).
4. The system has memory: changes that occur in one phase carry over into the other phase.

Some of the optimization algorithms we discussed incorporate elements of DPE. Both the great deluge and simulated annealing exploit the phase change in their underlying network to switch from global to local search.

4.2 GA

Application of the DPE model of evolution in a landscape led to the *cellular genetic algorithm* (CGA) [49–51]. In CGA, the agents in the population are distributed across

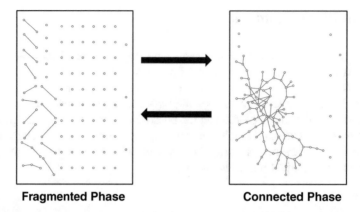

Fragmented Phase   **Connected Phase**

Fig. 3 Dual-phase evolution involves phase shifts back and forth between a connected phase and a fragmented phase. In a fragmented phase (shown at *left*), agents are either isolated or in small clumps, so interactions between agents play almost no role. In a connected phase (shown at *right*), agents are linked to others, so interactions have a major influence

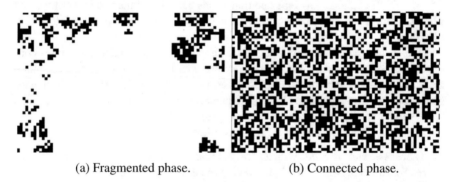

(a) Fragmented phase.   (b) Connected phase.

Fig. 4 Operation of the cellular genetic algorithm. Agents (marked by *black cells*) are scattered across a pseudo-landscape and breed only with agents in neighbouring cells. Intermittent disturbance wipes out all agents within randomly selected patches. These 'cataclysms' temporarily shift the population from a *connected phase*, in which local competition for space drives fitness selection, to a *fragmented phase*. This transition eliminates selection and allows agents to disperse freely until cleared space is reoccupied

a cellular automaton, which represents a landscape (**not** a fitness landscape). Within this space, agents can reproduce only with neighbours (local phase). This phase promotes the formation of local clumps of agents with similar genotypes and makes it possible for suboptimal genotypes to survive. Intermittent cataclysms create open territory (global phase), allowing agents to invade without being subject to selection and local competition (Fig. 4).

Lamarck proposed a theory of evolution that involved 'inheritance of acquired characters' [75]. That is, features acquired during an organism's lifetime become incorporated into its genes and can be passed on to its offspring. So if an animal habitually stretches its neck to reach leaves high on a tree, then its offspring will also have longer necks.

The biological theory was discredited during the twentieth century. However, the idea has been applied to genetic algorithms by having a reproductive (genetic) stage, followed by a learning phase in which each agent performs a hill climb [46, 78]. The new parameter values are then incorporated into its genes. Effective as it is, the method may fail because it is too efficient: during the hill climbing phase, each agent ascends to a local optimum. So the population can become trapped if agents cluster on local optima.

An alternative, which certainly does occur in nature, is the Baldwin effect [75]. This mimics the lifestyle of vertebrates, with a reproductive phase and a learning phase. Essentially, it operates like Lamarckian evolution, except that acquired characters are not incorporated into the genome. Thus, it retains greater genetic variability.

Memetic algorithms [57, 58] combine Dawkins' model of a meme (an idea that spreads from person to person) with the GA idea of an evolving population. The earliest versions achieved phase changes between search and selection at different levels by employing either Lamarck or Baldwin principles as described above. Many applications have addressed combinatorially complex problems, such as component deployment in embedded systems [1], software testing [2, 24] and various scheduling problems [29, 66].

4.3 Ant Colony Optimization

In nature, dual-phase evolution, combined with positive feedback, plays a part in food gathering by an ant colony. The underlying graph consists of points in the landscape, and pheromone trails created by foraging ants provide the edges linking points. Two processes operate on the edges: ants laying down pheromone trails and evaporation of pheromone trails. In a global phase, ants search for food, creating new paths as they go. As the ants explore at random, they can potentially link any point in the network they form. When food is found, the system switches to a local phase. Trails that lead to food are reinforced as more and more ants follow them to the food source and back to the nest. Less and less ants follow unproductive trails, so evaporation effectively deletes them. The twin effects of reinforcement and evaporation create positive feedback. Productive trails grow stronger, and more direct, whereas unproductive trails fade away.

Applications of ACO focus on the way trails evolve *after* ants discover food. The way in which paths from their nest to a food source become optimized provides a model for many combinatorial optimization problems. Typical applications include bin packing [52], assignment problems [3] or sequential processes, such as assembly lines [6]. In ACO, a network of states replaces points in a landscape, agents replace

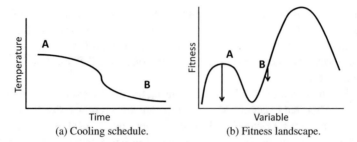

(a) Cooling schedule.   (b) Fitness landscape.

Fig. 5 Role of the cooling schedule in simulated annealing. The *cooling schedule* (**a**) controls the way the 'temperature' decreases over iterations of the algorithm. At point A, early on in the cooling schedule, large deteriorations in fitness are allowed (as illustrated by the *arrow* in (**b**)). So in the fitness landscape, this would allow the algorithm to escape from a low peak and move onto the main peak. At point B, late in the cooling schedule, only small variations are allowed, so the algorithm is now trapped on the peak

ants, and an agent's trail is a sequence of transitions between states. In nature, more evaporation of pheromones occurs on sections of a path that take longer to traverse. In ACO, the probability of choosing transitions of lower quality decreases between iterations of the algorithm, whereas direct transitions are reinforced, making them more attractive. Initially, agents can follow a great variety of paths. But the routes change as they find transitions that are stronger in pheromones. This results in sections of trails become disused, and the choices become increasingly limited. Eventually, the connectivity of the network of trails eventually falls below a critical threshold, after which the agents are confined to a small set of trails.

4.4 Simulated Annealing

As we saw earlier, the cooling schedule in simulated annealing determines the radius of neighbourhoods within the fitness landscape (Fig. 5). As that radius shrinks, the process crosses the critical threshold and becomes confined to a local region of the fitness landscape. In the example shown in the figure, the algorithm starts with the model located on a 'foothill' in the fitness landscape (point A). However, the cooling schedule starts with the process being 'hot', which means that a wide range of variation is permitted during iteration. This makes it possible for the algorithm to escape from the 'foothill' where it begins and move onto the large peak where the optimum lies (i.e. global search is possible). At the end of the cooling schedule, the process is 'cool', so only small variations around the current fitness are permitted. This means that the algorithm is trapped on the slope where it is and cannot move back onto the foothill. That is, only local search is now possible.

5 Evolutionary Dynamics

Evolutionary dynamics studies mathematically how the composition of a population changes over time as a result of natural selection. Most models are simple, aiming to capture the essential forces shaping natural selection in the real world. We will discuss how these simple models, and variations inspired by them, can also be used to understand NIAs.

Evolutionary dynamics comprises tools that include deterministic dynamical systems such as the replicator equation, as well as stochastic processes such as Markov chain models. Originating in biology and ecology, these simple models can be understood as models of computation [68]. They provide a useful, albeit simple, machinery to elucidate how and why such algorithms may work.

Evolutionary dynamics distinguishes two fundamentally different types of selection: *constant selection* and *frequency-dependent selection*. In constant selection, the fitness value of individuals is independent of the other individuals in the population. In frequency-dependent selection, the fitness values for an individual depend on the state of the population itself. The former corresponds to a static fitness landscape, while the latter represents a fitness landscape that changes along with the population exploring the space.

More precisely, if S is the search space of interest, constant selection deals with fitness functions of the form $F : S \to \mathbb{R}$. On the other hand, frequency-dependent selection deals with functions of the form $F : S^n \to \mathbb{R}$, where n is the population size. Algorithms based on frequency-dependent selection are commonly known in computer science as *coevolutionary algorithms*.

In this section, we discuss two main theoretical ideas from evolutionary dynamics, as well as their applications to NIAs. We will first show how Markov chain models can be used to describe constant selection. Then, we will discuss how game theoretical concepts, in particular the replicator equation, can be used to understand some problems arising from coevolutionary computation.

5.1 Markov Chain Models

A large part of constant selection models use stochastic processes, in particular discrete Markov chains [59]. A Markov chain is usually described in terms of a set of states $S = \{s_1, s_2, \ldots, s_n\}$. The dynamics is composed by a series of discrete *steps* between states of the chain. If the chain is in state s_j, it can transition into state s_i with probability p_{ji}. That is, future states depend, probabilistically, only on the current state of the system. The stochastic matrix $[p_{i,j}]$ is known as the transition matrix of the Markov chain.

Most of the applications of Markov chain theory to natural computation rely on the class of *ergodic* Markov chains. This concerns systems in which it is possible to go from every state to any other state (possibly in several steps) [37]. It can be shown that for this class of Markov chains, the long-term behaviour is characterized by a so-called stationary distribution $\mathbf{w}$. Let $\mathbf{P}$ be the transition matrix of an ergodic Markov chain, then $\mathbf{P}^n$ for $n \to \infty$ approaches a limiting matrix $\mathbf{W}$, where all the rows are given by the same vector $\mathbf{w}$. This quantity is a probability distribution over S, which is unique and does not depend on the initial state of the Markov chain [47]. Explicitly finding $\mathbf{w}$ or describing its properties is a standard method for studying the long-term behaviour of ergodic Markov chains.

5.1.1   Using Markov Chains to Understand GAs

In Markov chain analysis of genetic algorithms, the states of the Markov chain usually correspond to possible states of a population. If I is the set of all possible individuals, the state space S is made up of all possible n-tuples of elements in I. The Markov chain is formulated by considering how selection and mutation operators move the chain from one state to the other. The size of the state space is in this case $|I|^n$. The dynamics is also determined by an appropriately chosen fitness function that guides the process in the direction of higher fitness values.

One immediate apparent limitation of this approach is that the state space for meaningful problems is very large to be accounted for by this theoretical approach. This technique, however, is used to provide a model to help us understand the essential principles behind NIAs, rather than to solve meaningful problems outright. Notwithstanding the size of the space, the transition probabilities p_{ij} can often be presented in closed form [14, 77].

The main result in this particular area concerns the support of the stationary distribution. Once $\left[p_{i,j} \right]$ has been defined for a particular fitness function and set of operators, the asymptotic behaviour can be characterized by the properties of the stationary distribution $\mathbf{w}$. Nix and Vose [59] show in a stylized set-up that given enough time, the system will spend most of the time in a global maximum, if it exists, or in local optima, if there are several. A similar result is presented by Hernandez et al. [41], accounting for a strongly elitist selection operator as well as local mutations and non-local mutations resembling crossover.

Similar asymptotic studies, using Markov chain theory, exist for other search algorithms such as simulated annealing. For example, Granville et al. [30] prove asymptotic convergence to global optima in SA. Davis [12] discusses how convergence of GA and SA relates to each other using the Markov chain theory approach.

Theoretical results like the one presented above are yet to be bridged into the world of practical applications. Proofs of convergence are important, but it is still hard to quantify convergence time, which is the quantity that most practitioners would be interested in. Convergence guarantees that given enough time, the algorithm will confine itself to optimal states. But how long is enough?

One avenue that relies on Markov chain theory is based on the idea of predicting how long it will take for a Markov chain to reach stationarity. Hernandez et al. [42] use a statistical technique called coupling from the past to detect convergence time. Once the Markov chain associated with a GA has converged, no further generations will yield any improvement and it is sensible for the algorithm to stop. This is only the case because the stationary distribution is known to converge to optimal solutions.

Propp and Wilson [65] propose a sampling algorithm based on the idea of coupling. The cornerstone of the procedure is to build a Markov chain whose stationary distribution is a target distribution one wants to sample from. Theoretically, reaching stationarity requires infinite time, but Propp and Wilson provide an algorithm to figure out when stationarity has been reached. This so-called *burn-in* time is stochastic but finite. This idea is used by Hernandez et al. [42] to formulate a simple *self-stopping* algorithm. It uses a coupling technique to determine when stationarity has been reached and therefore when the algorithm should stop exploring solutions.

The approach outlined above is still far from practical applications. Coupling relies on running several instances of the same Markov chain. In most cases, the number of required instances is proportional to the size of the space. This renders the technique impractical for applications, but provides insight into how the timing for convergence may depend on the choice of the algorithm and its parameters. A number of statistical techniques have also been proposed to detect convergence, in what is an open and important question in the field [74].

5.2 The Replicator Equation

Chellapilla and Fogel [11] introduced a method of evaluating solutions based on comparing candidate solutions, rather than assessing the absolute qualities of potential solutions. This technique is known as *coevolution*. It promises to deliver the advantages of evolutionary computation in domains where expert knowledge is difficult to turn into a fitness function. Thus, solutions compete against each other. This means that the fitness of a solution depends on what other solutions are present in the population. The evolutionary dynamics equivalent of this set-up, discussed above, is known as frequency-dependent selection. This approach is also useful in modelling natural phenomena, as many complex scenarios in ecology, evolution and social science are examples of frequency-dependent selection [9].

Studying frequency-dependent selection adds significant complications to theory, as compared with constant selection. In these scenarios, the fitness landscape is dynamic and changes together with the population composition. This has led to a series of identified *biases* that sometimes prevent successful applications from straightforward implementations [64]. Such *biases* are inherent in the dynamic nature of the algorithms and highlight the importance of theory.

Theoretical analysis of coevolution is borrowed mostly from game theory. One particular technique of interest is the replicator equation. It describes the dynamics of selection acting on a population of individuals [73].

The main idea behind the replicator equation is that good solutions increase (or decrease) in numbers, proportional to the difference between their own fitness value and the average fitness of the population. This simple idea defines a system of ordinary differential equations that can be studied using standard mathematical techniques [71]. The replicator equation is general and has been used extensively across different fields, including economics [25], social science [70], biology [10] and computer science [21, 53].

Consider a set of phenotypes $\mathscr{P}$ of finite size n. The state of the population is represented by a vector $x \in \mathbb{R}^n$, such that $\sum_{i=1}^{n} x_i = 1$. The state space is thus the unit simplex of size n, which we denote $\mathscr{P}_n$. For a phenotype i, x_i is a number in [0, 1] that represents the proportion of that phenotype in the population.

The fitness function is defined as $F : \mathscr{P}_n \to \mathbb{R}^n$. It assigns a vector of fitness values to each phenotype, given a state of the population. Usually, the fitness of individuals is given by a game in normal form. The time evolution of the system is defined by:

$$\dot{x}_i = x_i(f_i(x) - \phi(x)), \tag{1}$$

where $\phi(x) = \sum_{i=1}^{n} x_i f_i(x)$ is the average population fitness.

This equation describes the deterministic dynamics of a process of selection. It is important to notice that this process ignores mutations and also assumes a infinitely large population. Those ingredients are part of extensions including the replicator–mutator equation [43] or stochastic processes that account for populations of finite size [23, 56, 79].

While the replicator equation aims to capture a learning process undergone by a population of agents, the long-run behaviour of the system is equivalent to that of well-known single-agent processes, such as reinforcement learning [8]. Thus, it provides general insights that may be applicable in learning theory, beyond population processes. In NIAs, the replicator equation provides a simple theoretical framework that can help us understand the forces of selection acting on a coevolving population.

5.2.1 GAs and the Replicator Dynamics

The use of the replicator equation, and more generally game theory, to study coevolution arises from the need to understand dynamics beyond what one can achieve by experiments only. In its original form, the replicator equation does not include mutation. Thus, it is particularly well suited to study selection mechanisms.

One way the replicator equation has been used is as a benchmark to assess whether dynamic features of evolutionary dynamics also arise in evolutionary algorithms. One such feature is the selection of polymorphic populations. Maynard Smith [54] introduced the so-called hawk–dove game. This game was designed to study animal aggression. It involves two types of individuals fighting for a resource of value V. Hawk types act aggressively, while dove types refrain from confrontation. Conflict entails a cost C. Thus, when two hawk types meet, they split the value assuming a cost, which leads to a fitness $(V - C)/2$. A hawk meeting a dove will result in the hawk earning the resource without the cost of confrontation, while the dove gets nothing (pay-offs V and 0). Doves against doves will split the resource without conflict, each earning $V/2$. The replicator dynamics in this case leads to the coexistence of aggressive and passive types. Such polymorphic equilibria are important in evolutionary ecology, but according to Ficici and Pollack [22], fail to materialize when using selection as given by standard selection mechanisms from evolutionary computation.

Ficici et al. [21] also compare the dynamics of different selection mechanisms used in GAs with the dynamics induced by the replicator equation. They conclude that when standard selection methods are placed in a frequency-dependent context, the resulting dynamics significantly differs from what is expected from the replicator equation. This result highlights the difficulties that arise from using techniques devised for constant selection in inherently dynamic scenarios.

5.2.2 ACO and the Replicator Dynamics

The replicator equation has also been used as a means to understand other algorithms for which the evolution metaphor is perhaps less direct. Pelillo [63] presented applications to classic optimization problems. In particular, they perform experiments in which the replicator equation is adapted to explore solutions of the maximum clique problem. The trajectories of the replicator equation are shown to asymptotically approach optimal solutions. Since the max clique problem is known to be NP-complete, this suggests that algorithms inspired on frequency-dependent selection may successfully solve difficult problems.

Another applications presented by Ammar et al. [4] study the dynamics of ant colony optimization. An equation inspired by the replicator equation is used to describe the rate of change in pheromone levels.

Other applications include artificial immune systems. Velez et al. [76] study the asymptotic properties of an immune network by modelling the system as a set of differential equations inspired by the replicator equation. These examples show that the replicator dynamics can be used to study a variety of algorithms including swarm processes and optimization algorithms based on networks.

The replicator equation is useful to describe the deterministic dynamics of an evolving population. It is, however, restricted by the size of the phenotype space. The dynamics are often solved numerically, but this method falters as the size of the space increases. This limitation confines results to theoretical contributions. In most computationally meaningful problems, the space is large enough that using the replicator on the whole phenotype space is prohibitive and not very insightful. Thus, the value of the replicator equation is best put to use on simple models that aim to capture essential ingredients.

The replicator equation relies on the assumption of an infinite population. Further work in applying evolutionary dynamics to problems in natural computation can benefit from exploring recent advances in evolutionary dynamics dealing with finite populations [27, 55, 60]. This shows that the community in evolutionary computation may stand to benefit not just from adapting the metaphors inspired by nature, but also from borrowing and tinkering the theoretical toolset that has been developed in biology and ecology.

6 Generalized Local Search Machines

Generalized local search machines (GLSMs) [45] are a formal framework for representing the different components of NIAs. GLSM distinguishes between search and search control: the search is implemented by simple procedures and represented by nodes, whereas the control mechanism is realized by a non-deterministic finite state machine. The main idea is based on the assumption that most efficient NIAs can be described as a combination of more simple and 'pure' search strategies using a control mechanism. Using this unifying framework to represent search methods helps understand the underlying similarities and differences between different algorithms.

6.1 The Model

The GLSM models NIAs as operating on two levels: the simple search strategies are executed at the lower level, whereas their activation and the transition between different search strategies are controlled at a higher level in the form of a finite state machine. Formally, a GLSM can be represented by a tuple $\mathcal{M} := (Z, z_0, M, m_o, \Delta, \sigma_Z, \sigma_\Delta, \tau_Z, \tau_\Delta)$, where

- Z is a set of states ('pure' search procedures),
- z_0 is the initial state,
- M is a set of memory states,
- $m_0 \in M$ is the initial memory state,
- $\Delta \subseteq Z \times Z$ is the transition relation between the states (describes the strategies used to switch between the 'pure' search procedures),
- σ_Z is the set of state types (surjective, not injective, different state types represent different search strategies),
- σ_Δ is the set of transition types (surjective),
- τ_Z is the function associating states with their corresponding types, and
- τ_Δ is the function associating transitions with their corresponding types.

The current search position is only changed by the search methods associated with states, not as a result of machine state transitions. The memory state, on the other hand, is only changed by state transitions. Examples of state transitions are deterministic (DET), conditional (COND), unconditional probabilistic (PROB), and conditional probabilistic (CPROB). The GLSM model allows adequate and uniform representation of nature-inspired algorithms and provides the conceptual basis for some of the best-known methods.

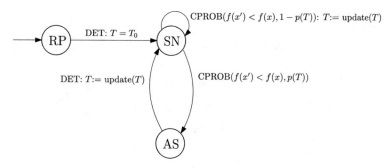

Fig. 6 GLSM of simulated annealing

6.2 SA

We model the SA algorithm with three states, $Z := \{RP, SN, AS\}$ (as shown in Fig. 6), where RP stands for random picking, SN is the selection of neighbour, and AS refers to the acceptance of solution.

The initial state (RP) is the initialization of the search with a randomly selected solution. This state can be defined procedurally as follows:

procedure RP(π, $S(\pi)$)
    input: *problem instance $\pi \in \Pi$, solution space $S(\pi)$*
    output: *candidate solution $s \in S(\pi)$*

    $s :=$ SELECTRANDOM(S);
    return *s*
end RP

After the completion of RP, the initial temperature is set to T_0 and the transition to the next state is performed deterministically. The selection of a neighbour (SN) and acceptance of the solution (AS) are executed iteratively until a stopping criterion. These two states can be defined by the following procedures:

procedure *SN*(π, *s*)
    input: *problem instance $\pi \in \Pi$,*
        solution $s \in S(\pi)$
    output: *solution $s' \in S(\pi)$*

    $s' :=$ SELECTRANDOM($N(s)$);
    return *s'*
end SN

procedure *AS*(s', *s*)
    input: *solutions $s, s' \in S(\pi)$*
    output: *solution $s \in S(\pi)$*

    $s := s'$;
    return *s*
end AS

SN selects one of the neighbours of s, whereas AS changes the search position to the neighbouring solution given the Metropolis acceptance criterion, which is modelled in the transitions as conditional probabilities. In essence, the neighbour is accepted as the new search position if it is better than the current solution. Otherwise, the temperature is used to probabilistically accept worse solutions.

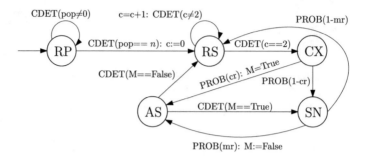

Fig. 7 GLSM of genetic algorithm

Initially, the temperature is high, which means that the algorithm has a higher probability of making deteriorating moves. As the search progressed, the temperature is decreased through update(T), leading the algorithm towards the local optimum and as a result ensuring convergence.

6.3 GA

The five states of the GA ($Z := \{RP, RS, CS, SN, AS\}$) are shown in Fig. 7. The procedure of RP is similar to the one described in Sect. 6.2. The transitions, however, are different. While SA operates on one solution, GA has a population of solutions; hence, the random picking procedure is executed n times, where n is the size of the population.

   RS (random solution) selects a solution at random from the population. This is executed twice (controlled by the deterministic condition) before the transition to the next state CX, which implements the crossover operator. RS is defined procedurally as follows:

procedure RS(S')
   input: *population of solutions* $S' \in S$
   $s :=$ SELECTRANDOM(S');
   return s
end RS

   While in RP any solution has equal probability of being selected, in RS, only solutions that are part of the population can be chosen. This biases the search towards a specific area of the fitness landscape. Next, the crossover operator is applied based on the crossover rate (cr) and followed by the mutation operator, which can be represented by the SN (select neighbour) state, defined in Sect. 6. The only difference at this stage is the transition type to accept solution (AS). In simulated annealing, the neighbour is accepted according to the Metropolis acceptance criterion, which is

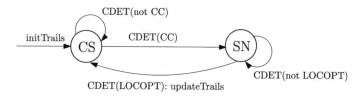

Fig. 8 GLSM of ant colony optimization

a conditional probabilistic transition type, whereas in genetic algorithms the mutant is accepted probabilistically, based on the mutation rate.

6.4 ACO

ACO has two main states: construction of solution (CS) and local search, which is depicted as the state 'selection of neighbour' (SN) in Fig. 8. The trails are initialized (initTrails) to the same value; hence, the first execution of the state CS is similar to RP described in Sect. 6, where solutions are initialized uniformly at random. After a solution is constructed, a local search procedure is applied. This is represented as the SN state (described in Sect. 6) and the conditional deterministic transition type, which checks whether the current solution is a local optimum.

The update of trails changes the pheromone matrix, introducing bias towards the selection of higher quality solution components in the next solution construction. This is where CS is differentiated from the RP procedure.

6.5 Discussion

All three algorithms are different in terms of both the search procedures involved and the search control. All of them, however, have a local search procedure incorporated into their structure, which is used to intensify the search and converge to a local optimum. This local search procedure is at the core of simulated annealing, and it is the temperature that controls when this intensification happens. Higher temperatures may lead the algorithm to explore different basins of attraction, while lower temperatures allow the method to approach the local optimum.

In GAs, it is the mutation operator that behaves like a local search procedure. The difference, however, is that mutation is not a 'hill climber' and stops after one neighbourhood search. The population of solutions is evolved in parallel, allowing the explorations of multiple basins of attraction.

Different selection mechanisms have been applied in the literature, which have been found to have different effects on search intensification. An elitist strategy, for instance, makes sure that the best solutions are always propagated to the next generation. If no diversification strategy is in place, elitist strategies may result in premature convergence. A careful balance between exploration and exploitation can be ensured through tuning the different parameter values, such as the selection strategy, population size, mutation rate and crossover rate.

Ant colony optimization is known as a constructive algorithm, because it constructs a solution one component at a time. The first ACO algorithms did not have a local search procedure in their implementation. It was later experimentally realized that its performance would improve when combined with local search.

The pheromone matrix in ACO acts like the temperature of SA. Initially, pheromone values are equal, which means that all solution components have equal probability of being realized. This is the exploration phase of the method, similar to high temperatures in SA. As new solutions are created, and the 'trails' in the pheromone matrix are updated, the search is intensified in high-quality areas of the fitness landscape. The local search procedure speeds up this intensification process, since pheromone values may take a long time to converge, due to the evaporation rates.

7   Conclusion

In this account, we have used three different paradigms from theoretical computer science—dual-phase evolution, evolutionary dynamics and generalized local search machines—to highlight relationships between different algorithms and to explore their workings. These paradigms show that there are deep relationships between many different algorithms and that there are sound theoretical reasons why these algorithms succeed.

Unfortunately, it is not feasible to explore every nature-inspired algorithm in detail. The exercise would be both tedious and needlessly repetitive; hence, our focus is on three of the best-known algorithms as typical examples. However, we have also discussed briefly several other nature-inspired algorithms to show that the models and principles do extend to many other algorithms.

The models that we have described not only show how NIAs work, but also provide insights about processes in many natural systems. However, we must conclude by noting that the methods we borrow from nature do not *duplicate* natural processes; they are only models. As with all models, they omit many details, and sometimes, the differences can be crucial. Also, it is important to realize that nature can provide more lessons than ways to design algorithms. It is also instructive to study the ways they are used in nature. These provide some important lessons.

Sometimes, a process that is indispensable in nature is seen as a problem in computing. As an example, consider genetic algorithms. Premature convergence in GAs refers to the phenomenon in which all members of the population acquire the same genetic make-up, thus stopping the search before a solution has been reached. In real populations, however, convergence plays a crucial role: it fixes alleles in a population, so enabling it to acquire new characters.

Perhaps, the biggest difference between problem-solving in nature and that in computer algorithms is that the algorithms virtually always treat problems as one-off. That is, they are treated as closed systems, isolated from any other problems that may be put to them. Natural systems, on the other hand, solve problem after problem, often in parallel, so they treat problem-solving in a different way.

One consequence of treating problems as closed systems is that computational methods do not learn from experience. They usually approach each problem de novo. Even in machine learning, algorithms learn only local knowledge from cases that are presented to them in the context of a particular problem. They do not retain that knowledge to use later on other problems.

Natural systems, in contrast, usually do retain knowledge gained from experience. Young animals learn from each experience, building up a repertoire of 'solutions' over time. Likewise, populations are shaped by their environment, which selects for characteristics best suited to solve future problems.

A notable example of the above is the different way in which optimization is adopted in computation versus in nature. Many human problems are posed as optimization. Moreover, the problems are posed within a closed context. Nothing outside is considered. So the aim becomes to find the absolute best possible solution, i.e. the topmost point of the tallest mountain. In fact, most of the theoretical investigations carried out in previous work are about proving convergence of NIAs (see Sect. 3.4).

In contrast, living systems rarely evolve to a global maximum. Many natural contexts involve trade-offs between competing aims. Usually, the problem boils down to how an organism allocates resources. One of the most fundamental is the tension between survival and reproduction (r vs. k selection). For instance, organisms can increase the chances of their offspring surviving by allocating resources in one of two ways: produce many offspring (e.g. locusts) or nurture few (e.g. birds and mammals).

One of the most important genetic trade-offs is between generalization and specialization [26]. Ecological generalists are species that survive and breed successfully in a wide range of conditions. However, in any specific situation, generalists may be outcompeted by specialists (quoted from Green et al. [33, p. 118]).

Mangroves, for instance, grow in estuaries where the water is increasingly brackish the closer they are to the sea. This environment forces a genetic trade-off between ability to tolerate high salt levels and growth rate, which limits when plants can reproduce [5]. The result is competitive exclusion. Faster growing trees restrict each species to zones at the extreme range of salinity they can tolerate (Green et al. [33, pp. 91–92]).

Climate is an important driver of genetic trade-offs. In a constant, predictable environment, plant and animal populations can adapt to achieve the optimum response to prevailing conditions. Examples of species that seem perfectly adapted to their environment helped to inspire the idea of a 'balance of nature', which suggests simple optimization.

However, many environments are not so predictable. Some are chronically disturbed by disasters, such as floods, droughts and fires. In such cases, adaptation to a single set of environmental conditions can be fatal. Plants that are perfectly adapted to a particular regime of soil, rainfall and temperature will quickly die when drought sets in.

In Australia, for example, the climate is so variable that many plants and animals have adapted to climatic variance, rather than the climatic mean. They need to be hardy enough to survive long droughts, but also they need to be able to take full advantage of short spells of rain. So they are not solving a single optimization problem, but rather coping with numerous problems, each of which requires an organism to use resources in different ways.

These examples show that finding the global optimum for a given set of conditions is not always a good strategy. Adapting too well to a particular problem leaves you vulnerable when conditions change. Expressing this in terms of machine learning, adapting too well to a particular set of conditions is akin to overfitting, wherein a model fits to local noise in the data, as well as the signal. Often, it is better in the long run to find a solution that is just adequate and maintain the flexibility to respond to other challenges that might arise.

Effectively, every problem is treated as multiobjective, where one objective is the problem to be solved now and another is to retain flexibility to solve unknown problems in the future. This can be seen as a tension between local and global solutions.

Acknowledgements This research was supported under Australian Research Council's Discovery Projects funding scheme, project number DE 140100017.

References

1. Aleti A (2014) Designing automotive embedded systems with adaptive genetic algorithms. Autom Softw Eng 22(2):199–240
2. Aleti A, Grunske L (2015) Test data generation with a kalman filter-based adaptive genetic algorithm. J Syst Softw 103:343–352
3. Aleti A, Grunske L, Meedeniya I, Moser I (2009) Let the ants deploy your software—an aco based deployment optimisation strategy. In: 24th IEEE/ACM International Conference on Automated Software Engineering ASE '09, pp 505–509
4. Ammar HB, Tuyls K, Kaisers M (2012) Evolutionary dynamics of ant colony optimization. In: Timm IJ, Guttmann C (eds) Multiagent system technologies,Oct. Lecture notes in computer science, vol 7598. Springer, Berlin Heidelberg, pp 40–52. doi:10.1007/978-3-642-33690-4_6
5. Ball MC (1988) Ecophysiology of mangroves. Trees 2:129–142
6. Blum C (2008) Beam-ACO for simple assembly line balancing. INFORMS J Comput 20(4):618–627
7. Blum C, Roli A (2003) Metaheuristics in combinatorial optimization: overview and conceptual comparison. ACM Comput Surv 35(3):268–308
8. Börgers T, Sarin R (1997) Learning through reinforcement and replicator dynamics. J Econ Theor 77(1):1–14 Nov
9. Brenner T (1998) Can evolutionary algorithms describe learning processes? J Evol Econ 8(3):271–283
10. Broom M, Rychtář J (2013) Game-theoretical models in biology. Chapman and Hall/CRC
11. Chellapilla K, Fogel DB (2001) Evolving an expert checkers playing program without using human expertise. IEEE Trans Evol Comput 5(4):422–428

12. Davis TE (1991) Toward an extrapolation of the simulated annealing convergence theory onto the simple genetic algorithm. Ph.D. thesis, University of Florida
13. Davis TE, Príncipe JC (1991) A simulated annealing like convergence theory for the simple genetic algorithm. In: ICGA, pp 174–181
14. Davis TE, Principe JC (1993) A markov chain framework for the simple genetic algorithm. Evol comput 1(3):269–288
15. Deneubourg J-L, Aron S, Goss S, Pasteels JM (1990) The self-organizing exploratory pattern of the argentine ant. J Insect Behav 3(2):159–168
16. Dorigo M (1992) Optimization, learning and natural algorithms. Ph. D. Thesis, Politecnico di Milano, Italy
17. Dorigo M, Birattari M (2010) Ant colony optimization. In: Encyclopedia of machine learning. Springer, pp 36–39
18. Dueck G (1993) New optimization heuristics: the great deluge algorithm and the record-to-record travel. J Comput phys 104(1):86–92
19. Erdős P, Rényi A (1959) On random graphs Publ Math (Debrecen) 6:290–297
20. Erdős P, Rényi A (1960) On the evolution of random graphs. Mat Kutato Int Koz 1(5):17–61
21. Ficici S, Melnik O, Pollack J (2000) A game-theoretic investigation of selection methods used in evolutionary algorithms. In: Proceedings of the 2000 congress on evolutionary computation, vol 2, pp 880–887
22. Ficici SG, Pollack JB (2000) A game-theoretic approach to the simple coevolutionary algorithm. In: Schoenauer M, Deb K, Rudolph G, Yao X, Lutton E, Merelo JJ, Schwefel H-P (eds) Parallel problem solving from nature PPSN VI. Lecture notes in computer science, 1917. Springer, Berlin Heidelberg, pp 467–476. doi:10.1007/3-540-45356-3_46
23. Fisher RA (1958) Polymorphism and natural selection. J Ecol 46(2):289–293 Jul
24. Fraser G, Arcuri A, McMinn P (2015) A memetic algorithm for whole test suite generation. J Syst Softw 103:311–327
25. Fudenberg D, Levine D (1998) The theory of learning in games. MIT Press, Cambridge MA
26. Futuyma DJ, Moreno G (1988) The evolution of ecological specialization. Annu Rev Ecol Syst 207–233
27. García J, Traulsen A (2012) The structure of mutations and the evolution of cooperation. PloS one 7(4):e35287
28. Goldberg DE, Segrest P (1987) Finite markov chain analysis of genetic algorithms. In: Proceedings of the 2nd international conference on genetic algorithms, vol 1, p 1
29. González MA, Vela CR, Varela R (2012) A competent memetic algorithm for complex scheduling. Nat Comput 11(1):151–160
30. Granville V, Křivánek M, Rasson J-P (1994) Simulated annealing: a proof of convergence. IEEE Trans Pattern Anal Mach Intell 16(6):652–656
31. Green DG (1982) Fire and stability in the postglacial forests of southwest nova scotia. J Biogeogr 9:29–40
32. Green DG (1994) Connectivity and complexity in ecological systems. Pac Conserv Biol 1(3):194–200
33. Green DG, Klomp NI, Rimmington GR, Sadedin S (2006a) Complexity in landscape ecology. Springer, Amsterdam
34. Green DG, Leishman TG, Sadedin S (2006b) Dual phase evolution—a mechanism for self-organization in complex systems. InterJournal, 1–8
35. Green DG, Liu J, Abbass H (2014) Dual phase evolution. Springer, Amsterdam
36. Green DG, Newth D, Kirley MG (2000) Connectivity and catastrophe—towards a general theory of evolution. In: Bedau M, McCaskill JS, Packard NH, Rasmussen S, McCaskill J, Packard N (eds) Proceedings of the 7th international conference on the synthesis and simulation of living systems (ALife VII)
37. Grinstead CM, Snell JL (2012) Introduction to probability. American Mathematical Soc
38. Gutjahr WJ (2000) A graph-based ant system and its convergence. Future Gener Comput Syst 16(8):873–888

39. Gutjahr WJ (2002) Aco algorithms with guaranteed convergence to the optimal solution. Inf Process Lett 82(3):145–153
40. Hajek B (1988) Cooling schedules for optimal annealing. Math Oper Res 13(2):311–329
41. Hernandez G, Nino F, Garcia J, Dasgupta D (2004) On geometric and statistical properties of the attractors of a generic evolutionary algorithm. In: Congress on Evolutionary Computation, CEC2004, vol 2. IEEE, pp 1240–1247
42. Hernandez G, Wilder K, Nino F, Garcia J (2005) Towards a self-stopping evolutionary algorithm using coupling from the past. In: Proceedings of the 2005 conference on genetic and evolutionary computation. ACM, pp 615–620
43. Hofbauer J, Sigmund K (1988) The theory of evolution and dynamical systems: mathematical aspects of selection. Cambridge University Press, Cambridge
44. Holland JH (1973) Genetic algorithms and the optimal allocation of trials. SIAM J Comput 2(2):88–105
45. Hoos HH, Stützle T (2004) Stochastic local search: foundations & applications. Elsevier
46. Houck CR, Joines JA, Kay MG (1996) Utilizing Lamarckian evolution and the Baldwin effect in hybrid genetic algorithms. NCSU-IE Technical Report 96-01
47. Kemeny JG, Snell JL (1960) Finite markov chains, vol 356. van Nostrand Princeton, NJ
48. Kirkpatrick S, Gelatt CD, Vecchi MP et al (1983) Optimization by simulated annealing. Science 220(4598):671–680
49. Kirley M, Green DG, Newth D (2000) Multi-objective problem, multi-species solution: an application of the cellular genetic algorithm. In: Mohammadian M (ed) Proceedings of international conference on advances in intelligent systems: theory and applications. IOS Press, pp 129–134
50. Kirley MG (2002) A cellular genetic algorithm with disturbances: optimisation using dynamic spatial interactions. J Heuristics 8:242–321
51. Kirley MG, Li X, Green DG (1999) Investigation of a cellular genetic algorithm that mimics landscape ecology. Lect Notes Comput Sci 1585(1999):90–97
52. Levine J, Ducatelle F (2004) Ant colony optimization and local search for bin packing and cutting stock problems. J Oper Res Soc 55(7):705–716
53. Liu W-B, Wang X-J (2008) An evolutionary game based particle swarm optimization algorithm. J Comput Appl Math 214(1):30–35 Apr
54. Maynard Smith J (1982) Evolution and the theory of games. Cambridge University Press, Cambridge
55. McAvoy A (2015) Stochastic selection processes. arXiv preprint arXiv:1511.05390
56. Moran PAP (1962) The statistical processes of evolutionary theory. Clarendon Press, Oxford
57. Moscato P (1989) On evolution, search, optimization, genetic algorithms and martial arts: towards memetic algorithms. Technical Report 826, California Institute of Technology, Pasadena, California, USA
58. Moscato P (1999) Memetic algorithms: a short introduction. In: Corne D, Dorigo M, Glover F (eds) New ideas in optimization. McGraw-Hill, pp 219–234
59. Nix AE, Vose MD (1992) Modeling genetic algorithms with markov chains. Ann Math Artif Intell 5(1):79–88
60. Nowak MA (2006) Evolutionary dynamics. Harvard University Press
61. Paperin G, Green DG, Leishman TG (2008) Dual phase evolution and self-organisation in networks. In: Li X, Kirley M, Zhang M, Green D, Ciesielski V, Abbass H, Michalewicz Z, Hendtlass T, Deb K, Tan KC, Branke J, Shi Y (eds) Proceedings of the 7th international conference on simulated evolution and learning. Springer, Melbourne, Australia
62. Paperin G, Green DG, Sadedin S (2011) Dual phase evolution in complex adaptive systems. J R Soc Interface
63. Pelillo M (1999) Replicator equations, maximal cliques, and graph isomorphism. Neural Comput 11(8):1933–1955
64. Popovici E, Bucci A, Wiegand RP, De Jong ED (2012) Coevolutionary principles. In: Handbook of natural computing. Springer, pp 987–1033

65. Propp JG, Wilson DB (1996) Exact sampling with coupled markov chains and applications to statistical mechanics. Random Struct Algorithms 9(1–2):223–252
66. Rager M, Gahm C, Denz F (2015) Energy-oriented scheduling based on evolutionary algorithms. Comput Oper Res 54:218–231
67. Rashedi E, Nezamabadi-pour HSS (2009) Gsa: a gravitational search algorithm. Inf Sci 179(13)
68. Savage JE (1998) Models of computation: exploring the power of computing
69. Schmitt LM (2001) Theory of genetic algorithms. Theor Comput Sci 259:1–61
70. Skyrms B (2003) The stag-hunt game and the evolution of social structure. Cambridge University Press, Cambridge
71. Strogatz S (1994) Nonlinear dynamics and chaos: with applications to physics, biology, chemistry, and engineering. Perseus Books, Cambridge, Massachusetts
72. Stützle T, Dorigo M (2002) A short convergence proof for a class of ant colony optimization algorithms. IEEE Trans Evol Comput 6(4):358–365
73. Taylor PD, Jonker LB (1978) Evolutionary stable strategies and game dynamics. Math Biosci 40(1):145–156 Jul
74. Trautmann H, Wagner T, Naujoks B, Preuss M, Mehnen J (2009) Statistical methods for convergence detection of multi-objective evolutionary algorithms. Evol Comput 17(4): 493–509
75. Turney P, Whitley D, Anderson R (1997) Evolution, learning and instinct: 100 years of the baldwin effect. Evolu Comput 4(3):
76. Velez M, Nino F, Alonso OM (2004) A game-theoretic approach to artificial immune networks. In: Artificial immune systems. Springer, pp 372–385
77. Voset MD, Liepinsl GE (1991) Punctuated equilibria in genetic search. Complex Syst 5:31–44
78. Whitley LD, Gordon VS, Mathias KE (1994) Lamarckian evolution, the baldwin effect and function optimization. In: Proceedings of the international conference on evolutionary computation. The 3rd conference on parallel problem solving from nature. Lecture notes in computer science. Springer, pp 6–15
79. Wright S (1931) Evolution in mendelian populations. Genetics 16:97–159
80. Yang XS, Karamanoglu M, He X (2014) Flower pollination algorithm: a novel approach for multiobjective optimization. Eng Optim 46(9):1222–1237
81. Yang XS (2008) Nature-inspired metaheuristic algorithms frome. Luniver Press
82. Yang XS (2010) A new metaheuristic bat-inspired algorithm
83. Yang X-S (2012) Flower pollination algorithm for global optimization. In: Unconventional computation and natural computation. Lecture notes in computer science, vol 7445, pp 240–249

Multimodal Function Optimization Using an Improved Bat Algorithm in Noise-Free and Noisy Environments

Momin Jamil, Hans-Jürgen Zepernick and Xin-She Yang

Abstract Modern optimization problems in economics, medicine, and engineering are becoming more complicated and have a convoluted search space with multiple minima. These problems are multimodal with objective functions exhibiting multiple peaks, valleys, and hyperplanes of varying heights. Furthermore, they are nonlinear, non-smooth, non-quadratic, and can have multiple satisfactory solutions. In order to select a best solution among several possible solutions that can meet the problem objectives, it is desirable to find many such solutions. For these problems, the gradient information is either not available or not computable within reasonable time. Therefore, solving such problems is a challenging task. Recent years have seen a plethora of activities to solve such multimodal problems using non-traditional methods. These methods are nature inspired and are becoming popular due to their general applicability and effective search strategies. In this chapter, we assess the ability of an improved bat algorithm (IBA) to solve multimodal problems in noise-free and additive white Gaussian noise (AWGN) environments. Numerical results are presented to show that the IBA can successfully locate multiple solutions in both noise-free and AWGN environments with a relatively high degree of accuracy.

M. Jamil (✉)
Harman/Becker Automotive Systems GmbH,
Becker-Goering Str. 16, 76307 Karlsbad, Germany
e-mail: momin.jamil@harman.com

H.-J. Zepernick
Blekinge Institute of Technology, 371 79 Karlskrona, Sweden
e-mail: hans-jurgen.zepernick@bth.se

X.-S. Yang
Middlesex University, School of Science and Technology, London NW4 4BT, UK
e-mail: x.yang@mdx.ac.uk

© Springer International Publishing AG 2017   29
S. Patnaik et al. (eds.), *Nature-Inspired Computing and Optimization*,
Modeling and Optimization in Science and Technologies 10,
DOI 10.1007/978-3-319-50920-4_2

1 Introduction

Most of the global optimization (GO) problems exhibit a highly rugged landscape. The challenge for any optimization algorithm is to find the highest or lowest point in such a landscape. In GO problems, the lowest or highest points refer to solution points in the function landscape. Over the years, nature-inspired population-based metaheuristic algorithms have been proven as an effective alternative to conventional optimization methods to solve modern engineering and scientific problems with complex search spaces and multiple optima. In case of multimodal problems with multiple optima, ideally, it is desirable to find all or as many as possible optima mainly for two reasons. First, to find all the optima can be dictated by a number of factors such as unavailability of some critical sources, or satisfaction of codal properties, etc. [37]. If such a situation arises, it is required to find or work on other available solutions. Second, the knowledge of the multiple solutions in a problem search space may provide valuable insight into the nature of the design space and, potentially, suggest alterative innovative solutions.

Population-based algorithms, such as differential evolution (DE), evolutionary strategies (ES), genetic algorithm (GA), and particle swarm optimization (PSO), have been extensively used to solve such problems. The intrinsic parallel nature of these algorithms enables them to simultaneously search multiple solution points in a search space. However, these algorithms tend to loose diversity and converge to a global best solution due to genetic drift [37]. Therefore, the two main challenges in these algorithms are (i) to maintain adequate population diversity so that multiple optimum solutions can be found, and (ii) how to preserve and maintain the discovered solution from one generation to another.

Iterative methods such as tabu search [9], sequential niche technique [2], and different niching methods [5, 10, 11, 20, 22, 25, 26, 28, 31, 40, 45] have been proposed to solve multimodal problems. Niches could aid differentiation of the species and maintain the diversity. These methods use various techniques that prevent algorithms to converge to the same solutions by prohibiting the algorithm from exploring those portions of the search space that have been already explored.

In recent years, Lévy flights also have been proposed within the context of meta-heuristics algorithms to solve optimization problems [19, 29, 30, 35, 42]. It should be noted that by incorporating Lévy flights, the probability of visitation to new areas in a function landscape is increased. This, in turn, reduces the probability of search particles returning to the previously visited points in a search space. Thus, it can be argued that Lévy flights can be used as a search strategy in stochastic algorithms to solve complicated problems.

Animal foragers in nature hunt or search for food in dynamical environments without a priori knowledge about its location. Many studies show that search or flight behavior of animals and insects demonstrates the typical characteristics of Lévy flights [3, 4, 34]. These studies show that animals search for food in a random or quasi-random manner comprising active search phases and randomly alternating with phase of fast ballistic motion. The foraging path of an animal is random walk, in which the next move is based on the current location and the transition probability to the next location.

In this work, we investigate the effectiveness of replacing the dynamics of the bats within the originally proposed bat algorithm (BA) [44] by random sampling drawn from a Lévy distribution. This improved version of the BA also known as IBA is used to solve multimodal optimization problems. Its performance is experimentally verified on a well-chosen, comprehensive set of complex multimodal test problems varying both in difficulty and in dimension in both noise-free and additive white Gaussian noise (AWGN) environments. Without loss of generality, we will consider only minimization problems, with an objective, given a multimodal function, to locate as many global minima as possible.

The rest of the chapter is organized as follows. Section 2 presents an overview of the improved bat algorithm (IBA). In Sect. 3, experimental results on the performance of IBA on multimodal functions are presented. In Sect. 4, we compare the performance of IBA with other metaheuristic algorithms. In Sect. 5, results on the performance of IBA in a noisy environment are presented. Finally, Sect. 6 concludes the chapter.

2 Improved Bat Algorithm

In this section, an improved version of the BA [44] is formulated with the aim of making the method more practical for a wider range of problems, but without loosing the attractive features of the BA. The BA, mimicking the echolocation behavior of certain species of bats, is based on the following set of rules and assumptions [44]:

1. It is assumed that the bats know the difference between food/prey, background barriers, and use echolocation to sense the proximate distance from the prey;
2. The virtual bats are assumed to fly randomly with a frequency f_{min} with velocity $\mathbf{v}_i$ at position $\mathbf{x}_i$ by varying wavelength λ (or frequency f) and loudness A_0. Depending on the proximity from the target, the wavelength (or frequency) of emitted pulses and the rate of pulse emission $r \in [0, 1]$ can be adjusted by the bats;
3. It is further assumed that the loudness varies from a large (positive A_0) to a minimum value of A_{min};
4. Furthermore, ray tracing is not used in estimating the time delay and three-dimensional topography;
5. The frequency f is considered in the range $[0, f_{max}]$ corresponding to the range of wavelengths $[0, \lambda_{max}]$;

By making use of the above rules and assumptions, the standard bat algorithm (SBA) will always find the global optimum [44]. However, in SBA, bats rely purely on random walks drawn from a Gaussian distribution; therefore, speedy convergence may not be guaranteed [44].

In the improved version of the BA, the random motion of bats based on Gaussian distribution is replaced by Lévy flight. The motivation is that the power-law behavior of the Lévy distribution produces some members of the random population in the distant regions of the search space, while other members are concentrated around the mean of the distribution. The power-law behavior of the Lévy distribution also helps to induce exploration at any stage of the convergence, making sure that the system does not get trapped in local minima. The Lévy distribution also reduces the probability of returning to the previously visited sights, while the number of visitations to new sights is increased.

Secondly, we incorporate an exponential model of frequency tuning and pulse emission rate compared to a fixed setting of these parameters in BA. This incorporation into this model has two advantages: (i) frequency tuning and dynamic control of exploitation; and (ii) exploration by automatic switching to intensive exploitation. The frequency-based tuning and pulse emission rate change to imitate the behavior of a bat lead to good convergence. Pulse emission rate and loudness control mechanism in the BA help in maintaining a fine balance between exploration and exploitation. In fact, an automatic switch to more extensive exploitation instead of exploration can be achieved when bats are in the vicinity of the optimal solution. The basic steps of this improved version of the BA known as IBA are summarized in the pseudo-code shown in Procedure 1 [16].

The position and velocity of bat i are represented by the vectors $\mathbf{x}_i$ and $\mathbf{v}_i$, respectively. The pulse frequency for bat i at position $\mathbf{x}_i$ is denoted by f_i. The initial pulse emission rate, updated pulse emission rate after each iteration, initial loudness, and updated loudness after each iteration for bat i are represented by symbols $r_i^{t_0}$, r_i, $A_i^{t_0}$, and A_i in (1) and (2), respectively. The symbols γ and α in (1) and (2) are constants. The symbols $\beta \in [0, 1]$, f_{min}, and f_{max} in (3) denote a random number drawn from a uniform distribution, minimum, and maximum frequency of the emitted pulse, respectively. In (4), the symbol $\mathbf{x}_i^{\text{best}}$ represents the current best solution found by bat i by comparing all the solutions among all the NB bats and 'rand' is a random number drawn from a uniform distribution. The parameter δ in (6) scales the random step size S and is related to the problem dimension [43]. For most of the cases, $\delta = 1$ can be used; however, it can be set to smaller values when the dimension of the problem is small [43]. A smaller value of δ might hold for simple unimodal problems. However, for multimodal problems with multiple global minima scattered in the search space, a smaller step size might hinder the search process. The operator $\odot$ denotes entrywise multiplication.

Procedure 1: Pseudo-code of IBA based on Lévy Flights [16]

1: Initialize the positions $\mathbf{x}_i$, $i = 1, 2, ..., NB$, and velocity $\mathbf{v}_i$ of the NB bats in D-dimensional problem space.
2: Define the pulse frequency f_i at position $\mathbf{x}_i$.
3: Initialize the pulse emission rate r^{t_0} and the loudness A^{t_0}.
4: **for** all $\mathbf{x}_i$
5:     Calculate objective function of all NB bats $F_i = f(\mathbf{x}_i)$

6: **end for**

7: Rank the position of the bats according to their fitness values.

8: Find the current best bat position $\mathbf{x}_i^{\text{best}}$.

9: $t = 1$

10: **while** (t < Maximum number of iterations)

11:    $t = t + 1$

12:    Update r_i and A_i as

$$r_i^{t+1} = r_i^{t_0}[1 - \exp(-\gamma t)] \tag{1}$$

$$A_i^{t+1} = \alpha A_i^{t_0} \tag{2}$$

13:        **for** (loop over all NB bats and all D dimensions)

14:            Adjust frequency as

$$f_i = f_{min} + (f_{max} - f_{min})\beta \tag{3}$$

15:        Update velocities as

$$\mathbf{v}_i^t = \mathbf{v}_i^{t-1} + (\mathbf{x}_i^{t-1} - \mathbf{x}_i^{\text{best}})f_i \tag{4}$$

16:        Update locations or new solutions as

$$\mathbf{x}_i^t = \mathbf{x}_i^{t-1} + \mathbf{v}_i^t \tag{5}$$

17:        **if** (rand > r_i)

18:            Generate new solution using Lévy Flight as

$$\mathbf{x}_i^t = \mathbf{x}_i^{\text{best}} + \delta \odot \mathbf{L}_\lambda(S) \tag{6}$$

19:        **end if**

20:        Check if the new solution (5) or (6) is within the problem domain.

21:        **if** (True)

22:            Move to next step.

23:        **else**

24:            Apply problem bound constraints.

25:        **end**

26:        Evaluate the objective function at new location (solution) generated as

$$F_k = f(\mathbf{x}_i) \tag{7}$$

27:        **if** ((rand < A_i) or ($F_k < F_i$))

28:            Accept the new solution.

29:        **end if**

30:        Rank the bats and find the current best $\mathbf{x}_i^{\text{best}}$.

31:        **end for**
32: **end while**

In (6), the random step size S is drawn from a Lévy distribution which has infinite mean and variance:

$$L_\lambda(S) \sim \frac{1}{S^{\lambda+1}} |S| >> 0 \qquad (8)$$

The probability of obtaining a Lévy random number in the tail of the Lévy distribution is given by (8). The choice of λ in (8) can have a significant impact on the search ability of the optimization algorithm based on Lévy flights. Therefore, some experimentation is required to find a proper value of λ [35, 43]. In order to determine a proper value of λ, five different benchmark functions were selected. These problems include unimodal (Beale), multimodal with a few local minima (Carrom table and Holder table), and multimodal functions with many local minima (Griewank, Eggholder) functions. For each of these problems, 50 independent trails were conducted for a fixed number of iterations with $\lambda = 1.3$, 1.4, 1.5, and 1.6. In order to determine an effective value of λ, the best values produced by the IBA for each landscape were recorded and averaged over the number of trails. The only value of λ that performed well on all landscapes was found to be 1.5. This value was subsequently chosen for all the tests presented later in this chapter. The value of $\lambda = 1.5$ has also been used in other metaheuristic algorithms based on Lévy flights [35, 43].

3   IBA for Multimodal Problems

In this section, the results obtained by applying IBA on a set of multimodal functions with multiple solutions are presented. The effectiveness of the IBA to handle multimodal problems is validated on a set of benchmark functions with different characteristics.

3.1   Parameter Settings

The universal set of initial values of parameters such as $NB, r_i^{t_0}, r_i, A_i^{t_0}$, and A_i that are applicable to all the benchmark functions used in this study do not exist. Therefore, finding effective values of these parameters requires some experimentation. The parameter settings listed in Table 1 are obtained from the trial experiments on a limited set of benchmark functions with two and higher dimensions. We have used different values of $NB = 50, 100, 150, 200$ and found that $NB = 100$ seems to be sufficient for most multimodal optimization problems considered in this work. For

Table 1 Parameter settings for IBA

Parameter	Value
Number of bats (population size) NB	100
Number of generations G	5000
Number of runs R	100
Initial loudness A^{t0}	0.1
Initial pulse emission rate r^{t0}	0.1
Constants $\alpha = \gamma$	0.9
Lévy step length λ	1.5
Minimum frequency f_{min}	0
Maximum frequency f_{max}	Dependent on problem domain size

tougher problems, larger NB can be used only if there is no better alternative, as it is more computationally expensive [43]. In summary, the parameter settings listed in Table 1 were used for all the experiments, unless we mention new settings for one or other parameters.

3.2   Test Functions

In various applications, the objective function exhibit multiple global minima. In order to experimentally evaluate the strengths and weaknesses of IBA, we have used a set of well-chosen test functions [12, 15, 21, 23], representing different characteristics and various levels of difficulty. The test suite includes functions having evenly and unevenly spaced multiple global optima, multiple global optima in the presence of multiple local minima, and deceptiveness. The test functions are listed in Table 2 with their respective domain sizes, number of minima (multimodal functions), and value of global minimum.

3.3   Numerical Results

Over the years, several different performance measures to evaluate the performance of stochastic algorithms have been reported in the literature [33]. However, there are no standard performance measures and in practice the results are reported as averages from certain number of independent runs for stochastic algorithms. Therefore, we will adapt commonly used performance measures and report results in terms of mean and standard deviation (SD) attained with a population size and number of generations.

Table 2 List of test functions

Function	Domain size	No. of minima	Global minima
f_1	$[-10, 10]$	1	-176.1375
f_2	$[0, pi]$	1	-1.0813
f_3	$[0, pi]$	1	0.9
f_4	$[0, 10]$	2	-0.6737
f_5	$[-5, 5]$	2	-4.5901
f_6	$[-5, 5]$	2	-0.8510
f_7	$[-5, 5]$	2	-1.0316
f_8	$[-10, 10]$	2	-10.8723
f_9	$[-5, 5]$	2	0
f_{10}	$[-5, 0; 10, 15]$	3	0.3979
f_{11}	$[-10, 10]$	4	-24.1568
f_{12}	$[-10, 10]$	4	-2.0626
f_{13}	$[-5, 5]$	4	0
f_{14}	$[-10, 10]$	4	-19.2085
f_{15}	$[-10, 10]$	4	-26.9203
f_{16}	$[-10, 10]$	4	-0.9635
f_{17}	$[-1, 1]$	6	-1
f_{18}	$[-10, 10]$	9	-176.541
f_{19}	$[-10, 10]$	9	-24.0624
f_{20}	$[-20, 20]$	12	-1
f_{21}	$[-10, 10]$	18	-186.7309
f_{22}	$[-7, 7]$	20	0
f_{23}	$[0, 1]$	25	-1
f_{24}	$[0, 1]$	25	-1

f_1: Levy 5, f_2: Michaelwicz, f_3: Periodic, f_4: Keane, f_5: Modified Ackley, f_6: Multimodal Yang, f_7: Six Hump Camel, f_8: Test Tube Holder, f_9: Trecanni, f_{10}: Branin, f_{11}: Carrom Table f_{12}: Cross-in-Tray, f_{13}: Himmelblau, f_{14}: Holder Table 1, f_{15}: Holder Table 2, f_{16}: Pen Holder f_{17}: Root 6, f_{18}: Hansen, f_{19}: Henrik-Madsen, f_{20}: Butterfly, f_{21}: Shubert f_{22}: Parsopoulos, f_{23}: Deb 1, f_{24}: Deb 2

The best fitness value produced by the IBA after each run was recorded. The mean fitness value and SD are presented in Table 3. From the results presented in Table 3, it can be seen that IBA performs equally well on almost all of the functions tested. The only two exceptions are f_{17} and f_{21} which show small deviation of $10E-4$ and $10E-5$ from the known global minima.

The minimum, maximum, and mean number of function evaluations (NFE) required by the IBA to converge to a solution averaged over the number of independent runs are also reported. From the results presented in Table 3, it can be seen that IBA can locate global minima with relatively few NFE (lower mean NFE values) with the exception of f_{20}. This function has 760 local minima and 18 global minima.

Table 3 Statistical results of 100 runs obtained by IBA for 2D functions (global minima ≥ 2). NFE$_{min}$: Minimum number of function evaluations; NFE$_{max}$: Maximum number of function evaluations; NFE$_{mean}$: Mean of number of function evaluations; Mean: Mean of fitness value; SD: Standard deviation of the fitness values

Function	NFE$_{min}$	NFE$_{max}$	NFE$_{mean}$	Mean	SD
f_4	800	2900	1817	−0.6737	3.3474E−16
f_5	3300	6700	5417	−4.5901	8.9265E−16
f_6	3300	11800	7294	−0.8512	1.2274E−15
f_7	1400	5500	3851	−1.0316	1.3390E−15
f_8	3700	12100	6572	−10.8723	1.7853E−15
f_9	2100	5500	3952	0	0
f_{10}	2800	8200	4899	0.3979	5.5791E−17
f_{11}	2500	5400	3774	−24.1568	1.7853E−14
f_{12}	3300	6400	4766	−2.0626	8.9265E−16
f_{13}	2600	6500	4343	0	0
f_{14}	2600	7500	4282	−19.2085	2.8656E−14
f_{15}	2400	5700	3910	−26.9203	4.9989E−14
f_{16}	2700	7700	5166	−0.9635	2.23116E−15
f_{17}	2200	77500	7244	−0.9995	1.4824E−04
f_{18}	4100	10100	6559	−176.5417	5.7130E−14
f_{19}	2600	10700	6343	−24.0624	1.0712E−14
f_{20}	7650	238800	65250	−1	0
f_{21}	4300	500000	125481	−186.7309	2.8468E−05
f_{22}	2900	8100	5322	0	0
f_{23}	1800	8400	3614	−1	1.3530E−06
f_{24}	1800	1800	8400	−1	0

The 18 global minima are classified into 9 different clusters. Each cluster consists of two solutions that are separated by small distance 0.98, while the distance between two clusters is 5.63 [13]. This function is difficult to optimize for any algorithm [13, 14] and IBA is no exception. Box plots are shown in Figs. 1 and 2 depicting the NFE required by the IBA to converge to a solution. The median and mean values of the NFE are depicted by the notch and 'o,' whereas the whiskers above and below the box give additional information about the spread of the NFE. Whiskers are vertical lines that end in a horizontal stroke. The '+' markers in these plots represent the outliers that are beyond the lower and upper whiskers, i.e., NFE$_{min}$ and NFE$_{max}$ in Table 3.

Figures 3, 4, 5, 6, and 7 show the contour plots of functions f_{20}, f_{21}, f_{22}, f_{23}, and f_{24}, respectively. The number of global minima for f_{20} and f_{22} depends on the problem domain size. For a domain size of $[-20, 20]^D$ and $[-7, 7]^D$, where D represents the problem dimension, functions f_{20} and f_{22} have 12 and 20 global

Fig. 1 Box plot depicting
the NFE for function f_4 to
f_{13}

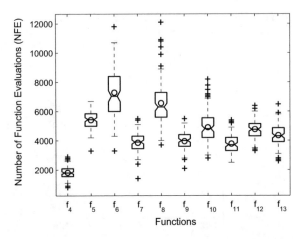

Fig. 2 Box plot depicting
the NFE for function f_{14} to
f_{24}

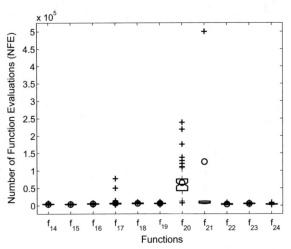

minima, respectively. The functions f_{23} and f_{24} have 25 global minima that are evenly and unevenly spread in the function landscape [3]. The '+' markers in these figures represent the global minima (solutions) located by the IBA in 100 independent runs with the parameter settings in Table 1. From the results presented in these figures, we can see that IBA is able to locate global minima with relatively high degree of accuracy. Due to the complex landscape of f_{20}, the number of bats $NB = 100$ is found to be not sufficient to locate all the global minima. We have found that with $NB = 125$, IBA was able to locate all the global minima for function f_{20}.

According to [36], for practical reasons, it is desirable to find all global minima. The ability and effectiveness of the IBA to accurately locate all of the multiple minima, a level of accuracy, $\epsilon \in (0, 1]$ is defined. If the Euclidean distance of a computed solution to a known global optimum is less than ϵ, then the solution is

Fig. 3 Contour plot of
function f_{20} (12 global
minima indicated by '+'
found by IBA after 100
independent runs)

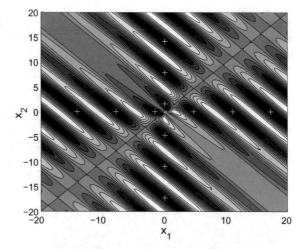

Fig. 4 Contour plot of
function f_{21} (18 global
minima indicated by '+'
found by IBA after 100
independent runs)

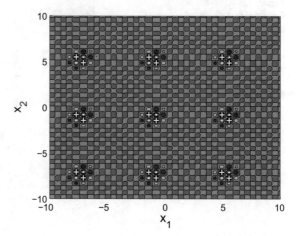

considered as a global optimum. The ability of IBA to consistently locate all the solutions for each function for a given set of parameters is also measured. The number of runs converged on solutions in % in Tables 4 and 5 signifies the number of the independent runs converged on the solutions. From the results presented in these tables, it can be seen that IBA could find all the solutions for a specified value of ϵ. These results also demonstrate the performance consistency of the IBA which measures the ability of the algorithm to consistently locate all the solutions for each function for a given set of parameters.

Fig. 5 Contour plot of function f_{22} (25 global minima indicated by '+' found by IBA after 100 independent runs)

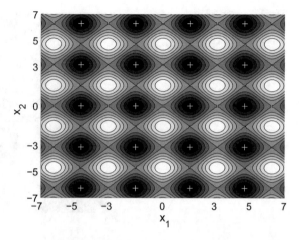

Fig. 6 Contour plot of function f_{23} (25 global minima indicated by '+' found by IBA after 100 independent runs). The minima are located at $\{x_1, x_2\} = 0.1, 0.3, 0.5, 0.7, 0.9$

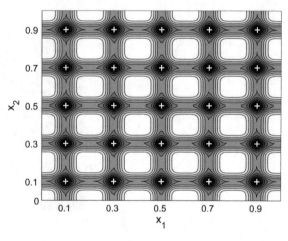

Fig. 7 Contour plot of function f_{24} (25 global minima indicated by '+' found by IBA after 100 independent runs). The minima are located at $\{x_1, x_2\} = 0.08, 0.246, 0.45, 0.681, 0.934$

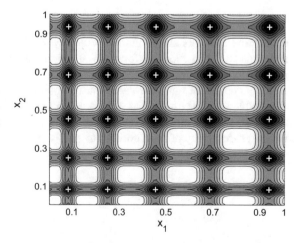

Table 4 IBA performance on test functions for $\epsilon = 10^{-5}$

Function	Solutions	% Number of runs converged on solutions								
		1	2	3	4	5	6	7	8	9
f_4	2	45	55	–	–	–	–	–	–	–
f_5	2	49	51	–	–	–	–	–	–	–
f_6	2	43	57	–	–	–	–	–	–	–
f_7	2	50	50	–	–	–	–	–	–	–
f_8	2	47	53	–	–	–	–	–	–	–
f_9	2	57	43	–	–	–	–	–	–	–
f_{10}	3	33	30	37	–	–	–	–	–	–
f_{11}	4	28	23	28	21	–	–	–	–	–
f_{12}	4	25	21	26	28	–	–	–	–	–
f_{13}	4	40	11	21	28	–	–	–	–	–
f_{14}	4	23	21	29	27	–	–	–	–	–
f_{15}	4	23	31	21	25	–	–	–	–	–
f_{16}	4	29	26	17	28	–	–	–	–	–
f_{17}	6	24	19	21	19	11	6	–	–	–
f_{18}	9	5	11	8	12	13	16	10	14	11
f_{19}	9	5	16	11	13	19	15	6	8	7

4 Performance Comparison of IBA with Other Algorithms

In this section, we compare IBA with other metaheuristic algorithms, such particle swarm optimization (PSO) [18], differential evolution (DE) [32], and evolutionary strategy (ES) [38]. We abstain to introduce these algorithms and did not make a special effort to fine-tune different parameters of these algorithms. For DE, we have used the strategy DE/best/2/bin [32] with the weighting factor $F = 0.8$ and crossover constant $CR = 0.7$. We have used the standard version of PSO with global learning, i.e., no local neighborhoods, an inertial constant $\omega = 0.3$, a cognitive constant $c_1 = 1$, and a social constant for swarm interaction $c_2 = 1$. For ES, we have used an offspring of $\lambda = 10$ for each generation and standard deviation $\sigma = 1$ for changing solutions. In order to have a fair comparison, we have used the settings of Table 1 for the population size NB, fixed NFE, and R for these algorithms.

In Table 6, the number of minima found by each algorithm in 100 independent runs is shown in parentheses. The numerator in the parentheses represents the total number of minima found by each algorithm, while the denominator represents the actual number of minima. The boldface terms represent the best results. For example, in the case of the function f_{18}, IBA has found 9 out of 9 minima, i.e., (**9/9**), whereas the other algorithms found either fewer or none of the minima. These results show that IBA has an ability to find multiple solutions for all the functions tested.

Table 5 IBA performance on test functions for $\epsilon = 10^{-5}$

Solutions	% Number of runs converged on solutions				
	f_{20}	f_{21}	f_{22}	f_{23}	f_{24}
1	8	7	4	2	4
2	5	5	3	4	3
3	2	6	2	6	2
4	3	9	5	3	5
5	10	5	2	3	2
6	22	6	2	3	2
7	19	5	2	5	2
8	11	3	3	8	3
9	2	6	2	5	2
10	1	3	4	4	4
11	4	8	1	3	1
12	13	4	5	3	5
13	–	3	9	2	9
14	–	2	7	5	7
15	–	6	3	3	3
16	–	4	2	6	2
17	–	6	1	7	1
18	–	4	6	3	6
19	–	2	6	2	6
20	–	6	8	4	8
21	–	–	3	6	3
22	–	–	3	4	3
23	–	–	8	4	8
24	–	–	4	3	4
25	–	–	5	2	5

5 IBA Performance in AWGN

Optimization in a noisy environment occurs in various applications such as experimental optimization. The problem of locating either minima or maxima of a function is vital in applications such as design of wireless systems [8], spectral analysis and radio-astronomy [27], and semiconductor modeling and manufacturing [39]. Traditionally, the simplex method by Nelder and Mead [7, 24] has been used for optimization in noisy environments. However, in order to overcome the drawbacks and the deficiencies of simplex methods [17], different variants of the simplex method have been proposed [41]. More sophisticated methods and extensive studies in this direc-

Table 6 Comparison of IBA with DE, ES, and PSO. Mean: mean of best value; SD: standard deviation of the best value; SEM: standard error of mean

Function		IBA	DE	ES	PSO
f_4	Mean	−0.67	−0.67	−0.67	−0.6737
	SD	0	0	$2.13E−07$	$4.44E−09$
	Minima	**(2/2)**	**(2/2)**	(1/2)	(1/2)
f_8	Mean	−10.87	−10.87	−10.87	−10.87
	SD	0	$1.07E−14$	$2.50E−03$	$7.23E−04$
	Minima	**(2/2)**	**(2/2)**	(1/2)	(1/2)
f_{14}	Mean	−19.20	−19.20	−19.20	−19.20
	SD	0	0	$2.54E−05$	$9.50E−03$
	Minima	**(4/4)**	**(4/4)**	(2/4)	(2/4)
f_{16}	Mean	−0.96	−0.96	−0.96	−0.96
	SD	0	0	$5.41E−05$	$3.42E−06$
	Minima	**(4/4)**	**(4/4)**	(3/4)	(3/4)
f_{17}	Mean	−0.99	−1	−0.99	−0.99
	SD	$2.80E−07$	0	$9.57E−05$	$1.19E−04$
	Minima	**(6/6)**	**(6/6)**	(1/6)	(1/6)
f_{18}	Mean	−176.54	−176.54	−162.41	−176.39
	SD	0	0	11.33	$1.40E−01$
	Minima	**(9/9)**	**(9/9)**	(0/9)	(3/9)
f_{19}	Mean	−24.07	−24.06	−22.51	−24.05
	SD	0	0	1.18	$8.20E−03$
	Minima	**(9/9)**	**(9/9)**	(0/9)	(6/9)
f_{22}	Mean	0	0	$1.3723E−05$	$1.66E−05$
	SD	0	0	$1.41E−05$	$1.85E−05$
	Minima	**(12/12)**	(11/12)	(6/12)	(5/12)

tion are discussed in [1]. Different population-based algorithms, e.g., PSO, have also been used to optimize functions in AWGN and multiplicative noise environments [27].

Information about the function $f(x)$ is obtained in the form of $\widehat{f}(x)$, where $\widehat{f}(x)$ is an approximation of the true function value $f(x)$, corrupted by AWGN. The influence of AWGN on the values of the objective functions was simulated according to [6] and is given as

$$\widehat{f}(x) = f(x) + \eta, \quad \eta \sim N\left(0, \sigma^2\right) \tag{9}$$

where $\eta \sim N(0, \sigma^2)$ is a Gaussian distributed random variable with zero mean and variance σ^2.

5.1 Numerical Results

In this section, we study the ability of IBA to handle the two-dimensional multimodal function optimization in AWGN. However, these problems can also be extended to dimensions greater than 2. Due to space limitations, we selected 9 functions from those listed in Table 1. These include two functions with single global minimum and 7 functions with two or more than two global minima. Experiments were carried out for four different levels of $\sigma^2 = 0.01, 0.05, 0.07,$ and 0.09. At each function evaluation,

Table 7 Statistical results of 100 runs obtained by IBA on selected functions in AWGN. Mean: mean of fitness value; SD: standard deviation of the fitness values; MAPE: mean absolute percentage error

Function		$\sigma^2 = 0.01$	$\sigma^2 = 0.05$	$\sigma^2 = 0.07$	$\sigma^2 = 0.09$
$f_1(-176.1375)$	Mean	−176.1830	−176.3646	−176.5443	−176.5442
	SD	0.0025	0.0121	0.0171	0.228
	MAPE	0.0258	0.1289	0.1799	0.2309
$f_2(-1.8013)$	Mean	−1.8468	−2.0283	−2.1206	−2.2095
	SD	0.0025	0.0122	0.0193	0.0242
	MAPE	2.5281	12.6022	17.7255	22.6617
$f_3(0.9)$	Mean	0.8556	0.6838	0.6053	0.5330
	SD	0.0027	0.0254	0.0386	0.0529
	MAPE	4.9308	24.0261	32.7394	46.5648
$f_5(-4.59012)$	Mean	−4.6349	−4.8137	−4.9029	−4.9957
	SD	0.0029	0.0118	0.0169	0.0258
	MAPE	0.9753	4.8701	6.8144	8.8361
$f_6(-0.851)$	Mean	−0.8950	−1.0741	−1.1595	−1.2473
	SD	0.0025	0.0153	0.0184	0.0234
	MAPE	5.1666	26.2138	36.2536	46.5684
$f_{11}(-24.1568)$	Mean	−24.2016	−24.3806	−24.4665	−24.5567
	SD	0.0026	0.0140	0.0156	0.0215
	MAPE	0.1856	0.9266	1.2822	1.6556
$f_{14}(-19.20850)$	Mean	−19.2530	−19.4329	−19.5236	−19.6095
	SD	0.0027	0.0137	0.0183	0.0229
	MAPE	0.2319	1.1683	1.6403	2.0875
$f_{17}(-1)$	Mean	−1.0424	−1.2112	−1.2946	−1.3851
	SD	0.0027	0.0137	0.0211	0.0310
	MAPE	4.2370	21.1208	29.4648	38.5136
$f_{18}(-176.5418)$	Mean	−176.5849	−176.7617	−176.8452	−176.9367
	SD	0.0025	0.0134	0.0160	0.0216
	MAPE	0.0244	0.1245	0.1719	0.2237

noise was added to the actual function according to (9) for different values of σ^2. For each variance value, 100 independent runs of IBA were performed.

Based on the mean absolute percentage error (MAPE) results presented in Table 7, it can be seen that increasing the value of σ^2 deteriorates the ability of IBA to locate global minimum/minima for unimodal (f_2 and f_3) and multimodal (f_6 and f_{17}) functions. For these functions, a performance degradation of more than 20 and 40% can be observed for $\sigma^2 = 0.09$ that may adversely effect the ability of IBA to accurately locate global optima for these functions. The term enclosed in the brackets in front of the function names represents the known value of the global minimum (unimodal) and minima (multimodal) for these functions.

Table 8 IBA performance on selected test function in AWGN with different σ^2

Function	Mean Euclidean distance of computed solution from known solution				
	σ^2	1	2	3	4
f_1	0.01	$8.2301E-04$	–	–	–
	0.05	0.0019	–	–	–
	0.07	0.0022	–	–	–
	0.09	0.0023	–	–	–
f_2	0.01	0.0540	–	–	–
	0.05	0.0132	–	–	–
	0.07	0.0160	–	–	–
	0.09	0.0201	–	–	–
f_3	0.01	0.0282	–	–	–
	0.05	0.0830	–	–	–
	0.07	1.1067	–	–	–
	0.09	2.2026	–	–	–
f_5	0.01	0.0119	0.0117	–	–
	0.05	0.0332	0.0294	–	–
	0.07	0.0288	0.0311	–	–
	0.09	0.0319	0.0379	–	–
f_6	0.01	0.0607	0.0708	–	–
	0.05	0.1403	0.1759	–	–
	0.07	0.1633	0.1727	–	–
	0.09	0.2278	0.2046	–	–
f_{11}	0.01	0.0048	0.0057	0.0060	0.0044
	0.05	0.0151	0.0162	0.0112	0.0133
	0.07	0.0179	0.0160	0.0149	0.0151
	0.09	0.0206	0.0158	0.0153	0.0178
f_{14}	0.01	0.0107	0.0094	0.0088	0.0081
	0.05	0.0215	0.0207	0.0199	0.0241
	0.07	0.0238	0.0262	0.0242	0.0210
	0.09	0.0291	0.0289	0.0243	0.0238

Next, in Tables 8, 9, and 10, we compare the mean Euclidean distance (the difference between the obtained and actual global minimum) of the computed solution from known a priori solution for four levels of σ^2. From the results presented in Tables 8, 9, and 10, we can see that as the value of σ^2 increases, the ability of the IBA to locate global minima decreases significantly for f_3 (periodic) and f_6 (Yang

Table 9 IBA performance for function f_{18} in AWGN with different σ^2

Solutions	$\sigma^2 = 0.01$	$\sigma^2 = 0.05$	$\sigma^2 = 0.07$	$\sigma^2 = 0.09$
1	$9.9842E-04$	0.0026	0.0017	0.1838
2	$6.9885E-04$	0.0018	0.0021	0.0025
3	$8.4551E-04$	0.0020	0.0020	0.0026
4	$8.6135E-04$	0.0017	0.0024	0.0023
5	$9.8587E-04$	0.0020	0.0023	0.0024
6	$7.9620E-04$	0.0014	0.0020	0.0029
7	$8.1328E-04$	0.0016	0.0021	0.0029
8	$9.2927E-04$	0.0021	0.0021	0.0020
9	0.0012	0.0012	0.0027	0.0020

Table 10 IBA performance for function f_{22} in AWGN with different σ^2

Solutions	$\sigma^2 = 0.01$	$\sigma^2 = 0.05$	$\sigma^2 = 0.07$	$\sigma^2 = 0.09$
1	0.0291	0.1353	0.0813	0.1324
2	0.0169	0.0692	0.0699	0.2266
3	0.0264	0.0488	0.1290	0.1272
4	0.0297	0.0623	0.1169	0.1436
5	0.0462	0.0804	0.0701	0.1254
6	0.0074	0.1318	0.1765	0.0758
7	0.0196	0.0686	0.0930	0.0554
8	0.0439	0.0775	0.0826	0.1054
9	0.0237	0.0655	0.0676	0.0512
10	0.0243	0.0870	0.1073	0.0927
11	0.0284	0.0927	0.0562	0.1019
12	0.0173	0.0856	0.1072	0.0889
13	0.0241	0.1015	0.1081	0.1140
14	0.0335	0.0971	0.0781	0.1124
15	0.0303	0.0782	0.1083	0.0626
16	0.0443	0.0647	0.0658	0.1866
17	0.0313	0.0291	0.0540	0.1225
18	0.0349	0.0649	0.0896	0.0873
19	0.0421	0.0986	0.0966	0.1733
20	0.0365	0.0825	0.1145	0.0874

multimodal) functions. For these two functions, the largest difference between the obtained and actual global minimum can be observed.

Interestingly, in the case of functions with more than one global minimum (with the exception of f_6 and f_{17}) such as f_5, f_{11}, f_{14}, f_{18}, and f_{23} with few or many local minima, IBA seems to perform relatively well in AWGN even for the highest value of σ^2. It seems that the noise plays a positive role for these multimodal functions by helping the IBA to escape local minima of the objective function. For these multimodal functions, the performance degradation is significantly low for the highest value of σ^2. On the contrary, in the case of unimodal problems such as f_2 and f_3, significant performance deterioration can be observed for the highest value of σ^2. One can hypothesize that for the unimodal functions high values of σ^2 can create deep troughs (for minimization problems) or crests (for maximization problems) in the function landscape that are mistaken by the IBA or any other optimization algorithm as a global minimum or maximum.

6 Conclusions

In this work, IBA has been presented as a viable alternative to existing numerical optimization methods for unimodal and multimodal problems. In this chapter, the ability of IBA to solve unimodal and multimodal problems in non-noise and AWGN has been investigated. Performance results have been reported for a set of test functions with varying levels of difficulty, different number of minima, and different level of noise variances. The experimental results indicate that IBA is very stable and efficient in the presence of noise. It is a very noise-tolerant method and can be used for minimization or maximization of noisy functions. It has performed exceptionally well even in the presence of noise with high variance. Conclusively, IBA appears to be a viable alternative technique for solving global optimization problems and may offer an alternative where other techniques fail. However, further research may be required to fully comprehend the dynamics and the potential limits of the IBA.

References

1. Arnold DV (2001) Local performance of evolution strategies in the presence of noise. PhD thesis, University of Dortmund, Germany
2. Beasley D, Bull DR, Martin RR (1993) A sequential Niche technique for multimodal function optimization. Evol Comput 1(2):101–125
3. Brits R, Engelbrecht AP, van den Bergh F (2007) Locating multiple optima using particle swarm optimization. Appl Math Comput 189(12):1859–1883
4. Brown C, Liebovitch LS, Glendon R (2007) Lévy flights in Dobe Ju/Hoansi foraging patterns. Hum Ecol 35(1):129–138
5. De Jong KA (1975) An analysis of the behavior of a class of genetic adaptive system. PhD thesis, University of Michigan, USA

6. Elster C, Neumaier A (1997) A method of trust region type for minimizing noisy functions. Computing 58(1):31–46
7. Fletcher R (1987) Practical methods of optimization, 1st edn. Wiley, Chichester
8. Fortune SJ, Gay DM, Kernighan BW, Landron O, Valenzulea RA, Wright MH (1995) WISE design of indoor wireless systems. IEEE Comput Sci Eng 2(1):58–68
9. Glover F (1989) Tabu search—part I. ORSA J Comput 1(3):190–206
10. Goldberg DE, Richardson J (1987) Genetic algorithms with sharing for multimodal function optimization. In: Proceedings of international conference on genetic algorithms and their application, USA, pp 41–49
11. Harik GR (1995) Finding multimodal solutions using using restricted tournament selection. In: Proceedings of international conference on genetic algorithms and their application, USA, pp 24–31
12. http://www-optima.amp.i.kyoto-u.ac.jp/member/student/hedar/Hedar_files/TestGO_files/Page364.htm
13. Iwamatsu M (2006) Multi-species particle swarm optimizer for multimodal function optimization. IEICE Trans Inf Syst (English Edition) E89-D(3):1181–1187
14. Iwamatsu M (2006) locating all the global minima using multi-species particle swarm optimizer: the inertia weight and the constriction factor variants. In: Proceedings of IEEE congress on evolutionary computation, Canada, pp 816–822
15. Jamil M, Yang XS (2013) A literature survey of benchmark functions for global optimisation problems. Int J Math Model Numer Optim 4(2):150–194
16. Jamil M, Zepernick HJ, Yang XS (2015) Synthesizing cross-ambiguity function using improved bat algorithm (Ed Xin-She Yang). Springer, Heidelberg, pp 179–202
17. Kelley CT (1995) Iterative methods for optimization. SIAM
18. Kennedy J, Eberhart RC (1995) Particle swarm optimization. In: Proceedings of IEEE international conference on neural networks, Australia, pp 1942–1948
19. Lee C-Y, Yao X (2004) Evolutionary programming using mutations based on Lévy probability distribution. IEEE Trans Evol Comput 8(1):1–13
20. Li J-P, Balaz ME, Parks GT (2002) A species conserving genetic algorithm for multimodal function optimization. Evol Comput 10(3):207–234
21. Madsen K, Žilinskas J Testing branch-and-bound methods for global optimization. http://www2.imm.dtu.dk/documents/ftp/tr00/tr05_00.pdf
22. Mahfoud SW (1995) Niching methods for genetic algorithms. PhD thesis, University of Illinois at Urbana-Champaign, USA
23. Mishra S Some new test functions for global optimization and performance of repulsive particle swarm method. http://mpra.ub.uni-muenchen.de/2718
24. Nelder JA, Mead R (1965) A simplex method for function minimization. Comput J 7(4):308–313
25. Parsopoulos K, Plagianakos V, Magoulas G, Vrahatis M (2001) Objective function "Stretching" to alleviate convergence to local minima. Nonlinear Anal 47(5):3419–3424
26. Parsopoulos KE, Vrahatis MN (2001) Modification of the particle swarm optimizer for locating all the global minima. In: Kurova V et al (eds) Artificial neural networks and genetic algorithms. Springer, Berlin, pp 324–327
27. Parsopoulos K, Vrahatis M (2002) Recent approaches to global optimization problems through particle swarm optimization. Nat Comput 1(2–3):235–306
28. Parsopolos K, Vrahatis M (2004) On the compuation of all minima through particle swarm optimization. IEEE Trans Evol Comput 8(3):211–224
29. Pavlyukevich I (2007) Non-local search and simulated annealing Lévy flights. J Comput Phys 226(2):1830–1844
30. Pavlyukevich I (2007) Cooling down Lévy flights. J Phys A: Math Theor 40(41):12299–12313
31. Petrowski A (1996) A clearing procedure as a Niching method for genetic algorithms. In: IEEE international conference on evolutionary computation, Japan, pp 798–803
32. Price K, Storn RM, Lampinen JA (2005) Differential evolution: a practical approach to global optimization. Natural computing series. Springer, New York

33. Rahnamayan S, Tizhoosh HR, Salama MMA (2008) Opposition-based differential evolution. IEEE Trans Evol Comput 12(1):64–79
34. Reynolds AM, Frey MA (2007) Free-flight odor tracking in dorsophila is consistent with an optimal intermittent scale-free search. PLoS One 2(4):e354
35. Richer TJ, Blackwell TM (2006) The Lévy particle swarm. In: Proceedings of IEEE congress on evolutionary computation, Canada, pp 808–815
36. Rönkkönen J (2009) Continuous multimodal gloabal optimization with differential evolution-based methods. PhD thesis, Lappeenranta University of Technology, Finland
37. Saha A, Deb K (2010) A bi-criterion approach to multimodal optimization: self-adaptive approach. In: Proceedings of international conference on simulated evolution and learning, India, pp 95–104
38. Schwefel H-P (1995) Evolution and optimum seeking. Wiley, New York
39. Stoneking D, Bilbro G, Trew R, Gilmore P, Kelley CT (1992) Yield optimization Using a GaAs process simulator coupled to a physical device model. IEEE Trans Microw Theory Tech 40(7):1353–1363
40. Tasoulis D, Plagianakos V, Vrahatis M (2005) Clustering in evolutionary algorithms to effectively compute simultaneously local and global minima. In: IEEE congress on evolutionary computation, Scotland, pp 1847–1854
41. Torczon V (1991) On the convergence of the multidimensional search algorithm. SIAM J Optim 1(1):123–145
42. Yang X-S, Deb S (2009) Cuckoo search via Lévy flights. In: Proceedings of world congress on nature and biological inspired computing, India, pp 210–214
43. Yang X-S, Deb S (2010) Engineering optimization by Cuckoo search. Int J Math Model Numer Optim 1(4):330–343
44. Yang XS (2010) A new metaheuristic bat-inspired algorithm. In: Gonzalez JR et al (eds) Nature inspired cooperative strategies for optimization. Studies in computational intelligence. Springer, Berlin, pp 65–74
45. Zaharie D (2004) Extension of differential evolution algorithm for multimodal optimization. In: International symposium on symbolic and numeric algorithms for scientific computing, Romania, pp 523–534

Multi-objective Ant Colony Optimisation in Wireless Sensor Networks

Ansgar Kellner

Abstract Biologically inspired ant colony optimisation (ACO) has been used in several applications to solve NP-hard combinatorial optimisation problems. An interesting area of application for ACO-based algorithms is their use in wireless sensor networks (WSNs). Due to their robustness and self-organisation, ACO-based algorithms are well-suited for the distributed, autonomous and self-organising structure of WSNs. While the original ACO-based algorithm and its direct descendants can take only one objective into account, multi-objective ant colony optimisation (MOACO) is capable of considering multiple (conflicting) objectives simultaneously. In this chapter, a detailed review and summary of MOACO-based algorithms and their applications in WSNs is given. In particular, a taxonomy of MOACO-based algorithms is presented and their suitability for multi-objective combinatorial optimisation problems in WSNs is highlighted.

1 Introduction

Wireless sensor networks (WSNs) consist of spatially distributed autonomous sensors that are used to monitor physical phenomena or environmental conditions including temperature, humidity and noise. Often, hundreds or even thousands of sensor nodes are deployed to form a self-organised wireless network. As it is generally desirable to use small devices at low costs, the sensor nodes are heavily constrained in their resources, such as memory, computation, communication bandwidth and energy. Such sensor nodes are usually equipped with a battery that must last the sensors' lifetime. To preserve energy, the radio range of each node is limited so that the sensor nodes have to work cooperatively to forward network packets via multiple hops from a source to a remote destination. Today's application areas of WSNs include area surveillance, industrial monitoring, wildlife tracking and health care.

A. Kellner (✉)
Institute of System Security, Technische Universität Braunschweig,
Rebenring 56, 38106 Braunschweig, Germany
e-mail: a.kellner@tu-bs.de

© Springer International Publishing AG 2017
S. Patnaik et al. (eds.), *Nature-Inspired Computing and Optimization*,
Modeling and Optimization in Science and Technologies 10,
DOI 10.1007/978-3-319-50920-4_3

Due to the fragile and dynamic structure of WSNs, adaptive and robust algorithms are required to provide the basic network services such as the routing of network packets or a certain quality of service (QoS). Biologically inspired algorithms match this requirements in several areas of application; in particular, ant colony optimisation (ACO)-based approaches have shown good results in the area of networking. While ACO-based algorithms only consider one objective, their extension to multi-objective ant colony optimisation (MOACO) takes multiple, even contrary, objectives into account.

The remainder of this chapter is organised as follows: in Sect. 2, the basic idea of multi-objective combinatorial optimisation problems is explained. Section 3 introduces the multi-objective ant colony optimisation heuristic that allows to solve multi-objective combinatorial optimisation problems. Subsequently, Sect. 4 discusses the different applications of MOACO-based algorithms in the area of WSNs. Section 5 concludes and summarises the chapter.

2 Multi-objective Combinatorial Optimisation Problems

In the following, the basic idea of combinatorial optimisation problems is defined as starting point of this research area. Subsequently, the related concept of Pareto optimality and decision-making and the basic ideas on solving multi-objective combinatorial problems are discussed.

2.1 Combinatorial Optimisation Problems

Combinatorial optimisation problems (COPs) are a special subset of global optimisation problems that are concerned with finding optimal solutions for discrete problems. An 'optimal solution' hereby refers to the best among all feasible solutions of a given problem according to a given cost function. The set of solutions is finite, and any solution has some combinatorial property such as a permutation, an arrangement of objects or a tree/graph which indicates a relationship between those objects [65].

Formally, a combinatorial optimisation problem can be defined as follows:

Definition 1 (*Instance of a combinatorial optimisation problem*) Let P be an *instance of a combinatorial optimisation problem*, $P = (S, \Omega, f)$, where S is a finite set of feasible solutions $\{s_1, s_2, \ldots, s_k\}$, $k \in \mathbb{N}$, $\Omega(t)$ is the set of constraints among the variables, and $f : S \rightarrow \mathbb{R}$ is the objective function, which assigns a real-valued cost $f(s)$ to each feasible solution $s \in S$.

The purpose of the optimisation process is to find the global optimum of this instance $s^* \in S$, which minimises the function f and satisfies all constraints in Ω:

$$s^* = \underset{s \in S}{\operatorname{argmin}} f(s) \tag{1}$$

The constraints $\Omega(t)$ as well as the objective function f are dependent on time t.

Definition 2 (*Combinatorial optimisation problem*) A *combinatorial optimisation problem (COP)* is a set of instances of an optimisation problem.

Definition 3 (*Compact representation of COP instance*) Let $P_c = < C, \Omega, J >$ be the *compact representation of an instance of a combinatorial problem*, where C is the finite set of variables (decisions) to select, the so-called solution components, $C = \{c_1, c_2, \ldots, c_{n_C}\}, n_C \in \mathbb{N}$, Ω is the finite set of constraints, and $J(S)$ is the real-valued cost function.

The set of feasible solutions S is a subset of solution components that are satisfying the relations in Ω: $S \subseteq \mathscr{P}(C) \cap \Omega(C)$. The instance of such a combinatorial problem can be solved by finding the element s^* such that

$$s^* = \arg \underset{s \in \{\mathscr{P}(C) \cap \Omega(C)\}}{\min} J(s) \tag{2}$$

2.2 Multi-objective Combinatorial Optimisation Problems

In many real-world applications, optimisation problems are dependent on multiple, even conflicting, objectives. For instance, in the networking domain, a route between two network nodes can be optimised regarding end-to-end delay, jitter and energy consumption. In traditional approaches, often all objectives are combined in a single-objective function that results in a single cost value. However, such a condensed representation is rather artificial and limited. Different objectives may have different representations and meanings such that the overall problem cannot adequately be represented. Alternatively, consider this sort of problem as multi-objective combinatorial optimisation problem (MCOP) that is capable of taking two or more (conflicting) objectives into account. An MCOP can then be optimised simultaneously regarding multiple constraints.

Definition 4 An *MCOP* can be defined as follows (cf. [15]):

$$\underset{\mathbf{x}}{\operatorname{argmin}} \mathbf{f}(\mathbf{x}) \qquad \mathbf{x} \in X \subseteq \mathbb{R}^k, \mathbf{f} \in Y \subseteq \mathbb{R}^n, n \geq 2 \tag{3}$$

$X \subseteq \mathbb{R}^k$ is the k-dimensional *decision space*, a multidimensional space of feasible solutions; $\mathbf{x} = (x_1, x_2, \ldots, x_k) \in X$, the *decision vector*, a vector of decision variables from X; $Y \subseteq \mathbb{R}^n$, the n-dimensional *objective space*, a multidimensional space constituted by the objective functions from $\mathbf{f}$; and $\mathbf{f} = (f_1(\mathbf{x}), f_2(\mathbf{x}), \ldots, f_n(\mathbf{x})) \in Y$, the *objective vector*, a vector of objective functions. The main goal of an MCOP is to optimise the objective vector $\mathbf{f}$.

Fig. 1 Mapping from *decision space* to *objective space*

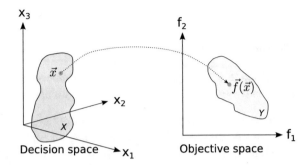

Figure 1 depicts an example of the mapping between the k-dimensional decision vector and the n-dimensional objective vector. For MCOPs, there is typically no single best solution. Instead, a set of solutions is determined that represents the best possible trade-off among the considered objectives. Often, *Pareto optimality* is used as quality criterion to find the best solutions. This related concept is discussed in the next section.

2.3 Pareto Optimality

Pareto optimality, named after the Italian engineer and economist Vilfredo Pareto (1848–1923), is a state in which it is impossible to make any individual property better without making at least one other property worse. Formally, Pareto optimality can be described as follows:

Definition 5 (*Pareto domination*) For two objective vectors $\mathbf{f}^*, \mathbf{f} \in Y$, $\mathbf{f}^*$ dominates $\mathbf{f}$ in terms of *Pareto's domination*, in short $\mathbf{f}^* \succ \mathbf{f}$, if it is better for one criterion and better or equal for the others. Formally this can be denoted as follows:

$$\forall i \in \{1, 2, \ldots, n\} : f_i^* \leq f_i \wedge \exists j \in \{1, 2, \ldots, n\} : f_j^* < f_j \tag{4}$$

Accordingly, it can be stated that the decision vector $\mathbf{x}^* \in X$ *dominates* the decision vector $\mathbf{x} \in X$, in short $\mathbf{x}^* \succ \mathbf{x}$, if $f(\mathbf{x}^*) \succ f(\mathbf{x})$.

Definition 6 (*Pareto optimal*) A decision vector $\mathbf{x}^* \in X$ is called *Pareto optimal*, *efficient* or *non-dominated*, if and only if, there is no other vector $\mathbf{x} \in X$ such that $\mathbf{x}$ dominates $\mathbf{x}^*$.

Definition 7 (*Weakly Pareto optimal*) A decision vector $x^* \in X$ is *weakly Pareto optimal* if there is no other decision vector $\mathbf{x} \in X$ for which all the components are better. Formally this can be denoted as follows:

$$\nexists \mathbf{x} \in X, \forall i \in \{1, 2, \ldots, n\} : f_i(\mathbf{x}) < f_i(\mathbf{x}^*) \ \mathbf{x}^* \in X \tag{5}$$

Fig. 2 Example of a *Pareto front* in the objective space

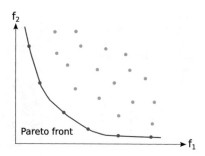

Definition 8 (*Pareto set*) The set of all Pareto optimal solutions, which is a subset of the decision space, is named *Pareto set* or *efficient set*. The Pareto set can be formally denoted as follows:

$$X^* \subseteq X, X^* = \{\mathbf{x}^* \in X | \nexists \mathbf{x} \in X : \mathbf{x} \succ \mathbf{x}^*\} \tag{6}$$

Definition 9 (*Pareto front*) The set of elements in the objective space that is corresponding to the set of elements of the Pareto set in the decision space is named *Pareto front, Pareto frontier* or *Non-dominated set*. Formally denoted as follows:

$$Y^* = f(X^*) \subseteq Y \tag{7}$$

Figure 2 shows an example of a mapping between the Pareto set in the decision space to the Pareto front in the objective space.

2.4 Decision-Making

Each element of the Pareto front is a valid solution to the MCOP, as a compromise among all objective functions. Since only the single best solution should be considered, the question of how to choose the final solution arises.

In the simplest case, an ordinary selection algorithm is applied that automatically finds the final solution. Due to the fact that no additional preferences are provided, this approach is also referred to as *no-preference method*. For more complex approaches, additional information is required that is not part of the objective functions. To this end, a *decision maker* component is introduced to take these into account. There are three major design choices to be made for such a decision component (see Fig. 3):

1. **A priori approaches**: specify the relative importance of each criteria before the optimisation process is started. Only a single optimisation is required to obtain a single solution, which makes this approach really fast.

Fig. 3 Decision maker
approaches: *a priori
decision*, *progressive
decision* and *a posteriori
decision*

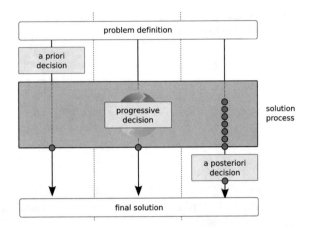

2. **Progressive approaches**: express the importance of the different criteria during
   the optimisation process; i.e. expressed preferences are used to guide the search.
   Frequent interaction is necessary for good results.
3. **A posteriori approaches**: obtain a set of (non-dominated) solutions by the chosen
   optimisation method and subsequently select the most suitable solution from the
   set. While the modelling phase before the optimisation process can be omitted, a
   higher computational effort is required to pick the final solution.

Depending on the specific problem, one or multiple decision maker approaches
must be chosen carefully. For MCOP, multi-objective metaheuristics are usually
applied with an a posteriori decision maker approach.

2.4.1 Multi-criteria Decision-Making

Multi-criteria decision-making (MCDM) is the corresponding research area that is
concerned with the assistance of one or multiple decision makers in taking an optimal
decision. The general idea of MCDM models is to rank the performance of a finite set
of solutions by decision criteria, i.e. finding the optimal solution among all solutions
found.

A comprehensive overview is given by Triantaphyllou [81] who introduces
MCDM methods and provides a comparative study of such models. Due to its wide
applicability, MCDM is a fast-growing research area that has spawned various areas
of applications. Since the multi-objective metaheuristics mainly focus on *a posteriori approaches*, the most prominent approaches will be briefly presented in the
following:

Weighted Sum

One of the most commonly used MCDM methods is the *Weighted Sum Model* (*WSM*),
originally proposed by Fishburn [32], Triantaphyllou [81]. In this model, first, a

weighting coefficient is assigned to each objective function, and then, the corresponding weighted sum of objectives is minimised.

The best solution satisfies the following equation:

$$A^*_{WSM-score} = \operatorname*{argmin}_i \sum_{j=1}^{n} w_j f_{ij}, \quad i = 1, 2, 3, \ldots, m \tag{8}$$

where n is the number of decision criteria, m the number of alternatives, w_j the relative weight of importance of the jth criterion with $0 \le w_j \le 1$; $\sum_{j=1}^{n} w_j = 1$ and f_{ij} the objective of the jth criterion and the ith alternative. The $A^*_{WSM-score}$ represents the alternative with the best WSM score.

In the case of a single dimension (all units of measurements are identical), the weighted sum can be easily applied. In contrast, in a multidimensional problem with different units, the additivity utility assumption would be violated. This, in turn, implies that the total value of each alternative would be equal to the sum of the products given in Eq. (8).

Weighted Product

A similar method is the *Weighted Product Model* (*WPM*). The main difference between both models is the operator used for combining the objectives: while WSM uses the addition, WPM makes use of the product. The first reference to this model can be found in Bridgman [8] as well as Miller et al. [63].

The best solution satisfies the following equation:

$$P(A_i) = \prod_{j=1}^{n} (f_{ij})^{w_j}, \quad i = 1, 2, 3, \ldots, m \tag{9}$$

where n is the number of criteria, m the number of alternatives, w_j the weight of importance of the jth criterion with $0 \le w_j \le 1$, $\sum_{j=1}^{n} w_j = 1$, and f_{ij} the objective function of the jth criterion and the ith alternative. The outcome $P(A_i)$ then is the (absolute) performance value of alternative A_i considering all criteria.

In contrast to WSM, WPM is dimensionless; i.e. all units of measure are eliminated by the structure of the formula. WPM can hence also be applied to multidimensional decision problems.

Weighted Metrics

The *Weighted Metrics Method* (*WMM*) [62] attempts to minimise the distance between the ideal criterion vector z^* and solutions in the criteria space. In the criteria space, the entries of z^* consists of the optima of the objective functions. If the ideal criterion vector originates from the admissible solutions space, the best solution will be the ideal criterion vector itself. However, this is only possible in case of non-conflicting criteria. The weighted approach is also referred to as *comprise programming* [87].

A multi-objective problem with m objectives can be transformed to a single-objective problem by applying the following L_p-metric:

$$s_p^* = \underset{x \in S}{\text{argmin}} \left(\sum_{i=1}^{m} w_i |f_i(x) - z_i^*|^p \right)^{1/p} \tag{10}$$

where m is the number of objectives, S the feasible decision variable space, $z^* \in \mathbb{R}^n$ the ideal objective vector, obtained by minimising each objective z_i^* separately (z^* is not a feasible vector because the objectives are conflicting), w_i the ith component of the weight vector with $0 \leq w_j \leq 1$, $\sum_{j=1}^{n} w_j = 1$, and f_i the objective function of the ith criterion. p ($1 \leq p < \infty$) defines the used metric; commonly used values are $p = 1$ (*Manhattan*), $p = 2$ (*Euclidean*) and $p = \infty$ (*Tchebycheff*). The solution of Eq. (10) is Pareto optimal, if all weights are positive or the solution is unique.

For $p = \infty$, the equation can also be rewritten as follows:

$$s_\infty^* = \underset{x \in S}{\text{argmin}} \max_{i=1,2,3,..,m} \left[w_i |f_i(x) - z_i^*| \right] \tag{11}$$

The latter is also referred to as *Weighted Tchebycheff Metric Method* (*WTM*). The solution of (11) is weakly Pareto optimal for positive weights. Furthermore, there is at least one Pareto optimal solution.

Achievement Scalarising Function Approach

The *achievement scalarising function approach* (*ASFA*) [83, 84] utilises a special type of scalarising functions. Such *achievement (scalarising) functions* are of the form $s_{\bar{z}} : Z \rightarrow \mathbb{R}$, where $\bar{z} \in \mathbb{R}^n$ is an arbitrary reference point. Due to the fact that Z is not explicitly known, the function $s_{\bar{z}}(f(\mathbf{x}))$ with $\mathbf{x} \in S$ is minimised. The underlying idea of the approach is to project the reference point $\bar{z}$ consisting of desirable aspiration levels onto the set of Pareto optimal solutions.

Achievement functions can be formulated in different ways, for example:

$$s(f(\mathbf{x})) = \max_{i=1,...,n} \left[w_i (f_i(\mathbf{x}) - \bar{z}_i) \right] + \rho \sum_{i=1}^{n} w_i (f_i(\mathbf{x}) - \bar{z}_i) \tag{12}$$

where w_i is the fixed normalising factor and ρ the augmentation multiplier (small positive scalar). An advantage of this approach is that the decision maker can obtain different Pareto optimal solutions by merely moving the reference point.

ε-Constraint Method

The *ε-Constraint Method* (*eCM*) [42] transforms a multi-objective problem to several single-objective problems with constraints. For each single-objective problem, one objective function is optimised as follows:

$$\begin{aligned}
\text{argmin} \quad & f_l(x), && l \in \{1, 2, 3, \dots, k\} \\
& f_j \leq \varepsilon_j, \ \forall j = 1, 2, 3, \dots, k \quad j \neq l, && (13) \\
& x \in S
\end{aligned}$$

with ε_j as upper bounds for the objective l and S as feasible region. To ensure Pareto optimality, either k different problems have to be solved or a unique solution needs to be found. Due to the fact that uniqueness of solutions is not easily verifiable, Vira and Haimes [82] propose systematic ways of perturbing the upper bounds to obtain different Pareto optimal solutions.

MCDM Summary

Each presented MCDM approach has its very own advantages and disadvantages: WSM is simple to implement and hence often used. However, it can only be applied when the units of all measures are identical. WSM can find Pareto optimal solutions if the solution is unique or all weights are positive. Furthermore, a small change in weights may have a great impact on the solution. Additionally, when evenly distributed weights are used, the representation of the Pareto optimal set may not be evenly distributed. For WPM on the contrary, all units of measure are eliminated by the product. Thus, measures can easily be combined. WMM produces a Pareto optimal solution, if it is unique or all weights are positive. However, not all Pareto optimal solutions may be found. ASFA can guarantee Pareto optimal solutions when lexicographical ordering is used. For eCM, a unique solution as well as each solution x^*, fulfilling $\varepsilon_j = f(x^*)$ $(i = 1, \dots, k \wedge j \neq l)$ for all objectives to be minimised, is Pareto optimal. However, for eCM, it is often difficult to find the upper bounds.

As a result, no general-purpose recommendation for MCDM can be given, but instead, depending on the context of the application, the appropriate model must be carefully selected.

2.5 Solving Combinatorial Optimisation Problems

Many (multi-objective) combinatorial optimisation problems are NP-hard and thus have at least an exponential worst-case time complexity. Famous NP-hard problems that are frequently studied in researches include the travelling salesman problem (TSP) [2], the assignment problem (AP) [54] or the knapsack problem (KP) [76].

While for small-scale problems, exact methods may be applied, for larger instances one reverts to approximations that are capable of obtaining near-optimal solutions at low computational costs. Such *heuristics* are used to find feasible solutions in reasonable time.

While heuristics are problem specific, *metaheuristics* provide a more generalised view on problems that require an abstract sequence of steps. A metaheuristic thus can be seen as 'general-purpose heuristic' that can be applied on various optimisation problems to find near-optimal solutions. Normally, only minor modifications have to be made to adapt it to a specific problem. There are several well-known metaheuristics that are used to find near-optimal solutions to difficult combinatorial optimisation problems: Iterated Local Search (ILS) [55], Variable Neighbourhood Search (VNS) [64], Simulated Annealing (SA) [47], shortest path finding algorithm A^* [44], Tabu Search (TS) [40, 41], ant colony optimisation (ACO) [28, 38], etc. A particular interesting subclass of metaheuristics are biologically inspired ant colony optimisation-based algorithms which are discussed in more detail in the next section.

3 Multi-objective Ant Colony Optimisation

In this section, the basic idea of multi-objective ant colony optimisation (MOACO) as application of metaheuristics is discussed. To this end, its origin in ant colony optimisation is presented in Sect. 3.1. Subsequently, Sect. 3.2 describes its extension to multi-objective ant colony optimisation by elaborating a taxonomy of MOACO algorithms and its applications.

3.1 Origins

Ant Colony Optimisation (ACO) belongs to the class of biologically inspired algorithms that can be used to approximate combinatorial optimisation problems. The origin of ACO lies in the observation of real ant colonies in nature, in which highly organised ants work in colonies together in a cooperative manner to solve difficult tasks, for instance finding the shortest path between a food source and the nest. Various experiments have been conducted to figure out how ant colonies are organised and how individuals communicate to solve complex tasks. Inspired by these biological observations, the ACO metaheuristic has been developed by imitating the behaviour of real ant colonies to solve combinatorial optimisation problems.

The main characteristic of ant colonies leveraged by ACO algorithms can be summarised as follows (cf. [28, 38]):

Colony of cooperating individuals ACO algorithms make use of artificial ant colonies that consist of autonomous agents. Those agents cooperate to solve difficult optimisation problems. Just as in real ant colonies, a single artificial ant is likely not able to find a reasonably good solution to a problem, while in cooperation with the other ants, a high-quality solution can be obtained.

Pheromone trail and stigmergy One of the key factors of cooperation is the communication between actors, for instance, using stigmergy: information is indirectly exchanged by modifying aspects of the environment. While real ants make use of pheromones to mark places they have visited, artificial ants modify a numeric value (artificial pheromone) at the states they have visited. Any other ant that visits the same state can access this numeric value. The value can be taken into account in the decision process and, subsequently, updated and stored back to the state. In nature, pheromones decay over time due to evaporation, removing inefficient paths from an ant colony's routes. A similar decaying mechanism is used for artificial ant colonies to allow the exploration of new problem solutions and to avoid local minima.

Shortest path searching and local movement   Both real and artificial ants try to find the shortest path between a source and a destination based on some cost function. While real ants explore the surrounding environment to find the shortest path between the nest and the food source, artificial ants try to iteratively construct a minimum cost solution between a start and an ending state. Depending on the application, the explicit meaning of 'state' and 'adjacent' may differ.

Stochastic decision policy for state transition   In natural as well as in artificial ant colonies, a probabilistic decision policy is used to transit from one to another adjacent state. The decision policy only incorporates local information without any look-ahead functionality regarding subsequent states. For the artificial ants, the probabilistic decision policy can be formally seen as a function of a priori defined information given by the problem specification and an a posteriori modification of the environment.

In comparison with real ants, artificial ones have additional characteristics without natural counterpart:

Heuristic information   Additionally to the pheromone information, artificial ants can make use of heuristic information to influence the probabilistic decision process. The heuristic information is problem specific.

Discrete world   While real ants operate in a continuous environment, artificial ants are moving in a discrete world of problem states. The movement of an artificial ant is the transition between two adjacent problem states. The transition rule considers the locally available pheromone information, heuristic information, the ant's internal state and the problem constraints.

Internal state   Each artificial ant has an internal state to store information about the visited problem states. This memory can be used to build feasible solutions, evaluate generated solutions and trace back the path from the current to the initial state.

Quality of solutions   Artificial ants can apply a weighting to prefer high-quality solutions by adapting the amount of pheromone they deposit on certain problem states.

Timing of pheromone deposit   While real ants continuously deposit pheromone on their walks, artificial ants are capable of depositing pheromone depending on the problem. As a result, artificial ants can first try to find a feasible solution and then deposit the pheromone depending on the quality of the solution found.

Extra capabilities   Artificial ants can make use of additional capabilities such as look-ahead, local optimisation or backtracking. However, these mechanisms are problem specific and the cost–benefit ratio must be taken into account.

Ant Colony Optimisation Algorithms

Based on the presented features of artificial ant colonies, several ACO-based algorithms have been proposed. The fundamental ones will be discussed in the following:

Ant System

The first ACO-based algorithm, *Ant System (AS)*, was proposed by Dorigo et al. [25, 27] for the travelling salesman problem (TSP). The proposed approach drew a lot of attention to ant-based algorithms so that various researches and applications appeared in this and related areas.

The fundamental idea of the original AS algorithm is that artificial ants should build a solution for a given TSP concurrently by traversing the construction graph, thereby making probabilistic decisions from vertex to vertex. The AS algorithm is subdivided into two different phases: the construction and the pheromone update phase. In the construction phase, each ant is assigned to a random state (city). In each step, each ant has to decide which state should be visited next. The decision utilises the *random proportional rule* that defines the kth ant moving from state i to state j. To avoid ants visiting the same state twice, each ant stores a list of visited cities in its internal memory.

The *random proportional rule* can be formally defined as follows:

$$p_{ij}^k = \begin{cases} \frac{[\tau_{ij}]^\alpha [\eta_{ij}]^\beta}{\sum_{l \in \mathcal{N}_i^k} [\tau_{il}]^\alpha [\eta_{il}]^\beta} & \text{if } j \in \mathcal{N}_i^k \\ 0 & \text{otherwise} \end{cases} \tag{14}$$

where m is the number of ants, τ_{ij} is the pheromone value that is changing during construction of solutions, and $\eta_{ij} = 1/d_{ij}$ is a heuristic value that is know a priory (d_{ij} is the distance between the cities i and j). The two parameters α and β define the relative influence of pheromone and heuristic information, while $\mathcal{N}_i^k$ is the feasible neighbourhood of ant k, i.e. the set of cities that can be reached from state i and that have not been visited yet by ant k.

The probability of choosing a certain edge (i, j) increases with the value of the associated pheromone information τ_{ij} and the value of the associated heuristic information η_{ij}.

When all ants have completed the construction of a solution, the pheromone update phase is triggered. In comparison with the later discussed ACO-based algorithms, in AS, all ants participate in the pheromone updating process. The pheromone update can be formally denoted as follows:

$$\tau_{ij} \leftarrow (1 - \rho)\tau_{ij} + \sum_{k=1}^{m} \Delta\tau_{ij}^k \tag{15}$$

where m is the number of ants, $\rho \in (0, 1]$ is the evaporation rate, and $\Delta\tau_{ij}^k$ is the quantity of pheromone on the edge (i, j).

The pheromone evaporation rate ρ is uniformly applied on all edges to avoid local minima. The $\Delta\tau_{ij}^k$, the quantity of pheromone on edge (i, j), is defined as follows:

$$\Delta\tau_{ij}^k = \begin{cases} \frac{Q}{L_k} & \text{if ant } k \text{ used edge } (i, j) \text{ on its tour} \\ 0 & \text{otherwise} \end{cases} \tag{16}$$

where Q is a constant (in most cases $Q := 1$) and L_k is the length of the tour built by the k-th ant, i.e. the sum of all passed edges on the tour.

By applying the presented rules, edges that are part of short ant tours will obtain a larger amount of pheromone than the edges that are part of longer tours. As a result, the pheromone is increased proportionally to a tour's quality. This impact is intensified by the fact that following ants are biased against edges with higher pheromone concentration and place additional pheromone on the used edges.

Elitist Ant System

The *Elitist Ant System* (*EAS*), proposed by Dorigo et al. [22, 27], is one of the earliest improvements of the original AS. EAS extends the original AS algorithm by applying an additional reinforcement mechanism that emphasises the *best-so-far tour* found by depositing additional pheromone on those edges. The additional pheromone depositing is carried out by a group of *elitist ants* which is sent along the best-so-far tour to place an additional amount of pheromone along these edges.

Formally, the original pheromone updating equation from AS (see Eq. (15)) is adapted to:

$$\tau_{ij} \leftarrow (1 - \rho)\tau_{ij} + \sum_{k=1}^{m} \Delta\tau_{ij}^k + e\Delta\tau_{ij}^{best} \tag{17}$$

where e is the number of elitist ants and $\Delta\tau_{ij}^{best}$ is the quantity of pheromone on the edge (i, j) of the best-so-far tour. The elitist ants deposit the additional amount of pheromone as follows:

$$\Delta\tau_{ij}^{best} = \begin{cases} \frac{Q}{L_{best}} & \text{if edge } (i, j) \text{ is part of the best-so-far tour} \\ 0 & \text{otherwise} \end{cases} \tag{18}$$

where Q is a constant (in most cases $Q := 1$) and L_{best} is the length of the best-so-far tour found.

Due to this reinforcement of the best-so-far tour, all other ants are influenced with a certain probability to use those edges in their own solution. The simulation results of Dorigo et al. [22, 27] show that with a well-balanced amount of elitist ants, better tours can be found more quickly than with the original AS algorithm.

Rank-Based Ant System

With AS_{rank}, Bullnheimer et al. [9] propose another variation of the AS algorithm, in which the amount of deposited pheromone depends on a weight that is linked to an ant's rank. An ant's rank is defined by sorting the ants by the length of their tours in descending order. Similar to the EAS algorithm, the ant that has found the best-so-far tour has an distinct position and, thus, can deposit an additional amount of pheromone. Formally, the pheromone updating equation for AS_{rank} is defined as follows:

$$\tau_{ij} \leftarrow (1 - \rho)\tau_{ij} + \sum_{r=1}^{w-1}(w - r)\Delta\tau_{ij}^r + w\Delta\tau_{ij}^{best} \qquad (19)$$

where r is the rank of the ant, w is the number of ants that are updating the pheromone information, and $\Delta\tau_{ij}^r$ is the quantity of pheromone on edge (i, j) of the r-ranked ant. The quantity of pheromone that is deposited by the r-ranked ant is defined as follows:

$$\Delta\tau_{ij}^r = \begin{cases} \frac{Q}{L_{rank}} & \text{if edge } (i, j) \text{ is part of the r-ranked ant's tour} \\ 0 & \text{otherwise} \end{cases} \qquad (20)$$

where Q is a constant (in most cases $Q := 1$) and L_{rank} is the length of the tour chosen by the r-ranked ant. Hence, additionally to the best-so-far ant, each of the $(w - 1)$ ants take part in the pheromone update process according to its weight $(w - r)$.

In the conducted experiments, Bullnheimer et al. [9] experimentally show that AS_{rank} provides a good compromise between exploitation (reinforcement of good path) and exploration (reinforcing several good path instead of concentrating on the best path). AS_{rank} significantly outperforms AS and slightly improves over EAS.

Max-Min Ant System

The *Max-Min Ant System* (*MMAS*), proposed by Stützle [77, 79], is the first major revision of the original AS algorithm. The proposed modifications target at two main objectives: (1) a better exploitation of the best-so-far tours and (2) a weakening of premature convergence towards initial solutions to avoid local optima.

To achieve these goals, the following changes have been made: To emphasise the best solutions, the best-so-far solution is heavily exploited, since only the best ant is allowed to deposit pheromone to its path. The 'best ant' can be either the ant that performed best in the current iteration (*iteration-best*) or the ant that has found the overall best solution so far (*best-so-far*). Depending on the application context, also hybrid approaches are possible. As a result of the modification, the algorithm may quickly stagnate in a state in which only good but not optimal paths are used. Such local minima are avoided by introducing pheromone value limits so that pheromone values have to be within the range $[\tau_{min}, \tau_{max}]$. Finally, to achieve a high exploration of different solutions during bootstrapping, the initial pheromone value is set to τ_{max}. In MMAS, the pheromone value is updated as follows:

$$\tau_{ij} \leftarrow (1 - \rho)\tau_{ij} + \Delta\tau_{ij}^{best} \qquad (21)$$

with

$$\Delta\tau_{ij}^{best} = \begin{cases} \frac{Q}{L_{best}} & \text{if edge } (i, j) \text{ is part of the best tour} \\ 0 & \text{otherwise} \end{cases} \qquad (22)$$

where Q is a constant (in most cases $Q := 1$) and L_{best} is the length of the best tour of the *iteration-best* or *best-so-far* ant.

While the *best-so-far* ant pheromone updating rule rapidly concentrates on the search space close to the best-so-far tour, the *iteration-best* ant pheromone updating rule provides a more greedy search by updating different good edges. Experiments have shown that the best results are obtained when using both pheromone update rules with a gradual increase in the *iteration-best* ant's rule [77, 79].

Stützle and Hoos recommend to determine the lower and upper bound of τ experimentally for each problem individually. However, they also present an approach to calculate these boundaries analytically [79].

In MMAS, occasionally the pheromone trails are reinitialised to foster the exploration of new solutions, i.e. $\tau_{ij} := \tau_{max}$. This is triggered when the algorithm enters the stagnation phase identified by analysing additional statistics or approximated as a certain number of iterations did not lead to significant improvement in the best-so-far solution [77, 79]. After each iteration, it is guaranteed that τ_{ij} is within the defined range $[\tau_{min}, \tau_{max}]$. The following rule is applied:

$$\tau_{ij} = \begin{cases} \tau_{min} & \text{if } \tau_{ij} < \tau_{min} \\ \tau_{ij} & \text{if } \tau_{min} \leq \tau_{ij} \leq \tau_{max} \\ \tau_{max} & \text{if } \tau_{ij} > \tau_{max} \end{cases} \tag{23}$$

Since MMAS significantly outperforms the original AS, it become one of the most studied algorithms in the research area of ACO algorithms and is used in various applications.

Ant Colony System

While the previously discussed methods describe incremental improvements to the original AS algorithm, *Ant Colony System (ACS)*, proposed by Dorigo and Gambardella [24, 35], can be rather seen as new ACO-based algorithm, instead of a direct descendant of AS. ACS is based on the earlier algorithm Ant-Q [34] that has been abandoned due to its high complexity. ACS uses three main steps: a pseudo-random proportional rule, a global pheromone update and a local pheromone update.

The *pseudo-random proportional rule* specifies how an ant k, currently located at state i, can move to state j:

$$j = \begin{cases} \text{argmax}_{l \in \mathcal{N}_i^k} \{\tau_{il}[\eta_{il}]^\beta\} & \text{if } q \leq q_0 \text{ exploitation} \\ J & \text{otherwise (biased exploration)} \end{cases} \tag{24}$$

where q is a random variable uniformly distributed in $[0, 1]$, q_0 is a parameter $(0 \leq q_0 \leq 1)$, determining the relative importance of exploitation versus exploration, and J is a random variable selected according to the probability distribution in Eq. (14).

Based on the pseudo-random proportional rule, an ant can make the best possible move with a probability of q_0 or it can explore new arcs with a probability of $(1 - q_0)$. The parameter q_0 hence specifies whether the ants should concentrate on finding better tours near the best-so-far tour or should rather explore new routes instead.

The *global pheromone update* in ACS is triggered after each iteration by the best-so-far ant exclusively. Formally, the pheromone update in ACS can be denoted as follows:

$$\tau_{ij} \leftarrow (1 - \rho)\tau_{ij} + \rho \Delta \tau_{ij}^{best} \tag{25}$$

with

$$\Delta \tau_{ij}^{best} = \begin{cases} \frac{Q}{L_{best}} & \text{if edge } (i, j) \text{ is part of the best tour} \\ 0 & \text{otherwise} \end{cases} \tag{26}$$

where Q is a constant (in most cases $Q := 1$) and L_{best} is the length of the best-so-far tour.

In contrast to AS, in ACS, the pheromone update and evaporation do solely affect the best-so-far tour which dramatically reduces the computational complexity for updating the pheromone trails. Furthermore, in ACS, the new pheromone trail is a weighted average between the old pheromone value and the newly deposited amount of pheromone.

Apart from the global pheromone update, ACS uses a *local pheromone update* that is triggered during the construction of a tour each time an ant passes an edge (i, j). The pheromone value of the used edge is reduced by applying the following formula:

$$\tau_{ij} \leftarrow (1 - \xi)\tau_{ij} + \xi \tau_0 \tag{27}$$

with τ_0 as the initial value of pheromone trails (normally a small constant) and ξ $(0 < \xi < 1)$ as the pheromone decay coefficient that controls the ants' degree of exploration. As a result, the pheromone concentration of traversed edges is decreased to encourage ants to choose a different edge than the previous ant. In this way, the exploration of different solutions is supported, while stagnation is avoided.

Applications of Ant Colony Optimisation Algorithms

As described, the ACO algorithms belong to the class of metaheuristics and, thus, can be applied to diverse combinatorial optimisation problems. Table 1 shows an overview of ACO algorithms in selected applications in various domains.

3.2 Multi-objective Ant Colony Optimisation

While ACO algorithms can solely consider one objective at a time, multi-objective ant colony optimisation (MOACO) algorithms can take multiple, even contradictory, objectives into account. The idea of most MOACO algorithms is derived from well-known single-objective ACO algorithms such as ACS or MMAS. To classify these algorithms, a taxonomy and its used criteria are discussed in the following.

Table 1 Selected applications of ACO algorithms

Problem type	Problem name	Authors
Routing	Travelling salesman	Dorigo et al. [26] Dorigo and Gambardella [23] Stützle and Hoos [78, 79]
	Vehicle routing	Gambardella et al. [37] Reimann et al. [69]
	Sequential ordering	Gambardella and Dorigo [36]
	Communication networks	Caro and Dorigo [13]
	Routing in MANETs	Ducatelle et al. [29] Caro et al. [14]
Assignment	Quadratic assignment	Stützle and Hoos [79] Maniezzo [56]
	Course timetabling	Socha et al. [72, 73]
	Graph colouring	Costa and Hertz [17]
Scheduling	Project scheduling	Merkle et al. [61]
	Total weighted tardiness	Den Besten et al. [19] Merkle and Middendorf [60]
	Open shop	Blum [6]
Subset	Set covering	Lessing et al. [51]
	k-cardinality trees	Blum and Blesa [7]
	Multiple knapsack	Leguizamon and Michalewicz [50]
	Maximum clique	Fenet and Solnon [30]
Other	Constraint satisfaction	Solnon [74, 75]
	Classification rules	Parpinelli et al. [66] Martens et al. [58]
	Bayesian networks	de Campos et al. [10], De Campos et al. [11]
	Protein folding	Shmygelska and Hoos [71]
	Protein–ligand docking	Korb et al. [49]

Taxonomy

Several parameters influence the performance of MOACO algorithms. Based on these parameters, a taxonomy of MOACO algorithms can be created considering the following criteria:

- Pheromone and heuristic information;
- Number of ant colonies;
- Aggregation of pheromone and heuristic information;
- Pheromone update;
- Evaluation of solutions; and
- Archival of solutions.

The different criteria will be discussed in more detail in the following:

Pheromone and Heuristic Information

To store pheromone and heuristic information in MOACO, two schemes are widely used:

Single matrix   A single matrix is used to store pheromone or heuristic information, similar to the traditional ACO algorithms. Since multiple objectives should be considered, those objectives need to be aggregated before storing them in the matrix.

Multiple matrices   However, typically each objective is stored in a separate matrix. Each matrix then reflects the solution component's value regarding a certain objective.

Number of Ant Colonies

Another important criteria for a MOACO algorithm is the number of used ant colonies. Since the term 'ant colonies' is not used consistently in the literature, here ant colonies are, in accordance with Iredi et al. [45], defined as multiple instances of single-colony algorithms. Each ant colony acts independently and, thus, manages its own pheromone and heuristic information.

There are two options:

One colony   One ant colony is used to optimise multiple objectives at the same time.

Multiple colonies   Multiple ant colonies are used to optimise multiple objectives. Usually, each colony is responsible for the optimisation of one objective. In most cases, each ant colony manages its own pheromone and heuristic matrices. A disadvantage of this approach is the increased overhead for each additional ant colony.

In the latter case, the ant colonies can communicate with each other to exchange information. To this end, either a common archive of solutions is used or the matrices of the other colonies are modified correspondingly.

Aggregation of Pheromone and Heuristic Information

When using multiple pheromone or heuristic matrices, these matrices usually have to be aggregated. There are different aggregation strategies to combine the matrices, and the most common are as follows:

Weighted sum   The matrices are aggregated by a weighted sum: $\sum_{l=1}^{n} \lambda_l \tau_{ij}^{l}$;
Weighted product   The matrices are aggregated by weighted product: $\prod_{l=1}^{n} (\tau_{ij}^{l})^{\lambda_l}$; and
Random.   In each construction step, one of the matrices is chosen of random.

The number of used weights normally corresponds either to the number of ants or the number of objectives. These weights λ_l may be either *fixed*, i.e. the weights for each objective are set a priori based on prior knowledge and remain unchanged the entire time; or the weights are assigned *dynamically*, i.e. different objectives can be weighted differently at different times. In the latter case, it is conceivable that ants assign different weights at different iterations.

Pheromone Update

In most ant-based algorithms, only the backward ants are allowed to update pheromone information. The most common ways of updating the pheromone information are as follows:

Elite solution   When only a single matrix is used for all objectives, the *best-so-far* or the *iteration-best* ant is chosen to update the pheromone information.

Best-of-objective solutions   When multiple matrices are used, the *best-so-far* or the *iteration-best* ant with respect to each objective is chosen.

Best-of-objective-per-weight solutions.   The *best-so-far* or the *iteration-best* ant with respect to each objective is chosen, while considering the weight λ.

Non-dominated solutions   Only the ants that have found a Pareto optimal solution are allowed to update the pheromone information. Additionally, these solutions may be stored in a set, and then, an elite solution is chosen from the set when the additional information is available.

All solutions   All ants that have found a valid solution are allowed to update the pheromone information.

Of course, combinations of the described approaches are possible. An example of such an hybrid solution would be that all ants deposit pheromone on their way, while the *best-so-far* ant adds an additional amount of pheromone to the path of the best solution.

Evaluation of Solutions

For multi-objective combinatorial optimisation problems, there are several methods to evaluate a solution with respect to different criteria. In case of MOACO algorithms, it is particularly important to distinguish Pareto optimal and non-Pareto optimal solutions:

Pareto solutions   For finding Pareto (non-dominated) solutions, all objectives are equally taken into account by applying the domination criterion.

Non-Pareto solutions   For finding non-Pareto solutions, either only one of the objectives or a combined value of all aggregated objectives is considered.

Archival of Solutions

There are different options for the storing of solutions for later use:

Offline storage   When a new solution is found, it is used to update the pheromone information and it is then stored in an offline storage: the archive. When specifically looking for Pareto solutions, only these are stored, and all non-Pareto solutions are discarded. The archive is only used to record historic data in a 'hall of fame' and may be returned at the end of the algorithm.

Online storage When a new solution is found, it is added to the population of solutions. A change in this triggers the pheromone update procedure using the improved solution set. In the case of Pareto solutions, all dominated solutions are removed. Consequently, the population of solutions always contains the best solutions found. The last population of solutions is returned at the end of the algorithm as set of final solutions.

Elite storage Only a single solution, the elite solution, is stored. The elite solution is used to update the pheromone information and can be returned as best-so-far solution at the end of the algorithm.

No storage New solutions are used to update the pheromone information, but are discarded subsequently. At the end of the algorithm, the last solution found is considered as final solution.

MOACO Algorithms Summary

Table 2 shows an overview of all components that are derived from the characteristics presented in the previous section. Due to the fact that the components are almost freely combinable, there are various possible MOACO algorithms that can be chosen depending on the considered problem.

Based on the identified components, the most common MOACO algorithms can be classified as shown in Table 3.

Performance Metrics

When applying different MOACO algorithms, the question arises how they can be compared regarding their performance. In contrast to ACO algorithms, in which only a single best solution is determined, MOACO algorithms return a set of solutions.

A simple option is the use of *unary quality metrics*, also referred to as *unary quality indicators*, to compare solution sets. In unary quality metrics, each solution set is

Table 2 Taxonomy-based components of MOACO algorithms

Component	Options
Pheromone information ($[\tau]$)	One matrix, multiple matrices
Heuristic information ($[\eta]$)	One matrix, multiple matrices
Number of ant colonies	Single colony (*single*), multiple colonies (*multiple*)
Aggregation of pheromone/heuristic information	Weighted sum ($\sum$), weighted product ($\prod$), random (*rand*)
Pheromone update	Elite solution (*es*), best-of-objective solutions (*boo*), best-objective-per-weight solutions (*bopw*), non-dominated solutions (*nd*), all solutions (*all*)
Evaluation of solutions	Pareto solutions (*p*), non-Pareto solutions (*np*)
Archival of solutions	Offline storage (*offline*), online storage (*online*), elite storage (*elite*), no storage (*none*)

Table 3 Taxonomy of MOACO algorithms with d as number of objectives and m number of ants

Algorithm	$[\tau]$	$[\eta]$	#Colonies	Aggregation	Weighting	τ-Update	Evaluation	Archival
MOAQ [57]	1	d	multiple	Π,Σ	d	nd	p	offline
MACS-VRPTW [37]	d	1	multiple	–	–	boo	np	elite
BicriterionAnt [45]	d	d	single	Π	m	nd	p	offline
COMPETants [20]	d	d	multiple	Σ	$d+1$	boo	p	none
ACOAMO [59]	1	1	single	–	–	all	np	elite
SACO [80]	1	1	single	–	–	elite	np	none
MACS [4]	1	d	single	Π	m	nd	p	online
MONACO [12]	d	1	single	Π	d	all	np	offline
PACO-MO [33]	d	d	single	Σ	m	all	p	online
P-ACO [21]	d	d	single	Σ	m	boo	np	offline
M3AS [68]	1	d	single	Π	d	nd	p	offline
MOA [39]	1	1	single	Π	d	nd	p	offline
CPACO [2]	d	d	single	Σ	d	nd	p	online
MOACSA [85]	1	1	single	Π	d	elite	np	none
mACO-1 [1]	d	d	multiple	rand	$d+1$	bopw	p	offline
mACO-2 [1]	d	d	multiple	Σ	$d+1$	bopw	p	offline
mACO-3 [1]	1	1	single	–	–	nd	p	offline
mACO-4 [1]	d	1	single	rand	d	boo	p	offline

assigned a quality value that allows to compare different solution sets. Although the basic idea is simple, the development of such unary quality metrics is difficult. Zitzler et al. [89] highlight that the development of such a metric is an MCOP itself, involving multiple objectives: (a) the distance of the resulting non-dominated set to the Pareto optimal front should be minimised; (b) a good (in most cases uniform) distribution of solutions should be targeted (e.g. distance metric-based); and (c) the extent of the obtained non-dominated front should be maximised; i.e. for each objective, a wide range of values should be covered by the non-dominated solutions.

Zitzler et al. [90] and Knowles et al. [48] present several unary quality indicators and discuss their limitations and the problems that can occur and may lead to false or misleading assessments of the quality of solution sets. Based on their findings, Knowles et al. [48] recommend three widely accepted unary quality indicators that can be applied to assess the quality of Pareto sets:

1. The hypervolume indicator I_H [88];
2. The unary Epsilon Indicators I_ε^1 and $I_{\varepsilon+}^1$ [90]; and
3. The I_{R2}^1 and I_{R3}^1 indicators [43].

Due to the fact that each quality indicator takes different preference information into account, no single 'best' quality indicator can be proposed. Knowles et al. [48] recommend to consider multiple quality indicators at the same time to get the most information.

4   Applications of MOACO Algorithms in WSNs

There is a wide range of researches using MOACO-based algorithms in various fields of application. In this chapter, use cases of such methods in the area of WSNs are highlighted:

Sett and Thakurta [70] propose to apply multi-objective optimisation (MOO) on quality of service (QoS) metrics to cluster a wireless sensor network to improve the QoS. Simultaneously, the number of cluster heads (CHs) is minimised, while the number of members in the cluster is maximised. This allows to reduce the delay in routing requests and to improve the coverage of the network. An ant colony optimisation (ACO)-based routing algorithm with additional network status checks for CHs is used. Simulation results show that the proposed algorithm outperforms the traditional approaches.

Kellner and Hogrefe [46] propose to use multi-objective ant colony optimisation (MOACO) algorithms for routing in wireless sensor networks (WSNs). Kellner and Hogrefe propose MARFWSN, a multi-objective ant colony optimisation routing framework for WSNs that applies MOACO algorithms for the routing in WSNs. Due to its modularised structure, different MOACO algorithms can be used through the provided interface. Additionally, to mitigate insider attacks, the trust of the sensor nodes is considered as one of the objectives in the route optimisation process.

Fidanova et al. [31] use multi-objective ant colony optimisation in wireless sensor networks to minimise the number of sensors needed to fully cover the monitoring region, while minimising the network's energy consumption at the same time. Furthermore, Fidanova et al. particularly study the influence of the number of ants on the performance of MAX-MIN Ant System (MMAS) ACO algorithm in the WSN.

Yazdi [86] present an multi-objective ant colony-based routing protocol with coloured pheromones and clustering to improve the QoS in WSNs. Four class types are introduced: real-time (red), streaming (blue), interactive (green) and best effort (black) to distinguish different traffic classes. Simulations show that the proposed algorithm outperforms EAQR and AntSensNet in terms of packet delivery ratio, end-to-end delay, throughput and lifetime. The algorithm additionally ensures the QoS differentiation based on the proposed classes.

Berre et al. [5] consider the deployment of WSNs as multi-objective problem. Three objectives are taken into account: the coverage of the network, its lifetime and the financial costs in terms of number of deployed sensor nodes. Different algorithms (NSGA-II, SPEA-II and MOACO) are applied to solve the multi-objective problem. Experimental results show that regarding the C measure, NSGA-II outperforms SPEA-II and MOACO.

Deepalakshmi and Radhakrishnan [18] propose an ant-based multi-objective on-demand QoS routing algorithm for mobile ad hoc networks, called AMQR, which is particularly suitable for real-time traffic. The authors describe their algorithm as highly adaptive, efficient and scalable, while reducing the end-to-end delay in high mobility environments. Using ns-2, AMQR was compared to AODV and AntHocNet in terms of delay, throughput and jitter. Experiments show an improvement in packet delivery ratio as well as shorter delay and jitter for AMQR. This however comes at the cost of an increased routing overhead.

Persis and Robert [67] discuss the use of multi-objective unicast MANET routing that considers network performance measures such as delay, hop distance, load, cost and reliability. Two algorithms are proposed, a multi-objective variant of the Ad hoc On-Demand Vector (AODV) routing protocol and an ant-based routing protocol. Experiments with ns-2 were conducted, and the results show improvements in delay, packet delivery ratio and throughput for the proposed protocols, in comparison with the original AODV protocol.

Constantinou [16] proposes the use of a multi-objective ACO approach for the routing in MANETs considering five power-aware metrics. The found routes should result in energy-efficient routes and a maximised lifetime of the overall network. Five algorithms are proposed to solve the multi-objective problem in the routing domain: the multi-pheromone algorithm (EEMACOMP) and the multi-heuristic algorithm (EEMACOMH) are based on a multi-objective version of ACS, while the multi-pheromone algorithm (EEMMASMP) and the multi-heuristic algorithm (EEMMASMH) are based on a multi-objective version of the MAX-MIN ant system. The last algorithm, multi-colony algorithm (EEMACOMC) uses multiple ant colonies in its algorithm.

Apart from the direct applications of MOACO algorithms in WSNs, there are various works that attempt to assist in choosing a suitable algorithm and to find meaningful parameter settings for solving the underlying problem:

One of the first in-depth studies about multi-objective ant colony optimisation has been presented by Angus and Woodward [3] in 2008. In their paper, a taxonomy of multiple objective ant colony optimisation algorithms is presented based on the existing approaches at that time. Furthermore, they derive guidelines for the development and the use of multiple-objective ant colony optimisation algorithms.

López-Ibáñez and Stützle [53] compare several MOACO algorithms in an experiment setting. For each algorithm, different parameter settings are tested to figure out how particular design choices affect the quality and the shape of the Pareto front. The empirical analysis provides some insights on how the different MOACO algorithms work and how the design of those algorithms may be improved in the future.

In a following work by the same authors [52], a formulation of algorithmic components that can describe most MOACO algorithms so far is proposed. A flexible algorithmic framework presented from existing MOACO algorithms can be initiated, but also new combinations of different components to obtain novel MOACO algorithms. The proposed framework enables the automatic configuration of MOACO algorithms such that several combinations can be easily tested. Simulation results show that the automatically configured algorithms outperform the MOACO algorithms that have inspired the framework. Their suitability for the routing in MANETs was shown. All five algorithms provide substantially better results than the non-dominated sorting genetic algorithm (NSGA-II) in terms of the quality of the solution. EEMACOMP outperforms the other four ACO algorithms as well as the NSGA-II algorithm in terms of the number of solutions, closeness to the true Pareto front and diversity.

The use of MOACO-based algorithms for applications in WSNs is an emerging research area. Most approaches apply MOACO algorithms on routing and quality of service, and for the deployment of sensors. In the future, one may expect improvements in the performance of MOACO-based algorithms for WSNs and the discovery of new areas of application.

5 Conclusion

In this chapter, the suitability of biologically inspired MOACO-based algorithms for their use in WSNs has been discussed. At first, the basics of multi-objective combinatorial problems have been presented and their approximation using the MOACO-based metaheuristics as applied in other application domains has been discussed. Afterwards, the current use of MOACO-based algorithms in the area of WSNs has been presented. The considered research shows the aptitude of MOACO-based algorithms for their use in WSNs, e.g. for the routing or deployment of sensors, due to their distributed, autonomous and self-organising nature. As a result, it is likely that the adaptive, scalable and robust MOACO-based algorithms will play an important role in the upcoming era of pervasive and ubiquitous computing.

References

1. Alaya I, Solnon C, Ghédira K (2007) Ant colony optimization for multi-objective optimization problems. In: 19th IEEE international conference on tools with artificial intelligence (ICTAI 2007), October 29–31, 2007, Patras, Greece, vol 1, pp 450–457
2. Angus D (2007) Crowding population-based ant colony optimisation for the multi-objective travelling salesman problem. In: IEEE symposium on computational intelligence in multicriteria decision making, MCDM 2007, Honolulu, Hawaii, USA, April 1–5, 2007, pp 333–340
3. Angus D, Woodward C (2009) Multiple objective ant colony optimisation. Swarm Intell 3(1):69–85
4. Barán B, Schaerer M (2003) A multiobjective ant colony system for vehicle routing problem with time windows. In: The 21st IASTED international multi-conference on applied informatics (AI 2003), February 10–13, 2003. Innsbruck, Austria, pp 97–102
5. Berre ML, Hnaien F, Snoussi H (2011) Multi-objective optimization in wireless sensors networks. In: 2011 international conference on microelectronics (ICM). IEEE, pp 1–4
6. Blum C (2005) Beam-aco—hybridizing ant colony optimization with beam search: an application to open shop scheduling. Comput OR 32:1565–1591
7. Blum C, Blesa MJ (2005) New metaheuristic approaches for the edge-weighted k-cardinality tree problem. Comput OR 32:1355–1377
8. Bridgman PW (1922) Dimensional analysis. Yale University Press
9. Bullnheimer B, Hartl R, Strauß C (1997) A new rank based version of the ant system—a computational study. Central Eur J Oper Res Econ. Citeseer
10. De Campos LM, Fernández-Luna JM, Gámez JA, Puerta JM (2002) Ant colony optimization for learning bayesian networks. Int J Approx Reason 31(3):291–311
11. De Campos LM, Puerta J et al (2008) Learning bayesian networks by ant colony optimisation: searching in two different spaces. Mathware Soft Comput 9(3):251–268

12. Cardoso P, Jesus M, Márquez A (2003) Monaco-multi-objective network optimisation based on an aco. Proc X Encuentros de Geometría Computacional, Seville, Spain
13. Caro GD, Dorigo M (1998) Antnet: distributed stigmergetic control for communications networks. J Artif Intell Res (JAIR) 9:317–365
14. Caro GD, Ducatelle F, Gambardella LM (2005) Anthocnet: an adaptive nature-inspired algorithm for routing in mobile ad hoc networks. Eur Trans Telecommun 16(5):443–455
15. Coello CAC, Dhaenens C, Jourdan L (eds) (2010) Advances in multi-objective nature inspired computing. In: Studies in computational intelligence, vol 272. Springer
16. Constantinou D (2011) Ant colony optimisation algorithms for solving multi-objective power-aware metrics for mobile ad hoc networks. University of Pretoria, Thesis
17. Costa D, Hertz A (1997) Ants can colour graphs. J Oper Res Soc 48(3):295–305
18. Deepalakshmi P, Radhakrishnan S (2011) An ant colony-based multi objective quality of service routing for mobile ad hoc networks. EURASIP J Wirel Commun Netw 2011:153
19. Den Besten M, Stützle T, Dorigo M (2000) Ant colony optimization for the total weighted tardiness problem. In: Parallel problem solving from nature PPSN VI. Springer, pp 611–620
20. Doerner K, Hartl R, Reimann M (2001) Are COMPETants more competent for problem solving? The case of a multiple objective transportation problem. Report series SFB adaptive information systems and modelling in economics and management science
21. Doerner KF, Gutjahr WJ, Hartl RF, Strauss C, Stummer C (2004) Pareto ant colony optimization: a metaheuristic approach to multiobjective portfolio selection. Ann OR 131(1–4):79–99
22. Dorigo M (1992) Optimization, learning and natural algorithms (in Italian). PhD thesis, Politecnico di Milano, Italy
23. Dorigo M, Gambardella LM (1997a) Ant colonies for the travelling salesman problem. BioSystems 43(2):73–81
24. Dorigo M, Gambardella LM (1997b) Ant colony system: a cooperative learning approach to the traveling salesman problem. IEEE Trans Evol Comput 1(1):53–66
25. Dorigo M, Maniezzo V, Colorni A (1991) The ant system: an autocatalytic optimizing process. In: TR91-016, Politecnico di Milano
26. Dorigo M, Maniezzo V, Colorni A, Maniezzo V (1991b) Positive feedback as a search strategy. Technical report, Dipartimento di Elettronica, Politecnico di Milano
27. Dorigo M, Maniezzo V, Colorni A (1996) Ant system: optimization by a colony of cooperating agents. IEEE Trans Syst Man Cybern Part B Cybern 26(1):29–41
28. Dorigo M, Caro GD, Gambardella LM (1999) Ant algorithms for discrete optimization. Artif Life 5(2):137–172
29. Ducatelle F, Caro GD, Gambardella LM (2005) Using ant agents to combine reactive and proactive strategies for routing in mobile ad-hoc networks. Int J Comput Intell Appl 5(2):169–184
30. Fenet S, Solnon C (2003) Searching for maximum cliques with ant colony optimization. In: Applications of evolutionary computing, EvoWorkshop 2003: EvoBIO, EvoCOP, EvoIASP, EvoMUSART, EvoROB, and EvoSTIM, Essex, UK, April 14–16, 2003, Proceedings, pp 236–245
31. Fidanova S, Marinov P, Paprzycki M (2013) Influence of the number of ants on multi-objective ant colony optimization algorithm for wireless sensor network layout. In: Large-scale scientific computing—9th international conference, LSSC 2013, Sozopol, Bulgaria, June 3–7, 2013. Revised Selected Papers, pp 232–239
32. Fishburn P (1967) Additive utilities with incomplete product set: applications to priorities and sharings
33. Fonseca CM, Fleming PJ, Zitzler E, Deb K, Thiele L (eds) (2003) Evolutionary multi-criterion optimization. In: Second international conference, EMO 2003, Faro, Portugal, April 8–11, 2003, Proceedings, Lecture notes in computer science, vol 2632. Springer
34. Gambardella LM, Dorigo M (1995) Ant-q: a reinforcement learning approach to the traveling salesman problem. In: Machine learning, Proceedings of the twelfth international conference on machine learning, Tahoe City, California, USA, July 9–12, 1995, pp 252–260

35. Gambardella LM, Dorigo M (1996) Solving symmetric and asymmetric tsps by ant colonies. In: International conference on evolutionary computation, pp 622–627
36. Gambardella LM, Dorigo M (2000) An ant colony system hybridized with a new local search for the sequential ordering problem. INFORMS J Comput 12(3):237–255
37. Gambardella LM, Taillard É, Agazzi G (1999) Macs-vrptw: a multiple ant colony system for vehicle routing problems with time windows. New ideas in optimization. McGraw-Hill Ltd., UK, pp 63–76
38. García OC, Triguero FH, Stützle T (2002) A review on the ant colony optimization metaheuristic: basis, models and new trends. Mathware Soft Comput 9(3):141–175
39. Gardel P, Baran B, Estigarribia H, Fernandez U, Duarte S (2006) Multiobjective reactive power compensation with an ant colony optimization algorithm. In: The 8th IEE international conference on AC and DC power transmission, 2006. ACDC 2006, IET, pp 276–280
40. Glover F (1989) Tabu search—part I. INFORMS J Comput 1(3):190–206
41. Glover F (1990) Tabu search—part II. INFORMS J Comput 2(1):4–32
42. Haimes YY, Ladson L, Wismer DA (1971) Bicriterion formulation of problems of integrated system identification and system optimization. IEEE Trans Syst Man Cybern 1(3):296–297
43. Hansen M, Jaszkiewicz A (1998) Evaluating the quality of approximations to the non-dominated set. Department of Mathematical Modelling, Technical University of Denmark, IMM
44. Hart PE, Nilsson NJ, Raphael B (1968) A formal basis for the heuristic determination of minimum cost paths. IEEE Trans Syst Sci Cybern 4(2):100–107
45. Iredi S, Merkle D, Middendorf M (2001) Bi-criterion optimization with multi colony ant algorithms. In: Evolutionary multi-criterion optimization, first international conference, EMO 2001, Zurich, Switzerland, March 7–9, 2001, Proceedings, pp 359–372
46. Kellner A, Hogrefe D (2014) Multi-objective ant colony optimisation-based routing in WSNs. IJBIC 6(5):322–332
47. Kirkpatrick S (1984) Optimization by simulated annealing: quantitative studies. J Stat Phys 34(5–6):975–986
48. Knowles J, Thiele L, Zitzler E (2006) A tutorial on the performance assessment of stochastic multiobjective optimizers. TIK report 214
49. Korb O, Stützle T, Exner TE (2006) PLANTS: application of ant colony optimization to structure-based drug design. In: Ant colony optimization and swarm intelligence, 5th international workshop, ANTS 2006, Brussels, Belgium, September 4–7, 2006, Proceedings, pp 247–258
50. Leguizamon G, Michalewicz Z (1999) A new version of ant system for subset problems. In: Proceedings of the 1999 congress on evolutionary computation, 1999. CEC 99, vol 2. IEEE
51. Lessing L, Dumitrescu I, Stützle T (2004) A comparison between ACO algorithms for the set covering problem. In: 4th international workshop Ant colony optimization and swarm intelligence, ANTS 2004, Brussels, Belgium, September 5–8, 2004, Proceedings, pp 1–12
52. López-Ibáñez M, Stützle T (2012a) The automatic design of multiobjective ant colony optimization algorithms. IEEE Trans Evol Comput 16(6):861–875
53. López-Ibáñez M, Stützle T (2012b) An experimental analysis of design choices of multi-objective ant colony optimization algorithms. Swarm Intell 6(3):207–232
54. López-Ibáñez M, Paquete L, Stützle T (2004) On the design of ACO for the biobjective quadratic assignment problem. In: Ant colony optimization and swarm intelligence, 4th international workshop, ANTS 2004, Brussels, Belgium, September 5–8, 2004, Proceedings, pp 214–225
55. Lourenco H, Martin O, Stützle T (2003) Iterated local search. Handbook of metaheuristics, pp 320–353
56. Maniezzo V (1999) Exact and approximate nondeterministic tree-search procedures for the quadratic assignment problem. INFORMS J Comput 11(4):358–369
57. Mariano CE, Morales E (1999) A multiple objective ant-q algorithm for the design of water distribution irrigation networks. Instituto Mexicano de Tecnología del Agua, Technical report HC-9904

58. Martens D, Backer MD, Haesen R, Baesens B, Mues C, Vanthienen J (2006) Ant-based approach to the knowledge fusion problem. In: Ant colony optimization and swarm intelligence, 5th international workshop, ANTS 2006, Brussels, Belgium, September 4–7, 2006, Proceedings, pp 84–95

59. McMullen PR (2001) An ant colony optimization approach to addressing a JIT sequencing problem with multiple objectives. AI Eng 15(3):309–317

60. Merkle D, Middendorf M (2003) Ant colony optimization with global pheromone evaluation for scheduling a single machine. Appl Intell 18(1):105–111

61. Merkle D, Middendorf M, Schmeck H (2002) Ant colony optimization for resource-constrained project scheduling. IEEE Trans Evol Comput 6(4):333–346

62. Miettinen K (1999) Nonlinear multiobjective optimization, vol 12. Springer

63. Miller DW et al (1960) Executive decisions and operations research. Prentice-Hall

64. Mladenović N, Hansen P (1997) Variable neighborhood search. Comput Oper Res 24(11):1097–1100

65. Paquete L, Stützle T (2007) Stochastic local search algorithms for multiobjective combinatorial optimization: a review. In: Handbook of approximation algorithms and metaheuristics, vol 13

66. Parpinelli RS, Lopes HS, Freitas AA (2002) Data mining with an ant colony optimization algorithm. IEEE Trans Evol Comput 6(4):321–332

67. Persis DJ, Robert TP (2015) Ant based multi-objective routing optimization in mobile ad-hoc network. Indian J Sci Technol 8(9):875–888

68. Pinto D, Barán B (2005) Solving multiobjective multicast routing problem with a new ant colony optimization approach. In: 3rd international Latin American networking conference, LANC 2005, Sponsored by IFIP TC6 communication networks and ACM SIGCOMM, Organized by CLEI (Centro Latino-Americano de Estudios en Informática), Cali, Colombia, October 10–13, 2005, pp 11–19

69. Reimann M, Doerner K, Hartl RF (2004) D-ants: savings based ants divide and conquer the vehicle routing problem. Comput OR 31(4):563–591

70. Sett S, Thakurta PKG (2015) Multi objective optimization on clustered mobile networks: an aco based approach. In: Information systems design and intelligent applications. Springer, pp 123–133

71. Shmygelska A, Hoos HH (2005) An ant colony optimisation algorithm for the 2d and 3d hydrophobic polar protein folding problem. BMC Bioinform 6:30

72. Socha K, Knowles JD, Sampels M (2002) A MAX-MIN ant system for the university course timetabling problem. In: Ant algorithms, Third international workshop, ANTS 2002, Brussels, Belgium, September 12–14, 2002, Proceedings, pp 1–13

73. Socha K, Sampels M, Manfrin M (2003) Ant algorithms for the university course timetabling problem with regard to the state-of-the-art. In: Applications of evolutionary computing, EvoWorkshop 2003: EvoBIO, EvoCOP, EvoIASP, EvoMUSART, EvoROB, and EvoSTIM, Essex, UK, April 14–16, 2003, Proceedings, pp 334–345

74. Solnon C (2000) Solving permutation constraint satisfaction problems with artificial ants. In: ECAI 2000, Proceedings of the 14th European conference on artificial intelligence, Berlin, Germany, August 20–25, 2000, pp 118–122

75. Solnon C (2002) Ants can solve constraint satisfaction problems. IEEE Trans Evol Comput 6(4):347–357

76. Sotelo-Figueroa MA, Baltazar R, Carpio JM (2010) Application of the bee swarm optimization BSO to the knapsack problem. In: Soft computing for recognition based on biometrics, pp 191–206

77. Stützle T (1999) Local search algorithms for combinatorial problems—analysis, improvements, and new applications, DISKI, vol 220. Infix

78. Stützle T, Hoos H (1997) Max-min ant system and local search for the traveling salesman problem. In: IEEE International conference on evolutionary computation, 1997. IEEE, pp 309–314

79. Stützle T, Hoos HH (2000) MAX-MIN ant system. Fut Gener Comput Syst 16(8):889–914

80. T'Kindt V, Monmarché N, Tercinet F, Laügt D (2002) An ant colony optimization algorithm to solve a 2-machine bicriteria flowshop scheduling problem. Eur J Oper Res 142(2):250–257
81. Triantaphyllou E (2000) Multi-criteria decision making methods. In: Multi-criteria decision making methods: a comparative study. Springer, pp 5–21
82. Vira C, Haimes YY (1983) Multiobjective decision making: theory and methodology, vol 8. North-Holland, New York
83. Wierzbicki AP (1982) A mathematical basis for satisficing decision making. Math Model 3(5):391–405
84. Wierzbicki AP (1986) On the completeness and constructiveness of parametric characterizations to vector optimization problems. Oper Res Spektr 8(2):73–87
85. Yagmahan B, Yenisey MM (2010) A multi-objective ant colony system algorithm for flow shop scheduling problem. Expert Syst Appl 37(2):1361–1368
86. Yazdi FR (2013) Ant colony with colored pheromones routing for multi objectives quality of services in wsns. Int J Res Comput Sci 3(1):1
87. Zeleny M, Cochrane JL (1973) Multiple criteria decision making. University of South Carolina Press
88. Zitzler E, Thiele L (1999) Multiobjective evolutionary algorithms: a comparative case study and the strength pareto approach. IEEE Trans Evol Comput 3(4):257–271
89. Zitzler E, Deb K, Thiele L (2000) Comparison of multiobjective evolutionary algorithms: empirical results. Evol Comput 8(2):173–195
90. Zitzler E, Thiele L, Laumanns M, Fonseca CM, da Fonseca VG (2003) Performance assessment of multiobjective optimizers: an analysis and review. IEEE Trans Evol Comput 7(2):117–132

Generating the Training Plans Based on Existing Sports Activities Using Swarm Intelligence

Iztok Fister Jr. and Iztok Fister

Abstract Planning the sports training sessions by an evolutionary computation and swarm intelligence-based algorithms has been becoming an interesting topic for research. Recently, many methods and techniques were proposed in theory and practice in order to help athletes in sports training. In a nutshell, integrating these methods and techniques in the same framework has resulted in creating an artificial sports trainer with abilities similar to a human trainer. In this chapter, we intend to extend the artificial sports trainer with an additional feature which enables athletes to generate a training plan on the basis of existing training courses tracked by mobile sports trackers. Experimental results suggest the usefulness of the proposed method.

1 Introduction

Once upon a time, many people began enjoying lighter sports activities. These activities were usually focused on walking, short jogging, or even light cycling. Since the raising of mass sports advertisements and sports competitions/events, many people look at sports activities more seriously, and, therefore, this has resulted in an expansion of participants in mass sports events [23]. In other words, casual events with a smaller number of participants suddenly became mass sport events. Typically, the mass sports event assembles from hundreds or thousands more participants. For example, running city marathons can be counted in this category of sports that recently became very popular [24]. All over the world, small and big cities organize running marathons where professional, amateur, and newbie athletes compete or just participate together without any special racing goals. On the other hand, more

I. Fister Jr. (✉) · I. Fister
Faculty of Electrical Engineering and Computer Science,
University of Maribor, Smetanova 17, 2000 Maribor, Slovenia
e-mail: iztok.fister1@um.si

I. Fister
e-mail: iztok.fister@um.si

© Springer International Publishing AG 2017
S. Patnaik et al. (eds.), *Nature-Inspired Computing and Optimization*,
Modeling and Optimization in Science and Technologies 10,
DOI 10.1007/978-3-319-50920-4_4

complex sport disciplines as, for example, cycling and multisport disciplines such as triathlons, have also been becoming very popular for a wide range of participants.

Anyway, participation in mass sports competitions is just one side of the coin. The other side of the coin is the fact that people have to be prepared in order to participate in such sports events. Preparation demands mostly proper sports training. The sports training is always a very complex and intricate process where athletes try to prepare themselves optimally for a sports event or competition. In this process, the presence of a sports trainer was inevitable until recently. The trainer prescribed his/her trainee training plans that need to be performed by athletes in order to increase their performances. To conduct a sports training efficiently, it is a very long process if trainers do not know the characteristics of the trainees' bodies in detail. Unfortunately, the trainers are generally very expensive.

To help athletes with a proper sports training program, an artificial sport trainer was designed that is an inexpensive variant of the real trainer. The artificial sports trainer consists of algorithms from a computational intelligence, which supports a wide spectrum of tracking mobile devices and are based on the sports training theory [7]. At this time, it is able to plan sports training sessions very efficiently [7], detects an overtraining phenomenon [9], recommends the most efficient food during endurance competitions [8], etc.

In the past 20 years, computational intelligence (CI) algorithms have been applied in dozens and dozens of applications. The following families of algorithms belong to the CI:

- Artificial Neural Networks (ANN) [14],
- Artificial Immune Systems (AIS) [6],
- Fuzzy Logic (FL) [26],
- Evolutionary Algorithms (EAs) [28],
- Swarm Intelligence (SI) [13, 16].

Recently, the most popular families of CI are EAs and SI-based algorithms. Typically, these stochastic population-based algorithms are inspired by natural systems. Thus, EAs mimic a Darwinian evolution [5], where the fitter individuals have more chances to survive in the cruel struggle for existence, while in SI, there is a bunch of unintelligent agents capable of performing the simplest tasks, but acting together in a community they exhibit collective behavior that enables them to survive, e.g., searching for a food by ants or building the magnificent buildings in which they live by termites.

Using computer technology to help athletes in training was also a vibrant topic for research in the past 10–15 years. In line with this, many useful applications emerged in theory with practical applicability. At the beginning, researchers applied machine learning techniques in the sport domain [18, 20–22]. On the other hand, a rapid development has been started of pervasive computing in sports [1, 2, 15, 19].

Mainly, the aim of this chapter was to present a pioneering work in planning the sports courses by cycling for a some period of time (also cycle). For instance, to select a proper combination of training courses by cycling for a duration of 1 week is very hard for sports trainers. In this work, a solution based on a swarm intelligence algorithm is proposed which performs optimal selection of sports courses from pre-defined course clusters. Initial experiments suggest that our solution is very helpful for athletes with many possibilities for further upgrade.

The structure of the remainder of the chapter is as follows. Section 2 deals with a short description of the artificial sports trainer. In Sect. 3, the proposed SI-based algorithm for generating the training plans based on existing cycling courses is illustrated. The experiments and results are the subjects of Sect. 4. The chapter finishes with summarizing the performed work and outlines possible directions for a future work.

2 Artificial Sports Trainer

Recently, the job of a human personal sports trainer has been becoming interesting and very popular. People start to hire personal trainers to help in training tasks due to different promotional activities for a healthy and low-fat life in society. On the other hand, many amateur athletes tend to increase their performance in some sports disciplines. A good example of this is the case of the Ironman Triathlon, where thousands and thousands of competitors compete in various Ironman competitions every year to achieve places for the Ironman World Championships taking place in Hawaii. For qualifying on Hawaii, an athlete must be very fast and, consequently, should finish the race on the podium. To achieve this good result, he/she must be prepared brilliantly for one of the qualification races. Thus, many athletes hire one or more personal trainers who prepare them using the proper training plans. Some personal trainers are very good, but some are not so good. On the other hand, the trainers are pretty expensive, and most of the athletes could not afford them.

After identifying this problem 2 years ago, the development of an artificial sports trainer [7, 10] has been started. The main purpose of the artificial sport trainer was to allow all people to have their own inexpensive personal trainers with capabilities comparable with the real sports trainer. It is based on the algorithms of computational intelligence and is able to cover almost all phases of the sports training. The following phases complete a traditional sports training (Fig. 1):

- Planning,
- Realization,
- Control,
- Estimation.

The plan of the specific sports training is accomplished in the first phase. Usually, this training plan is made for some time period, a so-called cycle (e.g., competition season, month, week), and depends on a training strategy. The training strategy is based on goals that need to be achieved at the end of the training period. As a result, all planned training sessions must comply with the strategy. Each training session prescribes a training type (e.g., aerobic, aerobic–anaerobic, anaerobic), magnitude, intensity, and iterations, thus defining an exercise load that an athlete needs to have

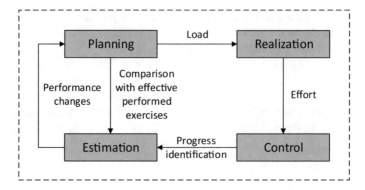

Fig. 1 Training cycle

accomplished during the training. In the realization phase, the training session is conducted by an athlete. All realizations of workouts, as demanded by the training session are, controlled by the trainer. In the past, the trainer measured an athlete's invested effort by tracking the workout with measures such as stopwatch for measuring the time and a measuring tape for measuring length. Recently, the quality of realized workouts has been measured by the mobile devices that produce a lot of data by tracking the athlete performing the sports session. This amount of data is suitable for identifying the progress of the athlete by the artificial sports trainer. At the end of the training period, the training cycle is estimated by the trainer. This estimation is performed by comparing the prescribed values and the obtained values during the workouts. However, performance changes are indicated by this comparison that has an impact on the training plan in the next training cycle.

The following CI algorithms can be used for the training period by the artificial sports trainer at the time:

- Generate sports training sessions [7] in the planning phase,
- Generate fitness sessions [11] in the realization phase,
- Predict food [8] in the control phase,
- Help prevent overtraining syndrome [9] in the estimation phase.

The efficiency of the artificial sport trainer was validated by a human sport trainer who confirmed that our solution is efficient enough. However, there are still many tasks for integration in the artificial sport trainer. One of these tasks is the generation of training plans based on existing sports activities as presented in the next section.

3   Generating the Training Plans

An artificial sports trainer is capable of replacing the real trainer, especially in training phases that demand the processing of a large amount of data and making decisions on their basis. The best suited for an automation of the real trainer tasks are individual

Fig. 2 Architecture of generating the sports training plans based on existing courses

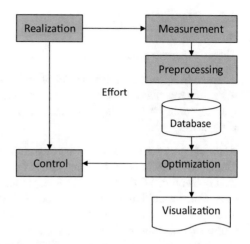

sports disciplines such as running, cycling, triathlon, where tracking of athletes in training is performed by mobile devices (i.e., smart phones and sports watches) automatically. These mobile devices generate a huge amount of data saved in a GPS exchange (GPX) or Training Center XML (TCX) dataset formats. Consequently, the artificial sports trainer is capable of classifying, clustering, analyzing, and data mining the training datasets and, on this basis, predicting those future training workouts that will most increase the current performances of athletes in training.

In line with this, generating the training plans based on an existing course is devoted to generate such courses that will comply the most with the training strategy on the basis of the current performance of an athlete. The architecture of the algorithm for generating the training plans is illustrated in Fig. 2.

Input of this kind of sports training generation presents a set of training datasets in GPX and/or TCX format that are obtained in the realization training phase by tracking mobile devices. At first, these datasets are identified and preprocessed, where the characteristic values about the training session are parsed. Then, the identified training sessions are clustered according to the characteristic values of the training session into clusters, and the obtained data are saved into a database. Finally, an optimization SI-based algorithm is launched that generates the optimal training course. On the one hand, this course is transferred to the control phase, while it can be visualized on the other hand.

Consequently, the proposed algorithm consists of the following stages:

- Preprocessing,
- Optimization,
- Visualization.

In the remainder of the chapter, the proposed stages are presented in detail.

3.1 Preprocessing

The preprocessing phase consists of four steps:

- Assembling and identifying a set of sports activities,
- Parsing the characteristics values in training data,
- Determining the training load indicators,
- Clustering.

A set of sport activity datasets was assembled and identified in the first step [25]. Then, the characteristics training values needed for the next step were parsed from the GPX and/or TCX datasets. Typically, these datasets are tracked by sport trackers (e.g., Garmin and Polar watches, or smart phones). Interestingly, athletes can upload a complete dataset on the server where it is archived. Additionally, there are many online services and applications to visualize these activities (Garmin Connect, Strava, Endomondo). A simple example of TCX dataset is presented in Table 1.

As can be seen from Table 1, each TCX dataset starts with the identification of the training session (i.e., activity). Then, the summary data about the activity follow, i.e.,

Table 1 Example of sport activity stored in a TCX file

```
⟨Activity Sport="Biking"⟩
 ⟨Id⟩2012-03-06T14:16:05.000Z⟨/Id⟩
 ⟨Lap StartTime="2012-03-06T14:16:05.000Z"⟩
  ⟨TotalTimeSeconds⟩1902.72⟨/TotalTimeSeconds⟩
  ⟨DistanceMeters⟩15042.79⟨/DistanceMeters⟩
  ⟨MaximumSpeed⟩16.774999618530273⟨/MaximumSpeed⟩
  ⟨Calories⟩273⟨/Calories⟩
  ⟨AverageHeartRateBpm⟩
   ⟨Value⟩119⟨/Value⟩
  ⟨/AverageHeartRateBpm⟩
  ⟨MaximumHeartRateBpm⟩
   ⟨Value⟩153⟨/Value⟩
  ⟨/MaximumHeartRateBpm⟩
  ⟨Intensity⟩Active⟨/Intensity⟩
  ⟨TriggerMethod⟩Manual⟨/TriggerMethod⟩
  ⟨Track⟩
   ⟨Trackpoint⟩
    ⟨Time⟩2012-03-06T14:16:05.000Z⟨/Time⟩
    ⟨Position⟩
     ⟨LatitudeDegrees⟩46.07827073894441⟨/LatitudeDegrees⟩
     ⟨LongitudeDegrees⟩14.516056096181273⟨/LongitudeDegrees⟩
    ⟨/Position⟩
```

a duration time, an overcame length, and an average heart rate. The dataset ends with a detailed description of the activity started with an XML tag ⟨Track⟩ and followed by a set of track points each started with ⟨Trackpoint⟩ an XML tag determining the current position of an athlete at a specific time. Track points are tracked in some time intervals, depending on the tracking device (e.g., each second).

In the parsing step, characteristic values are parsed from each activity dataset. These characteristics values are presented in bold in Table 1 and are needed for calculating the corresponding training load indicator. The training load is defined as the stress placed upon the body as a result of the training session [3]. There are more training load indicators. The basic TRIMP (TRaining IMPulse) was employed in this study. This training load indicator was at first proposed by Banister et al. in 1991 [3, 4] and it is simple, expressed as

$$\text{TRIMP} = t \cdot \overline{HR}, \tag{1}$$

where t denotes duration in minutes (min), and $\overline{HR}$ is an average heart rate in beats per minute (bpm). As can be seen from Eq. (1), the training load is defined as a product of duration and average heart rate. However, the main disadvantage of this indicator is that it is insensitive to the different levels of trainings. This means that long-term sports trainings of low intensity (i.e., with lower average HR) and the short-term sports trainings of high intensity (i.e., with higher average HR) have the similar TRIMP values.

Finally, the set of training sessions are clustered according to the identified TRIMP training load indicator in order to obtain groups of training sessions of the similar intensities. Here, k-means clustering [12] was taken into consideration. Thus, three TRIMP training zones are observed containing a low-, medium-, and high-intensive training sessions, where each point on Fig. 3 presents the TRIMP training load indicator of a specific training session. Let us notice that points with green color represent the low-intensive training activities, blue points the medium-intensive training activities, while the red points are the high-intensive training activities. Centers of clusters denoted as squares in the figure indicate the similar values of $HR = 130$ but of different duration. Therefore, the higher the duration, higher the TRIMP. The other training sessions are dispersed around these centers by the average HR and duration t. The most intensive training sessions in each cluster are characterized by the higher average HR and long duration t.

Intensity zones determine the intensity of the sports training session. However, the intensity load cannot be measured directly, because it is the result of complex chemical reactions carried out in an athlete's organism. Typically, it is estimated indirectly based on the average heart rate HR. Usually, two measures are employed to determine which stress is placed upon the athlete's body by performing some workout, i.e., HR zone and % max HR. In this study, four intensity zones were defined that cover the HR zones as illustrated in Table 2a. The k-means clustering splits identified sports training sessions into three clusters as presented in Table 2b.

Let us notice that the HR zones and % max HR are calculated for an athlete with the % max $HR = 200$ in Table 2a. Interestingly, the addressed set of sports activity

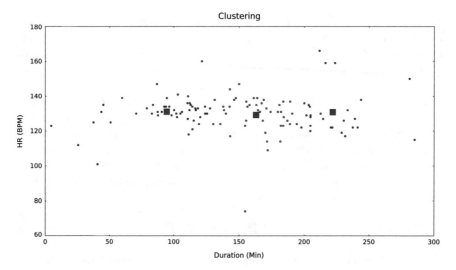

Fig. 3 Example of clustering

Table 2 Characteristics of the TRIMP clusters

(a) Training intensity zones

Intensity zone	*HR* zone	% max *HR* (%)
1	60–100	30–50
2	100–140	50–70
3	140–180	70–90
4	180–200	90–100

(b) TRIMP training zones

Cluster	Average *HR* (min)	Duration *t* (bpm)	TRIMP
1	130	90	11,700
2	130	160	20,800
3	130	215	27,950

datasets includes only training sessions in an aerobic *HR* zone (i.e., 50–70% max *HR*) and an aerobic–anaerobic *HR* zone (i.e., 70–90% max *HR*), while anaerobic *HR* zone (i.e., 90–100% max *HR*) is not presented in the current dataset. Training sports sessions of intensity 30–50% max *HR* indicate a resting phase. Interestingly, the training sessions of all observed intensity zones are observed in each cluster. As can be seen from Table 2b, the TRIMP training load indicator increases by raising the duration *t*.

The obtained clusters are saved into a database, from which they serve as an input for an optimization phase.

3.2 Optimization Process

The task of an optimization process is to accomplish a sports training plan for the duration of a training cycle (e.g., 1 week), in which training sessions of different intensities are distributed such that the less-intensive training sessions are planned at the start and the end of the cycle, while the more intensive are in the middle of the cycle. The clustering algorithm assembles the training sessions into three clusters where the intensity of training sessions increases from the first to the third cluster. As a result, the third cluster comprises the most intensive sports sessions. The optimization algorithm needs to include the proper training sessions from different clusters into the training plan so that the demanded intensity distribution is considered.

At the moment, we are witnessing the emerging effective computational intelligence algorithms which find their inspiration from nature. These so-called nature-inspired algorithms are very useful for solving the continuous, as well as discrete, optimization problems. A recent member of this family is a Bat Algorithm (BA) developed by Yang [29] that is, besides its simplicity, also capable of solving complex optimization problems efficiently. In this chapter, we took the BA for searching for the optimal training courses. It was used successfully for many tasks in design and development of the artificial sport trainer, and, thus, it is also appropriate for generating the training plans based on existing courses.

The original BA is presented in the remainder of the chapter. Then, the modified BA is discussed for the optimal training plan generation.

3.2.1 Bat Algorithm

The BA was developed by Yang in 2010 [29], where the phenomenon of echolocation arisen by some kind of microbats is explored as an inspiration for this stochastic population-based optimization algorithm. Mainly, the phenomenon is used by the microbats for orientation in the dark. In order to apply it to the algorithm, the author simplified the complex behavior of the natural bats with the following three rules:

- All bats use echolocation to sense distance to target objects.
- Bats fly randomly with the velocity v_i at position x_i, frequency $Q_i \in [Q_{min}, Q_{max}]$ (also wavelength λ_i), rate of pulse emission $r_i \in [0, 1]$, and loudness $A_i \in [A_0, A_{min}]$. The frequency (and wavelength) can be adjusted depending on the proximities of their targets.
- The loudness varies from a large (positive) A_0 to a minimum constant value A_{min}.

The BA maintains a swarm of the virtual bats, where each bat in the swarm represents a solution of the problem to be solved. Thus, each solution is represented as a real-coded vector

$$\mathbf{x}_i = [x_{i,1}, \ldots, x_{i,D}]^T, \quad \text{for } i = 1, \ldots, Np, \tag{2}$$

where t is a current generation, D denotes a dimensionality of the problem, and Np the number of virtual bats in the swarm. Actually, each vector determines a position of the virtual bat in a search space. Bats move toward the best bat position and thus explore the new regions of the search space. The BA supports two strategies of an exploration of the search space. The former moves the virtual bat according to the equation

$$Q_i^{(t)} = Q_{min}^{(t)} + \left(Q_{max}^{(t)} - Q_{min}^{(t)}\right) \cdot \beta,$$
$$\mathbf{v}_i^{(t+1)} = \mathbf{v}_i^{(t)} + \left(\mathbf{x}_i^{(t)} - \mathbf{x}_{best}^{(t)}\right) \cdot Q_i^{(t)}, \tag{3}$$
$$\mathbf{x}_i^{(t+1)} = \mathbf{x}_i^{(t)} + \mathbf{v}_i^{(t)}.$$

where a pulse frequency can vary in the interval $Q_i^{(t)} \in [Q_{min}, Q_{max}]$, a random number $\beta \in [0, 1]$ specifies the output pulse, and $\mathbf{x}_{best}^{(t)}$ presents currently the best solution. The latter improves the current bat position according to the equation

$$\mathbf{x}_i^{(t+1)} = \mathbf{x}_{best} + \varepsilon \cdot L(s, \alpha), \tag{4}$$

where $\varepsilon > 0$ is the step size scaling factor, $L(s, \alpha)$ the Lévy flight alpha-stable distribution with parameters scale s and exponent $\alpha \in (0, 2]$. The distribution reduces to Gaussian distribution for $\alpha = 2$ and to Cauchy distribution for $\alpha = 1$. The mentioned strategy is more exploitative than those illustrated in Eq. (4) and presents a kind of random walk that is primarily focused on exploring the vicinity of the current best solution.

Let us notice that both exploration strategies are balanced in the search process using the parameter pulse rate $r_i^{(t)} \in [0, 1]$. The closer the bat to the prey, the higher the pulse rate and vice versa. The evaluation function evaluates the quality of the generated solutions and models the optimization problem into the BA. Interestingly, the better trial solution replaces the on the same position laid solution in the current swarm only conditionally, i.e., according to the loudness $A \in [A_0, A_{min}]$. The motivation behind using this parameter lays in simulated annealing [17], where the better trial solution replaces the current solution only under some probability in order to avoid getting stacking into local optima.

The algorithm's pseudocode is presented in Algorithm 1, whose main algorithm components are summarized as follows:

- *Initialization* (lines 1–3): Initializing the algorithm parameters, generating the initial population, evaluating this, and, finally, determining the best solution $\mathbf{x}_{best}$ in the population,
- *Generate_the_new_solution* (line 6): Moving the virtual bats in the search space according to the physical rules of bat echolocation,
- *Local_search_step* (lines 7–9): Improving the best solution using the Random Walk Direct Exploitation (RWDE) heuristic,
- *Evaluate_the_new_solution* (line 10): Evaluating the new solution,
- *Save_the_best_solution_conditionally* (lines 12–14): Saving the new best solution under some probability A_i,
- *Find_the_best_solution* (line 15): Finding the current best solution.

Algorithm 1 Bat algorithm

Input: Bat population $\mathbf{x_i} = (x_{i1}, \ldots, x_{iD})^T$ for $i = 1 \ldots Np, MAX\_FE$.
Output: The best solution $\mathbf{x}_{best}$ and its corresponding value $f_{min} = \min(f(\mathbf{x}))$.
1: init_bat();
2: $eval$ = evaluate_the_new_population;
3: f_{min} = find_the_best_solution($\mathbf{x}_{best}$); {Initialization}
4: **while** termination_condition_not_meet **do**
5:   **for** $i = 1$ **to** Np **do**
6:     $\mathbf{y}$ = Generate_new_solution($\mathbf{x}_i$);
7:     **if** rand$(0, 1) > r_i$ **then**
8:       $\mathbf{y}$ = Improve_the_best_solution($\mathbf{x}_{best}$)
9:     **end if**{ Local search step }
10:     f_{new} = Evaluate_the_new_solution($\mathbf{y}$);
11:     $eval = eval + 1$;
12:     **if** $f_{new} \leq f_i$ **and** $N(0, 1) < A_i$ **then**
13:       $\mathbf{x}_i = \mathbf{y}$; $f_i = f_{new}$;
14:     **end if**{ Save the best solution conditionally }
15:     f_{min}=Find_the_best_solution($\mathbf{x}_{best}$);
16:   **end for**
17: **end while**

The original BA is devoted primarily for continuous optimization. When the algorithm is applied to another kind of problem, some modifications need to be applied into it. Therefore, necessary modifications for developing the BA for sports training course generation are presented in the next subsection.

3.2.2 BA for Generating the Sports Training Plan

Mathematically, generating the sports training plans based on existing sports activities can be defined as follows: Let a set of clusters $C = \{C_k\}$ for $k = 1, \ldots, n$ and an ordered list of D-tuples $(C_{\pi_1}, \ldots, C_{\pi_D})$, where n is the number of clusters and $\pi_j \in [1, n]$ for $j = 1, \ldots, D$ determines the cluster defined by j-th tuple, and D is dimension of the problem denoting the length of the training cycle. Thus, the tuple is defined as $C_k = \{s_{k,1}, \ldots, s_{k,n_k}\}$ where $s_{k,l}$ for $l = 1, \ldots, n_k$ denotes a specific training session and n_k is a size of k-th cluster. In addition, vectors $\mathbf{x}_i = [x_{i,1}, \ldots, x_{i,D}]^T$ for $i = 1, \ldots, Np$ are given, where each element $x_{i,j} \in [0, 1]$ is mapped to the corresponding training session. The training session $s_{\pi_j,l}$ is obtained from the element $x_{i,j}$ by determining the index l, as follows:

$$l = \lceil x_{i,j} \cdot n_{\pi_j} \rceil. \tag{5}$$

In fact, each element $s_{\pi_j,l}$ denotes the l-th training session in π_j-th cluster. As a result, the fitness function of the modified BA algorithm is expressed as follows:

Table 3 Distribution of clusters and their influence on the fitness function

Day	Cluster	Weight w_i	Cluster size	% max HR
Monday	–	–	–	–
Tuesday	1	0.1	27	30–50%
Wednesday	2	0.2	49	50–70%
Thursday	3	0.4	44	70–90%
Friday	–	–	–	–
Saturday	2	0.2	49	50–70%
Sunday	1	0.1	27	30–50%

$$f(\mathbf{x}_i) = \sum_{j=1}^{D} w_i \cdot \text{TRIMP}(s_{\pi_j,l}), \tag{6}$$

where w_i denotes the weight of the corresponding cluster and $\text{TRIMP}(s_{\pi_j,l})$ denotes the TRIMP training load indicator of the corresponding training session calculated by Eq. (5).

Let us suppose a training cycle of 1 week is given, where five days are devoted for the active training sessions (i.e., $D = 5$) and two days for resting. The distribution of clusters during the training cycle is defined as $C = \{C_1, C_2, C_3, C_2, C_1\}$. Consequently, each element of vector $\mathbf{x}_i$ can be mapped to a specific training session $x_{i,j} \mapsto s_{\pi_j,l}$. The weight values and distribution of clusters are presented in Table 3.

As can be seen from Table 3, the distribution of the intensity of the training sessions follows predefined demands, because the training sessions from the low-intensive cluster one are taken in the second and seventh training days, the more intensive cluster in the third and sixth training days, and the high-intensive cluster in the fourth training day (i.e., in the middle of the active training period). Two days in the training plan (i.e., Monday and Friday) are left for resting.

From the mathematical definition of the problem, it follows that the modified BA algorithm differs from the original BA in the interpretation of the solution. Indeed, each element of solution vector $x_{i,j}$ is mapped to the corresponding j-th training session in i-th cluster that represents a basis for the evaluation of the TRIMP training load indicator.

The task of the modified BA is to find the maximum value of the fitness function, in other words, the max $f(\mathbf{x})$ is searched for by the BA. Let us notice that the algorithm was implemented in Python programming language.

3.2.3 Visualization

The sports training courses proposed by the modified BA are visualized in this phase. Thus, the appropriate training course for each day of the week is selected, and the figure is drawn from the corresponding GPX or TCX dataset using Google Maps with the help of a visualizer [27]. In this manner, an athlete, as well as a potential trainer, obtains a feeling of how stressful the training session is without waiting for he/she to realize it during the training cycle. On the other hand, the athlete can allocate his/her strength properly along the entire training course.

4 Experiments

The goal of our experimental work was to show that the BA can be used as a generator of sports training plans based on existing courses. Thus, the quality of the generated sports training plans needs to be not worse than those planned by real trainers manually. In line with this, the archive of tracked TCX and GPX datasets obtained by the professional cyclists were taken into consideration during a period of 1 year. These datasets are clustered by k-means clustering into three groups according to the TRIMP training load indicator. Then, the BA was applied for generating the sports training plans. Finally, courses proposed in the optimal sports training plan were visualized. This method is useful for each sports discipline where the training course can be tracked by the mobile sports trackers (i.e., running, biathlon, duathlon, and triathlon).

During the experiments, the parameters of the BA were set as follows. The frequency was drawn from the Gaussian distribution in the interval $Q_i \in [0.0, 1.0]$, the loudness was fixed at $A_i^{(t)} = 0.7$, while the emission rate was $r_i^{(t)} = 0.2$. Let us notice that the population size was set to $Np = 70$ and algorithm can spend the maximum 10,000 function evaluations to terminate and that only one run of the BA was used for evaluation of the results.

The best result of the optimization with BA for generating the sports training plans obtained during the experimental work is illustrated Table 4. The total intensity load overcome by an athlete during this training cycle amounts to TRIMP $= 124, 758.56$.

Table 4 Generated sport training plan as proposed by the BA

Day	Cluster	Training session	Average *HR*	Duration *t*	TRIMP
Monday	–	–	–	–	–
Tuesday	1	9	130	124.04	16,125.20
Wednesday	2	2	128	184.18	23,575.04
Thursday	3	22	150	281.08	42,162.00
Friday	–	–	–	–	–
Saturday	2	27	137	189.15	25,913.55
Sunday	1	47	133	127.69	16,982.77

As can be seen from Table 4, the more intensive training courses from each cluster were incorporated into the training plan, i.e., thus the maximum duration is pursued by the maximum average *HR*. As matter of fact, the value of the TRIMP training load indicator increases from Tuesday to Thursday training sessions and decreases from Thursday to Sunday training sessions. As said before, Monday and Friday remain for resting. Although it seems that the duration t has higher impact on the TRIMP value, also average *HR* is above the value 130 in the last three training sessions. This training plan was also confirmed by two real trainers for cycling.

Visualization of the results from Table 4 is illustrated in Fig. 4, where specific maps are presented to an athlete in training to prepare themselves easily in mental

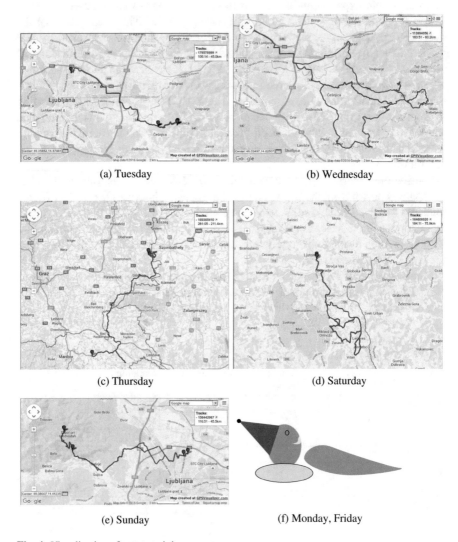

(a) Tuesday   (b) Wednesday

(c) Thursday   (d) Saturday

(e) Sunday   (f) Monday, Friday

Fig. 4 Visualization of sports training courses

preparation for the upcoming closures of training sessions in the next cycle. Training courses in these 5 maps are very vibrant, because they consists of small and medium hills, while there are also flat segments demanding the higher speeds.

Anyway, a problem has been discovered after the experiments, due to some courses that are not generated within the same geographic area (e.g., Central Slovenia or northeastern Slovenia). However, this is not a problem, when the competition (e.g., National Championship) is planned in another geographic region, and athletes would like to try it before the competition takes place. Unfortunately, this mode of training is often unusual for realizing the normal training sessions. As a result, the generator must be upgraded in the future with some geographical filter where all activities laying outside of particular geographical region are removed from an observation.

5 Conclusion with Future Ideas

This chapter presented an innovative solution for generating the sports training plans based on existing sports activities that expands the functionality of the artificial sports trainer. In particular, swarm intelligence algorithms were used to search for the best combination of training sessions within a cycle of duration 1 week. Indeed, the training sessions are clustered according to their TRIMP training load indicators into three clusters, from which the appropriate training sessions are included into the training plan such that obey the prescribed distribution of training intensities. Practical experiments showed that the proposed solution is very bright for use in such a domain.

In the future, there is still room for improvement and extension of this approach. Firstly, it would be good to take more training sessions into account. Additionally, the design of a prototype for generation of training plans based on existing sports activities in multisport disciplines would be a good way for future development. On the other hand, it might be very promising to study the influence of the other values extracted from GPX or TCX datasets besides duration and average heart rate, e.g., altitude, power meters, and cadence. Moreover, considering the other training types, such as intervals, hills, power, should be outlined in more detail.

References

1. Baca A, Dabnichki P, Heller M, Kornfeind P (2009) Ubiquitous computing in sports: a review and analysis. J Sports Sci 27(12):1335–1346
2. Bächlin M, Förster K, Tröster G (2009) Swimmaster: a wearable assistant for swimmer. In: Proceedings of the 11th international conference on ubiquitous computing. ACM, pp 215–224
3. Banister EW (1991) Modeling elite athletic performance. Physiol Test Elite Athl 403–424
4. Banister EW, Carter JB, Zarkadas PC (1999) Training theory and taper: validation in triathlon athletes. Eur J Appl Physiol Occup Physiol 79(2):182–191

5. Darwin C (1859) On the origins of species by means of natural selection. Murray, London, p 247
6. DasGupta D (1999) An overview of artificial immune systems and their applications. Springer
7. Fister I, Rauter S, Yang X-S, Ljubič K, Fister I Jr (2015) Planning the sports training sessions with the bat algorithm. Neurocomputing 149:993–1002
8. Fister I Jr, Fister I, Fister D, Ljubič K, Zhuang Y, Fong S (2014) Towards automatic food prediction during endurance sport competitions. In: 2014 international conference on soft computing and machine intelligence (ISCMI). IEEE, pp 6–10
9. Fister I Jr, Hrovat G, Rauter S, Fister I (2014) Am I overtraining? A novel data mining approach for avoiding overtraining. In: Computer science research conference, pp 47–52
10. Fister I Jr, Ljubič K, Suganthan PN, Perc M, Fister I (2015) Computational intelligence in sports: challenges and opportunities within a new research domain. Appl Math Comput 262:178–186
11. Fister I Jr, Rauter S, Fister KL, Fister D, Fister I (2015) Planning fitness training sessions using the bat algorithm. In: 15th conference ITAT 2015 (CEUR workshop proceedings, ISSN 1613-0073, vol 1422), pp 121–126
12. Hartigan JA, Wong MA (1979) Algorithm as 136: a k-means clustering algorithm. Appl Stat 100–108
13. Hassanien AE, Alamry E (2015) Swarm intelligence: principles, advances, and applications. CRC Press
14. Hassoun MH (1995) Fundamentals of artificial neural networks. MIT Press
15. Hey J, Carter S (2005) Pervasive computing in sports training. IEEE Perv Comput 4(3):54
16. Kennedy J, Eberhart RC, Shi Y (2001) Swarm intelligence. Morgan Kaufmann
17. Kirkpatrick S, Gelatt CD, Vecchi MP (1983) Optimization by simulated annealing. Science 220(4598):671–680
18. Lan T-S, Chen P-C, Chang Y-J, Chiu W-C (2015) An investigation of the relationship between performance of physical fitness and swimming ability based on the artificial neural network. In: Innovation in design, communication and engineering: proceedings of the 2014 3rd international conference on innovation, communication and engineering (ICICE 2014), Guiyang, Guizhou, PR China, October 17–22, 2014. CRC Press, p 181
19. Michahelles F, Schiele B (2005) Sensing and monitoring professional skiers. IEEE Perv Comput 4(3):40–45
20. Novatchkov H, Baca A (2012) Machine learning methods for the automatic evaluation of exercises on sensor-equipped weight training machines. Proc Eng 34:562–567
21. Novatchkov H, Baca A (2013) Artificial intelligence in sports on the example of weight training. J Sports Sci Med 12(1):27
22. Perl J, Baca A (2003) Application of neural networks to analyze performance in sports. In: Proceedings of the 8th annual congress of the European College of Sport Science, S, vol 342
23. Rauter S, Topič MD (2011) Perspectives of the sport-oriented public in Slovenia on extreme sports. Kineziologija 43(1):82–90
24. Rauter S, Topič MD (2014) Runners as sport tourists: the experience and travel behaviours of ljubljana marathon participants. Coll Antropol 38(3):909–915
25. Rauter S, Fister I, Fister I Jr (2015) How to deal with sports activity datasets for data mining and analysis: some tips and future challenges. Int J Adv Perv Ubiquitous Comput (IJAPUC) 7(2):27–37
26. Ross TJ (2009) Fuzzy logic with engineering applications. Wiley
27. Schneider A (2009) GPS visualizer. http://www.gpsvisualizer.com
28. Simon D (2013) Evolutionary optimization algorithms. Wiley
29. Yang X-S (2010) A new metaheuristic bat-inspired algorithm. In: Nature inspired cooperative strategies for optimization (NICSO 2010). Springer, pp 65–74

Limiting Distribution and Mixing Time
for Genetic Algorithms

S. Alwadani, F. Mendivil and R. Shonkwiler

Abstract We use methods from Markov chain theory to analyze the performance of some simple GA models on a class of deceptive objective functions. We consider the invariant distribution, the average expected hitting time, the mixing rate, and limits of these quantities as the objective function becomes more and more highly skewed.

1 Introduction

The most extensive use to which genetic algorithms (GAs) have been put has been to solve optimization problems. Yet despite a great deal of empirical work, very little is known about how the parameters of the algorithm actually impact the efficacy of the search.

What is known is that, mathematically, a GA is an instance of a Markov chain, often taken to be time homogeneous. The casting of a GA as a Markov chains seems to have first been done independently in [2–4] and [10, 14]. These papers studied two different models of GAs as Markov chains and investigated their convergence properties, as either a control parameter went to zero [4] or population size becomes unbounded [10]. Since then, there has been a wealth of other work on convergence properties of GAs and other evolutionary algorithms (e.g., [5, 7, 11, 12] and the references therein).

One of the most fundamental properties of such a chain is the existence of a stationary distribution to which the chain converges as it undergoes iteration. Other than knowing that the convergence rate to the stationary distribution is geometric

S. Alwadani · F. Mendivil (✉)
Acadia University, Wolfville, NS, Canada
e-mail: franklin.mendivil@acadiau.ca

S. Alwadani
e-mail: 122349a@acadiau.ca

R. Shonkwiler
Georgia Tech, Atlanta, Georgia
e-mail: shenk@math.gatech.edu

© Springer International Publishing AG 2017
S. Patnaik et al. (eds.), *Nature-Inspired Computing and Optimization*,
Modeling and Optimization in Science and Technologies 10,
DOI 10.1007/978-3-319-50920-4_5

95

based on general principles, this is a largely unexplored aspect of GA research. What characterizes that distribution? What sort of stationary distribution should be striven for? How do the specific mechanisms of the GA affect the stationary distribution? How quickly does the chain tend to it and how do the mechanisms of the GA affect this? Even more fundamentally, is the convergence rate a good way to gauge the efficacy of genetic algorithm?

These are important concepts for Markov Chain-based optimizers because they are related to the degree to which this kind of search methodology wastes cycles trapped in a suboptimal basins. In particular, this applies to genetic algorithm methodologies. In this study, we investigate these questions by working on a range of small but representative problems. Problems for which there is a characteristic size parameter which can be adjusted to allow, in many cases, exact computations.

The paper is organized as follows. Section 2 presents some background and establishes notation. This section also introduces the simple class of problems we use as a benchmark. Section 3 discusses expected hitting time as a means of comparing two different search strategies and uses two different (and simple) random walk models as illustrations and as a baseline to which to compare the GAs of the following sections. Section 4 presents some results on a style of GA as originally described by Holland. Section 5 presents results on two variations of a simple GA (strongly related to, but not the identical to, the "simple GA" studied by others). Section 6 introduces a new, very straightforward GA which uses random shuffles for "crossover" and analyzes the performance of this algorithm. Finally, Sect. 7 contains some general conclusions and a brief discussion of further directions for investigation.

2 Preliminaries

We consider the problem of optimizing (usually maximizing) the function $f : \mathbb{X} \rightarrow \mathbb{R}$, where $\mathbb{X}$ is finite. We make no assumptions as to the nature of the set $\mathbb{X}$, but for many practical problems, $\mathbb{X}$ has some extra structure (such as being a subset of $\mathbb{R}^d$). The algorithms we examine are all *iterative* in that they produce a sequence (x_n) of potential solutions, hopefully getting better as the iterations proceed. The iterations are usually *local* in the sense that x_{n+1} is chosen as a perturbation from some *neighborhood* of x_n. In this way, each such search algorithm defines a *topology* (via neighborhood systems) on $\mathbb{X}$ which gives $\mathbb{X}$ the structure of a graph. Thus, given $x \in \mathbb{X}$ the *neighborhood*, $N(x)$, of x is the set of all $y \in \mathbb{X}$ which are adjacent to x. By the convention in graph theory, $x \notin N(x)$, and $N[x] = N(x) \cup \{x\}$ is the *closed neighborhood* of x.

Once the state space $\mathbb{X}$ has a topology, then the objective function f establishes local maxima (and of course minima) as follows. A *local maximum* is a real value v^* such that for some $x^* \in \mathbb{X}$

$$f(x^*) = v^*, \quad \text{and} \quad f(x) \le v^* \text{ for all } x \in N(x^*).$$

We refer to x^* as a *local maximizer*. If $N(x^*)$ can be taken as $\mathbb{X}$ in the above, then v^* is the *global maximum* and x^* is a *global maximizer*. Given a local maximizer $x^* \in \mathbb{X}$, its *basin*, B_{x^*} is defined as the set of all $x \in \mathbb{X}$ for which there exists a sequence $x = x_0, x_1, x_2, \ldots, x_n = x^*$ with $x_i \in N(x_{i-1})$ and $f(x_i) \ge f(x_{i-1})$. Note that the set of all basins forms a covering of $\mathbb{X}$ but if x and y are two local maxima, it is possible for $B_x \cap B_y \ne \emptyset$.

The basin structure of a search plays a key role in its performance. It can fairly be said that the central problem of optimization is to avoid becoming trapped in suboptimal basins. This is seen as follows.

Let $f : \mathbb{X} \to \mathbb{R}$ be a minimization problem for which there is a single basin, a *uni-basin* search. Then, for any starting point x_0, its neighborhood $N(x_0)$ contains a point x_1 such that $f(x_1) \le f(x_0)$. Now, repeat these steps with x_1. Eventually, this process will lead to the *global* minimizer. In other words, there is a downhill path leading to the global minimizer. Of course, this is only of value if the neighborhoods are small enough that a downhill improvement can be quickly found. That is not case if, for example, the entire space $\mathbb{X}$ happens to be the neighborhood of each point. We refer to this as the *maximal topology*.

On the other hand, if neighborhoods are small in a practical way, then being uni-basin is a tremendous help in locating the problem's global minimizer.

At this point, we introduce a family of example problems which have the property that they are "deceptive" in the sense that they tend to force a search algorithm toward suboptimal basins. Their detailed definition is that for an integer $n > 1$, let the function $s_n : [0, n-1] \to \mathbb{R}$ be defined according to

$$s_n(x) = \begin{cases} n+1-nx, & \text{if } x \in [0,1], \\ x, & \text{if } x \in [1, n-1]. \end{cases}$$

We call this the *sandia-n* function. Since all our examples have finite state spaces, we generally sample the function at equally spaced points. For methodologies whose states are bit sequences, sandia-4 is a 2-bit problem with domain $\{0, 1, 2, 3\}$, thought of in binary as $\{00, 01, 10, 11\}$. Similarly, sandia-8 is a 3-bit problem on the domain $\{000, 001, \ldots, 111\}$. In the event that we restrict its domain, we will still refer to it as the sandia-n function. The goal of a search applied to the sandia-n function is to find its global maximum, which occurs at $x = 0$. As defined above, sandia-n is a 2-basin function and it's graph is in Fig. 1a. For a more complicated example, we can add another basin (hilltop) to the basic sandia function. This is shown in Fig. 1b.

We will use these sandia functions often in the sequel because they are simple examples as well as deceptive objective functions.

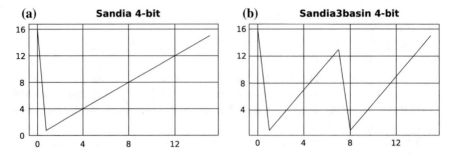

Fig. 1 **a** The basic sandia-n function. **b** The sandia 3-basin function

2.1 Random Search and Markov Chains

The iterative search methods that we examine will all incorporate some element of randomness and so can be thought of as (biased or guided) random searches on $\mathbb{X}$. In addition, all of these strategies will have the property that x_{n+1} depends only on x_n, and thus, they can be modeled as Markov chains. If a method has no parameters that vary with iteration, then the chain is *(time) homogeneous* and (almost always) has an associated *invariant distribution*. This is a probability distribution π on $\mathbb{X}$. Under mild conditions, from any given starting point, as the number of iterations increases, the proportion of time spent in any given state $x \in \mathbb{X}$ will converge to π_x.

For any two states x and y, let p_{xy} be the probability of a transition from x to y in one iteration and let the *transition matrix* be given by $P = (p_{xy})$. The invariant distribution satisfies $\pi P = \pi$ (here π is considered as a row vector). The matrix P characterizes the search. From it one can calculate various measures of the efficacy of the search, two of these are the *rate of convergence* and the *expected hitting time*. Powers of P show how easy or difficult it is to get between states in a stochastic sense. For instance, the entries of P^2 give the probability of getting between two states in two iterations. By itself, P shows how likely it is to get between any two states in one iteration. In this way, it defines the neighborhood of each state, namely those for which the transition probability is positive.

If P is *irreducible* and *aperiodic* (see [1]), as $t \to \infty$ the rows of P^t converge to the invariant distribution π. The rate of this convergence is the *mixing time* of the search. In Fig. 2, we show the invariant distributions of two of the algorithms we study in this article.

(a) **Invariant Distribution: Sandia 3-bits** **(b)** **Invariant Distribution: Sandia 3-bits**

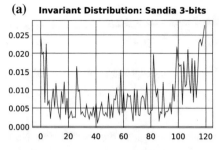 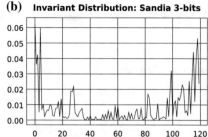

Fig. 2 **a** Holland algorithm. **b** Simple GA

2.2 Boltzmann Distribution and Simulated Annealing

Simulated annealing (SA) is an optimization method inspired by statistical mechanics. It seeks to minimize "energy" E over the state space, for a positively valued function $E : \mathbb{X} \to [0, \infty)$. At the heart of SA is the *Metropolis–Hastings algorithm* in which states are proposed for the next iteration and either accepted or not. Thus, an iteration consists of two processes, the proposal process or generation process, in which from a given state x, a state y is selected from $N(x)$ and proposed for the next iteration. This process may be defined by a matrix (g_{xy}) where the element g_{xy} is the probability of proposing y from x. If $E(y) < E(x)$, then y is accepted and becomes the state for the next iteration. However, if $\Delta E = E(y) - E(x) > 0$, then y is accepted on the basis of a side experiment with acceptance probability $e^{-\frac{\Delta E}{kT}}$. In this, k is a constant and T a parameter of the process (which usually changes with iteration count in SA). If y is not accepted, then x is retained to start the next iteration.

For the purposes of our discussion here, we will take $T = k = 1$ for simplicity. Of course, the choice of the combination kT will influence the convergence properties of the resulting Markov chain.

If a_{xy} denotes the probability of accepting a transition from x to y, then the transition probability matrix becomes

$$p_{xy} = g_{xy}a_{xy} \text{ and } p_{xx} = 1 - \sum_{y \neq x} p_{xy}.$$

Now, suppose the proposal process is symmetric, $g_{xy} = g_{yx}$. Then, the following relationship holds between every two states x and y

$$e^{-E(x)}p_{xy} = e^{-E(y)}p_{yx}. \tag{1}$$

This relationship is called *detailed balance*. To prove it, we may assume without loss of generality that $E(x) < E(y)$, then $a_{xy} = e^{-E(y)+E(x)}$ while $a_{yx} = 1$. Hence,

$$e^{-E(x)}p_{xy} = e^{-E(x)}g_{xy}a_{xy} = e^{-E(x)}g_{yx}\frac{e^{-E(y)}}{e^{-E(x)}}$$
$$= e^{-E(y)}g_{yx}a_{yx} = e^{-E(y)}p_{yx}. \tag{2}$$

From detailed balance, it follows that the invariant distribution is proportional to

$$\pi_x \propto e^{-E(x)}.$$

In fact, $\pi_x = e^{-E(x)}/\sum_{y\in\mathbb{X}} e^{-E(y)}$. This is called the *Boltzmann distribution*. Note that the sum appearing in the denominator is generally not known.

If now the purpose is to maximize a function $f : \mathbb{X} \to \mathbb{R}$, we may still use the Metropolis–Hastings algorithm by defining energy $E(x)$ in such a way that

$$f(x) = e^{-E(x)}.$$

Then, the acceptance process becomes as follows: If $f(y) > f(x)$ so that $E(y) < E(x)$, then y is accepted (with probability 1). Otherwise, y is accepted with probability

$$e^{-(E(y)-E(x))} = \frac{e^{-E(y)}}{e^{-E(x)}} = \frac{f(y)}{f(x)}.$$

In this, we continue to assume a symmetric proposal matrix. Thus, the invariant distribution of the process is proportional to the exponential of the objective function, $\pi_x \propto \exp(-f(x))$.

Even if the proposal process is not symmetric, we may still recover the Boltzmann distribution if detailed balance can be enforced,

$$f(x)g_{xy}a_{xy} = f(y)g_{yx}a_{yx}. \tag{3}$$

But now put

$$a_{xy} = \min\left(1, \frac{f(y)g_{yx}}{f(x)g_{xy}}\right). \tag{4}$$

If the ratio is bigger than 1, then $a_{xy} = 1$, and since then the inverse ratio will be less than 1, it follows that $a_{yx} = (f(x)g_{xy}/f(y)g_{yx})$. Substituting these quantities into the detailed balance equation (1) confirms it continues to be valid.

What is remarkable is that the Metropolis–Hastings method allows one to take a very general process that given by g_{xy} and modify it to have *exactly* the distribution you desire by adding an acceptance phase. We use this idea in Sect. 6 in order to control the invariant distribution of the genetic algorithm described in that section. Notice, however, that Eq. (3) requires that $g_{xy} \neq 0 \Leftrightarrow g_{yx} \neq 0$. Of course, this is true if g is symmetric, but this condition often prevents one from using this mechanism to enforce a desired invariant distribution for a given search process.

3 Expected Hitting Time as a Means of Comparison

In employing a search methodology, we are interested in how long it might take to find a suitable solution, be it a global maximizer or perhaps just something good enough for our purposes. The most direct answer to this is the *expected hitting time*. Very simply, this is the expected number of iterations that will be required to reach whatever goal we set. Since the methodology is stochastic, we must settle for the probabilistically expected time.

Were it possible to know the transition probability matrix, $P = (p_{ij})$, of the method, then we could proceed as follows. For any goal state, g, the expected hitting time when starting in this state is evidently 1, $E_g = 1$. For any other state x, the time is 1 plus whatever time is expected for any state y, we may transition to from x, thus

$$E(x) = 1 + \sum_{y \in \mathbb{X}} p_{xy} E(y).$$

This leads to a system of equations,

$$
\begin{array}{ccccc}
(1 - p_{11})E_1 & -p_{12}E_2 & \cdots & -p_{1n}E_n & = 1 \\
-p_{21}E_1 & (1 - p_{22})E_2 & \cdots & -p_{2n}E_n & = 1 \\
\vdots & \vdots & \ddots & \vdots & \vdots \\
-p_{n1}E_1 & -p_{n2}E_2 & \cdots & (1 - p_{nn})E_n & = 1
\end{array}
\tag{5}
$$

Note that the pattern above is for non-goal rows and that for them the sum of the coefficients in any row is 0.

The solution of this system gives the expected hitting time from each state. If we also take into account the starting process, for example, starting equally likely over all possible states, the weighted average with the $E's$ gives the expecting hitting time for the methodology. As in [1, 8, 13], we can use matrix methods to find these quantities. Suppose that we wish to find the hitting time to the subset $G \subseteq \mathbb{X}$. Let $\hat{\alpha}$ and $\hat{P}$ be the restriction of the starting distribution α and transition matrix P to $\mathbb{X} \setminus G$. Furthermore, let $\mathbb{1}$ denote a vector (of appropriate length) with all entries having the value 1. Then, we see that (5) yields

$$\text{EHT} = \text{average expected hitting time} = \hat{\alpha}^T (I - \hat{P})^{-1} \mathbb{1}. \tag{6}$$

For an example, consider the domain $D = \{0, \frac{1}{2}, 1, \frac{3}{2}, 2, \ldots, n - 1\}$ along with the sandia-n function $s_n(\cdot)$. In this topology, s_n has two basins, $B_0 = \{0, \frac{1}{2}, 1\}$ and $B_n = \{1, \frac{3}{2}, 2, \ldots, n - 1\}$. For solving this problem, we try a *random walk* search in which the neighborhood of x is the two states $x \pm 1/2$, or the single adjacent state at the end points 0 and $n - 1$. The walk selects a point from the neighborhood of its present iteration with probability p if the function value is uphill and $q = 1 - p$ if downhill. At $x = 1$, both adjacent points are uphill so the next iterate is selected from these equally likely.

Fig. 3 Expected iterations to find the goal $x = 0$ starting from x, with $p = 0.6$ and $q = 0.4$ in the sandia-8 problem

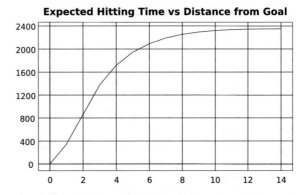

Decomposing on the first iterate, the expectation E_x from an interior point in B_n satisfies $E_x = 1 + q E_{x-1} + p E_{x+1}$. Writing the similar equation for each $x \in D$ leads to a system of $2n - 2$ linear equations that is easily solved. Figure 3 shows the expected hitting times for this problem with the choices $p = 0.6$ and $q = 1 - p = 0.4$. Starting the search from any point equally likely gives an overall expected hitting time to the global minimizer at $x = 0$ of 1784 iterations. To gauge the effect of basins, starting from a point in B_0, the expected hitting time to the global minimizer at $x = 0$ is 574, and from a point in B_n, it is 1979. Even more dramatically, switching things around and regarding $x = 7$ as the goal gives the basin expectations as 67 from B_0 and 34 from B_n.

Interestingly, for this example problem, the minimal overall expected hitting time occurs when $p \approx 0.2564$, significantly less than the balanced case of $p = 1/2$. The explanation for this is that if we start in any state $x > 1$, we need to take many "downhill" steps to get to the other side where the global maximum occurs. To obtain this value for the minimal overall expected hitting time, we computed (6) explicitly with $\alpha = (1/15)\mathbb{1}$ to obtain an expression for EHT:

$$\frac{7\,p^{12}-3\,p^{11}+171\,p^{10}-745\,p^9+2450\,p^8-5616\,p^7+9224\,p^6-10892\,p^5+9207\,p^4-5451\,p^3+2155\,p^2-513\,p+56}{15p(1-p)^{11}}.$$

Notice that the numerator and denominator are polynomials of the same degree and that the expected hitting time becomes unbounded as $p \to 0$ or $p \to 1$.

Under a small modification, this random walk search becomes similar to a simulated anneal (see Sect. 2.2). If an uphill neighbor is chosen, then make that move with certainty. For a downhill neighbor, make the move with probability q, otherwise stay put. This is similar to a fixed temperature anneal; a true anneal is realized by having q tend to zero (in some fashion) with iteration count. Again, by decomposing on the first iteration, we can develop the system of expectations for calculating the hitting time. For a general state x, we have

$$E_x = 1 + \frac{1}{2}E_u + \frac{1}{2}q E_d + \left(\frac{1}{2} - \frac{q}{2}\right)E_x,$$

where u is the uphill neighbor of x and d the downhill neighbor. In this, $1/2$ is the chance of picking each of the neighbors, and $1/2 - q/2$ is the chance of staying at x. Incredibly, even though there are only 15 states in $\mathbb{X}$, taking $q = 0.4$ gives an expected hitting time of $800, 962$ for equally likely starting. This shows how basins play a key role in optimal search. As a point of interest, it is easy to show that EHT is a decreasing function of q for this example, and thus, the minimum value for EHT of $385/3 = 128.333$ occurs for $q = 1$. In fact, we can compute that

$$EHT = \frac{385}{3} + \frac{3549}{5}(1 - q) + O((1 - q)^2),$$

and so the average expected hitting time increases fairly rapidly as we decrease q from 1. Of course, for $q = 0$, the expected hitting time starting from any state $x > 1$ is infinite. Again, we see that for this sandia-n problem, it better to take the downhill steps to avoid getting trapped.

Finally, for $q = 1$, the process is a symmetric random walk on $\mathbb{X}$ with reflecting barriers at both ends. This problem is fairly elementary and also very classical. One can easily compute (or lookup in a reference) that when $\mathbb{X}$ has n points with a maximum at $x = 0$, we have $EHT = (n - 1)(4n - 5)/6$. Recall that $q = 1$ gives a minimum for EHT, and so EHT grows (at least) quadratically as a function of the size of $\mathbb{X}$.

For $q \in (0, 1)$, the Markov chain is irreducible and aperiodic and thus has a unique invariant distribution which we can easily calculate as

$$\pi_q = \frac{1}{S}(q^{10}, 2q^{11}, 2q^{12}, 2q^{11}, 2q^{10}, 2q^9, \ldots, 2q^4, 2q^3, 2q^2, 2q^1, 1),$$

where S is a normalizing constant. Notice that π_q places the greatest mass near the local maximum at the right-hand end point. This is clearly undesirable. When $q = 0$ or $q = 1$, the chain does not have a unique invariant distribution. From general theory of Markov chains, the size of the second largest (in modulus) eigenvalue controls the rate at which the process converges to the stationary distribution (see [9]). When $q = 0$, there are two absorbing states, and thus, two eigenvalues equal to 1, but then, the next largest eigenvalue is $1/2$; thus, the convergence is rather fast in this case. However, of course, the process is most likely to converge to the local maximum, so it is normally converging to the wrong value! When $q = 1$, the chain has period two and thus again has no unique stationary distribution. In this case, there are again two eigenvalues of absolute value 1, and then, the next largest eigenvalue (in modulus) is approximately 0.901, so the convergence, to the period two behavior, is a bit slower but still not so bad.

The lesson to be drawn from this example is that while it is nice to have fast convergence, we need to be careful to what we are converging. For this reason, we focus on the expected hitting time rather than on mixing times or convergence rates.

3.1 "No Free Lunch" Considerations

The famous "No Free Lunch" theorem (see [16]) might seem to indicate that it is pointless to discuss differences between search algorithms since when averaged over all possible objective functions, they all do equally well (or poorly). Of course, as is discussed in [15], another interpretation is that the theorem emphasizes that we should match a problem with the correct algorithm.

As originally framed in [16], the result applies only to algorithms which do not revisit states. Since implementing this restriction requires either some method of generating elements of $\mathbb{X}$ with no repetition or keeping a list of all states which have been visited and then using something like rejection sampling, this result is of more theoretical than practical importance for our considerations. In particular, almost no random search methods satisfy this restriction.

In the world of random search methods, it is clear that there are some Markov chains which do worse, even when averaged over all possible locations for the maximum, than others. For instance, if the Markov chain is reducible, then the expected hitting time from some x to some other y can be infinite. Not surprisingly, the smallest EHT comes from a search which does not revisit states; this type of random search is more like a random permutation of $\mathbb{X}$ or random sampling without replacement. Again, implementing such an algorithm is usually completely impractical.

Theorem 1 *Let $|\mathbb{X}| = N$. Then we have that $\frac{N+1}{2} \leq EHT$ for any random search method on $\mathbb{X}$.*

Proof Let X_n be an Ω-valued process where, with no loss of generality, we assume $EHT < \infty$. Define $\widetilde{X}_n$ by $\widetilde{X}_1 = X_1$ and for $n \geq 1$

$$\widetilde{X}_{n+1} = X_m, \quad m = \inf\{t \geq 1 : X_t \notin \{\widetilde{X}_1, \widetilde{X}_2, \ldots, \widetilde{X}_n\}\}$$

(so that $\widetilde{X}_n$ is the process X_n with all repeated states removed).

For each $x \in \Omega$ and any process Y_n on Ω, let $HT(Y_n, x) = \inf\{n : Y_n = x\}$ be the first hitting time to state x for the process Y_n. Now, clearly $HT(\widetilde{X}_n, x) \leq HT(X_n, x)$ for all x and

$$\sum_{x \in \Omega} HT(\widetilde{X}_n, x) = \sum_{i=1}^{N} i = \frac{N(N+1)}{2}$$

and thus

$$\frac{N(N+1)}{2} = \sum_{x \in \Omega} HT(\widehat{X}_n, x) \leq \sum_{x \in \Omega} HT(X_n, x).$$

Taking expectations, we get

$$\frac{N(N+1)}{2} = E\left[\sum_{x \in \Omega} HT(\widehat{X}_n, x)\right] \leq E\left[\sum_{x \in \Omega} HT(X_n, x)\right].$$

Therefore,

$$\frac{N+1}{2} \leq \frac{1}{N} \sum_{x \in \Omega} E\left[HT(X_n, x)\right] = EHT. \tag{7}$$

$\square$

The quantity $(N + 1)/2$ is the number of samples you expect to need if you are looking for something and you randomly sample without replacement. The inequality (7) is the simple fact that the fastest way to perform a random search is to sample without replacement.

The random target lemma (see [9]) can be interpreted as giving some guidance if we have some prior knowledge about the distribution of where in $\mathbb{X}$ we expect the maximum to occur. Specifically, if we believe that the distribution π on $\mathbb{X}$ describes likely occurrences of the maximum for our class of functions, then we should try to use a search process whose behavior mimics π.

4 The Holland Genetic Algorithm

We now move on to a discussion of basins and hitting times for genetic algorithms. We start with the genetic algorithm as John Holland defined it (approximately, the inversion operator there will be disposed of) [6]. A *population* $\mathscr{P}$ is a fixed-size collection of states from the state space $\mathbb{X}$, say of cardinality Z. The collection is not ordered and can contain duplications. The elements of a population are often called *chromosomes* or *individuals*. Further, by the *fitness* of a chromosome, we mean the (strictly positive) objective value $f(x)$. A genetic algorithm is designed to maximize fitness. As a search methodology, a GA iteration consists of 3 steps: crossover, mutation, and selection.

Crossover is accomplished by selecting two chromosomes, y_1, y_2, from $\mathscr{P}$ by the technique known as *roulette wheel*. This means the probability of selecting $x \in \mathscr{P}$ is proportional to its fitness. If $\Sigma_{\mathscr{P}}$ is the sum of the fitnesses of the chromosomes of $\mathscr{P}$, the probability for x is $f(x)/\Sigma_{\mathscr{P}}$. Since some, or all, of the chromosome of $\mathscr{P}$ could be the same, we make no effort to assure the two selected are distinct. With y_1 and y_2 as arguments, a binary operation is performed on them to give a result y. This operation is called *crossover* and is often stochastic. If the states are given as strings of bits, and this is the classical situation, then crossover means to divide both arrays between the kth and $k + 1$st elements and interchange the bits between the arguments from $k + 1$ on. Here, k is an integer from 1 up to the length of the bit arrays minus 1. One of the resulting two bit arrays is selected as the *offspring y*.

The next operation in the Holland algorithm is *mutation*. This is a unary stochastic operation on y producing the result y'. If y is a bit array, the Holland algorithm "flips" each bit with some small probability p_m. If a given bit is 0, it is changed to 1 with probability p_m, and if a given bit is 1, it is changed to 0.

Finally, *selection* in this algorithm means to select one of the original population chromosomes equally likely, say x_j where $j \in \{1, 2, \ldots, Z\}$. Once x_j is selected, it is replaced by y' completing the iteration to produce the next *generation*. In this algorithm, it would be more appropriate to call this operation *replacement*.

One sees that the search space for the GA is the set of all populations from $\mathbb{X}$ of size Z; call this set Ω. The neighborhood of a any $\mathscr{P} \in \Omega$ is defined by the details of crossover, mutation, and selection. The next iterate $\{x_1, \ldots, y', \ldots, x_Z\}$ is by the definition in this neighborhood.

To study this algorithm, we chose a very small example: the sandia-4 problem using a population size 3 GA. We use $\mathbb{X} = \{0, 1, 2, 3\}$ which we represent in bits, so $\mathbb{X} = \{00, 01, 10, 11\}$ (alternatively as a shorthand, we also use $\mathbb{X} = \{A, B, C, D\}$). The fitnesses given by the sandia-4 function are $f(0) = 4$, $f(1) = 1$, $f(2) = 2$, and $f(3) = 3$. Populations can be represented as a multiset, for example, $\{00, 00, 10\}$ or equivalently $\{A, A, C\}$ or as a *type-count array*. In the latter, for an array indexed by the original states (of $\mathbb{X}$), enter the multiplicity of that state in the population. For typical problems, this array has a length in the millions or more, but here, it has length 4. For example, using this type-count array representation, the population above becomes $(2, 0, 1, 0)$ since $\{A, A, C\}$ contains two copies of A, one copy of C, and no copies of either B or D.

In the Holland algorithm for the sandia-4 problem, crossover has only a bit length of 2 to work on. Therefore, there is only 1 possibility for the position of crossover, namely $k = 1$. The choice of the two resulting "offspring" is 50–50. For mutation, let the mutation probability be p_m, for example, 0.2. Since every bit is subject to a possible flip, it means any state can be the result of mutation. Finally, selection (replacement) is equally likely, and thus, there is a $1/3$ chance of replacing any one chromosome.

By well-known combinatorics, namely the "stars and bars" formula, the size of Ω is $\binom{4+3-1}{3} = 20$. Given all this information, it is straightforward to find explicit expressions for all the entries of the 20×20 transition matrix P for this chain.

To illustrate, we show the calculation for one row of the table, the row indexed by AAB. The sum of fitnesses for this population is 9, and therefore, the chance of choosing A is 8/9 and B is 1/9. If the pair chosen for crossover is AA, with probability $(\frac{8}{9})^2$, then the result of crossover will be A or 00. Letting $q_m = 1 - p_m$, under mutation, this becomes A with probability q_m^2, or B or C each with probability $q_m p_m$ or D with probability p_m^2. Finally considering replacement, an A in the original population has $2/3$ chance of being replaced. Furthermore, replacing it with an A, via this line of reasoning, probability is $(\frac{8}{9})^2 q_m^2$, with B or C probability is $(\frac{8}{9})^2 p_m q_m$, and with D the probability is $(\frac{8}{9})^2 p_m^2$. In summary,

$$Pr(AAB \rightarrow AAA | \mathscr{A}) = q_m^2, \qquad Pr(AAB \rightarrow AAB | \mathscr{A}) = p_m q_m,$$
$$Pr(AAB \rightarrow AAC | \mathscr{A}) = p_m q_m, \quad Pr(AAB \rightarrow AAD | \mathscr{A}) = p_m^2,$$

where $\mathscr{A} = \{$the two parents chosen for crossover are $AA\}$. Using this reasoning, the transitions from state AAB are (with $r = \frac{8}{9}$ and $s = \frac{1}{9}$)

Table 1 Allowable transitions for the Holland algorithm

	A	A	A	A	A	A	A	A	A	A	B	B	B	B	B	B	C	C	C	D
	A	A	A	A	B	B	B	C	C	D	B	B	B	C	C	D	C	C	D	D
	A	B	C	D	B	C	D	C	D	D	B	C	D	C	D	D	C	D	D	D
AAA	■	■	■	■																
AAB	■	■	■	■	■	■	■													
AAC	■	■	■	■		■		■	■											
AAD	■	■	■	■			■		■	■										
ABB		■			■	■	■				■	■	■							
ABC		■	■		■	■	■	■	■			■		■	■					
ABD		■		■	■	■	■		■	■			■		■	■				
ACC			■			■		■	■					■			■	■		
ACD			■	■		■	■	■	■	■					■			■	■	
ADD				■			■		■	■						■			■	■
BBB					■						■	■	■							
BBC					■	■					■	■	■	■	■					
BBD					■		■				■	■	■		■	■				
BCC						■		■				■		■	■		■	■		
BCD						■	■		■			■	■	■	■	■		■	■	
BDD							■			■			■		■	■			■	■
CCC								■						■			■	■		
CCD								■	■					■	■		■	■	■	
CDD									■	■					■	■		■	■	■
DDD										■						■			■	■

$$Pr(AAB \rightarrow AAA) = \tfrac{1}{3}q(rq + sp), \quad Pr(AAB \rightarrow AAC) = \tfrac{1}{3}p(rq + sp),$$
$$Pr(AAB \rightarrow AAD) = \tfrac{1}{3}p(rp + sq), \quad Pr(AAB \rightarrow ABB) = \tfrac{2}{3}q(rp + sq),$$
$$Pr(AAB \rightarrow ABC) = \tfrac{2}{3}p(rq + sp), \quad Pr(AAB \rightarrow ABD) = \tfrac{2}{3}p(rp + sq),$$
$$Pr(AAB \rightarrow AAB) = \tfrac{2}{3}q(rq + sp) + \tfrac{1}{3}q(rp + sq).$$

Other transitions with exactly two distinct chromosomes are similar to this only with different values for r and $s = 1 - r$. We will leave the further calculations to the reader.

Turning to the neighborhood system, the calculation is simple: Since the mutation step can produce any of the chromosomes, to get the neighbors, it is merely a matter of replacing each of the starting chromosomes with the states, A, B, C, D, one at a time. In Table 1, we give the possible transitions. A "■" in the entry means the indicated transition is possible, and therefore, this population is in the neighborhood of the antecedent; otherwise, an empty cell indicates that the transition is impossible.

Table 2 Population fitness of the 3-sized population for the sandia objective

A	A	A	A	A	A	A	A	A	A	B	B	B	B	B	B	C	C	C	D
A	A	A	A	B	B	B	C	C	D	B	B	B	C	C	D	C	C	D	D
A	B	C	D	B	C	D	C	D	D	B	C	D	C	D	D	C	D	D	D
4	4	4	4	4	4	4	4	4	4	1	2	3	2	3	3	2	3	3	3

Table 3 Expected hitting time for the Holland algorithm

	2-bit	3-bit	4-bit
Original sandia	5.7	7.0	10.2
Sandia 3-basin	6.1	7.6	10.9

Looking at Table 1, we see that more diverse populations have larger neighborhoods; thus, for the purpose of mixing, it is desirable to have diverse populations. This is true generally independent of any objective.

In order to define basins, it is necessary for us to assign a fitness to populations. Since every chromosome is evaluated, it makes natural sense to make the population fitness equal to the maximum of the fitnesses of its chromosomes. In Table 2, we show the population fitnesses. From the table, we see a profound effect, and the global basin has expanded out to encompass one-half of the search space (in this small example, at least). In addition to that there are no local minima since the mutation operation can create any chromosome, the chromosome A will always be in the neighborhood of any population.

From the transition matrix, we can compute the expected number of iterations of the GA in order to find the global maximum. Assuming equally likely starting, i.e., chose A, B, C, D equally likely for each of the 3 population chromosomes, the hitting time for the 2-bit sandia-n problem is given in Table 3. This is a substantial improvement over the random walk search. The table also shows hitting times for the 3-bit and 4-bit problems. These increase the population size to 120 and 816, respectively. By adding another basin (hilltop), the times increase a little.

Figure 4 shows the relationship between EHT and the probability of a bitmutation, p_m, for the sandia-16 function using four bits. The plots show results for both the standard sandia function (with two basins) and a three basin variations. It is interesting to observe that the minimum value for EHT occurs at $p_m \approx 0.73$ for both test functions.

Fig. 4 Expected iterations for the Holland algorithm to find the goal $x = 0$ starting uniformly versus the mutation probability p_m for the 4-bit sandia problem. In both graphs, the minimum occurs at $p_m = 0.73$

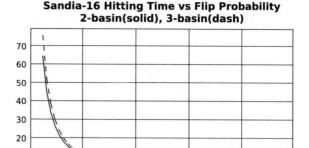

5 A Simple Genetic Algorithm

The second genetic algorithm we investigate is what we will call the *simple GA*. The generation cycle in our simple GA consists of four stages: (1) crossover, (2) selection, (3) mutation, and (4) selection again. These are given as follows.

For crossover, we choose two members of the population equally likely and perform a standard one-point crossover of their bit strings. Again, the crossover point k is also chosen equally likely. One of the offspring is chosen equally likely and added to the *augmented* population.

We bring the population back to its original size by a selection operation, or more accurately a select-out operation. This is roulette wheel based with the probability of selecting out a given individual in proportion to the negative exponential of its fitness,

$$Pr(\text{select-out } x) \propto \exp(-\mu \cdot \text{fitness}(x)).$$

In this, μ is a parameter of the algorithm; we call it the *fitness multiplier*. For μ less than 1, the differences between fitnesses are mollified, while for μ bigger than 1, fitnesses are exaggerated. For example, if $f_A = \text{fitness}(A) = 4$ and $f_B = \text{fitness}(B) = 1$, then $\exp(-f_A) = 0.0183$ and $\exp(-f_B) = 0.3679$ for a ratio B to A of 20. But for $\mu = 0.5$, the ratio is 4.48.

Next, a member of the resulting population is selected equally likely for bitmutation; from this, one a bit is selected equally likely and flipped. This is followed once again by the same select-out process as above.

To study this algorithm, we chose the sandia-n problem using 2, 3, and 4 bits together with a population size of 3. In this algorithm, there are genuine population basins because only one bit is mutated in each generation. As with the objective itself, there are two basins. The goal basin is any population containing $A = (0, 0, \ldots, 0)$. The other basin is any population not containing A. In the 4-bit problem, of the 816 populations, 136 contain the allele A. In the general m-bit problem, out of the $\binom{2^m + 3 - 1}{3}$

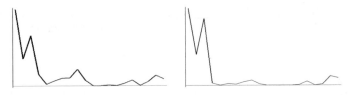

Fig. 5 Invariant distribution for simple GA (original and modified) on sandia-4 problem

possible populations, $2^{m-1}(2^m + 1)$ contain the allele A. Clearly, the proportionate size of the "goal" basin decays to zero very quickly as $m \to \infty$.

We can compute the transition probability matrix and figure how likely it is to get from the "farthest" population from the goal to the goal. In the 2-bit problem, to get from the population $\mathscr{P} = \{D = (1, 1), D, D\}$ to any population containing A is a two-step process because only one bit can be flipped in a generation. Taking the fitness multiplier $\mu = \ln(2)$ (so that $e^{-\mu f} = 2^{-f}$), the probability of a transition from $\{D, D, D\}$ to $\{C, D, D\}$ is 3/10 and to $\{B, D, D\}$ is 3/14. From $\{C, D, D\}$, it is possible to get to populations containing A in one step, namely $\{A, C, C\}$, for which the transition occurs with probability 2/297, or to $\{A, C, D\}$ with probability 976/13365, or to $\{A, D, D\}$ with probability 56/1215. Thus, the probability of transitioning in one step from $\{C, D, D\}$ to some population containing the individual A is 1682/13365 ≈ 0.126. Similarly, the probability of a single-step transition from $\{B, D, D\}$ to any population containing A is 550/5187 ≈ 0.106. With this type of reasoning, we can compute that the probability of a transition from $\{D, D, D\}$ to any population containing the goal state in two steps is approximately 0.064. The invariant distribution is in Fig. 5. In addition, Table 4 shows the allowable transitions. Comparing this table with Table 1, we see that the simple GA allows many more transitions than the Holland-style GA did.

Figure 2b shows the invariant distribution for the 3-bit version (so the function is sandia-8), where Ω contains 120 distinct populations. From this figure, one can see that the algorithm spends very little time in the states between basins. Therefore, it takes a great deal of time to get out of the local maximum basin.

Using the expected hitting time to a population containing A, we can make some comparisons. First, the expected hitting time for uniform start-up for the 2-bit, 3-bit, and 4-bit problems is given in Table 5. In this, we have taken the fitness multiplier at $\mu = 0.2$. In Fig. 6, we show the dependence of the hitting time on μ. As shown in the figures, the hitting time is much lower for μ taking small values. These correspond to the selection process evening out the differences in fitnesses between population members. This results in allowing greater diversity in the population.

We also compare with the three basin problems as shown in Fig. 1b. The results are given in Table 5 where it is seen that the extra basin resulted in a proportionately greater hitting time for the 4-bit version of the problem.

Having an explicit symbolic form for the transition matrix enables us to do some analytical experiments. For instance, modify the standard sandia-4 function on $\mathbb{X} = \{00, 01, 10, 11\}$ to be $f(00) = 4n, f(01) = n, f(10) = 2n, and f(11) = 3n$ (so n

Table 4 Allowable transitions for the simple GA

	AAA	AAB	AAC	AAD	ABB	ABC	ABD	ACC	ACD	ADD	BBB	BBC	BBD	BCC	BCD	BDD	CCC	CCD	CDD	DDD
AAA	■	■	■																	
AAB	■	■	■	■	■	■	■				■	■	■							
AAC	■	■	■	■		■		■	■				■				■	■		
AAD	■	■	■	■	■	■	■	■	■	■			■		■	■		■	■	
ABB	■	■	■	■	■	■	■				■	■	■							
ABC	■	■	■	■	■	■	■	■	■	■	■	■	■	■	■	■	■	■	■	■
ABD	■	■	■	■	■	■	■	■	■	■	■	■	■	■	■	■		■	■	■
ACC	■	■	■	■		■		■	■				■		■	■				
ACD	■	■	■	■	■	■	■	■	■	■		■	■	■	■	■	■	■	■	■
ADD		■	■	■	■	■	■	■	■	■			■		■	■		■	■	■
BBB				■							■		■							
BBC	■	■		■	■	■	■	■			■	■	■	■	■	■		■	■	
BBD				■		■			■	■	■	■			■	■			■	■
BCC	■	■		■	■	■	■	■			■	■	■	■	■	■		■	■	
BCD	■	■	■	■	■	■	■	■	■	■	■	■	■	■	■	■	■	■	■	■
BDD			■		■			■	■	■	■				■	■			■	■
CCC						■											■	■		
CCD						■	■	■					■	■	■	■	■	■	■	■
CDD						■	■	■					■	■	■	■	■	■	■	■
DDD															■				■	■

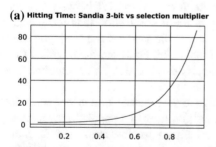

(a) Hitting Time: Sandia 3-bit vs selection multiplier

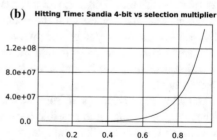

(b) Hitting Time: Sandia 4-bit vs selection multiplier

Fig. 6 **a** For the 3-bit problem. **b** For the 4-bit problem

Table 5 Expected hitting time for the simple GA

	2-bit	3-bit	4-bit
Original sandia	3.1	6.4	38.8
Sandia 3-basin	3.2	7.1	48.2

times the standard sandia-4 function values). Then, taking the limit as $n \to \infty$, one expects to get absorbing states at the two populations $\{00, 00, 00\}$ and $\{11, 11, 11\}$. However, it is very interesting that the state $\{00, 11, 11\}$ is also an absorbing state for the limiting chain. Of course, the limiting chain does not arise from any fitness function, but it indicates how the chain behaves for a very skewed objective.

If instead we use the function values $f(00) = n$, $f(01) = 2n$, $f(10) = 3n$, and $f(11) = 4n$, then the population $\{11, 11, 11\}$ is the only absorbing state as expected.

Varying the selection in the simple GA

As another slight variation on the simple GA, we keep the same crossover and mutation but modify the selection mechanisms. For the first selection (after crossover), we deterministically remove the least-fit individual; if there is a tie, we choose equally likely. For the second selection (after mutation), we use a roulette wheel selection but instead of "selecting out," we "select in" without replacement. More specifically, given the four individuals after mutation, we select one proportional to fitness and add it to the new population. Then, we select another one proportional to fitness but now conditioned to the subpopulation of those which have not yet been chosen. We do this again, now choosing one of the two remaining individuals. This results in a more involved algorithm to implement and in much more complicated formulas for the entries of the transition matrix. The question is: Does this significantly change the performance of the GA?

First, we see from a plot of the invariant distribution (second image in Fig. 5) that the modification causes more of the mass to be on the "good" states (compare two plots in Fig. 5). However, surprisingly, we calculate EHT for the modification at 8.74 which is quite a bit higher than EHT of 3.1 from Table 5. This is another indication that at times EHT and the invariant distribution are competing and measuring two completely different things. It is likely that the difference in EHT between the two variations has more to do with the change in the first selection than the second. The problem is that if you on the "wrong" side of the sandia-4 function, you need to move "downhill' in order to find the optimal point. However, the modified first selection discourages this, and so on average, it takes longer with the modification than without.

Another unexpected result from the change in the first selection is that when the fitness values change, it is possible for a transition from one population to another to appear or disappear. For example, there are fifteen possible transitions from the population $\{00, 00, 11\}$ when the fitness values are all equal, but only six with the sandia-4 objective function (compare Tables 4 and 6). In fact, 56 out of the 172 allowable transitions with a constant objective function disappear when we change to the sandia-4 objective function. (For reference, the allowable connections for the constant objective are the same as given in Table 4.) This is a significant difference in the structure of the directed graph which underlies the Markov chain.

As in the (unmodified) simple GA, taking a limit with ever more skewed sandia-4 function values results in the same three absorbing states, the popula-

Table 6 Allowable transitions for modified simple GA for sandia-4

	AAA	AAB	AAC	AAD	ABB	ABC	ABD	ACC	ACD	ADD	BBB	BBC	BBD	BCC	BCD	BDD	CCC	CCD	CDD	DDD
AAA	■	■	■																	
AAB	■	■	■	■	■	■	■													
AAC	■	■	■	■		■		■	■											
AAD	■	■	■	■			■		■											
ABB	■	■	■	■	■	■	■				■	■	■							
ABC	■	■	■	■	■	■	■	■	■	■		■		■	■		■	■	■	
ABD		■	■	■	■	■	■	■	■	■		■		■	■			■	■	
ACC	■	■	■	■		■		■	■				■		■	■				
ACD		■	■	■		■	■	■	■	■				■	■			■	■	
ADD				■		■	■									■		■		
BBB				■							■		■							
BBC	■	■		■	■	■	■	■			■	■	■	■	■			■	■	
BBD				■		■					■	■	■	■	■	■			■	■
BCC		■			■		■	■						■	■		■	■	■	
BCD		■	■		■	■	■	■	■		■	■	■	■	■		■	■	■	■
BDD					■		■				■			■	■			■	■	
CCC					■										■	■				
CCD					■	■	■						■	■	■	■		■	■	■
CDD					■	■								■	■		■	■	■	■
DDD																■			■	■

tions {00, 00, 00}, {00, 11, 11}, and {11, 11, 11}. Thus, this feature of the algorithm remains unchanged.

Non-uniform invariant distribution under equal fitness

A very interesting observation is that even when the objective function is constant, so that all individuals in $\mathbb{X}$ have the same fitness value for the GA, the invariant distribution is not uniform over Ω, the set of populations. Even without selection pressure, the details of the crossover and mutation give constraints on the allowable transitions which, in turn, create bottlenecks for the chain and lead to a non-uniform invariant distribution.

As an example of a non-uniform distribution, for the simple GA, the invariant distribution for equal fitness values is

$$(\alpha, \beta, \beta, \gamma, \beta, \delta, \delta, \beta, \delta, \gamma, \alpha, \gamma, \beta, \gamma, \delta, \beta, \alpha, \beta, \beta, \alpha), \tag{8}$$

where we have $\alpha = 4061/243656$, $\beta = 17694/243656$, $\gamma = 3195/243656$, and $\delta = 18270/243656$, and we order the states in the same way as in Tables 1 and

4. Notice that there are only four different values in this probability distribution. The populations which have a given probability under the invariant distribution are as follows:

$$Pr(\text{ three copies of one individual}) = \alpha,$$
$$Pr(\text{ three different individuals}) = \delta,$$
$$Pr(\text{ two the same, different children}) = \gamma,$$
$$Pr(\text{ two the same, same children}) = \beta.$$

The last two categories require a small explanation. For the population $\{00, 00, 10\}$, we notice that no matter which two individuals are chosen as the "parents" for crossover, the two "children" will just be copies of the two parents. However, for the population $\{00, 00, 11\}$, this is not true; the two parents 00 and 11 produce the offspring 01 and 10. This simple difference results in different behavior and thus different values for their probability under the invariant distribution. This naturally leads to the question of how to predict which states will have the same probability under the invariant measure.

We have that $\mathbb{X} = \{0, 1\}^2$ and Ω consists of populations of elements of $\mathbb{X}$. Consider the mapping $\phi_1 : \mathbb{X} \to \mathbb{X}$ given by

$$\phi_1(00) = 10, \quad \phi_1(01) = 11, \quad \phi_1(10) = 00, \quad \text{and } \phi_1(11) = 01. \qquad (9)$$

We denote by $\Phi_1 : \Omega \to \Omega$ the function which is induced by ϕ_1. It is easy to show that Φ_1 commutes with the crossover, mutation, and selection operations on Ω, and thus, when the fitness values are all constant, ω and $\Phi_1(\omega)$ will have the same probability under the invariant distribution. There are four functions, ϕ_i, similar to ϕ_1 as given in (9) (including the identity, that is), and each of them induces a symmetry in the resulting Markov chain on populations.

This type of reasoning will work for arbitrarily large $\mathbb{X}$ and arbitrarily large population sizes; of course, the details will be more complicated and there will be many more "cases." That is, anytime we have a bijection $\Psi : \Omega \to \Omega$, which commutes with crossover, mutation, and selection, we get symmetries of the invariant distribution when the fitness values are constant. If Ψ also preserves the level sets of the fitness function, then we get symmetries of the invariant distribution in the general case as well. This idea is helpful if we want to find explicit expressions for the entries of the transition matrix as a function of the fitness values or other parameters of the GA.

The modified simple GA has the same pattern as given in (8), except with different values for $\alpha, \beta, \gamma, \delta$. This is clear because the functions Φ_i commute with all the operators in the modified algorithm. In particular, this means that just the fact that the operations commute with all the Φ_i does not specify the invariant distribution but only the pattern of which values are the same.

The influence of the connection structure on the invariant measure can be seen even when the objective function is not constant. This structure of the chain has a secondary influence on the invariant distribution of the chain. For example, for the modified simple GA with objective functions $f(00) = 1$, $f(01) = 2$, $f(10) = 3$, and $f(11) = 4$, the state $\{11, 11, 11\}$ has lower probability under the invariant distribution than the state $\{10, 11, 11\}$ (0.274 versus 0.390). By using the Metropolis–Hastings idea (Sect. 2.2), we can usually "engineer" the invariant distribution as desired; however, this usually has strong implications for the hitting times of the resulting chain since the only way to "correct" for the connection structure is to delay the chain in the current state sufficiently long.

6   Shuffle-Bit GA

We now introduce a completely new genetic algorithm, one for which we can engineer the invariant population distribution as desired, for example, to be the Boltzmann distribution. As in simulated annealing, we achieve this through the use of an acceptance phase (as in the Metropolis–Hastings algorithm). What makes it possible is a kind of symmetry in the proposal scheme; namely, if population $j \in N(i)$, then also $i \in N(j)$. Overall, the new GA is given by

> For the proposal process, from the population i at the start of an iteration, we (1) mutate (flip) one of bits of the combined "gene pool" and then (2) make a permutation of the resulting combined bits.

In more detail, the first operation consists in amalgamating all the bits of the population into a single array, the *gene pool*. In length, it will be the number of bits per chromosome times the population size s. Now, select 0 or 1 with probability $1/2$ each and choose, equally likely, a bit of that type from the gene pool and flip it. If there are only bits of one type, choose one uniformly at random and flip it. For the second operation, make a uniform permutation of the gene pool, that is, a random shuffle. Now, divide it back into individual chromosomes, s in number. This is the population presented to the acceptance phase of the iteration. The allowable transitions from this are in Table 7.

Notice that it does not matter if instead we performed the two operations in the reverse order; the resulting Markov transition matrix would be identical. Of course, given a specific permutation and a specific bit location to be flipped, the order of the operations does matter. However, in a probabilistic sense, they do not.

For the acceptance phase, we use the Metropolis–Hastings algorithm, in particular Eq. (4). In order to implement Eq. (4), we must be able to compute, on demand, any term g_{ij} of the population proposal scheme. Let z be the number of zeros in the gene pool of j. (As a check, z must be either one more or one less than the number of zeros in the gene pool of i; otherwise, $g_{ij} = 0$). If n is the size of the gene pool, then the number of ways of selecting the placement of z zeros is the number of combinations $\binom{n}{z}$. Having selected one of these possibilities uniformly at random, divide up the

Table 7 Allowable transitions for the shuffle GA

	A A A	A A B	A A C	A A D	A B B	A B C	A B D	A C C	A C D	A D D	B B B	B B C	B B D	B C C	B C D	B D D	C C C	C C D	C D D	D D D
AAA		■	■																	
AAB	■			■	■	■		■												
AAC	■			■	■	■		■												
AAD		■	■			■		■			■	■		■			■			
ABB		■	■			■		■			■	■		■			■			
ABC		■	■			■		■			■	■		■			■			
ABD				■	■	■		■		■			■		■			■		
ACC		■	■			■		■			■	■		■			■			
ACD				■	■	■		■		■			■		■			■		
ADD						■		■			■	■		■		■	■		■	
BBB				■	■	■		■		■			■		■			■		
BBC				■	■	■		■		■			■		■			■		
BBD						■		■			■	■		■		■	■		■	
BCC				■	■	■		■		■			■		■			■		
BCD						■		■			■	■		■		■	■		■	
BDD								■					■		■			■		■
CCC				■	■	■		■		■			■		■			■		
CCD						■		■			■	■		■		■	■		■	
CDD								■					■		■			■		■
DDD															■			■		

gene pool back into individual chromosomes. In doing so, there may be several sets of identical chromosomes. Let their sizes be $k_1, k_2, \ldots, k_m$. This is equivalent to identifying the population in type-count format. For a population consisting of just one type of chromosome, then $m = 1$ and $k_1 = s$. The chance that this population is the permutation result is $1/\binom{n}{z}$ or

$$\frac{k_1!/k_1!}{\binom{n}{z}}.$$

If the population consists of sall distinct members, then $k_1 = k_2 = \cdots = k_s = 1$. Since there are $s!$ permutations of this population, it follows that the chance it is obtained is

$$\frac{s!}{\binom{n}{z}}.$$

In between these two extreme cases, and including them, the chance of obtaining a given population is given by the multinomial coefficient

$$\frac{(k_1 + k_2 + \cdots + k_m)!/(k_1!k_2!\ldots k_m!)}{\binom{n}{z}}.$$

As an example, consider a population of size 3, each individual consisting of two bits. As usual, denote A for 00, B for 01, C for 10, and D for 11. The population AAD is a neighbor of AAB since the former has one more "1" bit than the latter. The probability of proposing AAD from AAB is

$$\frac{1}{2} \times \frac{\frac{(2+1)!}{2!1!}}{\binom{6}{2}} = \frac{1}{10}.$$

6.1 Results

We implemented the shuffle GA on our test problems with excellent results. In Fig. 7a, we show the invariant distribution on the sandia-2-bit problem, and in Fig. 7b, we show the invariant distribution on the 3-bit problem. The horizontal line on the left is the common probability for every population that has an A chromosome. (Recall the fitness of a population is the maximum of its chromosomes.) There are 10 such population; their common probability under the invariant distribution is 0.06154. The next series of plateaus is for the populations that have a D chromosome but no A. This value is 0.04615 for all 6 of them. For 3 populations that have a C but no A or B, the value is 0.03077. And for the one population BBB, the invariant probability is 0.01538. Note the ratios: $Axx/BBB = 0.06154/0.01538 = 4.00$, $Dxx/BBB = 3.00$, and $Cxx/BBB = 2.00$. Thus, the invariant distribution is indeed Boltzmann. In Table 8, we show the expected hitting times for the shuffle GA.

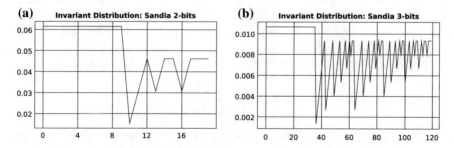

Fig. 7 **a** For the 2-bit problem. **b** For the 3-bit problem

Table 8 Expected hitting time for the shuffle GA

	2-bit	3-bit	4-bit
Original sandia	3.1	3.0	3.55
Sandia 3-basin	3.0	3.0	3.55

6.2 Estimate of Expected Hitting Time

Suppose that each chromosome contains ℓ bits and there are s chromosomes in the population. Then, there are $m = 2^\ell$ possible individuals, and the "gene pool" has size $r = s\ell$.

Let us say, without loss of generality, that the goal state is all zero bits and let us calculate the probability that a random permutation of r bits will make the first ℓ of them all zeros. (Thus, the first chromosome of the population hits the goal.)

For example, take $\ell = 4$ so there are 16 possible individuals: $0000 = A, 0001 = B, \ldots, 1111 = P$, and take $s = 2$, so the number of populations is

$$\binom{s + m - 1}{m - 1} = 136.$$

We number the populations as: pop1 = AA, pop2 = AB, ..., pop136 = PP. The gene pool size is $r = 8$.

Letting H be the chance that a permutation will result in the first ℓ specified to be 0, we see that it is a function of how many zero bits there are in the gene pool. Let z be the number of 0's and y be the number of 1's, $y = r - z$.

If $z < \ell$, no permutations will work, and so $H(z) = 0$ if $z < \ell$. If $z = \ell$, then only one permutation will work, in the example 00001111. The number of permutations of r distinct symbols is $r!$, but if k of them are the same, then the number of distinct permutations is $r!/k!$. In our case, there are z 0's and y 1's, so the number of distinct permutations is $r!/(z!y!) = \binom{r}{z}$. In particular, $H(\ell) = 1/(\ell!\ell!)$. In the example, this is $(24 \cdot 24)/(40320) = 1/70$.

In general, for $z \geq \ell$, we have that $H(z)$ is $(r - \ell)!/(z - \ell)!y!$ divided by $r!/z!y!$ or

$$H(z) = \frac{(r - \ell)!z!}{r!(z - \ell)!}.$$

This is seen as follows. The total number of distinct permutations was derived above. For the numerator, we count among their number those having the first ℓ to be 0. There are $r - \ell$ remaining positions in the permutation, and there are $z - \ell$ remaining zeros (and y 1's); therefore, the count is $(r - \ell)!/(z - \ell)!y!$.

If the ℓ 0's can be in any of the s chromosomes, then the chance of a hit in any one of them is bounded from above by $sH(z)$.

6.2.1 Upper Bound on the Generation of ℓ Zeros

As we saw above, there is no chance of hitting the goal chromosome unless there is the minimal number of the correct bits present. Here, we want to find an upper bound on that happening. Again, we assume the correct bits are ℓ zeros. In each iteration of the search, one bit, 0 or 1 equally likely, is flipped, except if all bits are 0, then one is flipped to 1, and if all are 1, then one is flipped to 0.

For the worst case, assume that initially all the bits are 1's. Let E_z be the expected number of iterations to reach ℓ zeros given there are z presently where $z < \ell$. If there are $\ell - 1$ zeros, then with one-half chance, there will be ℓ in 1 iteration and with 1/2 chance it will take 1 plus the expectation from $\ell - 2$ zeros, so

$$E_{\ell-1} = \frac{1}{2}1 + \frac{1}{2}(1 + E_{\ell-2}) = 1 + \frac{1}{2}E_{\ell-2}.$$

On the other hand, if there are no 0's, then in 1 iteration, there will be one with certainty; hence,

$$E_0 = 1 + E_1.$$

In general, we have

$$E_z = 1 + \frac{1}{2}E_{z-1} + \frac{1}{2}E_{z+1}.$$

The solution system is given by the $\ell \times \ell$ matrix

$$\begin{pmatrix} 1 & -\frac{1}{2} & 0 & 0 & 0 & \dots & 0 & 0 \\ -\frac{1}{2} & 1 & -\frac{1}{2} & 0 & 0 & \dots & 0 & 0 \\ 0 & -\frac{1}{2} & 1 & -\frac{1}{2} & 0 & \dots & 0 & 0 \\ \vdots & & \ddots & \ddots & \ddots & & \vdots & \vdots \\ 0 & 0 & 0 & 0 & 0 & \dots & -1 & 1 \end{pmatrix} \begin{pmatrix} E_{\ell-1} \\ E_{\ell-2} \\ E_{\ell-3} \\ \vdots \\ E_0 \end{pmatrix} = \begin{pmatrix} 1 \\ 1 \\ 1 \\ \vdots \\ 1 \end{pmatrix}.$$

Carrying out elimination on this matrix gives

$$\begin{pmatrix} 1 & -\frac{1}{2} & 0 & 0 & 0 & \dots & 0 & 0 \\ 0 & 1 & -\frac{2}{3} & 0 & 0 & \dots & 0 & 0 \\ 0 & 0 & 1 & -\frac{3}{4} & 0 & \dots & 0 & 0 \\ 0 & 0 & 0 & 1 & -\frac{4}{5} & \dots & 0 & 0 \\ \vdots & \vdots & \vdots & \ddots & \ddots & \ddots & & \vdots \\ 0 & 0 & 0 & 0 & 0 & \dots & 1 & (1-\ell)/\ell \\ 0 & 0 & 0 & 0 & 0 & \dots & 0 & 1/\ell \end{pmatrix} \begin{pmatrix} E_{\ell-1} \\ E_{\ell-2} \\ E_{\ell-3} \\ E_{\ell-4} \\ \vdots \\ E_1 \\ E_0 \end{pmatrix} = \begin{pmatrix} 1 \\ 2 \\ 3 \\ 4 \\ \vdots \\ \ell - 1 \\ \ell \end{pmatrix}.$$

From this, it follows that $E_0 = \ell^2$, $E_1 = (\ell - 1)(\ell + 1)$, ..., $E_k = (\ell - k)(\ell + k)$, $E_{\ell-1} = 2\ell - 1$.

6.2.2 Expected Hitting Time from Any Given Starting Point

Suppose the algorithm is started having exactly n zeros. Let E_n be the expected hitting time in this event. The probability that in 1 iteration the goal will be found is, from above, bounded by $sH(n)$. Since we are looking for an upper bound on EHT, we will just take it to be $sH(n)$. If the goal is not found on the next iteration, then either there will be $n+1$ zeros or $n-1$ for the next step. The former case occurs with probability $1/2$, and in that case, the expected hitting time is E_{n+1}. In the latter, the expected hitting time is the expected time to return to the case with n zeros plus the expected hitting time from that state. So

$$E_n = sH(n) + 1 + \frac{1}{2}E_{n+1} + \frac{1}{2}\left(2n - 1 + E_n\right).$$

When there are $n+1$ zeros, we have

$$E_{n+1} = 1 * sH(n+1) + \frac{1}{2}E_n + \frac{1}{2}E_{n+2}.$$

If there are fewer than s 1's, that is $z > r - s$, then it is necessarily the case that some chromosome has n zeros and the goal is found. Hence,

$$E_{r-s-1} = 1 + \frac{1}{2}E_{r-s-2} + \frac{1}{2}.$$

This reasoning leads to the system of equations

$$\begin{pmatrix} \frac{1}{2} & -\frac{1}{2} & 0 & 0 & \dots & 0 & 0 \\ -\frac{1}{2} & 1 & -\frac{1}{2} & 0 & \dots & 0 & 0 \\ \dots & & \dots & & \dots & \vdots & \vdots \\ 0 & 0 & 0 & 0 & \dots & -\frac{1}{2} & 1 \end{pmatrix} \begin{pmatrix} E_{r-s-1} \\ E_{r-s-2} \\ \vdots \\ E_n \end{pmatrix} = \begin{pmatrix} \frac{3}{2} \\ 1 \\ \vdots \\ \frac{1}{2} + n + sH(n) \end{pmatrix}.$$

Solving for E_n, we have

$$E_n \leq \frac{r-s-n}{r-s-n+1}\left(1 + 2n + sH(n) + r - s - n - 1 + \frac{1}{r-s-n}\right).$$

7 Discussion and Future Work

There are several conclusions which can be drawn from our study. It is desirable to have some idea, if possible, of how the invariant distribution distributes the asymptotic mass of the process. For a GA, any symmetries of crossover and mutation can be used to determine the structure in the invariant distribution. While one might think that fitness is the major factor in determining the invariant distribution, the "connection structure" of the GA, which is induced via the genetic operators, is equally important and might result in unexpected states having high probability under the limiting distribution. This is particularly evident if one considers a GA with constant objective function where the invariant distribution is not uniform and favors those populations with many connections (and also a high probability of remaining in the same population).

The rate at which the Markov chain converges to stationary behavior (the invariant distribution, if it exists) might be interesting, but at times, it can be deceptive if the stationary behavior is not the desired behavior. For this reason, for random search processes, we do not suggest to investigate the mixing time in general. Our primary focus in this paper was the average expected hitting time (EHT), where we use the uniform starting distribution to average the hitting times.

Having many connections is desirable, but sometimes, having too many connections is a handicap. This can be seen by comparing the values of EHT in Tables 3, 5, and 8 to the connection structures in Tables 1, 4, and 7. The shuffle GA does a good job on the 2-bit, 3-bit, and 4-bit versions of the sandia problem, and we speculate that it is the regular structure of the underlying connection graph which is largely responsible for this.

Like most work on this topic, our study has raised more questions than we could possibly answer and only represents another small step toward the resolution of the overall question of how to match optimization problems with algorithms. Clearly, it would be ideal to extend our analysis to very large versions of the sandia-n problem and possibly other, more involved deceptive problems. There are no known techniques to estimate EHT for a general objective function, even for our very simple models for GAs. Our results indicate that changing the objective function can significantly change EHT, but exactly how this relationship works is unknown.

Acknowledgements The work of FM is partially supported by a Discovery Grant from the Natural Sciences and Engineering Research Council of Canada (238549-2012). The paper contains some results from the MSc thesis of the first author, and her work is supported by the Saudi Cultural Bureau.

References

1. Brémaud P (1999) Gibbs fields, Monte Carlo simulation, and queues. Texts in Applied Mathematics, vol 31. Springer, New York
2. Davis TE (1991) Toward and extrapolation of the simulated annealing convergence theory onto the simple genetic algorithm. PhD Thesis, University of Florida
3. Davis TE, Principe JC (1991) A simulated annealing-like convergence theory for the simple genetic algorithm. In: Proceedings of the 4th international conference on genetic algorithms, San Diego, July 1991, pp 174–181
4. Davis TE, Principe JC (1993) A Markov chain framework for the simple genetic algorithm. Evol Comput 1(3):269–288

5. Eiben AE, Aarts EHL, van Hee KM (1991) Global convergence of genetic algorithms: a markov chain analysis. In: Schwefel HP, Männer R (eds) Parallel problem solving from nature, pp 4–12. Springer, Berlin and Heidelberg
6. Holland J (1992) Adaptation in natural and artificial systems. MIT Press, Cambridge, MA
7. Jansen T (2013) Analyzing evolutionary algorithms: the computer science perspective. Springer, Heidelberg
8. Kemeny JG, Snell JL (1976) Finite Markov chains. (Reprint of the 1960 original) undergraduate texts in mathematics. Springer, New York-Heidelberg
9. Levin DA, Peres Y, Wilmer EL (2009) Markov chains and mixing times. American Mathematical Society, Providence, RI
10. Nix AE, Vose MD (1992) Modeling genetic algorithms with Markov chains. Ann Math Artif Intell 1(5):79–88
11. Rudolph G (1986) Finite Markov chain results in evolutionary computation: a tour d'horizon. Fund Inform 35(1–4):67–89
12. Rudolph G (2010) Stochastic convergence. In: Rozenberg G, Bäck T, Kok J (eds) Handbook of natural computing, pp 847–869. Springer, Berlin
13. Takacs C (2006) On the fundamental matrix of finite state Markov chains, its eigensystem and its relation to hitting times. Math Pannon 17(2):183–193
14. Vose MD, Liepins GE (1991) Punctuated equilibria in genetic search. Complex Syst 1(5):31–44
15. Wolpert D (2012) What the no free lunch theorems really mean; how to improve search algorithms. SFI Working Paper 2012-10-017, Santa Fe Institute
16. Wolpert D, Macready WG (1997) No free lunch theorems for optimization. IEEE Trans Evol Comput 1(1):76–82

Permutation Problems, Genetic Algorithms, and Dynamic Representations

James Alexander Hughes, Sheridan Houghten and Daniel Ashlock

Abstract Permutation problems are a very common classification of optimization problems. Because of their popularity countless algorithms have been developed in an attempt to find high quality solutions. It is also common to see many different types of search spaces reduced to permutation problems as there are many heuristics and metaheuristics for them due to their popularity. This study incorporates the travelling salesman problem, bin packing problem, and graph colouring problem. These problems are studied with multiple variations and combinations of heuristics and metaheuristics with two distinct types of representations. The majority of the algorithms are built around the Recentering-Restarting Genetic Algorithm. The algorithm variations were effective on all problems examined although the variations between the different algorithms in this particular study were for the most part not statistically different. This observation led to the conclusion that algorithm success is dependent on problem instance; in order to fully explore the search space, one needs to study multiple algorithms for a given problem.

1 Introduction

Evolutionary algorithms are popular search strategies that are commonly used to generate competitive candidate solutions to computationally intractable problems. Many evolutionary algorithms involve modifying permutations of elements with the

J.A. Hughes · S. Houghten (✉)
Brock University, St. Catharines, ON, Canada
e-mail: shoughten@brocku.ca

J.A. Hughes
e-mail: jh08tt@gmail.com

D. Ashlock
University of Guelph, Guelph, ON, Canada
e-mail: dashlock@uoguelph.ca

© Springer International Publishing AG 2017
S. Patnaik et al. (eds.), *Nature-Inspired Computing and Optimization*,
Modeling and Optimization in Science and Technologies 10,
DOI 10.1007/978-3-319-50920-4_6

objective of ordering them in a way to optimize some criteria. It is important to study this style of problem as they are a common type of optimization problem. In fact, many other types of optimization problems are very easily reduced to these permutation problems. Fortunately, there are many problems that exist which are inherently of this classification, the best known being the travelling salesman problem (TSP). There are many other problems which can be considered of this classification by developing clever representations and evaluations, two examples being the bin packing problem and vertex colouring variant of the graph colouring problem. These three problems are ideal for the study of novel algorithms as they are well studied and provide a large collection of benchmarks.

The *no free lunch theorem* [31] tells us no one algorithm will always dominate a search space. A large number of studies use one specific algorithm and tune it to work optimally for the given problem domain. Unfortunately, this approach is extremely limiting considering the large search spaces of the intractable problems. Ideally, one should study a large variety of algorithms in addition to conducting a simple parameter sweep. These algorithms should be exploited in every way possible to learn their strengths and weaknesses. By combining algorithms, one can also greatly improve the effectiveness of the traversal through the search space. This idea is particularly important for evolutionary algorithms as there are many effective modular enhancements which can be easily incorporated into the search.

A Genetic Algorithm (GA) variation called the Recentering-Restarting Genetic Algorithm (RRGA) has been shown to work well on permutation problems with high quality results [1, 3, 4, 18, 19, 21]. This collection of studies analyzed the effectiveness of this maturing approach on varying problems including epidemic networks, error correction, and DNA fragment assembly. This chapter builds upon previous work by introducing new ideas into the RRGA, analyzing additional evolutionary algorithm variations and other search algorithms, and studying a collection of popular benchmark problems.

In this chapter, we use the RRGA on all problems to learn its strengths and understand how to exploit its search through the fitness landscape. Other search algorithms studied include *nearest neighbour*, *minimal spanning tree*, *2-Opt*, and *Lin-Kernighan*. *Ring species* and the *island model*—two popular genetic algorithm modular variations—are also studied. Combinations of all these strategies will be used to explore the search spaces and to gain an understanding of how to utilize all the benefits of each algorithm. Although within this chapter it will become clear that the RRGA does perform well on its own, we want to emphasize that *it is best used in combination with other modular enhancements*; a major theme in this chapter is that one of the many advantages of evolutionary search is the ease with which enhancements can be incorporated.

2 Problem Descriptions

2.1 Bin Packing Problem

The one-dimensional bin packing problem (BPP) is defined as follows: given a set of n objects of sizes $s_1, s_2, ..., s_n$, pack the objects into containers (bins) such that the number of containers is minimized [26]. The BPP is known to be NP-Hard [23, 26].

Many algorithms have been applied to large instances of this problem successfully. Additionally, many heuristics have been created which produce reasonably good results in a relatively short amount of time. One popular heuristic for this problem is the *first fit algorithm* which places each object into the first bin in which it fits. If no bin has sufficient size, then an additional bin is added. This algorithm provides a solution that is no worse than twice the ceiling of the sum of the sizes of the objects [32]. This basic heuristic is particularly noteworthy as it is a simple and ideal algorithm for calculating the fitness values of a given permutation of objects; the quality of an ordering can be evaluated with this algorithm. Although a solution to the problem is dependent on the algorithm *and* the ordering of the objects, we will fix the algorithm and only alter the ordering, since we are studying the ability of the algorithm to evolve a permutation.

Variations of the BPP increase the number of dimensions (two/three-dimensions). Only the one-dimensional BPP is considered in this study as it is effective for demonstrating and evaluating algorithms.

2.2 Graph Colouring Problem

The graph colouring problem (GCP), specifically the vertex colouring problem, is very similar to the BPP as it is a permutation problem with fitness value dependent on all genes before a given gene (although this does depend on the algorithm being used for evaluation). Because of this, algorithms that work well on the BPP tend to also be applied to the GCP successfully [11]. The GCP is identified as being NP-Hard [7, 13].

The problem is defined as follows: given a graph, colour each vertex such that no two adjacent vertices have the same colour. The goal is to minimize the *chromatic number*—number of colours used to colour the graph.

Similar to the BPP, there are many sophisticated algorithms that can produce optimal solutions for large instances of the GCP. In particular, there is a heuristic very similar to the greedy algorithm used for the BPP: given a set of vertices $v_1, v_2, ..., v_n$, assign the smallest available colour (based on some ordering) to v_i such that none of v_i's neighbours has that colour. If no colour is available, add an additional colour. In addition to the vertex colouring problem, the GCP may refer to Chromatic Colouring

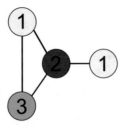

Fig. 1 An example instance of a graph with a chromatic number of 3. The vertices of these graphs are coloured with a *number* for simplicity. Note that no two vertices have the same colour. There are numerous possible colourings for this particular instance, however, this is an example of an optimal colouring

(counting the number of colourings of a graph with a given chromatic number), Edge Colouring, and Total Colouring (a combination of vertex and edge colouring).

An example of a GCP vertex colouring problem can be seen in Fig. 1.

2.3 Travelling Salesman Problem

The travelling salesman problem (TSP) is a well known combinatorial optimization problem identified as NP-Hard [9]. The problem is defined as follows: given a complete undirected graph $G = (V, E)$ with a non-negative cost $c(u, v)$ associated with each edge $\in E$, find a Hamiltonian cycle of G such that the total distance (cost) is minimized. In this problem, the vertices are referred to as *cities*.

The TSP has become widely used as a benchmark problem for combinatorial optimization algorithms. There exist many approximation algorithms that produce high quality solutions for the TSP including 2-Opt and Lin–Kernighan (LKH). These algorithms will be briefly discussed further in Sect. 4. Along with the approximation approaches, many evolutionary algorithms have been applied to the TSP with success.

In 2001 Discrete Mathematics and Theoretical Computer Science (DIMACS) held an implementation challenge for the TSP, and the website [10] archives many problem instances along with a results page summarizing the best results obtained by many different types of algorithms. The problem instances are in Travelling Salesman Problem Library (TSPlib) format and can ultimately be obtained from TSPlib [27]. The information gathered throughout this implementation challenge was recorded and published in [24] which can serve as a reference for benchmarks when exploring new optimization techniques.

3 Previous Work on Small Travelling Salesman Problem Instances

Previous work studied a collection of small TSP instances varying from 12 to 20 cities on a variety Genetic Algorithm variations [18]. Figures 2 and 3 summarize the results. These figures show the obvious trend of the instances becoming more difficult as the number of cities increases. Specifically, Fig. 2 suggests that as the number of cities increases one algorithm begins to dominate.

There are a number of reasonable explanations for why this happened, but the most sound argument is that due to the restarts and recentres of the RRGA variations, it is likely that these algorithms never had the chance to fully explore the search space before a restart. These restarts and recentres are very destructive operations for small problems as they would remove any significant gains made in the search space. This phenomenon would probably not be an issue if the RRGA variations were given enough generations to fully explore the search trajectories taken.

This hypothesis led to the addition of experiments which increased the number of generations by an order of 10. Figure 3 summarizes the results. These results clearly show that no one algorithm dominates all solutions with 4/8 of the algorithms studied achieving the best result on at least half of the 20-city problems. These improvements were not gained by disadvantaging the basic GA or giving the RRGA an advantage, but rather, by removing the advantage the basic GA originally had over the RRGA variations. It is expected that if one were to increase the number of generations further, the RRGAs would have an unfair advantage as the RRGA variations excel at

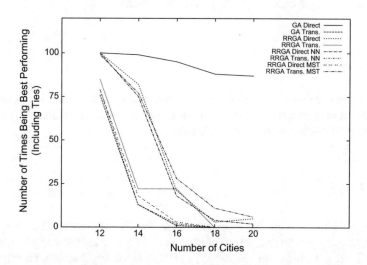

Fig. 2 Summary of algorithm performance on the small travelling salesman problem instances. As the number of cities increases, the number of times an algorithm achieves the optimal result decreases. One algorithm begins to dominate, namely the basic GA with a direct representation

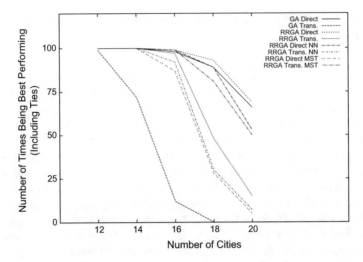

Fig. 3 Summary of algorithm performance on the small travelling salesman problem instances with *additional generations*. Unsurprisingly, as the number of cities increases, the number of times an algorithm achieves the optimal result decreases; however, no one algorithm dominates in this case

avoiding both local optima and early convergence. This would allow these algorithms to continue searching the space, while the basic GA would get stuck early.

This demonstrated phenomenon led to the idea of studying multiple algorithm variations to ensure a thorough analysis of the search space.

4   Algorithms

This section discusses the algorithms used, including the heuristics and a detailed explanation of the set of genetic algorithm variations used. In addition, this section includes a description of the representations used.

4.1   2-Opt

The 2-Opt algorithm was developed as an effective heuristic for the TSP but can be applied to many similar problems. The success of 2-Opt has led to the implementation of 3-Opt, the generalization into K-Opt, and other algorithms including the dynamic approach of Lin–Kernighan. The basic idea behind 2-Opt is to take a particular location within a permutation and reverse the subpermutation following the specified

Fig. 4 Illustration of 2-Opt [15]. In this example, half of the circle's ordering is removed and then replaced in the reverse order

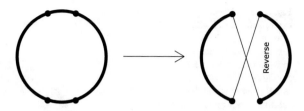

location. This has the effect of crossing the permutation on itself. This strategy can create all possible permutations. Figure 4 depicts this process.

4.2 Lin–Kernighan

The Lin–Kernighan heuristic (LKH) was developed for the TSP and has been empirically shown to be one of the most effective and efficient algorithms for this particular problem [15]. It also can be extended to other similar problems. The LKH involves applying multiple K-opts—a generalization of 2-opts—iteratively while intelligently determining the value of K for each step.

There have been many variations of the LKH, but K. Helsgaun's implementation consistently produces high quality results [15]. The LKH currently holds the record for all instances with a known optimum used in the *DIMACS TSP Implementation Challenge* [10, 15].

4.3 Genetic Algorithm Variations

A collection of Genetic Algorithm (GA) variations were selected for use: Recentering-Restarting GA (RRGA), Island Model GA (IM), and a GA which uses ring species (RS). The RRGA was used on all problems, and the IM and RS were used for the TSP. These variations were used in combination for the TSP to better explore the search space to maximize the significance of results.

4.3.1 Ring Species

The motivation for ring species (RS) in evolutionary algorithms is based on a natural phenomenon: the differentiation of a population when expanding around geographical barriers. These populations tend to develop differences, and fertility rates drop as the distance between two populations increases until some threshold distance is met at which the two populations cannot produce offspring [2].

Fig. 5 Ring Species
example for $d = 3$ and
starting index 1. The *arrows*
represent and cover the
maximum distance which
candidate solution 1 can
travel to mate with other
candidate solutions. Since
the population is thought of
as a ring, and because the
distance is 3, candidate
solution 1 has the
opportunity to mate with
candidate solutions 22, 23,
24, 2, 3, and 4

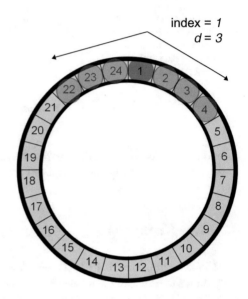

A simpler variation of ring species can be implemented where the population is treated as a ring. Consider the first and last members of the population as adjacent, with mating only allowed to occur between chromosomes that fall within some parametrized distance d of each other. This idea is demonstrated in Fig. 5 where chromosome 1 can only mate with chromosomes within a distance of 3, namely chromosomes 22, 23, 24, 2, 3, and 4.

4.3.2 Island Model

The Island Model (IM), developed by Whitley, is a popular evolutionary algorithm variation. It has been used successfully for many problems and different types of evolutionary algorithms including Genetic Algorithms and Genetic Programming [29, 30].

The IM is a parallelized approach that evolves multiple, isolated, subpopulations (islands) concurrently to encourage genetic diversity. Each subpopulation is allowed to traverse the search space along its own trajectory without being affected by the other subpopulations until periodic migrations occur between islands. These migrations promote unique offspring by combining very genetically different parents. Figure 6 depicts the subpopulations and migrations.

4.3.3 Recentering-Restarting Genetic Algorithm

The Recentering-Restarting Genetic Algorithm (RRGA) has been applied to epidemic networks [1], Side Effect Machines for decoding [19], DNA Fragment Assem-

Fig. 6 Example of 3
subpopulations evolving
separately in parallel.
Migrations happen
periodically between
subpopulations to encourage
genetic diversity. The
amount of time required
before a migration can be
arbitrarily defined

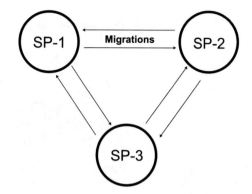

bly [21], and the travelling salesman problem [20]. In all of these, the RRGA was
able to produce high quality significant results.

One of the RRGA's major strengths is its tendency to avoid fixating on local
optima—a shortcoming of many metaheuristics. The RRGA provides a sequential
traversal through the search space for a global optimum. This sequential search, along
with the nature of the required dynamic representation (see Sect. 4.4 for a description
of the representations) allows the algorithm to avoid local optima.

Before the algorithm can begin, a centre must be selected. This centre is typically
a direct representation of a potential solution to the problem. This centre can be randomly selected, or more ideally, with some heuristic; the RRGA excels at polishing
high quality solutions.

Once the centre is selected, the population can be created. Depending on the
representation used, some alterations may be required. Two different representations
are studied in this chapter: a direct representation and an indirect transposition-based
representation (see Sect. 4.4). For the direct representation, a string of n randomly
generated transpositions are applied to the centre to generate each member of the
population. For the indirect representation, strings of n randomly generated transpositions are simply placed into the population to be used as the chromosomes. The
value of n is parametrized at initialization and is used to generate a population similar
to, but at a distance from, the centre. Seeding a GA with a high quality solution for
a problem sometimes results in stagnation as no further learning can be found due
to fixating on a local optimum. The RRGA avoids this due to the nature in which
the population is derived from the centre; the population is dynamically altered as
restarts occur (Figs. 7 and 8).

After the population is created, the evolutionary search is run. An additional step is
required when using the transposition representation to translate the chromosome into
a candidate solution that the fitness function can evaluate. This translation takes the n
transpositions of the chromosome and applies them to the current centre to generate a
permutation which can be analyzed. Figure 9 depicts the translation process in detail.

Once the GA completes a run, the best chromosome in the final population is
compared to the current centre. If this chromosome has a better fitness value than

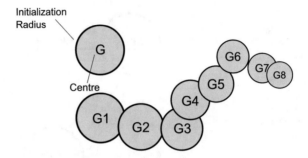

Fig. 7 G represents a typical initialization area and G1 through G8 show an example traversal of the search space by the RRGA where an increase in number signifies a restart transition. As restarts occur, the centre is being updated to the best candidate solution found. In this example, the search radius is decreased on each improvement from a restart

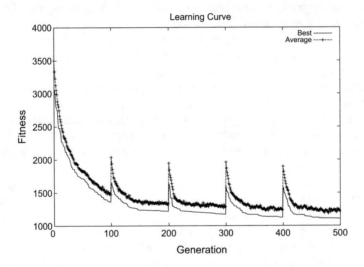

Fig. 8 Typical RRGA learning curve. High peaks of poor fitness occur when the algorithm recenters, since the system has yet to have an opportunity to improve with the new population. The overall trend of the curve is an improvement in fitness

Fig. 9 Demonstration of the translation algorithm required to translate a candidate solution with the transposition representation into something that can be evaluated. This algorithm takes a chromosome of neighbouring transpositions and applies them to some centre

the current centre, then it will become the new centre and the whole process repeats but with the number of transpositions n increased (or decreased). If the best chromosome's fitness does not improve upon the current centre's fitness, then the current centre is left unchanged, the final population is discarded, and the process is repeated with the number of transpositions altered in the opposite way than if the centre had been replaced. In other words, if the number of transpositions was increased when the centre was replaced, then the number of transpositions would decrease if the centre was left unchanged, and vice versa. One determines at initialization through a parameter whether n is increased or decreased on an improvement. Once these alterations are made, the GA will be run again, and the whole process will repeat multiple times. By altering the centre and the n transpositions used, the search will be different from its previous run allowing for a very dynamic traversal through the fitness landscape with the intention of finding the global optimum.

The number of restarts that are used for an actual run can be defined in two different ways. Either there is a fixed, static number of restarts or the number of restarts is set to be dynamic: restart after y generations of no learning.

One may choose to alter the n transpositions after each completion of the GA by a fixed value. For example, if there is an improvement, then decrease the number of transpositions by 5, and if no improvement, then increase by 10. In this approach, one needs to be careful to avoid using only *even* or only *odd* permutations [5]. This requirement comes from the fact that compositions of two even permutations always results in an even permutation, the composition of two odd permutations always results in an even permutation, and the composition of an even and an odd permutation always results in an odd permutation [22]. If one were to ignore this fact, then not all permutations would be explored by the search algorithm resulting in a detrimental loss of potential high quality results. An alternative approach is to use a percentage paradigm, for example multiply the number of transpositions n by 0.95 on an improvement or by 1.10 with no improvement. This approach will result in different compositions of even and odd permutations on every restart, eliminating the concerns of the fixed approach.

Due to the dynamic nature of this algorithm, the number of transpositions used may increase or decrease rapidly. To curb this problem, upper and lower limits are set for the number of transpositions. If either limit is passed, the number of transpositions is reset to the initial value. The limits can be set to $1/10th$ and $10*$ the initial value.

Figure 7 depicts this traversal through the search space. In this particular example, the number of transpositions used for each GA run decreases as better solutions are found, resulting in the radius of the circles decreasing as the algorithm continues through evolution.

Figure 8 depicts a typical learning curve for the Recentering-Restarting Genetic Algorithm.

The Recentering-Restarting Genetic Algorithm process can be seen in Algorithm 1.

Algorithm 1 Recentering-Restarting Genetic Algorithm [19]

Assign Random Centre σ
Set Flag
while Stopping Criteria is Not Met **do**
    if Flag is Set **then**
        Centre = σ
        Initialize Population Through use of Transpositions
    end if
    Run Genetic Algorithm
    if Solution τ is Better than σ **then**
        $\sigma = \tau$
        Set Flag
    else
        Reset Flag
    end if
    Increase/Decrease Number of Transpositions to be Generated
end while

A two point mutation was used for both representations. A single point crossover was used with the indirect representation due to its less destructive nature. Partially mapped crossover was used with the direct representation in order to maintain the integrity of the chromosomes; a city must exist once and only once within a chromosome.

4.4 Representation

Direct representations are frequently used in search algorithms for their ability to generate quality solutions; however, it is important to explore other representations in order to exploit the search space completely. The goal of these new representations would be to have different/unique characteristics while still maintaining the beneficial qualities provided by a direct representation. Two different types of representations will be used for all variations of the evolutionary algorithms: a direct representation and an indirect representation based on transpositions.

4.4.1 Direct Representation

The direct representation is simply an ordered list which represents a solution to the problem. Depending on which problem is being studied, this representation will alter slightly. For the TSP, a chromosome is a permutation which represents the order in which to visit the cities. For the BPP, the permutation represents an order in which to pack the items into bins with the first fit rule. The GCP's direct representation is the order in which to start colouring the vertices.

4.4.2 Indirect Transposition Representation

The indirect transposition representation is an ordered string of transpositions that exchange *neighbouring* elements of the centre. With this representation, the chromosome is an ordered list of transpositions, and the genetic operators only alter the order of the transpositions and never directly alter the order of a candidate solution. This approach always exchanges neighbouring elements meaning this representation is formed from a set of local search operators; by adjusting the number of transpositions, one can alter the degree of locality of the search space.

For a complete mathematical description of the indirect transposition representation refer to [18].

5 Experimental Design

This section covers the details regarding the experimental design.

A large parameter sweep was performed for each problem and algorithm to help ensure a thorough analysis. Different values for the parameters were allowed on the unique problems. These values were determined empirically through preliminary testing. All experiments were run 30 times for statistical significance.

Parametrized values include the following: crossover rate, mutation rate, population size, number of restarts, number of transpositions, how to alter transpositions on restart, maximum distance for ring species, number of subpopulations for the island model, and the number of generations between migrations for the island model. Note that the number of generations is dependent on the population size ($Generations = x/PopSize$). This is to ensure a consistent number of mating events to give a fair analysis between algorithms and sets of system parameters.

Although a through analysis was performed with a larger suite of statistical tests, they were not included in this chapter for space considerations. The extensive set of results can be found in [16].

5.1 Bin Packing Problem

The purpose of studying the BPP is twofold: firstly, to analyze the effectiveness of the RRGA and the representations on this type of grouping permutation problem, and secondly, to demonstrate the areas where the RRGA excels over a conventional GA. A focus is being given to the latter. It is important to note that there are more ideal representations that have been shown to produce very high quality results with basic GAs. These representations are not utilized as they make it difficult to evaluate the pros and cons of the RRGA.

Two distinct data sets were used. The first set, titled *u1000*, was generated by Falkenauer [12] and was used in studies exploring grouping representations and appropriate genetic operators [12]. This data set contains 20 unique instances and was selected as it supplies a large range of problem instances that have multiple properties making them very inconsistent. The second data set is called the *hard28*. These 28 data sets were created by Schoenfield [28] and are known for their small size and high level of difficulty [6].

The dynamic restart paradigm was used for this problem with a fixed number of transpositions added or removed on a restart.

All GAs were seeded with either the identity permutation or a relatively good solution obtained with 2-Opt.

All results are analyzed in multiple ways. The mean, max, and standard deviations are recorded for each run and are used to analyze the consistency of the results within the 30 runs. These values are also used to determine which, if any, system parameters performed best, and to compare the results of the two representations. A paired t-test is used to determine if there is any statistical difference between the results obtained from the two representations. We pay special attention to the effect the seeds will have on the results of the RRGA versus the basic GA. The full set of results can be found in [17].

5.2 Graph Colouring Problem

The purpose of studying the GCP is very similar to the BPP; study the effectiveness, strengths, and weaknesses of the RRGA on this problem. Note that once again, the genetic operators discussed by Falkenauer [12] would likely produce the best quality results. Using this approach would defeat the purpose of this study as it would simply produce near optimal results making it difficult to analyze the effectiveness of the RRGA.

The data sets can be found at [14]. Selecting the data sets for the GCP was a complicated task as there is a wide range of difficulty. The goal was to use a group of data sets that were considered highly difficult but would run in a reasonable amount of time. One of the requirements for the selection of a problem instance was that the RRGA variation would not easily solve it in the majority of runs. Dozens of challenging instances were analyzed, but in the end, 18 of the most difficult problems were selected with the requirement that they would typically complete 30 runs in less than 20 h.

The dynamic restart paradigm and the factor approach to initializing the number of transpositions were used for the GCP with the percentage alteration of the number of transpositions. The identity permutation and a solution obtained by 2-Opt were used as seeds for all GAs.

The analysis performed on this problem is the same as the BPP. The full set of results can be found in [17].

5.3   Travelling Salesman Problem

The travelling salesman problem is revisited, studying large problem instances as opposed to a collection of small city problems.

A total of 16 benchmark data sets were selected for the large TSP problem instances. These problem instances were obtained from the DIMACS TSP implementation challenge [10] but can ultimately be obtained via TSPlib [27]. Similar to the GCP instances selected, these 16 were chosen for their difficulty in combination with reasonable computation time. All of the benchmarks selected have between 1000 and 2000 cities. These 16 benchmarks also have a large collection of very competitive results along with suspected optima which can be used to analyze the quality of the results obtained by the RRGA. Another reason for selecting many of these particular instances is that the reported results on the DIMACS TSP implementation's results page show that the LKH algorithm does not find the suspected optimal solution [10].

Once again, the system parameters used for the TSP are very similar to those for BPP and GCP. Three variations of RRGAs are used in studying this problem: RRGA, RRGA using the Island Model (RRGA + IM), and RRGA using ring species (RRGA + RS). There are some algorithm-specific system parameters. Variations of these algorithms using no restarts are implemented to study the non-Recentering-Restarting equivalents.

The dynamic restart approach and the percentage alteration of transpositions were used for the TSP instances. In addition, three seeds were used for this problem: identity permutation, 2-Opt solution, and LKH solution.

The analysis for the TSP is similar to those of the BPP and GCP.

The results are compared to those of other algorithms and the optimal solutions that are available at [10]. The optimal solutions were determined using the Concorde TSP Solver [8].

Unlike the previous experiments, the number of generations used was varied as a system parameter. This was to determine the effect of the number of generations on the RRGA variations.

Lastly, 2-Opt will be used as a postoptimization technique to determine how well a combination of heuristics and metaheuristics will perform. These results will be compared to the results obtained with no postoptimization technique.

The TSP experiments also include ring species and Island Model variations of the RRGA to expand the scope of the study due to the possibility that the problem instances might be too difficult for any one GA variation. Although this may not be the case, the small problem TSP instances previously studied demonstrate the necessity of using multiple search strategies [18, 20].

6   Results and Discussion

This section discusses the results obtained by all experiments performed and also includes summary tables of the best and average best results obtained by each algorithm. These tables are brief summaries and should not be assumed to carry any statistical significance. Although algorithms may achieve the same best and/or best average result, this does not indicate identical results in all cases. This section mentions comprehensive tables and plots of all experiments available in [17].

6.1   Bin Packing Problem

Tables 1 and 2 show the best results obtained by the basic GA and the RRGA for both representations for each problem instance, respectively. The RRGA is seen to perform reasonably well on the u1000 instances while performing very well on the hard28 instances, often achieving the optimum. Additionally, it is seen that the RRGA does prove to have an edge over the basic GA with both representations.

The overall results of the RRGA variations on the BPP instances are competitive. There are multiple cases where the algorithms obtained the optimum on the hard28 instances, and overall, the majority of the results were extremely close to optimal. In almost all cases, the best result found was seeded with 2-Opt.

When doing a quick analysis of all the results, one can see that the RRGA outperforms the basic GA and that the direct representation tends to perform better on average; however, these trends cannot clearly be observed in the summary table of the best performing results; Tables 1 and 2. Paired t-tests with 95% confidence were performed to compare both representations for each set of system parameters. For the u instances, there was a lot of inconsistency; some were statistically different where many were not. The hard28 problem instances' results had fewer inconsistencies, and in general, the direct representation outperformed the indirect representation. The summarized results in Table 2 and the full results in [17] show the direct representation performing marginally better; however, this trend has no statistical significance.

When studying the plotted results seen in [17], it is observed that as the likelihood of more restarts increased, so did the quality of the solutions. Many of the plots demonstrate this trend, but there are some cases where too many restarts did not benefit performance. In all cases, restarts provided an increase in performance or at the very least a tie with the basic GA. This trend appears to be amplified when considering the indirect representation.

In many runs, it appears that as the likelihood of more restarts increased the standard deviation and the range of the 95% confidence interval decreased.

On the hard28 instances, the basic GA performs better when starting with the random seed as opposed to the 2-Opt seed. This seems unusual as 2-Opt produces such a good starting point. A reasonable explanation is that the 2-Opt solution starts

Table 1 Summary of best and best average results found by the basic GA for BPP instances

Data set	Theo. opt.	Best direct	Avg. direct	Best trans.	Avg. trans.
u1000_00	399	413	415.5	414	416.3
u1000_01	406	420	422.7	422	423.5
u1000_02	411	425	427.4	427	428.4
u1000_03	411	425	428.3	428	429.3
u1000_04	397	411	414.0	413	414.2
u1000_05	399	413	415.7	414	416.0
u1000_06	395	408	410.7	409	410.6
u1000_07	404	417	419.5	419	420.3
u1000_08	399	412	414.8	415	417.1
u1000_09	397	411	413.8	413	414.8
u1000_10	400	413	416.0	415	416.5
u1000_11	401	416	418.5	418	419.6
u1000_12	393	405	407.7	407	408.4
u1000_13	396	409	412.0	410	411.2
u1000_14	394	409	411.5	411	412.3
u1000_15	402	416	418.7	419	420.2
u1000_16	404	417	419.3	419	420.3
u1000_17	404	417	419.9	420	421.0
u1000_18	399	413	415.1	415	417.1
u1000_19	400	413	415.7	415	416.5
h1	62	62	62.0	62	62
h2	60	61	61.0	61	61
h3	59	60	60	60	60
h4	62	63	63.0	63	63
h5	58	59	59.0	59	59
h6	64	65	65	65	65
h7	62	63	63.0	63	63
h8	63	64	64	64	64
h9	67	68	68.0	68	68
h10	64	65	65.0	65	65
h11	67	68	68.0	68	68
h12	68	69	69.0	69	69
h13	71	72	72.0	72	72
h14	72	73	73.0	73	73
h15	76	76	76.0	76	76
h16	71	72	72.0	72	72
h17	72	75	75.0	75	75
h18	76	76	76.0	76	76
h19	71	77	77.0	77	77

(continued)

Table 1 (continued)

Data set	Theo. opt.	Best direct	Avg. direct	Best trans.	Avg. trans.
h20	74	74	74.0	74	74
h21	76	73	73.0	73	73
h22	77	72	72.0	72	72
h23	75	76	76.0	76	76
h24	84	84	84.0	84	84
h25	80	81	81.0	81	81
h26	80	81	81.0	81	81
h27	83	84	84.0	84	84
h28	81	82	82.0	82	82

Table 2 Summary of best and best average results found by the RRGA for BPP instances

Data set	Theo. opt.	Best direct	Avg. direct	Best trans.	Avg. trans.
u1000_00	399	412	413.8	412	414.2
u1000_01	406	419	421	420	421.1
u1000_02	411	425	426	425	426.4
u1000_03	411	425	426.7	425	426.8
u1000_04	397	410	412	411	412.3
u1000_05	399	412	413.4	412	413.1
u1000_06	395	407	408.1	408	408.8
u1000_07	404	417	418.3	418	419
u1000_08	399	411	413.1	412	413.4
u1000_09	397	411	411.6	410	411.5
u1000_10	400	413	414.1	413	414.7
u1000_11	401	415	416.3	415	416.7
u1000_12	393	405	406.2	405	406.4
u1000_13	396	408	409.4	408	409.5
u1000_14	394	408	409.5	408	409.7
u1000_15	402	416	417.4	416	417.6
u1000_16	404	417	417.8	417	418
u1000_17	404	417	418.5	418	418.8
u1000_18	399	412	413.3	412	413.4
u1000_19	400	413	413.9	413	414.5
h1	62	62	62	62	62
h2	60	61	61	61	61
h3	59	60	60	60	60

(continued)

Table 2 (continued)

Data set	Theo. opt.	Best direct	Avg. direct	Best trans.	Avg. trans.
h4	62	63	63	63	63
h5	58	59	59	59	59
h6	64	65	65	65	65
h7	62	63	63	63	63
h8	63	64	64	64	64
h9	67	68	68	68	68
h10	64	65	65	65	65
h11	67	68	68	68	68
h12	68	69	69	69	69
h13	71	72	72	72	72
h14	72	73	73	73	73
h15	76	76	76	76	76
h16	71	72	72	72	72
h17	72	75	75	75	75
h18	76	76	76	76	76
h19	71	77	77	77	77
h20	74	74	74	74	74
h21	76	73	73	73	73
h22	77	72	72	72	72
h23	75	76	76	76	76
h24	84	84	84	84	84
h25	80	81	81	81	81
h26	80	81	81	81	81
h27	83	84	84	84	84
h28	81	82	82	82	82

the GA at a local optimum from which the basic GA cannot escape. This problem seems to disappear as restarts are introduced, which is one of the benefits of the RRGA. Many of the results obtained using restarts tie or surpass the results obtained when starting with a random seed.

These results demonstrate the quality of the results obtained by the RRGA. In addition, these results show the effectiveness of the restarts at avoiding stagnation on a local optimum.

6.2   Graph Colouring Problem

Tables 3 and 4 shows the best results obtained by the basic GA and the RRGA, respectively.

Overall, the algorithms were able to obtain strong results; however, it is difficult to determine how well these algorithms performed as many of the experiments are run on data with no known optimal solution. It is expected that results obtained by using the grouping representation and the accompanying genetic operators [11] would be of higher quality; however, one of the purposes of analyzing the GCP is to study the qualities and behaviours of the RRGA relative to a simple GA; maximizing results is secondary.

Unusually, when analyzing Table 4, one can even see that the best and best average results (to one decimal place) obtained by each representation with the RRGA are tied on all instances. In general, when studying the full results in [17], the difference between the results from the two representations on the GCP appears to be smaller than the difference between the results of the representations on the BPP. Statistical tests showed that in almost all of the cases, there was no significant difference between the results.

Table 3   Summary of best and best average results obtained by the basic GA on the GCP instances

Data set	Opt. chromatic number	Best direct	Avg. direct	Best trans.	Avg. trans.
DSJC125.1	5	6	7.1	7	7.1
DSJC125.5	17	20	21.8	21	23.2
DSJC125.9	44	44	49.1	48	49.3
DSJC250.1	8[a]	10	11.8	11	11.8
DSJC250.5	28[a]	35	40.5	37	39.7
DSJC250.9	72	81	89.3	87	90.7
DSJC500.1	12[a]	17	18.1	18	18.1
DSJC500.5	47[a]	64	70.5	66	68.2
DSJC1000.1	20[a]	28	30.1	29	29.8
flat300_28_0	28	39	44.5	42	43.5
queen10_10	11	12	13.4	12	13.4
queen11_11	11	13	16.1	14	15.5
queen12_12	12	15	16.5	15	18.3
queen13_13	13	16	17.8	16	17.5
queen14_14	14	17	19.7	17	20.4
queen15_15	15	19	20.7	19	20.1
r250.5	65	68	70.9	70	71.4

[a] Upper Bound

Table 4 Summary of best and best average results obtained by the RRGA on the GCP instances

Data set	Opt. chromatic number	Best direct	Avg. direct	Best trans.	Avg. trans.
DSJC125.1	5	6	6	6	6
DSJC125.5	17	19	20.1	19	20.1
DSJC125.9	44	44	45.6	44	45.6
DSJC250.1	8[a]	10	10.8	10	10.8
DSJC250.5	28[a]	35	36.2	35	36.2
DSJC250.9	72	80	82.1	80	82.1
DSJC500.1	12[a]	17	17	17	17
DSJC500.5	47[a]	64	65.3	64	65.3
DSJC1000.1	20[a]	28	28.8	28	28.8
flat300_28_0	28	39	40.6	39	40.6
queen10_10	11	12	12.4	12	12.4
queen11_11	11	13	13.9	13	13.9
queen12_12	12	14	15	14	15
queen13_13	13	16	16.4	16	16.4
queen14_14	14	17	17.9	17	17.9
queen15_15	15	18	19	18	19
r250.5	65	67	68.2	67	68.2

[a]Upper Bound

Many of the observations made when analyzing the BPP can also be made when studying the GCP. The RRGA outperforms the basic GA and the direct representation appears to perform best by a very small margin. As the likelihood of the number of restarts increases so does the quality of results obtained. In many of these cases, the differences are statistically significant, and there is minimal to no overlap between confidence intervals. This trend appears to be amplified when observing the results of the indirect transposition representation. Similar to the BPP, there are cases where too large a number of restarts does not benefit the quality of results. Even though the transposition representation does not perform as well as the direct representation, the actual difference between these results is very small.

The results for the GCP demonstrate that the RRGA variants surpass the quality of results obtainable by a basic GA. The results also demonstrate the degree to which an increased number of restarts positively affects the performance of the indirect transposition representation.

6.3   Travelling Salesman Problem

Tables 5 and 6 show the best results obtained by all studied algorithms for each problem instance. Note that these tables include results for the RRGA when coupled with ring species and the island model. No summary of the basic GA results are presented in this chapter; however, once again, the full results can be seen in [17].

The LKH algorithm was used as one of the seeds for these experiments to demonstrate the effectiveness of the LKH algorithm for these permutation problems, but experiments using this as the seed were halted, and 2-Opt was used instead. Unlike the results shown in the DIMACS TSP Implementation Challenge results page, the LKH algorithm does in fact achieve the suspected optimum on all problem instances selected. This was unexpected as part of the reason these instances were selected because the LKH was not supposed to be able to find the optimum. As one would expect, when these values were used as seeds no learning occurred.

The effect of the initial seed on the results for these large TSP instances is substantial. 2-Opt has an extremely large benefit on every result. Surprisingly, this trend is reversed when 2-Opt was used as postoptimization, i.e. using a random initial seed produces better final results when 2-Opt is used as a postoptimization technique. It is clear that postoptimization substantially improves results.

In comparing the best performing results, one can observe that these are close to their respective optima and comparable to results obtained by many other competitive algorithms [10]. This is true when considering both the results with no postoptimization and those with postoptimization.

Unlike the BPP and GCP, in most cases, the value of the standard deviation was 0. With few outliers, in many of the few situations where this value was not 0, the standard deviations were relatively small, leading to small confidence intervals.

When no postoptimization was performed, it is clear that the effect of restarts on the overall results is minimal. Dissimilar to the BPP and GCP results, the transposition representation results tend to outperform the direct representation. A reasonable explanation for this might be that since these datasets are larger and arguably more complex, the transposition representation should perform well because it is built upon local search operations. Additionally, it is difficult to say which version of the RRGA actually performed best as the results are similar with the majority of cases having no statistically significant differences (as seen in [17]). Although in several cases the RRGA and RRGA + RS with the transposition representation out performed the others, this trend is not consistently seen on all instances. In a number of cases, it can be observed that the performance of the RRGA transposition representation deteriorates as the likelihood of more restarts increases.

Results that were postoptimized are substantially more consistent than those without postoptimization.

No one variation of the algorithm dominates on all large TSP instances, or even one instance. This corresponds to the observations made about the small TSP instances which suggests that no one algorithm can dominate on all instances within a problem domain [25].

Table 5 Summary of best and average results after postoptimization for TSP instances using direct representation. Bold entries are best performing. Bold entries carry into Table 6

Data set name	Suspected direct opt.	Best direct RRGA	Avg. direct RRGA	Best direct RRGA+RS	Avg. direct RRGA+RS	Best direct RRGA+IM	Avg. direct RRGA+IM
d1291	50801	57709	57709	57709	57709	57709	57709
d1655	62128	69235	70594	70594	70594	69559	70594
dsj1000	18659688	20384354	20883123.2	20309097	20866084.4	20335251	20890961.8
fl1400	20127	20939	21672	20850	21672	21044	21672
fl1577	22249	23964	23964	23964	23964	23964	23964
nrw1379	56638	62228	62317	62083	62317	62170	62317
pcb1173	56892	64116	64116	64116	64116	64116	64116
pr1002	259045	284844	284844	284844	284844	284844	284844
rl1304	252948	277989	287163	282632	287163	277583	287163
rl1323	270199	295860	307723.4	299083	307805	296089	307805
rl1889	316536	349719	361240.3	349152	361827.5	**348189**	361092.2
u1060	224094	248394	248394	248394	248394	248394	248394
u1432	152970	169575	169575	169575	169575	169575	**169575**
u1817	57201	65378	65378	65378	65378	65378	65378
vm1084	239297	**257655**	269165	260765	268455	262160	269407.3
vm1748	336556	367356	378619.5	370201	379006.9	368218	377854.4

Table 6 Summary of best and best average results after postoptimization for TSP instances using indirect representation. Bold entries are best performing. Bold entries carry from Table 5

Data set name	Suspected opt.	Best trans. RRGA	Avg. trans. RRGA	Best trans. RRGA+RS	Avg. trans. RRGA+RS	Best trans. RRGA+IM	Avg. trans. RRGA+IM
d1291	50801	**55005**	**57020.2**	55683	57164	55163	57226.2
d1655	62128	**68522**	**69464.8**	69068	70390.9	68938	70481.8
dsj1000	18659688	**20270323**	**20575104.8**	20390987	20856183.6	20366489	20868276.8
fl1400	20127	20864	**21524.7**	**20831**	21672	20885	21665.2
fl1577	22249	**23254**	**23568.2**	23964	23964	23753	23940.8
nrw1379	56638	**61587**	**62155.6**	61905	62317	62011	62317
pcb1173	56892	62264	63631.4	**62096**	**63496.8**	62428	63583.4
pr1002	259045	**280255**	**284600.7**	280826	284844	280271	284844
rl1304	252948	**273670**	**280076.9**	276012	287163	274838	287066.2
rl1323	270199	**294922**	**301065.3**	297374	307805	**294536**	306510.7
rl1889	316536	349806	355775.6	348947	360834	348719	**360070.2**
u1060	224094	**242114**	246192	245609	247701.5	244570	247921
u1432	152970	**167730**	**169773.4**	168591	**169575**	167974	**169575**
u1817	57201	**63819**	**64617.8**	65378	65378	64758	65350.7
vm1084	239297	259815	**266687.2**	260550	268797.4	258388	267986.6
vm1748	336556	**366519**	**372609.5**	369669	378293.1	370116	379045.5

7 Conclusions

Genetic algorithm variations were studied (Recentering-Restarting, ring species, and island model) in combination with each other with success. These algorithms were also combined with heuristics and other local search algorithms with significant results. These algorithms worked very well when combined with each other as they were able to take advantage of their respective strengths while avoiding their short-comings. An example of this is the strong performance of the RRGA when 2-Opt was used as an initial seed: heuristics can produce high quality solutions but often fall into local optima; by using the RRGA, which is known to be very effective at remov-ing itself from local optima, the heuristic solutions could be greatly improved upon. In addition to this specific benefit, in many cases, the RRGA was observed to pro-duce much better solutions than conventional heuristics, conventional metaheuristics, and even highly specialized heuristics and metaheuristics. When the Recentering-Restarting idea was implemented in combination with these other approaches, the quality of results was improved.

Two representations were analyzed: a direct representation and an indirect trans-position representation. Both of these approaches worked well in general, but there appeared to be an interesting and important trend. When working in any search space, the direct representation always performed relatively well but seemed to excel over the transposition representation when the starting location within the search space was randomly generated with no guarantee of the quality of the initial population. In contrast to the direct representation, the indirect transposition representation per-formed very well in very large search spaces and appeared to outperform the direct representation when the starting location was already of high quality. An example of this is the experiments conducted that used heuristics to seed the algorithm. This observation is not surprising as the transposition representation is built upon local search operators allowing it to search in more than one way at a time.

It was demonstrated that all algorithms performed reasonably well and became even more competitive when combined with one another. This supports the idea that if one wants to study a specific problem then one should not simply take one specific algorithm, tune it, and perform a parameter sweep. By doing this, one would severely limit the quality of results and miss the opportunity to draw significant conclusions. To avoid this problem, many algorithms should be studied and implemented. By combining multiple approaches, it is possible to thoroughly traverse a search space and achieve better results. This becomes particularly important when studying sensi-tive problems which have major real world implications. Scientific progress could be substantially inhibited if scientists were to limit themselves with this one algorithm approach.

Throughout the study, it was observed that the small trends being detected in the data rapidly disappeared as more problem instances, and more system parameters were studied. This phenomenon appeared on all three problems studied. This sug-gests that the minor trends were insignificant and dependent on the specific problem instances. This problem-specific observation corresponds to the one made in [25].

This further emphasizes the need to thoroughly explore our algorithms and use them together to find higher quality results.

Acknowledgements This research was supported in part by the Natural Sciences and Engineering Research Council of Canada.

References

1. Ashlock D, Lee C (2008) Characterization of extremal epidemic networks with diffusion characters. In: IEEE symposium on Computational Intelligence in Bioinformatics and Computational Biology, CIBCB'08. IEEE, pp 264–271
2. Ashlock D, McEachern A (2009) Ring optimization of side effect machines. Intell Eng Syst Artif Neural Netw 19:165–172
3. Ashlock D, Shiller, E (2011) Fitting contact networks to epidemic behavior with an evolutionary algorith. In: 2011 IEEE Symposium on Computational Intelligence in Bioinformatics and Computational Biology (CIBCB). IEEE, pp 1–8
4. Ashlock D, Warner E (2008) Classifying synthetic and biological dna sequences with side effect machines. In: IEEE Symposium on Computational Intelligence in Bioinformatics and Computational Biology, CIBCB'08. IEEE, pp 22–29
5. Barlow T (1972) An historical note on the parity of permutations. Am Math Month 766–769
6. Belov G, Scheithauer G (2006) A branch-and-cut-and-price algorithm for one-dimensional stock cutting and two-dimensional two-stage cutting. Eur J Oper Res 171(1):85–106
7. Brooks RL (1941) On colouring the nodes of a network. In: Mathematical proceedings of the cambridge philosophical society, vol 37. Cambridge Univ Press, pp 194–197
8. Concorde TSP Solver (2011). http://www.math.uwaterloo.ca/tsp/concorde/index.html. Accessed 19 Jan 2014
9. Cook W (2012) In pursuit of the traveling salesman: mathematics at the limits of computation. Princeton University Press
10. DIMACS TSP Challenge (2008). http://dimacs.rutgers.edu/Challenges/TSP/. Accessed 19 June 2014
11. Eiben ÁE, Van Der Hauw JK, van Hemert JI (1998) Graph coloring with adaptive evolutionary algorithms. J Heuris 4(1):25–46
12. Falkenauer E (1996) A hybrid grouping genetic algorithm for bin packing. J Heuris 2(1):5–30
13. Garey MR, Johnson DS, Stockmeyer L (1974) Some simplified np-complete problems. In: Proceedings of the sixth annual ACM symposium on theory of computing. ACM, pp 47–63
14. Gualandi S, Chiarandini M, Graph coloring benchmarks: vertex coloring. https://sites.google.com/site/graphcoloring/vertex-coloring. Accessed 2 Sept 2013
15. Helsgaun K (2000) An effective implementation of the lin-kernighan traveling salesman heuristic. Eur J Oper Res 126(1):106–130
16. Hughes JA (2012) Reanchoring, recentering & restarting an evolutionary algorithm. Undergraduate thesis, Brock University
17. Hughes JA (2014) A study of ordered gene problems featuring DNA error correction and DNA fragment assembly with a variety of heuristics, genetic algorithm variations, and dynamic representations. Master's thesis, Brock University
18. Hughes JA, Houghten S, Ashlock D (2014) Recentering and restarting a genetic algorithm using a generative representation for an ordered gene problem. Int J Hybrid Intell Syst 11(4):257–271
19. Hughes J, Brown JA, Houghten S, Ashlock D (2013) Edit metric decoding: representation strikes back. In: 2013 IEEE Congress on Evolutionary Computation (CEC). IEEE, pp 229–236
20. Hughes J, Houghten S, Ashlock D (2013) Recentering, reanchoring & restarting an evolutionary algorithm. In: 2013 World Congress on Nature and Biologically Inspired Computing (NaBIC). IEEE, pp 76–83

21. Hughes J, Houghten S, Mallen-Fullerton GM, Ashlock D (2014) Recentering and restarting genetic algorithm variations for DNA fragment assembly. In: 2014 IEEE conference on computational intelligence in bioinformatics and computational biology. IEEE, pp 1–8
22. Jacobson N (2012) Basic algebra I. Courier Dover Publications
23. Johnson DS, Garey MR (1979) Computers and intractability: a guide to the theory of NP-completeness. Freeman & Co, San Francisco, p 32
24. Johnson DS, McGeoch LA (2007) Experimental analysis of heuristics for the STSP. In: The traveling salesman problem and its variations. Springer, pp 369–443
25. Kanda J, Carvalho A, Hruschka E, Soares C (2011) Selection of algorithms to solve traveling salesman problems using meta-learning. Int J Hybrid Intell Syst 8(3):117–128
26. Martello S, Toth P (1990) Knapsack problems: algorithms and computer implementations. John Wiley & Sons, Inc
27. Reinelt G (2013) TSPlib. http://comopt.ifi.uni-heidelberg.de/software/TSPLIB95/. Accessed 13 Nov 2013
28. Schoenfield JE (2002) Fast, exact solution of open bin packing problems without linear programming. Draft, US Army Space & Missile Defense Command, p 45
29. Whitley D (1992) An executable model of a simple genetic algorithm. In: Foundations of generic algorithms, vol. 2, pp 45–62
30. Whitley D, Rana S, Heckendorn RB (1999) The island model genetic algorithm: on separability, population size and convergence. J Comput Inf Technol 7:33–48
31. Wolpert DH, Macready WG (1997) No free lunch theorems for optimization. IEEE Trans Evol Comput 1(1):67–82
32. Xia B, Tan Z (2010) Tighter bounds of the first fit algorithm for the bin-packing problem. Discret Appl Math 158(15):1668–1675

Hybridization of the Flower Pollination Algorithm—A Case Study in the Problem of Generating Healthy Nutritional Meals for Older Adults

Cristina Bianca Pop, Viorica Rozina Chifu, Ioan Salomie,
Dalma Szonja Racz and Razvan Mircea Bonta

Abstract This chapter investigates the hybridization of the state-of-the-art Flower Pollination Algorithm as a solution for improving its execution time and fitness value in the context of generating healthy nutritional meals for older adults. The proposed hybridization approach replaces the local and global pollination operations from the Flower Pollination Algorithm with Path Relinking-based strategies aiming to improve the quality of the current solution according to the global optimal solution or to the best neighbouring solution. We model the problem of generating healthy nutritional meals as an optimization problem which aims to find the optimal or near-optimal combination of food packages provided by different food providers for each of the meals of a day such that the nutritional, price, delivery time and food diversity constraints are met. To analyse the benefits of hybridization, we have comparatively evaluated the state-of-the-art Flower Pollination Algorithm, adapted to our problem of generating menu recommendations, versus the hybridized algorithm variant. Experiments have been performed in the context of a food ordering system experimental prototype using a large knowledge base of food packages developed in-house according to food recipes and standard nutritional information.

Keywords Flower Pollination Algorithm · Path relinking · Personalized menu recommendation · Meta-heuristic hybridization · Fitness function

C.B. Pop (✉) · V.R. Chifu · I. Salomie · D.S. Racz · R.M. Bonta
Computer Science Department, Technical University of Cluj-Napoca,
26-28 G. Baritiu Street, Cluj-Napoca, Romania
e-mail: Cristina.Pop@cs.utcluj.ro

V.R. Chifu
e-mail: Viorica.Chifu@cs.utcluj.ro

I. Salomie
e-mail: Ioan.Salomie@cs.utcluj.ro

© Springer International Publishing AG 2017
S. Patnaik et al. (eds.), *Nature-Inspired Computing and Optimization*,
Modeling and Optimization in Science and Technologies 10,
DOI 10.1007/978-3-319-50920-4_7

1 Introduction

Our daily life can be seen as a struggle to solve several optimization problems, whether we think of simple things such as finding the shortest route towards a destination, booking the cheapest holiday and organizing the daily programme or more complex things such as optimizing the energy consumption in our households to save money. For many of these problems, the current technologies offer us multiple solutions such as smart GPS-based devices for searching the shortest route, smart online systems for booking holidays, including transportation and accommodation, or smart home energy management systems, and many integrate heuristic algorithms to provide the most appropriate solutions. Searching for optimality continues to be the goal of many researchers as there still is place for improvement in the current state-of-the-art heuristics. Consequently, new sources of inspiration for designing heuristics have been searched, and one of the most promising is nature as it provides many metaphors for designing optimization algorithms since living organisms (e.g. animals, insects, plants) have been able to survive along time due to the efficient manner in which they search for food, reproduce, and adapt to the living environment. However, it is important to keep in mind that we can inspire from nature, but we should avoid copying it because the chances of success are limited as the flight pioneers have discovered in the early days [9]. As a result of inspiring from nature, many meta-heuristic algorithms have been proposed such as Ant Colony Optimization [4], Particle Swarm Optimization [13], Cuckoo Search [19], Honey Bees Mating Optimization [8] and others with successful applications in several domains. The search for better heuristic algorithms has not stopped here, and thus, the hybridization of nature-inspired meta-heuristics has emerged as a new research direction.

This chapter addresses the problem of hybridizing the state-of-the-art Flower Pollination nature-inspired meta-heuristic [22] with the Path Relinking search technique [7] in the context of generating personalized menu recommendations for older adults. The choice for this type of problem is motivated by the fact that many people nowadays are interested in adopting a healthy lifestyle which involves finding the most appropriate food to eat such that dietary, cost and personal preferences constraints are met. In particular, our solution addresses to the elders whose health could be seriously affected if their nutrition is not appropriate. The proposed solution consists of three steps. The first step implies gathering the older adult's personal health information, such as personal characteristics (weight, height, age, gender), information regarding the elder's health (previous diseases, chronic illness, family health history), personal dietary preferences (what types of food the elder likes or dislikes) and information regarding allergies towards certain types of foods in order to be able to automatically eliminate these from the generated recommendations. In the second step, a detailed user profile will be generated from the so far gathered information. This profile will include information such as the older adult's body mass index, weight diagnosis (underweight, normal weight, overweight or obese) and necessary daily caloric, nutrient, mineral and vitamin intakes. The recommended daily nutritional intake levels consist of the daily intake level of a nutrient that is considered to be sufficient to meet the requirements of 97–98% of healthy individuals. All of this information will be of help in developing the most appropriate dietary proposal. Then, in the third step, the proposed hybrid Flower Pollination-based Algorithm will be used to search for the combination of food packages that best satisfies the older adult's nutritional needs and the constraints imposed by personal preferences, health status and lifestyle habits. The search is performed in a large knowledge base of food packages provided by food providers which has been developed in-house.

This chapter is structured as follows. Section 2 introduces the theoretical background. Section 3 reviews the state-of-the-art approaches for generating healthy nutritional meals. Section 4 formally defines the problem of generating personalized menu recommendations as an optimization problem. Section 5 presents the Flower Pollination Algorithm's hybridization approach we proposed in the context of generating personalized menu recommendations. Section 6 discusses the experimental results, while Sect. 7 presents the chapter's conclusions.

2 Background

This section presents the theoretical background required for the development of the hybrid Flower Pollination Algorithm we proposed for generating personalized menu recommendations. Sect. 2.1 introduces the main concepts related to optimization problems. Sect. 2.2 presents an overview of meta-heuristics, focusing on describing the meta-heuristics that are used in this chapter.

2.1 Optimization Problems

Optimization problems involve searching for an optimal solution that minimizes or maximizes a set of given fitness functions under certain constraints. In what follows, we present the main concepts related to optimization problems definition as they have been introduced by Yang in [20, 21].

An optimization problem can be mathematically represented as follows:

$$OP = (S, F, C) \tag{1}$$

where S is the *set of candidate solutions*, F is the *set of objective functions* that need to be maximized or minimized, and C is a *set of constraints* that must be fulfilled.

The *set of candidate solutions*, S, is formally defined as follows:

$$S = \{x | x = (x_1, x_2, ..., x_n)\} \tag{2}$$

where x_i is a decision variable whose value can be real continuous, discrete or a combination of them. The domain of values of the decision variables forms the *search space* of the optimization problem. According to the value types of the decision variables, optimization problems can be classified in *discrete, continuous* and *mixed problems.*

The *set of objective functions, F*, is formally defined as follows:

$$F = \{f_1, f_2, ..., f_m\} \tag{3}$$

where f_i is a linear or nonlinear objective function that needs to be either maximized or minimized. The domain of values that the objective function values can take forms the *solution space* of the optimization problem. According to the number of objective functions, optimization problems can be classified in *single-objective* and *multi-objective problems.* Single-objective problems have only one objective function that needs to be minimized or maximized. Multi-objective (multi-criteria, multi-attributes) problems have multiple objective functions defined; some of these objective functions must be minimized, others maximized; and sometimes, they may be conflicting. There are five approaches to manage the multiple objective functions of a multi-objective optimization problem [5]: (1) ranking objectives according to their priority and turning the multi-objective optimization problem in a set of sequential single-objective problems, (2) associating a target (i.e. predefined threshold value) to each objective and aggregating the objective functions into a single function, (3) normalizing the objectives in the same interval and aggregating the objective functions into a single weighted function, (4) applying Pareto dominance, and (5) defining a minimal threshold for each objective and finding a solution that satisfies all these objectives.

The *set of constraints, C*, is formally defined as follows:

$$C = \{c_1, c_2, ..., c_p\} \tag{4}$$

where c_i is a constraint that must be satisfied by a solution of the optimization problem. According to the number of constraints, optimization problems can be classified in *constrained* and *unconstrained problems.*

2.2 Meta-Heuristic Algorithms

Optimization problems can be solved using *deterministic* and *stochastic* algorithms. Deterministic algorithms represent a class of algorithms that follow the same paths and provide the same solutions for the same initial state regardless the number of times and the moment in time when the algorithm is run [21]. As opposed to deterministic algorithms, stochastic algorithms such as meta-heuristic algorithms do not always follow the same path and do not have repeatable values when they are run multiple times for the same initial state, due to the randomness feature they incorporate [21]. The meta-heuristic algorithms are more suitable for solving optimization problems with large search spaces as they are able to provide an optimal or near-optimal solution in an acceptable execution time by processing the best solutions in each iteration and by avoiding stagnation in a local optimum through randomization [21]. As a result of inspiring from the processes encountered in nature which have ensured the survival of plants, insects and animals, a new class of meta-heuristics has emerged, namely the nature-inspired meta-heuristics.

To improve the performance of meta-heuristics, hybridization techniques have been introduced. These techniques combine (*i*) principles from two or more meta-heuristics, or (*ii*) meta-heuristics with cooperative search, or (*iii*) meta-heuristics with systematic methods [2].

In what follows, we briefly present the state-of-the-art meta-heuristics that we have used to develop a solution for generating personalized menu recommendations for older adults.

2.2.1 The Flower Pollination Algorithm

The Flower Pollination Algorithm [22, 23] is an optimization algorithm inspired by the flower pollination process encountered in nature. This process implies transferring the pollen from one flower to another either by insects, birds and animals (i.e. biotic pollination), or by wind and water diffusion (i.e. abiotic pollination). Some pollinators maintain the flower constancy as they prefer to pollinate only a specie of flowers, which leads to an intensification of this specie reproduction, process also called self-pollination. Other pollinators perform cross-pollination which implies transferring the pollen between different flower species. In the context of the state-of-the-art Flower Pollination Algorithm, the biological concepts are modelled as follows: (i) a flower is modelled as a solution of an optimization problem, (ii) biotic and cross-pollination are modelled as global pollination processes aiming to modify a solution according to the global optimal solution encountered so far, and (iii) abiotic and self-pollination are modelled as local pollination processes aiming to modify a solution based on two neighbour solutions.

The Flower Pollination Algorithm starts from an initial set of randomly generated solutions which are improved in an iterative process through global and local pollination-based strategies. In the global pollination-based strategy, a solution x_i is updated according to the following formula [22]:

$$x_i^{t+1} = x_i^t + L(x_i^t - g)$$ (5)

where x_i^{t+1} is the solution at the current iteration, x_i^t is the solution at the previous iteration, g is the global best solution encountered so far, and L is a parameter that represents the strength of the pollination.

In the local pollination-based strategy, a solution x_i is updated according to the following formula [22]:

$$x_i^{t+1} = x_i^t + \varepsilon(x_j^t - x_k^t)$$ (6)

where x_i^{t+1} is the solution at the current iteration, x_i^t is the solution at the previous iteration, ε is a constant value in the interval $[0, 1]$, and x_j^t and x_k^t are two solutions generated in the neighbourhood of the x_i^t solution.

The Flower Pollination Algorithm has been applied in various optimization problems such as linear antenna array optimization [16], economic load dispatch optimization [15] or fractal image compression [12].

2.2.2 Path Relinking Algorithm

Path Relinking [7] is a search technique which explores the paths between an initial solution and a guiding solution with a high fitness value. The main idea of this technique is to obtain the guiding solution by applying a set of modification strategies on the initial solution. In this way, each time when a modification strategy is applied on the initial solution, the obtained intermediate solution will be more similar with the guiding solution and less dissimilar with the initial one. By using this search technique, new paths between the initial and the guiding solution are generated, so that the solutions on these paths could be sources for generating new paths.

3 Literature Review

This section presents the state-of-the-art in the field of generating healthy lifestyle recommendations.

Merwe et al. propose in [14] an approach that combines an expert system with linear programming to generate eating plans for teenagers according to their nutritional requirements and food preferences. The main components of the expert system are as follows: (i) a user interface that is used for collecting information about users and for providing the generated eating plan, (ii) the working memory that translates the input data collected through the user interface into a specific format required by the inference engine, (iii) the inference engine that is able to produce an eating plan based on the data provided by the user, a set of production rules and a knowledge base and (iv) the knowledge base that contains relevant knowledge (i.e. list of foods grouped by the quantity of carbohydrate, protein and fat they contain) provided by nutritionists that is used for generating the eating plan. The inference engine integrates two interactive linear programming models, namely a goal programming model and a multi-objective linear programming model. The proposed approach returns the cheapest eating plan that satisfies the teenager's food preferences.

Chavez-Bosquez et al. propose in [3] an approach which combines mathematical programming with belief merging for planning menus such that the nutritionist's recommendation and the user food preferences are met as much as possible. Based on the user's caloric needs, a menu approximation for the meals of a day is generated using a linear programming-based approach by processing a list of foods. Then, a belief merging operator is applied upon the user's food preferences, the generated menu approximation and a set of nutritional rules resulting in the final menu.

Kale and Auti propose in [10] a decision tree-based approach for planning children food menus according to their health profile, food preferences and availability, as well as nutritional needs. In particular, the ID3 algorithm is used to generate a decision tree which will establish the food items that should be included in the menu. Once the food items are selected, their recommended quantities are computed based on nutritional constraints.

An ontology-based system for food and menu planning is proposed by Snae and Bruckner in [17]. The proposed system is composed of the following: (i) an ontology containing food and nutrition-related concepts such as ingredients, nutrients and recommended daily intakes and (ii) an expert system whose inference engine suggests meals and dishes using the knowledge stored in the ontology and the customer's food preferences.

Sivilai et al. propose in [18] a personalized nutrition and food planning system for older people. The system uses an inference engine to generate food menus per meal/day/week according to the user's personal health record and favourite/disliked food ingredients. The personal health record contains medical information such as weight, height, blood pressure, pulse rate or the chronic diseases the user is suffering of. Additionally, when generating menus, the inference engine takes into account the food and nutrition-related information stored in a knowledge base which comprises an ontology, a set of rules and conditions, and a database.

A decision support system for generating menu recommendations based on the rough set theory is proposed by Kashima et al. in [11]. The system generates menu recommendations according to the information collected from customers by means of questionnaires in which they must specify what they like/dislike from ten randomly selected types of menus. Based on this information, a preference rule is extracted and used for generating the customer's menu recommendation.

Bing et al. present in [1] a personalized recommender system based on multi-agents and the rough set theory. The proposed system consists of the following layers: (1) the viewer layer which enables the interaction between the learners and the system, (2) the control layer which transfers information from the user to the agent service layer, (3) the agent service layer which is composed of learner and recommender agents, the latter using the rough set theory to recommend learning resources based on the online behaviours of the user, (4) the lucene service layer which builds a personalized retrieval function to support recommendations, and (5) the data service layer which is responsible for storing data such as information about the learner's features, records of the learner's activities, learning resources.

Gaal et al. propose in [6] the use of genetic algorithms to design an automated menu generator for people suffering from cardiovascular diseases. A menu is generated such that its food items can be combined and the user's preferences are met. In this approach, an individual in the genetic algorithm is modelled as a single meal or as a daily menu consisting of meals with dishes associated. The individuals are evolved using the genetic crossover, mutation and selection operators and are evaluated using a penalty-based fitness function. The initial population consists of individuals that have been generated previously for other similar problems. Experiments have been performed on a commercial nutritional database for Hungarian lifestyle and cuisine which contains information about recipes and their ingredients and nutritional details.

4 Problem Definition

Given an older adult health and lifestyle profile, as well as his nutritionist recommended diet, our goal is to generate a personalized menu recommendation consisting of a set of food packages, one for each of the meals of a day, out of a large repository of food packages provided by different food providers. Due to the large number of available food packages, the number of personalized menu recommendations that can be generated based on them is also large and the identification of the most appropriate one requires a high computational time. Thus, the problem of generating these recommendations can be defined as an optimization problem which can be solved using heuristic approaches.

According to Formula 1, we formally define the problem of generating healthy menu recommendations for older adults as follows:

$$OP = (S, F, C) \tag{7}$$

where (*i*) S is the set of all possible menu recommendations for one day obtained by combining the food packages provided by a set of food providers, (*ii*) F is the set of fitness functions which evaluate a menu recommendation from the nutritionist's personalized diet, the price, the user culinary preferences and the delivery time perspectives, and (*iii*) C is the set of constraints referring to the allergies of the older adult that must be considered when selecting the food packages to be part of the menu recommendation. The following subsections detail each component of the formal definition of our optimization problem.

4.1 Search Space and Solution Representation

The search space consists of the entirety of available food packages provided by different food providers which can be combined to obtain menu recommendations. In our approach, we consider that a food package is a combination of food items corresponding to a meal of the day. Breakfast and snack packages will contain fitting foods. The lunch package will contain a starter dish (i.e. a soup), a main dish and possibly a side dish, while the dinner package will contain a main dish and possibly a side dish. The search space is represented as an n-dimensional space, where each food package is represented as a point with n coordinates. Each coordinate will indicate the value of one of the n nutritional components (e.g. carbohydrates and proteins) that will be taken into consideration. Thus, a food package represented as a point in

Table 1 Example of representation for a menu recommendation

Meal	Description	Representation
Breakfast	100 g of oatmeal bread with 10 g of butter	foodPack1 = (8.48, 48.506)
Snack 1	100 g of apple	foodPack2 = (0.26, 13.81)
Lunch	200 g of chicken soup with 200 g of cheese-filled ravioli	foodPack3 = (13.8, 28.78)
Snack 2	150 g of nonfat plain Greek yogurt with 50 g of corn flakes	foodPack4 = (18.54, 47.45)
Dinner	200 g of tuna salad	foodPack5 = (32.08, 18.82)

the search space is formally defined as follows:

$$foodPack = (nutrient_1, nutrient_2, ..., nutrient_n) \qquad (8)$$

where $nutrient_i$ is the total nutrient value corresponding to the food items part of the food package (e.g. the value of carbohydrates for a lunch package consisting of 200 g of chicken soup with 200 g of cheese-filled ravioli is 28.78 [24]), and n is the total number of nutrients that are taken into account.

A solution of our optimization problem represents a personalized menu recommendation, MR, for one day, formally defined as follows:

$$MR = \{foodPack_b, foodPack_{s1}, foodPack_l, foodPack_{s2}, foodPack_d\} \qquad (9)$$

where $foodPack_b$, $foodPack_{s1}$, $foodPack_l$, $foodPack_{s2}$ and $foodPack_d$ represent the food packages for breakfast, the first snack, lunch, the second snack and dinner, respectively. A menu recommendation can also be represented as a point in an n-dimensional space, whose coordinates are computed by summing up the coordinates of the individual food packages part of the recommendation.

Table 1 illustrates an example of menu recommendation for one day; for simplicity, we consider only the proteins and carbohydrates nutrients for each food package part of the menu recommendation (the nutrients' values are taken from [24]). The menu recommendation from Table 1 can be represented as a point in a two-dimensional space as follows: $MR = (73.16, 157.366)$.

4.2 Fitness Function

We model the problem of generating healthy menu recommendations as a multi-objective optimization problem for which we define four objective functions:

- FF_{rec}—objective function evaluating how much the candidate personalized menu recommendation (i.e. solution) adheres to the nutritionist's recommended diet.

- FF_{price}—objective function evaluating the price of the food packages part of a menu recommendation.
- FF_{pref}—objective function evaluating how much a menu recommendation fulfils the older adult's preferences.
- FF_{time}—objective function evaluating the delivery time of the food packages part of a menu recommendation.

We manage these four objective functions by normalizing them in the same interval and by aggregating them into a single weighted function:

$$FF(sol) = w_1 * FF_{rec}(sol) + w_2 * FF_{price} + w_3 * FF_{pref} + w_4 * FF_{time} \qquad (10)$$

where (i) *sol* is the personalized menu recommendation being evaluated; (ii) FF_{rec}, FF_{price}, FF_{pref} and FF_{time} are the fitness functions associated with each considered objective; and (iii) w_1, w_2, w_3 and w_4 are weights that specify the importance of each of the objectives, such that $w_1 + w_2 + w_3 + w_4 = 1$. The values of the weights are established by the older adult who decides the order of importance of the nutritionist's recommended diet, food price, food preferences and food delivery time.

In what follows, we detail each objective function considered.

The FF_{rec} objective function. FF_{rec} is defined as the sum of deviations of the menu recommendation from the nutritionist's recommended diet for each meal, scaled for each nutritional component:

$$FF_{rec}(sol) = \sum_{foodPack \in sol} \sum_{i=1}^{noNC} \alpha_i * NCD_i(foodPack) \qquad (11)$$

where *noNC* is the number of nutritional components considered, α_i is a weight indicating the importance of a nutritional component whose value is established by the nutritionist such that $\alpha_1 + \alpha_2 + \cdots + \alpha_n = 1$, and NCD_i is a function that evaluates the difference between the optimal value (i.e. the value recommended by the nutritionist) of a nutritional component *i* and the current value of the nutritional component *i* associated with the food package *foodPack*.

The function NCD_i is defined as follows:

$$NCD_i(foodPack) = \frac{abs(optimal_i - \sum_{foodItems \in foodPack} NCV_i(foodItem))}{optimal_i} \qquad (12)$$

where $optimal_i$ is the optimal value (i.e. the value recommended by the nutritionist) of the nutritional component *i*, and NCV_i computes the actual value of a nutritional component corresponding to a food item part of the evaluated food package.

The FF_{rec} objective function takes the 0 value if the candidate menu recommendation has the nutritional values recommended by the nutritionist. The aim is to minimize this objective function.

The FF_{price} objective function. FF_{price} evaluates the total price of the food packages part of the menu recommendation *sol* as follows:

$$FF_{price}(sol) = \frac{\sum_{foodPack \in sol} Price(foodPack) - preferedPrice}{preferedPrice} \quad (13)$$

where *Price* evaluates the price of a food package, and *preferredPrice* is the maximum price preferred by the user. The FF_{price} objective function takes the following values:

- 0—if the total price of the food packages' part of the evaluated menu recommendation is equal to the older adult's preferred maximum price;
- negative values—if the total price is lower than the older adult's preferred maximum price;
- positive values—if the total price is above the older adult's preferred maximum price.

The aim is to minimize this objective function.

The FF_{pref} objective function. FF_{pref} evaluates the number of food items that are preferred by the older adult and that are found among the menu recommendation's food packages:

$$FF_{pref}(sol) = -NumberOfPreferredFoodItems(sol, preferredFoodItems) \quad (14)$$

where *preferredFoodItems* represents the set of food items that are preferred by the older adult, and the minus sign is used to obtain negative values in case many food items part of the menu recommendation are also food items preferred by the older adult, as the aim is to minimize the FF_{pref} objective function.

The FF_{time} objective function. FF_{time} evaluates the total time required for preparing and delivering the food packages part of a menu recommendation *sol* as follows:

$$FF_{price}(sol) = \frac{\sum_{foodPack \in sol} Time(foodPack) - preferedTime}{preferedTime} \quad (15)$$

where *Time* evaluates the time of a food package computed as the sum of its preparation and delivery time, and *preferredTime* is the maximum time the older adult is willing to wait for the food.

4.3 Constraints

In our approach, the problem of generating menu recommendations is modelled as a constrained problem in which the constraints are related to the allergies the older adult might have. The older adult must specify in his personal profile whether he suffers from any allergies to certain groups of food products. Then, based on this information, all the food packages that contain any food items belonging to food groups that the older adult is allergic to will automatically be excluded from the search space. Once this constraint is imposed, we ensure the fact that no possible menu recommendation will contain any food items that the older adult is allergic to.

5 Hybridizing the Flower Pollination Algorithm for Generating Personalized Menu Recommendations

This section presents our approach for hybridizing the state-of-the-art Flower Pollination Algorithm in the context of generating personalized menu recommendations for older adults.

5.1 Hybrid Flower Pollination-Based Model

We define a Hybrid Flower Pollination-based Model which will be used in the design of the Hybrid Flower Pollination-based Algorithm for generating healthy menu recommendations for older adults. The model consists of the Flower Pollination-based core component and the Path Relinking-based hybridization component.

5.1.1 Flower Pollination-Based Core Component

The Flower Pollination-based core component is defined by mapping the concepts from the state-of-the-art Flower Pollination Algorithm to the concepts of the problem of generating personalized menu recommendations for older adults as follows: (i) a flower (i.e. solution) becomes a personalized menu recommendation defined using Formula 9, (ii) the pollen grain becomes a new menu recommendation obtained by applying a modification strategy on a menu recommendation, (iii) the local and global pollination processes are modelled as strategies based on the Path Relinking meta-heuristic, (iv) the flower characteristics become the nutritional, price and delivery time-related information associated with a menu recommendation, and (v) the evaluation of a flower's attributes is the evaluation of a menu recommendation performed with the fitness function defined in Formula 10.

5.1.2 Path Relinking-Based Hybridization Component

The aim of the Path Relinking-based hybridization component is to improve the search capabilities of the state-of-the-art Flower Pollination Algorithm using a strategy based on the Path Relinking meta-heuristic which explores the paths connecting

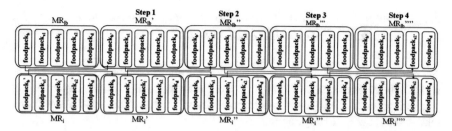

Fig. 1 An example of applying the path relinking-based strategy between two menu recommendations

the current solution with the local best solution also named guiding solution. Thus, in our approach, we aim to obtain the local best menu recommendation by applying a crossover-based modification strategy on the currently processed menu recommendation. This way, new paths are generated between the current and the local best menu recommendations from which the best one is selected for further processing. Figure 1 shows an illustrative example of applying the Path Relinking-based strategy between two menu recommendations.

As it is shown in Fig. 1, the current menu recommendation, MR_i, is updated until the local best menu recommendation, MR_{lb}, is obtained by iteratively performing a one-point crossover between the two menu recommendations. More exactly, the following steps are performed:

- Initial solution $MR_i = (foodPack'_b, foodPack'_{s1}, foodPack'_l, foodPack'_{s2}, foodPack'_d)$, guiding solution $MR_{lb} = (foodPack_b, foodPack_{s1}, foodPack_l, foodPack_{s2}, foodPack_d)$
- First Step—Changing first elements between the two recommendations. The obtained recommendations are as follows: $MR'_i = (foodPack_b, foodPack'_{s1}, foodPack'_l, foodPack'_{s2}, foodPack'_d)$, $MR'_{lb} = (foodPack'_b, foodPack_{s1}, foodPack_l, foodPack_{s2}, foodPack_d)$.
- Second Step—Changing the second elements between the two recommendations MR'_i and MR'_{lb}. The obtained recommendations are as follows: $MR''_i = (foodPack_b, foodPack_{s1}, foodPack'_l, foodPack'_{s2}, foodPack'_d)$, $MR''_{lb} = (foodPack'_b, foodPack'_{s1}, foodPack_l, foodPack_{s2}, foodPack_d)$.
- Third Step—Changing the third elements between the recommendations MR'''_i and MR_{lb}''. The obtained recommendations are follows: $MR'''_i = (foodPack_b, foodPack_{s1}, foodPack_l, foodPack'_{s2}, foodPack'_d)$, $MR'''_{lb} = (foodPack'_b, foodPack'_{s1}, foodPack'_l, foodPack_{s2}, foodPack_d)$.
- Fourth Step—Changing the fourth elements between the two recommendations MR''''_i and MR''''_{lb}. The obtained recommendations are follows: $MR'''_i = (foodPack_b, foodPack_{s1}, foodPack_l, foodPack_{s2}, foodPack'_d)$, $MR'''_{lb} = (foodPack'_b, foodPack'_{s1}, foodPack'_l, foodPack'_{s2}, foodPack_d)$.

5.2 Flower Pollination-Based Algorithms for Generating Personalized Menu Recommendations

We propose two Flower Pollination-based Algorithms, one which adapts and one which hybridizes the state-of-the-art Flower Pollination Algorithm in the context of generating personalized menu recommendations for older adults. Both algorithms take as input the following parameters:

- *UserProfile*—the personal profile of an older adult containing the following: (i) *personal information* including the age, gender, weight, height, the geographical area, where the older adult is living, and information regarding the physical activity performed by the older adult; (ii) *health information* including the diseases and allergies the older adult is suffering from; (iii) *price constraint* referring to the maximum amount of money allocated for food per day; (iv) *time constraint* referring to the maximum time the older adult is willing to wait for food delivery; and (v) importance given to factors such as the amount of money spent and the time needed to wait for the delivery of the food;
- *PopSize*—the population size;
- *NrIterations*—the maximum number of iterations;
- *DietRecom*—the personalized diet recommendation established by a nutritionist according to the older adult's profile;
- *FoodOffers*—the set of food packages offered by food providers;

Both algorithms return the optimal or a near-optimal personalized menu recommendation and consist of an initialization stage followed by an iterative stage. The initialization stage for the two algorithm variants is the same and includes the following steps:

- Generating the initial population of menu recommendations based on the food packages offered by food providers, the older adult's profile (i.e. personal profile), the older adult's preferences regarding price, delivery time and the nutritionist's diet recommendation.
- Identifying the optimal menu recommendation among the set of recommendations in the initial population.
- Generating a random probability between 0 and 1, which will decide for each iteration and for each menu recommendation whether local or global pollination-based processes will be performed.

The iterative stage is performed until a stopping condition (i.e. a predefined number of iterations) is satisfied. Since the iterative stage is different for the two approaches, they will be discussed separately in the following subsections.

5.2.1 The Iterative Stage of the Classical Flower Pollination-Based Algorithm for Generating Personalized Menu Recommendations

In the iterative stage of the Classical Flower Pollination-based Algorithm, at each iteration, for each menu recommendation, MR_i, in the population, the following operations are performed:

- A vector, L, of 0 s and 1 s having a length equal to the length of a menu recommendation (i.e. the length is given by the number of food packages in the recommendation) is generated;
- A random number in the interval [0,1] is generated, and based on this number, one of global or local pollination-based processes is performed upon MR_i resulting in a new recommendation MR'_i;
- If the new menu recommendation MR'_i resulting from global/local pollination has a better fitness than the current MR_i recommendation, then MR'_i will replace MR_i in the population;
- At the end of each iteration, the menu recommendation with the best fitness is identified among the set of recommendations in the population and the global optimum is updated.

In what follows, we detail how we have interpreted the local and global pollination processes from the state-of-the-art Flower Pollination Algorithm in the context of generating personalized menu recommendations. In both processes, we consider that the menu recommendations are represented as points in an n-dimensional space (see Sect. 4).

Global Pollination. The global pollination process is applied between the currently processed menu recommendation, MR_i, and the best menu recommendation obtained so far, MR_g, according to Formula 5 adapted in the context of generating menu recommendations.

First, a subtraction is made between the elements of MR_g and the elements of MR_i, and whenever the two elements are identical, a random element is generated. For example, the coordinates of the breakfast food package belonging to MR_g are subtracted from the coordinates of the breakfast food package belonging to MR_i (the same operation is performed for the snacks, lunch and dinner food packages). The resulting menu recommendation MR'_i is then multiplied with the vector L of 0 s and 1 s. For each element in the vector equal to 0, the corresponding element from MR'_i is replaced by the coordinates of a randomly chosen valid food package.

Second, an addition is made between the elements of MR_i and MR'_i. The addition will once again be performed element by element using the food package representation as a point in our n-dimensional space. The addition will be performed among the corresponding nutritional values as shown in the example below:

$$MR_i(P_b, P_{s1}, P_l, P_{s2}, P_d) + MR'_i(P'_b, P'_{s1}, P'_l, P'_{s2}, P'_d) = MR_r(P''_b, P''_{s1}, P''_l, P''_{s2}, P''_d)$$

$$(16)$$

where each P'' is obtained by adding, element by element, the nutritional values corresponding to the breakfast/snacks/lunch/dinner food packages corresponding to MR_i and MR_i'. However, since simply adding the coordinates will result in increasingly large values that would most likely be in conflict with the daily intake recommendation, we suggest first multiplying the coordinates with two adjusting weights aw_1 and aw_2 (see Formula 17) computed from the proximity of the menu recommendations to the ideal menu recommendation (the point representing the nutritionist's recommendation). That is, if we consider the addition of points P_1 and P_2 reported to the optimal solution P_{opt}, we start by computing the distances d_1 and d_2 from P_1 to P_{opt} and P_2 to P_{opt}, respectively. Then, the adjusting weights aw_1 and aw_2 will have the following form:

$$aw_1 = \frac{d_2}{d_1 + d_2}; \; aw_2 = \frac{d_1}{d_1 + d_2} \tag{17}$$

In conclusion, the result of this adapted form of addition will be as follows:

$$MR_r(aw_1 * P_b + aw_2 * P_b', aw_1 * P_{s1} + aw_2 * P_{s2}', aw_1 * P_l + aw_2 * P_l', aw_1 * P_{s2} + \\ aw_2 * P_{s2}', aw_1 * P_d + aw_2 * P_d') \tag{18}$$

All of the results of the operations made on these points in the search space will have to be converted back to the form of a food package. The closest package that corresponds to the same meal of the day will be considered to be the actual result of the operation (by measuring the distance from the point obtained to the point matching the representation of the given package in the search space).

Local Pollination. The local pollination process is applied between the current menu recommendation MR_i and the best neighbour of MR_i according to Formula 6 adapted in the context of generating menu recommendations. The subtraction and multiplication with a vector are performed in the same manner as in the case of the global pollination process.

5.3   The Iterative Stage of the Hybrid Flower Pollination-Based Algorithm for Generating Healthy Menu Recommendations

In the iterative stage of the Hybrid Flower Pollination-based Algorithm (see Algorithm 1), at each iteration, for each menu recommendation MR_i in the population, the following operations are performed:

- A random number in the interval [0,1] is generated;
- If the randomly generated number is higher than a probability p, then the Path Relinking-based strategy (see line 10) presented in Sect. 5.1.2 is applied between the current processed menu recommendation, MR_i, and the best menu recommendation found so far, MR_g; otherwise, the Path Relinking-based strategy is applied between MR_i and the best menu recommendation generated in the neighbourhood of MR_i (see line 12).
- If the fitness of the newly obtained menu recommendation is better than MR_g, then MR_g is updated. If not, a one-point mutation is performed on the newly obtained menu recommendation by replacing the worst food package with a randomly generated one. If this mutated menu recommendation is better than MR_g, it will replace it.
- The newly resulted menu recommendation will replace the initial one in the population.

Algorithm 1: Hybrid_Flower_Pollination_based_Algorithm

---

1 **Input:** *UserProfile, PopSize, NrIterations, DietRecom, FoodOffers*
2 **Output:** sol_{opt}
3 **begin**
4   Pop = **Randomly_Generate_Pop**(*PopSize, FoodOffers, UserProfile, DietRecom*)
5   sol_{opt} = **Get_Best_Solution**(*Pop*)
6   p = **Generate_Random_Probab**(0, 1)
7   **while** (*stopping condition is not satisfied*) **do**
8   **foreach** *flower* **in** *Pop* **do**
9   **if**(**RandomNumber**() $< p$)**then**
10   *newFlower* = **Path_Relinking**(*flower*, sol_{opt})
11   **else**
12   *newFlower* = **Path_Relinking**(*flower*, **Get_Best_Neighbor**(*flower*))
13   **if**(**GetFitness**(*newFlower*) $<$ **GetFitness**(sol_{opt}))**then**
14   sol_{opt} = *newFlower*
15   **else**
16   **Mutate_on_Worst_Index**(*newFlower*)
17   **if** (**GetFitness**(*newFlower*) $<$ **GetFitness**(sol_{opt}))**then**
18   sol_{opt} = *newFlower*
19   **end if**
20   **endfor**
21   **Replace_Flower**(*Pop, flower, newFlower*)
22   **end while**
23   **return** sol_{opt}
24 **end**

---

6 Performance Evaluation

This section presents the experimental prototype used to evaluate the Classical and Hybrid Flower Pollination-based algorithms for generating menu recommendations for older adults and discusses the experimental results.

6.1 *Experimental Prototype*

To evaluate the Classical and Hybrid Flower Pollination-based algorithms, we have developed the experimental prototype of a system (see Fig. 2) dedicated to older adults for ordering cooked food daily.

The experimental prototype is composed of the following modules and resources: *Ontology-driven Graphical User Interface*, *Personal Profile Generator*, *Food and User Ontology* and *Flower Pollination-Based Healthy Meal Generator*.

The *Ontology-driven Graphical User Interface* guides the older adult in specifying the food menu request and the constraints related to delivery time or cost. Additionally, this module allows a nutritionist to introduce information about the medical history and personalized diet recommendations of the older adults registered in the system.

The *Personal Profile Generator* generates the personal profile of an older adult, based on the information introduced by the older adult or by the nutritionist using the *Ontology-Driven Graphical User Interface*. The information contained in the personal profile of an older adult is stored in the *Food and User Ontology*.

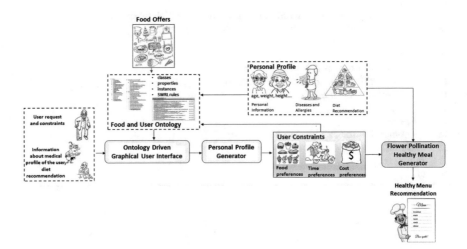

Fig. 2 Conceptual architecture of the experimental prototype

Fig. 3 A fragment of the Food and User Ontology

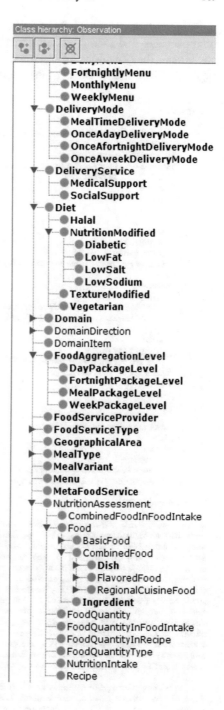

The *Food and User Ontology* contains information about food, food recipes, food packages and food providers as well as relevant information about the older adults using the system, such as personal information and health-related information. Figure 3 illustrates a fragment of the Food and User Ontology.

```
<recipe>
  <head>
    <title>1-Pot: Cheesy Turkey Casserole</title>
    <categories>
      <cat>Poultry etc</cat>
      <cat>Casseroles</cat></categories>
    <yield>4</yield></head>
  <ingredients>
    <ing-div>
      <title/>
      <ing>
        <amt>
          <qty>3/4</qty>
          <unit>cups</unit></amt>
        <item>Dry bread crumbs</item></ing>
      <ing>
        <amt>
          <qty>2</qty>
          <unit>tablespoons</unit></amt>
        <item>Parmesan, freshly grated</item></ing>
      <ing>
        <amt>
          <qty>1/4</qty>
          <unit>teaspoons</unit></amt>
        <item>Salt</item></ing>
      <ing>
        <amt>
          <qty>1/4</qty>
          <unit>teaspoons</unit></amt>
        <item>Pepper</item></ing>
      <ing>
        <amt>
          <qty>1/4</qty>
          <unit>teaspoons</unit></amt>
        <item>Dried basil</item></ing>
        ...............
    </ingredients>
    <directions>
<step> Eggplant Tomato Sauce: Peel and chop eggplant. In heavy saucepan,
heat oil over medium heat; cook eggplant, onion and garlic, stirring, for
about 5 minutes or until softened. Add tomatoes, mashing with fork, basil,
sugar,  oregano and pepper; bring to boil. Reduce heat; simmer, stirring
occasionally, for about 15 minutes or until thickened.In bowl, combine bread
crumbs, Parmesan, salt, pepper and basil. In separate bowl, whisk egg whites.
cut turkey into serving-size portions; dip into egg whites, then into crumb
mixture to coat on both sides. In large nonstick skillet, heat 1 ts each of
the oil and butter over medium-high heat; cook turkey in batches, adding
remaining butter and oil as necessary, for 4 minutes or until browned on both
sides. Pour sauce into shallow 10-cup casserole dish; arrange turkey in single
layer over sauce. Sprinkle with cheese. Cover with foil. Bake, covered, in 350F
180C oven for about 25 minutes or until turkey is tender and sauce is bubbly.
Remove foil; broil for 2 minutes or until cheese is bubbly.
.........
Per Serving: about 440 calories, 39 g protein, 15 g fat, 39 g carbohydrate,
high source fibre, excellent source iron, calcium
</step></directions></recipe></recipeml>
```

Fig. 4 Fragment of the recipe for 1-Pot Cheesy Turkey Casserole written in Recipe ML [25]

Rules:						
~~uivHasMaxironNeed(?d, 40.0)~~						
PersonalData(?p), hasDailyIntakeValues(?p, ?d), hasGender(?p, "male"^^string), hasAge(?p, ?a), greaterThanOrEqual(?a, 18), lessThan(?a, 50) -> divHasIronNeed(?d, 8.0)						
PersonalData(?p), hasDailyIntakeValues(?p, ?d) -> divHasVitaminBNeed(?d, 1.3)						
PersonalData(?p), hasAnthropometricMeasurements(?p, ?a), hasDailyIntakeValues(?p, ?d), hasWeight(?a, ?w), multiply(?rez, ?w, 0.75) -> divHasAvgProteinNeed(?d, ?rez)						
PersonalData(?p), hasDailyIntakeValues(?p, ?d), hasWeightDiagnostic(?p, Overweight), divHasBasicEnergyNeed(?d, ?be), multiply(?m, ?be, 0.15), subtract(?rez, ?be, ?m) -> divHasFinalEnergyNeed(?d, ?rez)						
PersonalData(?p), hasAnthropometricMeasurements(?p, ?a), hasBMI(?d, ?bmi), lessThan(?bmi, 18.5) -> hasWeightDiagnostic(?p, Underweight)						
PersonalData(?p), hasDailyIntakeValues(?p, ?d), hasGender(?p, "male"^^string), hasAge(?p, ?a), greaterThanOrEqual(?a, 14), lessThan(?a, 18) -> divHasIronNeed(?d, 11.0)						
PersonalData(?p), hasDailyIntakeValues(?p, ?d), hasGender(?p, "female"^^string) -> divHasVitaminANeed(?d, 700.0)						
PersonalData(?p), hasDailyIntakeValues(?p, ?d), hasAge(?p, ?a), greaterThanOrEqual(?a, 1), lessThan(?a, 70) -> divHasVitaminDNeed(?d, 15.0)						
PersonalData(?p), hasDailyIntakeValues(?p, ?d), hasGender(?p, "male"^^string) -> divHasVitaminANeed(?d, 900.0)						
PersonalData(?p), hasDailyIntakeValues(?p, ?d), hasAge(?p, ?a), greaterThanOrEqual(?a, 3),						

Fig. 5 A fragment of SWRL rules defined in the SWRL rules editor tab

We have used a repository of recipes [25] written in the Recipe ML language to define the instances of the *Recipe* concept. These recipes contain information about: (i) the name of the food recipe, (ii) optionally, the categories to which the food recipe belongs (e.g. snacks, meat), (iii) the ingredients, where for each ingredient the quantity and the unit of measure is specified, (iv) the recipe text and (v) the nutritional values of the food recipe. Figure 4 illustrates a fragment of such a food recipe for 1-Pot Cheesy Turkey Casserole.

The main reason behind choosing to use ontologies was their reasoning capabilities. In order to accomplish this, we have defined SWRL rules directly in the ontology. Figure 5 illustrates some examples of SWRL rules for computing the nutritional needs and body mass index for older adults.

The *Flower Pollination Healthy Meal Generator* generates personalized menu recommendations for older adults using either the Classical or Hybrid Flower Pollination-based optimization algorithms as well as the *Food and User Ontology*.

6.2　Test Scenarios

The Classical and Hybrid Flower Pollination-based algorithms have been tested on a set of older adults' profiles suffering from different diseases. In Table 2, we present the personal profiles of two older adults containing information about their age, gender, weight and height, geographical area in which they live, the physical activity they perform, diseases or allergies, money allocated for food per day, maximum meal expectation time and importance given to factors such as the amount of money they can spend on food and the time willing to wait for the delivery of the food.

Table 2 Examples of older adults profiles

Value	Older adult 1	Older adult 2
Name	John Smith	Mary Jenkins
Gender	Male	Female
Age	61	68
Weight	79	62
Height	180	167
Physical activity factor	1.2 (Sedentary)	1.375 (Mild activity)
Diseases	Knees arthritis	–
Allergies	Dairy products	–
Price/day (Euro)	50	45
Delivery time (min)	60	40
Ranks (from most important to least important)	Time, price	Time, price

Table 3 Daily nutritional need computed for the older adults profiles from Table 2

Value	Older Adult 1	Older Adult 2
Proteins (g)	49.2–172.19	42.57–149
Lipids (g)	98.39–172.19	85.14–154.96
Carbohydrates (g)	221.39–319.78	191.57–276.71
Energy (kcal)	1967.87	1702.84
Calcium (mg)	1200	1200
Iron (mg)	8	8
Sodium (mg)	1500	1500
Vitamin A (g)	900	700
Vitamin D (g)	15	15
Vitamin C (mg)	90	75
Vitamin B6 (mg)	6.095	6.095

Using the profiles presented in Table 2 and the defined SWRL rules, the older adults' nutritional need is computed (see Table 3). These values will be used to analyse the results and outcomes given by the two algorithms.

6.3 Setting the Optimal Values of the Algorithms' Adjustable Parameters

Both Classical and Hybrid Flower Pollination-based algorithms have the same adjustable parameters, namely the population size (*PopSize*) and the maximum number of iterations (*NrIterations*). To set the optimal values of the adjustable parameters, a trial-and-error strategy was adopted. In what follows, we present fragments of the best experimental results obtained while tuning the values of the adjustable parameters of both algorithms for the older adults profiles presented in Table 2. For the user profiles from Table 2, we have considered the values illustrated in Tables 4–5 for the weights associated with the fitness function components (see Sect. 4). The experiments have been performed on a number of 2643 food packages which are distributed as follows: 101 food packages for breakfast, 2001 packages for lunch, 429 packages for dinner and 112 packages for snacks. Hence, this gives a total number of possible combinations of $101 * 112 * 2001 * 112 * 429 = 1.087.581.470.976$. Tables 6, 7, 8, and 9 present a fragment of the best experimental results obtained while varying the values of the adjustable parameters for the considered older adult profiles (each table row represents an average value of the results obtained while running the algorithm for 100 times).

The first conclusion that we can draw by analysing the experimental results in the case of the Classical Flower Pollination-based Algorithm (see Tables 6 and 7) is the fact that increasing the population size is a much better option than increasing the number of iterations. For each number of iterations (e.g. 20, 30, 40, 50), when increasing the population size, the fitness function decreases in most of the cases. The plot in Fig. 6 shows a downward slope for each group of iteration numbers, meaning that the fitness function is always decreased. On the other hand, for each group of population sizes (15, 25, 35, 50), keeping the population size constant and

Table 4 Values for the weights associated to the fitness function components

Fitness function component	Weight value
FF_{rec}	$w_1 = 0.5$
FF_{price}	$w_2 = 0.25$
FF_{pref}	$w_3 = 0.1$
FF_{time}	$w_4 = 0.15$

Table 5 Nutritional components and the values associated to their weights in the FF_{rec} objective function

Nutritional component	Weight value
Proteins	$\alpha_1 = 0.12$
Lipids	$\alpha_2 = 0.02$
Carbohydrates	$\alpha_3 = 0.12$
Energy	$\alpha_4 = 0.3$
Calcium	$\alpha_5 = 0.08$
Iron	$\alpha_6 = 0.08$
Sodium	$\alpha_7 = 0.08$
Vitamin A	$\alpha_8 = 0.05$
Vitamin D	$\alpha_9 = 0.05$
Vitamin C	$\alpha_{10} = 0.05$
Vitamin B6	$\alpha_{11} = 0.05$

Table 6 Experimental results for the older adult profile 1 when running the Classical Flower Pollination-based Algorithm

Population size	Number of iterations	Average time	Average fitness
15	20	16.77	1.89
25	20	18.76	1.83
35	20	27.92	1.8
50	20	49.11	1.7
15	30	22.03	1.92
25	30	30.66	1.83
35	30	42.41	1.75
50	30	49.94	1.62
15	40	25.32	1.904
25	40	38.75	1.85
35	40	56.49	1.65
50	40	82.12	1.56
15	50	33.75	1.88
25	50	53.84	1.73
35	50	66.006	1.57
50	50	101.04	1.572

Table 7 Experimental results for the older adult profile 2 when running the Classical Flower Pollination-based Algorithm

Population size	Number of iterations	Average time	Average fitness
15	20	15.48	1.79
25	20	20.08	1.72
35	20	31.1	1.7
50	20	42.5	1.66
15	30	21.27	1.74
25	30	36.98	1.73
35	30	43.66	1.63
50	30	62.82	1.58
15	40	22.38	1.67
25	40	40.18	1.662
35	40	63.69	1.66
50	40	81.53	1.63
15	50	35.46	1.8
25	50	47.14	1.6
35	50	62.81	1.58
50	50	104.28	1.57

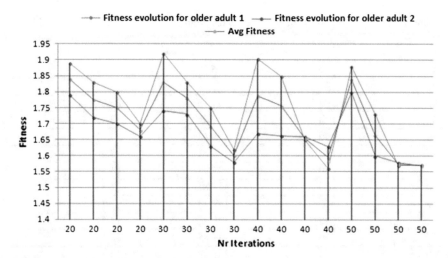

Fig. 6 Increasing population size while keeping the same number of iterations in the case of the Classical Flower Pollination-based Algorithm

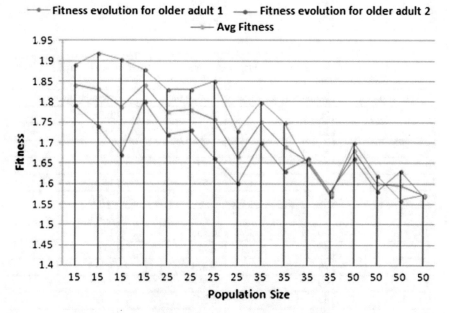

Fig. 7 Increasing the number of iterations while maintaining the same population size in the case of the Classical Flower Pollination-based Algorithm

increasing the number of iterations may lead to worse fitness function values as shown in Fig. 7. For example, for a constant population size of 15 menu recommendations, increasing the number of iterations from 40 to 50 leads to much worse results going from an average fitness of 1.78–1.84. Overall, the best results (i.e. good fitness value

Table 8 Experimental results for the older adult profile 1 when running the Hybrid Flower Pollination-based Algorithm

Population size	Number of iterations	Average time	Average fitness
15	20	5.704	1.72
25	20	7.27	1.61
35	20	14.13	1.45
50	20	12.95	1.51
15	30	9.61	1.45
25	30	8.81	1.66
35	30	17.47	1.36
50	30	17.82	1.53
15	40	9.15	1.542
25	40	10.72	1.52
35	40	11.85	1.506
50	40	15.97	1.5
15	50	6.57	1.492
25	50	18.64	1.442
35	50	20.782	1.434
50	50	26.892	1.39

Table 9 Experimental results for the older adult profile 2 when running the Hybrid Flower Pollination-based Algorithm

Population size	Number of iterations	Average time	Average fitness
15	20	5.59	1.55
25	20	7.22	1.64
35	20	7.0	1.525
50	20	11.47	1.5
15	30	5.67	1.72
25	30	9.52	1.578
35	30	10.258	1.606
50	30	14.45	1.51
15	40	6.87	1.64
25	40	8.528	1.53
35	40	14.25	1.618
50	40	16.56	1.49
15	50	7.77	1.57
25	50	11.48	1.49
35	50	14.35	1.406
50	50	20.142	1.48

in an acceptable execution time) obtained for the Classical Flower Pollination-based Algorithm were for a population size of 35 and a number of 40 iterations.

In the case of the Hybrid Flower Pollination-based Algorithm (see Tables 8 and 9), the best results (i.e. good fitness value in an acceptable execution time) were obtained for a population size of 35. Unlike in the case of the Classical Flower Pollination-based Algorithm, for the Hybrid Flower Pollination-based Algorithm, there is no clear pattern followed when keeping one of the two parameters constant and increasing the other one as shown in the charts from Figs. 8 and 9. However, the number of iterations also influences the end result, 30 and 40 being among the best candidates for this parameter.

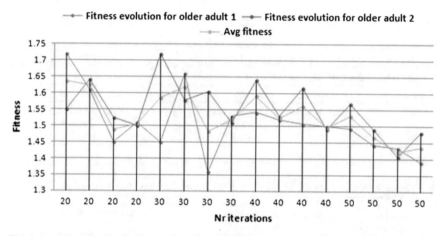

Fig. 8 Increasing population size while keeping the same number of iterations in the case of the Hybrid Flower Pollination-based Algorithm

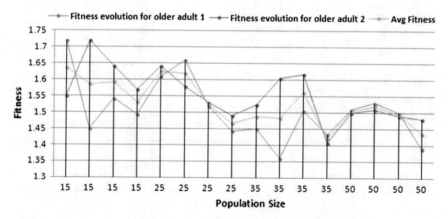

Fig. 9 Increasing the number of iterations while maintaining the same population size in the case of the Hybrid Flower Pollination-based Algorithm

Tables 10 and 11 present an example of the best personalized menu recommendations for the older adult profiles 1 and 2 generated with the Hybrid Flower Pollination-based Algorithm based on the available food packages, while Tables 12 and 13 present their personalized nutritional features. Additionally, the price and delivery time constraints for these menus are presented.

Table 10 Personalized menu recommendation for the older adult profile 1 when running the Hybrid Flower Pollination-based Algorithm

Breakfast	First snack	Lunch	Second snack	Dinner
Meat pie	Arancia genovese	Ark soup, Artichoke and Leek pizza	Apricot pear tart	Arroz con pollo Spanish

Table 11 Personalized menu recommendation for the older adult profile 2 when running the Hybrid Flower Pollination-based Algorithm

Breakfast	First snack	Lunch	Second snack	Dinner
Granola	Pear smoothie	Tomato soup, Rice Pilaf, Turkey Piccata	Apple smoothie	Aromatic green casserole

Table 12 Nutritional Features for the older adult profile 1 generated when running the Hybrid Flower Pollination-based Algorithm

Nutritional features and older adult preferences	Older adult needs	Generated value
Proteins (g)	49.2–172.1920	78.53
Lipids (g)	98.39–172.19	59.46
Carbohydrates (g)	221.39–319.78	200.12
Energy (kcal)	1967.87	1625.85
Calcium (mg)	1200	522.49
Iron (mg)	8	9.49
Sodium (mg)	1500	1815.27
Vitamin A (g)	900	566.24
Vitamin D (g)	15	2.92
Vitamin C (mg)	90	87.64
Vitamin B6 (mg)	6.095	0.84
Price	50	55.64
Delivery time	60	75

Table 13 Nutritional Features for the older adult profile 2 generated when running the Hybrid Flower Pollination-based Algorithm

Nutritional features and older adult preferences	Older adult needs	Generated value
Proteins (g)	42.57–149	76.45
Lipids (g)	85.14–154.96	63.24
Carbohydrates (g)	191.57–276.71	135.79
Energy (kcal)	1702.84	138.22
Calcium (mg)	1200	694.83
Iron (mg)	8	7.76
Sodium (mg)	1500	872.5
Vitamin A (g)	700	297.3
Vitamin D (g)	15	1.31
Vitamin C (mg)	75	65.14
Vitamin B6 (mg)	6.095	2.16
Price	45	51.47
Delivery time	40	45

6.4 Comparison Between the Classical and Hybrid Flower Pollination-Based Algorithms

After analysing the Classical and Hybrid Flower Pollination-based algorithms separately, we have made a comparison between them to evaluate the impact of hybridization. As far as execution time is concerned, hybridizing the Flower Pollination Algorithm with Path Relinking is a better choice as demonstrated by Fig. 10. In this figure, we illustrate how the average execution time varies in the case of both algorithms when increasing the population size. Thus, we calculated an average of 0.34-s execution time per iteration for the Hybrid Flower Pollination-based Algorithm. On the other hand, the Classical Flower Pollination-based Algorithm has an average execution time per iteration of 1.29, which is more than 3.5 times worse than the time previously mentioned. However, this difference in execution time between the two approaches derives from the fact that the Classical Flower Pollination-based Algorithm performs much more complex operations between two solutions (i.e. menu recommendations) at each iteration than does the hybridized algorithm variant.

As far as performance regarding fitness function values is concerned, the Hybrid Flower Pollination-based Algorithm produced slightly better results. Figures 11 and 12 show that regardless of increasing the population size or the number of iterations, the average fitness function values obtained for the hybrid algorithm were always better than the ones produced by the Classical Flower Pollination-based Algorithm. As opposed to the difference in execution times, the difference between the two algorithms is not so big as far as average fitness function values are concerned. The

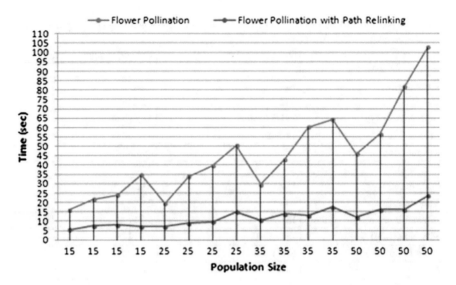

Fig. 10 Average execution time comparison between the Classical and Hybrid Flower Pollination-based Algorithms

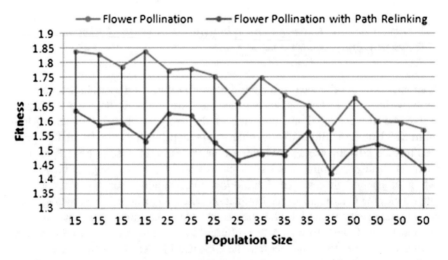

Fig. 11 Average fitness function values comparison when increasing the number of iterations

Classical Flower Pollination-based Algorithm produced an average fitness function value of 1.71, while the hybrid one yielded a value of 1.53, which is only 0.1 times better.

In conclusion, by running a series of tests on a set of older adults profiles and by varying the parameters for population size and iterations number, we have found that the Hybrid Flower Pollination-based algorithm yields better results in terms of both

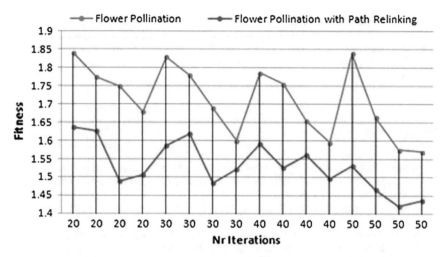

Fig. 12 Average fitness function values comparison when increasing the population size

execution time and fitness function value. While the difference between the quality of the solutions provided by the two algorithms is not that substantial, the difference in execution time is remarkable, considerably being in favour of the hybridization of the Classical Flower Pollination-based Algorithm with Path Relinking.

7 Conclusions

In this chapter, we have presented two algorithms for generating personalized menu recommendations for older adults, one adapting and the other hybridizing the state-of-the-art Flower Pollination Algorithm. The hybrid algorithm combines the Flower Pollination Algorithm with strategies based on the Path Relinking meta-heuristic to improve the search for the optimal or near-optimal personalized menu recommendation in terms of execution time and fitness value. Experimental results performed on a set of older adult profiles have proven that by hybridizing the Flower Pollination Algorithm with Path Relinking, an improvement in terms of execution time and fitness value is obtained in the context of generating personalized menu recommendations for older adults. As future work, we plan to develop a method for dynamically setting the values of the proposed algorithms' adjustable parameters (i.e. population size and number of iterations) and to integrate the proposed algorithms and experimental prototype into a commercial system connected to a real market place of food providers. Additionally, we intend to investigate the impact of hybridizing the state-of-the-art Flower Pollination Algorithm with Path Relinking on the fitness value and execution time in the context of other optimization problems.

Acknowledgements This work has been carried out in the context of the Ambient Assisted Living Joint Programme project DIET4Elders [http://www.diet4elders.eu] and was supported by a grant of the Romanian National Authority for Scientific Research, CCCDI UEFISCDI, project number AAL16/2013. This document is a collaborative effort. The scientific contribution of all authors is the same.

References

1. Bing W, Fei W, Chunming Y (2010) Personalized recommendation system based on multiagent and rough set. In: Proceedings of the 2nd international conference on education technology and computer, pp 303–307
2. Blum C, Puchinger J, Raidl GR, Roli A (2011) Hybrid metaheuristics in combinatorial optimization: a survey. Appl Soft Comput J 11(6):4135–4151
3. Chavez-Bosquez O, Marchi J, Pozos-Parra P (2014) Nutritional menu planning: a hybrid approach and preliminary tests. Res Comput Sci J 82:93–104
4. Dorigo M, Birattari M, Stutzle T (2006) Ant colony optimization—artificial ants as a computational intelligence technique. IEEE Comput Intell Mag 1(4):28–39
5. Floreano D, Mattiussi C (2008) Bio-inspired artificial intelligence—theories, methods, and technologies. The MIT Press
6. Gaal B, Vassanyi I, Kozmann G (2005) A novel artificial intelligence method for weekly dietary menu planning. Methods Inf Med 44(5):655–664
7. Glover F, Laguna M (2000) Fundamentals of scatter search and path relinking. Control Cybern J 29(3):653–684
8. Haddad OB, Afshar A, Mario MA (2006) Honey-Bees Mating Optimization (HBMO) algorithm: a new heuristic approach for water resources optimization. Water Resour Manage J 20(5):661–680
9. Hinchey MG, Sterritt R (2007) 99% (Biological) Inspiration.... In: Proceedings of the Fourth IEEE international workshop on engineering of autonomic and autonomous systems, pp 187–195
10. Kale A, Auti N (2015) Automated menu planning algorithm for children: food recommendation by dietary management system using id3 for indian food database. Proc Comput Sci 50:197–202
11. Kashima T, Matsumoto S, Ishii H (2011) Decision support system for menu recommendation using rough sets. Int J Innov Comput Inf Control 7(5):2799–2808
12. Kaur G, Singh D, Kaur M (2013) Robust and efficient RGB based fractal image compression: flower pollination based optimization. Int J Comput Appl 78(10)
13. Kennedy J, Eberhart R (1995) Particle swarm optimization. Proc IEEE Int Conf Neural Netw 4:1942–1948
14. van der Merwe A, Kruger H, Steyn T (2015) A diet expert system utilizing linear programming models in a rule-based inference engine. J Appl Oper Res 7(1):13–22
15. Prathiba R, Moses BM, Sakthivel S (2014) Flower pollination algorithm applied for different economic load dispatch problems. Int J Eng Technol 6(2)
16. Saxena P, Kothari A (2016) Linear antenna array optimization using flower pollination algorithm. SpringerPlus 5:306
17. Snae C, Bruckner M (2008) FOODS: a food-oriented ontology-driven system. In: Proceedings of the second ieee international conference on digital ecosystems and technologies, pp 168–176
18. Sivilai S, Snae C, Bruckner M (2012) Ontology-driven personalized food and nutrition planning system for the elderly. In: Proceedings of the 2nd international conference in business management and information sciences
19. Yang XS, Deb S (2009) Cuckoo search via Lvy flights. In: World congress on nature & biologically inspired, computing, pp 210–214

20. Yang XS (2010) Nature-Inspired metaheuristic algorithms, 2nd edn. Luniver Press
21. Yang XS (2010) Engineering optimization—an introduction with metaheuristic applications. Wiley
22. Yang XS (2012) Flower pollination algorithm for global optimization. In: Unconventional computation and natural computation, Lecture Notes in Computer Science, vol 7445, pp 240–249
23. Yang XS, Karamanoglu M, He X (2014) Flower pollination algorithm: a novel approach for multiobjective optimization. Eng Optim 46(9):1222–1237
24. National Nutrient Database for Standard Reference Release 28. http://ndb.nal.usda.gov/ndb/search/list
25. Squirrel's RecipeML Archive. http://dsquirrel.tripod.com/recipeml/indexrecipes2.html

Nature-inspired Algorithm-based Optimization for Beamforming of Linear Antenna Array System

Gopi Ram, Durbadal Mandal, S.P. Ghoshal and Rajib Kar

Abstract Nature-inspired algorithms have brought great revolution in all fields of electromagnetics where the optimization of certain parameters is highly complex and nonlinear. With the help of proper design of the cost function or the fitness function in terms of optimizing parameters, any type of problem can be solved. The nature-inspired algorithms play an important role in the optimal design of antenna array with better radiation characteristics. In this work, hyper-beamforming of linear antenna array has been taken as an example of nature- inspired optimization in antenna array system. An emerging nature-inspired optimization technique has been applied to design the optimal array to reduce the side lobes and to improve the other radiation characteristics to show the effect of the optimization on design via the nature-inspired algorithms. Various nature-inspired algorithms have been considered for the optimization. Flower pollination algorithm (FPA) is applied to determine the optimal amplitude coefficients and the spacing between the elements of the array of the optimized hyper-beamforming of linear antenna array. FPA keeps the best solution until it reaches the end of the iteration. The results obtained by the FPA algorithm have been compared with those of other stochastic algorithms, such as real-coded genetic algorithm (RGA), particle swarm optimization (PSO), differential evolution (DE), firefly algorithm (FFA), simulated annealing (SA), artificial immune system (AIS), and artificial bee colony (ABC). Optimal hyper-beamforming of the same obtained by FPA can obtain the best improvement in side lobe level (SLL) with fixed first null beam width (FNBW). Directivity of the array is calculated by using Simpsons 1/3 rule. The entire simulation has been done for 10-, 14-, and 20-element linear antenna arrays.

G. Ram (✉) · D. Mandal · S.P. Ghoshal · R. Kar
Department of ECE, NIT Durgapur, India
e-mail: gopi203hardel@gmail.com

D. Mandal
e-mail: durbadal.bittu@gmail.com

© Springer International Publishing AG 2017
S. Patnaik et al. (eds.), *Nature-Inspired Computing and Optimization,*
Modeling and Optimization in Science and Technologies 10,
DOI 10.1007/978-3-319-50920-4_8

185

1 Introduction

Antenna is a far-reaching element here and now with the constantly progressive technology. Antenna plays an imperative role in the wireless communication either for short- or for long-distance communication. For distinctive applications, different types of antenna are used. According to IEEE, the antenna is defined as "transmitting or receiving radio waves." The contemporary research is going on for the development of antenna with increased accuracy and efficiency. For the prolonged distance communication, the design of antenna should be highly directive with high gain and high directivity with less interference, but the radiation characteristics of a single type of antenna will have low directivity, low gain, and low efficiency. To accommodate these characteristics, the single antenna will have to have a large size which is essentially impractical. To overcome this problem, identical antennas are arranged in the form of an array of identical elements. The geometric arrangements of the arrays may be of different configurations depending on the applications. Array structures can be linear, circular, planner, conformal, and hybridization of the structures. The reduction of side lobes is a serious concern for the antenna research community. There are many conventional and numerical methods for the reduction of side lobes and the improvement of other radiation characteristics, but these have some disadvantages, such as high dynamic range ratio of current excitation weights. The great difficulty of optimal pattern synthesis is the simultaneous reduction of SLL and directivity. This is a very challenging task for the antenna designer. To overcome these difficulties and challenges, the authors have designed a novel and a unique cost function dealing with conflicting pattern characteristics such as SLL reduction and directivity improvement, modeled as a single objective optimization with suitable weighting factors for the objectives.

The classical gradient-based optimization methods are not suitable for optimal design of hyper-beamforming of linear antenna arrays due to the following reasons: (i) highly sensitive to starting points when the number of solution variables and hence the size of the solution space increase, (ii) frequent convergence to the local optimum solution or divergence or revisiting the same suboptimal solution, (iii) requirement of continuous and differentiable objective function (gradient search methods), (iv) requirement of the piecewise linear cost approximation (linear programming), and (v) problem of convergence and algorithm complexity (nonlinear programming). So, evolutionary methods have been employed for the optimal design of hyper-beamforming of linear antenna arrays with better parameter control.

Beamforming is a signal processing technique used to control the direction of the transmission and reception of the radio signals [1]. This beamforming processing significantly improves the gain of the wireless link over the conventional technology, thereby increasing range, rate, and penetration [2–4]. It has found numerous applications in radar, sonar, seismology, wireless communication, radio astronomy, acoustics, and biomedicine [5]. It is generally classified as either conventional (switched and fixed) beamforming or adaptive beamforming. Switched beamforming system [6, 7] is a system that can choose one pattern from many predefined patterns in order to enhance the received signals. Fixed beamforming uses a fixed set of weights and time delays (or phasing) to combine the signals received from the sensors in the array, primarily using only the information about the locations of the sensors in space and the wave direction of interest [8]. Adaptive beamforming is based on the desired signal maximization mode and interference signal minimization mode [9–11].

The major drawbacks of some stochastic techniques employed for optimization in electromagnetic such as simulated annealing (SA) [12–14], real-coded genetic algorithm (RGA) [15–20], firefly algorithm (FFA) [21–28], particle swarm optimization (PSO) [29–33], differential evolution (DE) [34–39], artificial immune system (AIS) [40–44], and artificial bee colony (ABC) [45–47] are premature convergence and entrapment to suboptimal solution. To enhance the performance of optimization algorithms in global and local searches, the authors suggest an alternative technique with Levy flight-based [48–54] flower pollination algorithm (FPA) [55–65] for the optimal hyper-beamforming of linear antenna array in this work. The main contribution of the authors in this work is the novel application of FPA in different problems of electromagnetics with new antenna designs. The statistical analysis and t-test [66] have also been done to prove the comparative better performance of FPA. There are also other researchers reported based on the improvement of particular antenna elements [67–69].

The rest of the paper is arranged as follows: In Sect. 2, the problem formulation of hyper-beamforming of linear antenna array is presented. Section 3 briefly discusses on FPA employed for the design. Section 4 describes the simulation results obtained by employing different algorithms. Convergence characteristics of different algorithms are explained in Sect. 5. Section 6 concludes the paper. Finally, future research topics based on the nature-inspired algorithms are described in Sect. 7.

2 Problem Formulation

Consider a broadside linear antenna array of N equally spaced isotropic elements as shown in Fig. 1. For broadside beams, the array factor is as follows [6].

$$AF(\theta) = \sum_{n=1}^{N} I_n e^{j(n-1)Kd[\sin\theta\cos\phi - \sin\theta_0\cos\phi_0]}$$ (1)

where θ = angle of arrival in the elevation plane; d = inter-element separation; K = propagation constant; N = sum of elements in the array; I_n = current excitation coefficient in the nth element.

Sum and the difference patterns can be given as (2), (3), respectively [8].

Sum pattern,

$$Sum(\theta) = |R_L| + |R_R|$$ (2)

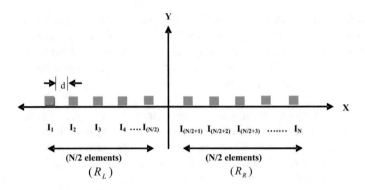

Fig. 1 Geometry of an N-element linear array along the x-axis

Difference pattern,

$$Diff(\theta) = |R_L - R_R| \tag{3}$$

where

$$R_L = \sum_{n=1}^{N/2} I_n e^{j(n-1)Kd[\sin\theta\cos\varphi - \sin\theta_0\cos\varphi_0]} \tag{4}$$

$$R_R = \sum_{n=N/2+1}^{N} I_n e^{j(n-1)Kd[\sin\theta\cos\varphi - \sin\theta_0\sin\varphi_0]} \tag{5}$$

The general equation of hyper-beam which is a function of hyper-beam exponent u is given by

$$AF_{Hyper}(\theta) = \left\{ (|R_L| + |R_R|)^u - (|R_L - R_R|)^u \right\}^{1/u} \tag{6}$$

where u ranges from 0.2 to 1. All the antenna elements are assumed to be isotropic. The directivity of linear antenna array is given by

$D_{AF_{Hyper}}$ = radiation intensity in the direction (θ_0, ϕ_0)/average radiation intensity

$$= \frac{|AF_{Hyper_0}(\theta_0, \phi_0)|^2}{\frac{1}{4\pi} \int_0^{2\pi} \int_0^{\pi} |AF_{Hyper}(\theta, \phi)|^2 \sin\theta d\theta d\phi} \tag{7}$$

Here $D_{AF_{Hyper}}$ is the directivity of the linear antenna arrays; θ_0 denotes the direction in which $AF_{Hyper}(\theta, \phi)$ has the maximum value.

The power radiated (P_0) by the time-modulated concentric ring circular array, P_0 at the fundamental (center) frequency is given by (8).

$$P_0 = \int_0^{2\pi} \int_0^{\pi} |AF_0(\theta, \phi)|^2 \sin\theta d\theta d\phi \tag{8}$$

The cost function (CF) for improving the SLL of radiation pattern of hyper-beam linear antenna array is given in (9)

$$
\begin{aligned}
CF = \quad & w_1 * Max \left| AF_{Hyper}(\theta_{msl1}, I_n) \right| / \left| AF_{Hyper}(\theta_0, I_n) \right| \\
+ & w_2 * Max \left| AF_{Hyper}(\theta_{msl2}, I_n) \right| / \left| AF_{Hyper}(\theta_0, I_n) \right| \\
+ & w_3 * 1 / D_{max\_hyper-beam}
\end{aligned}
\tag{9}
$$

where θ_0 is the angle where the maximum of central angle is attained in $\theta \in [-\pi/2, \pi/2]$. θ_{msl1} is the angle where the maximum side lobe $AF_{Hyper}(\theta_{msl1}, I_n)$ is attained in the lower band of hyper-beam pattern. θ_{msl2} is the angle where the maximum side lobe $AF_{Hyper}(\theta_{msl2}, I_n)$ is attained in the upper side band of hyper-beam pattern. $D_{max\_hyper-beam}$ is the maximum directivity. w_1, w_2, w_3 are the weighting factors to impart the different contributions to the cost function. In CF, both the numerator and the denominator are in absolute magnitude. Minimization of CF means the maximum reduction of SLL. FPA is employed for the minimization of CF by optimizing the current excitation weights of elements and the inter-element spacing. Optimal results for CF and SLL are discussed in Sect. 4.

3 Flower Pollination Algorithm [55]

Flower pollination is essentially the transmission of pollen from a male plant to a female plant. Some flowers contain male and female parts, and the pollination can occur by transferring pollen from the male organ to the female organ in the same plant. Pollination occurs naturally by animals, by wind, or by self-pollination. Sometimes, a person needs to intervene and pollinate flowers by hand. Pollination is the course of action of mating in the plants [55–65]. There are two types of pollination: self-pollination and cross-pollination [55–65]. To develop FPA, there are four rules which are summarized as follows:

FPA developed by Yang [55] is a new population-based meta-heuristic optimization technique inspired by the flower pollination process in nature. It mimics the flower pollination behavior. Pollination is a natural physiological process of mating in plants which is associated with transfer of pollens by pollinators such as insects. Pollination is further divided into cross-pollination and self-pollination. By establishing analogy between the natural flower pollination process and the solution of real-world optimization problems, certain rules have been established in [55–60] to help in formulating the FPA. The following four rules provide the guidance regarding the i) pollination process employed and ii) selection of step size.

Rule-1: For global pollination (also called as biotic and cross-pollination), pollinators travel over long distances and their movements are similar to Levy flights distribution [55–60].

Rule-2: The local pollination (abiotic or self-pollination) is conducted by means of random distribution.

Rule-3: The local pollination takes place among flowers of the same plant or flowers of the same species. Flower constancy may be regarded as an equivalent to a reproduction probability that is proportionate to the similarity of two flowers involved.

Rule-4: A switching probability $(0 < p < 1)$ decides whether a flower is pollinated by local or global pollination.

The entire FPA is divided into two major steps:

3.1 Global Pollination:

Now the rules are converted to the following mathematical equations: For the global pollination, Rule-1 and Rule-3 can be represented mathematically as follows [55–60]:

$$X_i^{iter+1} = X_i^{iter} + \gamma L(\lambda)(g_* - X_i^{iter}) \tag{10}$$

where X_i^{iter} is the pollen i or the solution vector X_i at iteration $iter$ and g_* is the current best solution found among all the solutions at the current generation/iteration; γ is a scaling factor to control the step size; $L(\lambda)$ is the Levy flight-based step size that corresponds to the strength of the pollination. Insects can fly over a long distance with different distance steps; this is drawn from a Levy distribution [55–60]. The parameter L is the strength of the pollination, which essentially is a step size. Since insects may move over a long distance with various distance steps, a Levy flight can be used to mimic this characteristic efficiently [55–60]. That is, L (> 0) drawn from a Levy distribution is given by

$$L \sim \frac{\lambda' \Gamma(\lambda') \sin(\pi \lambda'/2)}{\pi} \frac{1}{s^{1+\lambda}}, \quad (s \gg s_0 > 0) \tag{11}$$

where $\Gamma(\lambda')$ is the standard gamma function, and this distribution is valid for large steps s > 0. In all the simulations below, $\lambda' = 1.5$ is used.

3.2 Local Pollination:

For the local pollination, both Rule-2 and flower constancy (Rule-3) can be represented as [55–60]:

$$X_i^{iter+1} = X_i^{iter} + \varepsilon(X_j^{iter} - X_k^{iter}) \tag{12}$$

where X_j^{iter} and X_k^{iter} are the pollens from different flowers of the same plant species. This essentially mimics the flower constancy in a limited neighborhood. If X_j^{iter} and X_k^{iter} come from the same species or are selected from the same population, this becomes a local random walk equivalently if ε is drawn from a uniform distribution in [0, 1] [55–60].

Most flower pollination activities can occur at both the local and global scales. In practice, adjacent flower patches or flowers in the not-so-far-away neighborhood are more likely to be pollinated by local flower pollens. For this, a switch probability (Rule-4) or proximity probability P is used to switch between the global pollination and local pollination. Section 3.3 shows the pseudocode of FPA. Figure 2 shows the flowchart of flower pollination algorithm.

3.3   Pseudo-code for FPA:

Step 1: Initialize the variables, P11 = popul (npop, nvar), population size, npop = 120; ε [0, 1]; switching probability, P = 0.5; the minimum and the max-imum boundary values of (nvar) optimizing variables; nvar depends on the different number of current excitation weights of three linear antenna arrays; max-iter = 100.

Step 2: Assume the initial population and compute the CF values.

Step 3: *for* (cy = 1; cy $\leq$ max_iter; cy^{++})
    for (j = 1 , j $\leq$ np; j^{++})
    // calculate fitness
    $Cost_{cy,j}$ = fitness value;
    end *for*
    // gbest value calculation
    for (j = 1; j $\leq$ np; j^{++})
    for (j = 1; j $\leq$ np; j^{++})
    if ($Cost_{cy,j}$ = min (Cost))
    $P1gbest_{cy,j}$ = $P11_{j,jspg}$;
    break
    end *if*
    end *for*
    end *for*
    for (npop = 1; npop $\leq$ np; npop^{++})
    for (jspg = 1; jspg $\leq$ nvar; jspg^{++})
    if (rand(1)< P)
    // Global pollination
    // Draw a (n- dimensional) step vector L which obey the levy distribution

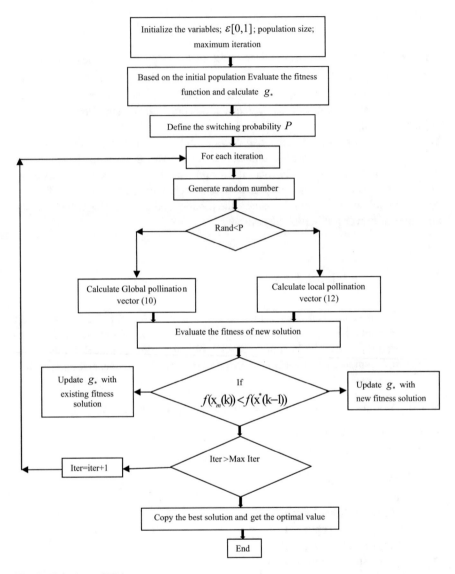

Fig. 2 Flowchart of FPA

(11)
$$T_{npop,jspg} = P11_{npop,jspg} + L * (P1gbest_{cy,jspg} - P11_{npop,jspg})$$
// check T against boundary limits.
if
$$T_{npop,jspg} \leq T_{jspg,\min}; T_{npop,jspg} = T_{jspg,\min}$$
end
if

$T_{npop,jspg} \geq T_{jspg,\max}; T_{npop,jspg} = T_{jspg,\max}$
end
else
// Local pollination
$\varepsilon = 1.5 * rand(1); HKJ = randperm\ (np);$
$T_{npop,jspg} = P11_{npop,jspg} + \varepsilon * (P11_{HJK(1),jspg} - P11_{HJK(2),jspg})$
// *check T against boundary limits.*
if
$T_{npop,jspg} \leq T_{jspg,\min}; T_{npop,jspg} = T_{jspg,\min}$
end if
$T_{npop,jspg} \geq T_{jspg,\max}; T_{npop,jspg} = T_{jspg,\max}$
end
end *if*
end *for*
end *for*
initialize $P11_{npop,jspg} = T_{npop,jspg}$
end *for cycle.*

Step 4: Calculate the CF and P1gbest. Repeat all the steps from step 3. P1best is the near-global optimal solution vector having the optimal optimizing variables as current excitation weights and one number of uniform inter-element spacing.

4 Simulation Results

This section gives the simulation results for various hyper-beams of linear antenna array designs. Three linear antenna array designs considered are of 10-, 14-, and 20-element sets, each maintaining a uniform inter-element spacing. Reduction of SLL can be achieved by varying the hyper-beam exponent value u, thereby obtaining different hyper-beam patterns. The results show that the SLL reduction increases as the exponent value u decreases. For 10-, 14-, and 20-element linear arrays, with $u = 1$, SLL are -19.91 dB, -20.10 dB, and -20.20 dB, respectively, whereas with $u = 0.5$, SLL reduces to -32.78 dB, -33.02 dB, and -33.20 dB, respectively, as shown in Figs. 3, 4, 5, 6, 7, and 8. Uniform linear array with $(I_n = 1)$ and $\lambda_A = 0.5$ shows the respective SLL and directivity values as -12.97, -13.11, -13.20 dB and 10, 11.4613, 13.0103 dB. Therefore, in comparison with uniform linear array, hyper-beam technique yields much more reduction of SLL. Table 1 shows the best chosen parameters for the optimization purpose.

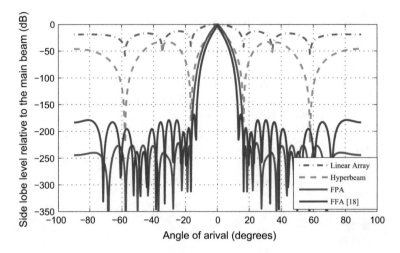

Fig. 3 Radiation pattern obtained by FPA for the 10-element array at $u = 0.5$

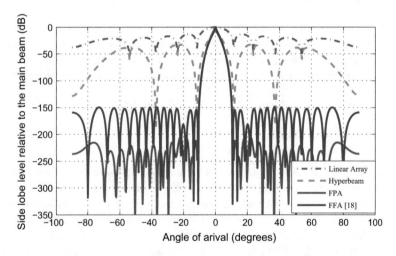

Fig. 4 Radiation pattern obtained by FPA for the 14-element array at $u = 0.5$

4.1 Optimization of Hyper-Beam by Using FPA

The optimization of SLL and directivity is applied in the radiation pattern obtained from the array factor of the hyper-beam pattern. From Table 2 and Figs. 3, 4, and 5, in which the exponent value $u = 0.5$, it is clear that the reduction of SLL by FPA-based optimization is much better as compared to those obtained by RGA, PSO, DE, FFA, SA, AIS, and ABC. From Table 3, it is confirmed that FPA proves its superiority in yielding better SLL with $u = 1.0$. The directivity values obtained by using FPA for all the arrays are better than those of conventional linear antenna array and un-optimized

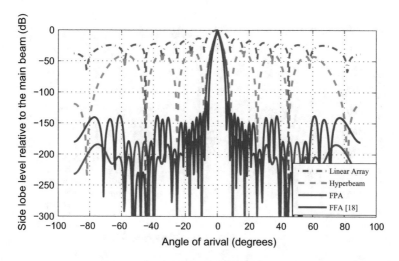

Fig. 5 Radiation pattern obtained by FPA for the 20-element array at $u = 0.5$

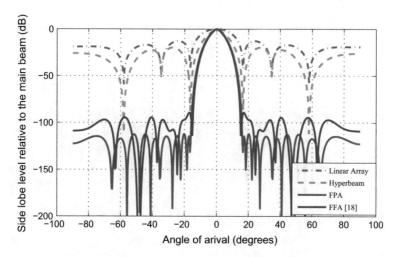

Fig. 6 Radiation pattern obtained by FPA for the 10-element array at $u = 1$

hyper-beam pattern. From Table 3, in which the exponent value $u = 1.0$, the same observations can be made with regard to SLL for the algorithms. In this case, also FPA proves its superiority in yielding better SLL and directivity as compared to those obtained by RGA, PSO, DE, FFA, SA, AIS, and ABC. FPA efficiently computes N number of near-global optimal current excitation weights and one number optimal uniform inter-element separation for each hyper-beam linear antenna array to have the maximum SLL reduction, an improved directivity and the fixed FNBW (Figs. 6, 7 and 8).

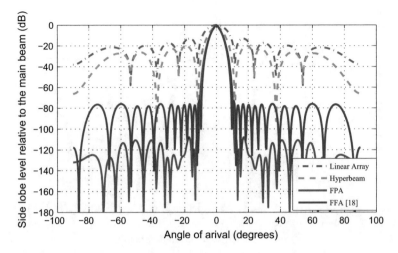

Fig. 7 Radiation pattern obtained by FPA for the 14-element array at $u = 1$

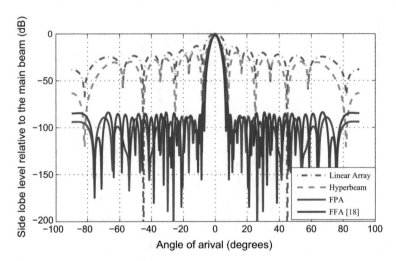

Fig. 8 Radiation pattern obtained by FPA for the 20-element array at $u = 1$

4.2 Comparisons of Accuracies Based on t test

Two independent sample: t test is a hypothesis testing method for determining the statistical significance of the difference between two independent samples of equal or unequal sample size [66]. The values of t test contain positive values and have certain values depending on the confidence interval as given in the critical table of [66]. Unequal population variance/standard deviation ($\sigma_1^2 \neq \sigma_2^2$) is assumed. Two-sample t test with equal sample size has been adopted with sample size of 51. Assuming unequal variances, the test statistic is calculated as follows:

Table 1 Best chosen parameters of various algorithms

Parameters	RGA	PSO	DE	FFA	FPA	SA	AIS	ABC
Population size	120	120	120	120	120	120	120	120
Iteration cycle	100	100	100	100	100	100	100	100
Crossover rate	0.8	–	–	–	–	–	–	–
Crossover	Two point cross over	–	–	–	–	–	–	–
Mutation rate	0.05	–	–	–	–	–	–	–
Mutation	Gaussian mutation	–	–	–	–	–	–	–
Selection, probability	Roulette wheel, 1/3	–	–	–	–	–	–	–
C_1, C_2	–	1.5, 1.5	–	–	–	–	–	–
v_i^{min}, v_i^{max}	–	0.01, 1.0	–	–	–	–	–	–
w_{max}, w_{min}	–	1.0, 0.4	–	–	–	–	–	–
C_r	–	–	0.3	–	–	–	–	–
F	–	–	0.5	–	–	–	–	–
α, γ, β_0	–	–	–	0.01, 0.2, 0.6	–	–	–	–
ε	–	–	–	–		–	–	–
P	–	–	–	–	0.5	–	–	–
λ'	–	–	–	–	1.5	–	–	–
s	–	–	–	–	1.25	–	–	–
n_o	–	–	–	–	–	–	–	50% of the colony
n_e	–	–	–	–	–	–	–	50% of the colony
n_s	–	–	–	–	–	–	–	1
Limit	–	–	–	–	–	–	–	$n_e.D$
Clone size factor	–	–	–	–	–	–	2	–
Maturation rate	–	–	–	–	–	–	0.40	–
T_0 (initial temperature)	–	–	–	–	–	1	–	–
T_f (final temperature)	–	–	–	–	–	10^{-4}	–	–
α (cooling rate)	–	–	–	–	–	0.95	–	–

n_o = onlooker number, n_e = employed bee number, n_s = scout number, D = dimension of the problem

Table 2 Optimal current excitation weights, optimal inter-element spacing, SLL, FNBW, and directivity for hyper-beam pattern of linear array with hyper-beam exponent ($u = 0.5$), obtained by various algorithms

N	Algorithms	Optimized current excitation weights and $[I_1, I_2, I_3, I_4 \ldots \ldots I_N]$	Optimal inter-element spacing (λ_A)	SLL of hyper-beam with optimization (dB)	FNBW of hyper-beam with optimization (deg)	Directivity (dB)
10	GA [18]	0.2844 0.5240 0.8813 0.9032 0.4231	0.5441	−100.6	41.04	NR*
		0.8425 0.4564 0.6402 0.3414 0.3853				
	PSO [18]	0.2398 0.6414 0.9123 0.9722 0.4312	0.5717	−117.2	39.60	
		0.9502 0.4327 0.6582 0.3571 0.3982				
	DE [18]	0.1029 0.3802 0.6258 0.9394 0.7907	0.8470	−151.9	34.56	
		1.0000 0.5211 0.5538 0.2118 0.2156				
	FFA [18]	0.2490 0.4210 0.1819 0.6099 0.8929	0.9460	−168.5	26.64	
		0.5685 0.6123 0.6745 0.3250 0.1579				
	SA	0.0231 0.3185 0.6509 0.9015 0.8179	0.8822	−176.9	33.12	18.2531
		0.9777 0.5934 0.4869 0.7933 0.6430				
	AIS	0.0389 0.2737 0.5865 0.7933 0.6430	0.8658	179	33.12	18.5097
		0.9531 0.4555 0.4488 0.3270 0.1919				

(continued)

Table 2 (continued)

N	Algorithms	Optimized current excitation weights and $[I_1, I_2, I_3, I_4, ...I_N]$	Optimal inter-element spacing (λ_A)	SLL of hyper-beam with optimization (dB)	FNBW of hyper-beam with optimization (deg)	Directivity (dB)
	ABC	0.1713 0.3291 0.4811 0.5951 0.9288	0.9910	−212	32.4	16.9523
		0.9488 0.9731 0.6248 0.2336 0				
	FPA	0.1790 0.3914 0.3183 0.6821 0.9705	0.9231	−221.7	33.12	15.4508
		0.9858 0.9872 0.9025 0.3700 0.1313				
14	GA [18]	0.3631 0.2555 0.4905 0.0043 0.6114	0.5878	−96.21	25.92	NR*
		0.5778 0.8634 0.5042 0.5782 0.5913				
		0.7502 0.5545 0.2878 0.3431				
	PSO [18]	0.2319 0.1857 0.6027 0.5089 0.7906	0.6036	−113	25.20	
		0.4163 0.6275 0.7212 0.9097 0.2907				
		0.2525 0.2755 0.5506 0.3615				
	DE [18]	0.2297 0.3701 0.3080 0.2229 0.6599	0.7949	−125.8	23.04	

(continued)

Table 2 (continued)

N	Algorithms	Optimized current excitation weights and $[I_1, I_2, I_3, I_4, \ldots, I_N]$	Optimal inter-element spacing (λ_{-A})	SLL of hyper-beam with optimization (dB)	FNBW of hyper-beam with optimization (deg)	Directivity (dB)
		0.9495 0.6941 0.8597 0.4157 0.7559				
		0.7305 0.2389 0.3759 0.0982				
	FFA [18]	0.3124 0.3709 0.5444 0.3167 0.8885	0.8760	−149.7	20.16	
		0.9115 0.9869 0.9916 0.9243 0.6967				
		0.9560 0.4002 0.1592 0.1990				
	SA	0 0.3186 0.2749 0.7167 0.8797	0.8854	−155.2	23.04	18.8780
		0.4955 0.8739 0.7213 0.9256 0.5171				
		0.1593 0.3630 0.4199 0.2121				
	AIS	0.1758 0.4332 0.4342 0.5007 0.6025	0.9820	−163.7	20.16	20.7098
		0.9815 0.9996 0.9781 0.8408 0.9440				
		0.6747 0.4594 0.0962 0.0863				
	ABC	0.0149 0.0676 0.3827 0.5842 0.9443	0.9312	−192.7	23.04	20.9416

(continued)

Table 2 (continued)

N	Algorithms	Optimized current excitation weights and $[I_1, I_2, I_3, I_4 \ldots\ldots I_N]$	Optimal inter-element spacing (λ_A)	SLL of hyper-beam with optimization (dB)	FNBW of hyper-beam with optimization (deg)	Directivity (dB)
		0.8909 0.9717 0.9740 0.8890 0.5431				
		0.5385 0.4064 0.3326 0.1646				
	FPA	0.1294 0.3195 0.4092 0.2984 0.5739	0.9889	−205	23.04	19.1510
		0.9705 0.9418 0.9163 0.7405 0.8670				
		0.7406 0.3445 0.2256 0.0576				
20	GA [18]	0.2505 0.3933 0.4881 0.4829 0.3027	0.5361	−83.69	19.44	NR*
		0.6697 0.3436 0.9551 0.5974 0.8952				
		0.5252 0.9773 0.4056 0.6612 1.0000				
		0.1577 0.8144 0.3284 0 0.5558				
	PSO [18]	0.1675 0.2453 0.2113 0.5168 0.6011	0.5353	−88.71	18.72	
		0.5661 0.7962 0.2148 0.8279 0.2476				
		0.9888 0.3429 0.8064 0.1836 0.2281				
		0.1792 0.4317 0.6579 0.2244 0.3467				

(continued)

Table 2 (continued)

N	Algorithms	Optimized current excitation weights and [$I_1, I_2, I_3, I_4 \ldots \ldots I_N$]	Optimal inter-element spacing (λ_A)	SLL of hyper-beam with optimization (dB)	FNBW of hyper-beam with optimization (deg)	Directivity (dB)
	DE [18]	0.1567 0.1345 0.5561 0.4817 0.9529	0.5852	−101.9	18	
		0.7651 0.9420 0.7511 0.6736 0.5927				
		0.9889 0.8862 0.4313 0.4025 0.2891				
		0.3316 0.4286 0.4649 0.4306 0.3195				
	FFA [18]	0.0737 0.3285 0.4033 0.5217 0.8113	0.9520	−115.1	11.52	
		0.9979 0.9665 0.7002 0.7260 0.7769				
		0.9957 1.0000 0.9467 0.4586 0.2044				
		0.2364 0.5734 0.6510 0.4416 0.2468				
	SA	0.1975 0.3803 0.4046 0.6823 0.4824	0.9989	−116.1	12.96	19.5717
		0.3973 0.9136 0.7014 0.9308 0.9736				
		0.8978 0.8516 0.9655 0.5969 0.9134				
		0.4990 0.3093 0.0245 0.0966 0				

(continued)

Table 2 (continued)

N	Algorithms	Optimized current excitation weights and $[I_1, I_2, I_3, I_4\ldots\ldots I_N]$	Optimal inter-element spacing (λ_A)	SLL of hyper-beam with optimization (dB)	FNBW of hyper-beam with optimization (deg)	Directivity (dB)
	AIS	0 0.0250 0.1335 0.3851 0.6344 0.8001 0.9121 0.9432 0.8958 0.9064 0.9619 0.9205 0.7311 0.5706 0.5059 0.4439 0.4529 0.4836 0.3543 0.1851	0.8042	−136.4	16.56	21.7815
	ABC	0 0.1784 0.2709 0.3304 0.3437 0.5403 0.4797 0.6846 0.8459 0.8455 0.9433 0.8439 0.8557 0.8385 0.4966 0.4670 0.1755 0.0063 0 0	0.9839	−156.6	16.56	21.1380
	FPA	0 0 0.0064 0.2110 0.3786 0.5921 0.7978 0.8638 0.8769 0.9311 0.9475 0.9212 0.7570 0.5873 0.5320 0.4865 0.3984 0.3611 0.1813 0.0608	0.9547	−170.8	16.56	19.3398

NR* = Not Reported

Table 3 Optimal current excitation weights, optimal inter-element spacing, SLL, FNBW, and directivity for hyper-beam pattern of linear array with hyper-beam exponent ($u = 1$), obtained by various algorithms

N	Algorithms	Optimized current excitation weights and $[I_1, I_2, I_3, I_4 \ldots \ldots I_N]$	Optimal inter-element spacing (λ_A)	SLL of hyper-beam with optimization (dB)	FNBW of hyper-beam with optimization (deg)	Directivity (dB)
10	GA [18]	0.2844 0.5240 0.8813 0.9032 0.4231 0.8425 0.4564 0.6402 0.3414 0.3853	0.5441	−100.6	41.04	NR*
	PSO [18]	0.2398 0.6414 0.9123 0.9722 0.4312 0.9502 0.4327 0.6582 0.3571 0.3982	0.5717	−117.2	39.60	
	DE [18]	0.1029 0.3802 0.6258 0.9394 0.7907 1.0000 0.5211 0.5538 0.2118 0.2156	0.8470	−151.9	34.56	
	FFA [18]	0.2490 0.4210 0.1819 0.6099 0.8929 0.5685 0.6123 0.6745 0.3250 0.1579	0.9460	−168.5	26.64	
	SA	0.0231 0.3185 0.6509 0.9015 0.8179 0.9777 0.5932 0.4869 0.3407 0.2439	0.8822	−176.9	33.12	18.2531
	AIS	0.0389 0.2737 0.5865 0.7933 0.6430 0.9531 0.4555 0.4488 0.3270 0.1919	0.8658	179	33.12	18.5097

(continued)

Table 3 (continued)

N	Algorithms	Optimized current excitation weights and [$I_1, I_2, I_3, I_4......I_N$]	Optimal inter-element spacing (λ_A)	SLL of hyper-beam with optimization (dB)	FNBW of hyper-beam with optimization (deg)	Directivity (dB)
	ABC	0.1713 0.3291 0.4811 0.5951 0.9288 0.9488 0.9731 0.6248 0.2336 0	0.9910	−212	32.4	16.9523
	FPA	0.1790 0.3914 0.3183 0.6821 0.9705 0.9858 0.9872 0.9025 0.3700 0.1313	0.9231	−221.7	33.12	15.4508
14	GA [18]	0.3631 0.2555 0.4905 0.0043 0.6114 0.5778 0.8634 0.5042 0.5782 0.5913 0.7502 0.5545 0.2878 0.3431	0.5878	−96.21	25.92	NR*
	PSO [18]	0.2319 0.1857 0.6027 0.5089 0.7906 0.4163 0.6275 0.7212 0.9097 0.2907 0.2525 0.2755 0.5506 0.3615	0.6036	−113	25.20	
	DE [18]	0.2297 0.3701 0.3080 0.2229 0.6599 0.9495 0.6941 0.8597 0.4157 0.7559	0.7949	−125.8	23.04	

(continued)

Table 3 (continued)

N	Algorithms	Optimized current excitation weights and [$I_1, I_2, I_3, I_4......I_N$]	Optimal inter-element spacing (λ_A)	SLL of hyper-beam with optimization (dB)	FNBW of hyper-beam with optimization (deg)	Directivity (dB)
		0.7305 0.2389 0.3759 0.0982				
	FFA [18]	0.3124 0.3709 0.5444 0.3167 0.8885	0.8760	−149.7	20.16	
		0.9115 0.9869 0.9916 0.9243 0.6967				
		0.9560 0.4002 0.1592 0.1990				
	SA	0 0.3186 0.2749 0.7167 0.8797	0.8854	−155.2	23.04	18.8780
		0.4955 0.8739 0.7213 0.9256 0.5171				
		0.1593 0.3630 0.4199 0.2121				
	AIS	0.1758 0.4332 0.4342 0.5007 0.6025	0.9820	−163.7	20.16	20.7098
		0.9815 0.9996 0.9781 0.8408 0.9440				
		0.6747 0.4594 0.0962 0.0863				
	ABC	0.0149 0.0676 0.3827 0.5842 0.9443	0.9312	−192.7	23.04	20.9416
		0.8909 0.9717 0.9740 0.8890 0.5431				

(continued)

Table 3 (continued)

N	Algorithms	Optimized current excitation weights and $[I_1, I_2, I_3, I_4......I_N]$	Optimal inter-element spacing (λ_A)	SLL of hyper-beam with optimization (dB)	FNBW of hyper-beam with optimization (deg)	Directivity (dB)
		0.5385 0.4064 0.3326 0.1646				
	FPA	0.1294 0.3195 0.4092 0.2984 0.5739	0.9889	−205	23.04	19.1510
		0.9705 0.9418 0.9163 0.7405 0.8670				
		0.7406 0.3445 0.2256 0.0576				
20	GA [18]	0.2505 0.3933 0.4881 0.4829 0.3027	0.5361	−83.69	19.44	NR*
		0.6697 0.3436 0.9551 0.5974 0.8952				
		0.5252 0.9773 0.4056 0.6612 1.0000				
		0.1577 0.8144 0.3284 0 0.5558				
	PSO [18]	0.1675 0.2453 0.2113 0.5168 0.6011	0.5353	−88.71	18.72	
		0.5661 0.7962 0.2148 0.8279 0.2476				
		0.9888 0.3429 0.8064 0.1836 0.2281				
		0.1792 0.4317 0.6579 0.2244 0.3467				

(continued)

Table 3 (continued)

N	Algorithms	Optimized current excitation weights and [$I_1, I_2, I_3, I_4 \ldots\ldots I_N$]	Optimal inter-element spacing (λ_A)	SLL of hyper-beam with optimization (dB)	FNBW of hyper-beam with optimization (deg)	Directivity (dB)
	DE [18]	0.1567 0.1345 0.5561 0.4817 0.9529 0.7651 0.9420 0.7511 0.6736 0.5927 0.9889 0.8862 0.4313 0.4025 0.2891 0.3316 0.4286 0.4649 0.4306 0.3195	0.5852	−101.9	18	
	FFA [18]	0.0737 0.3285 0.4033 0.5217 0.8113 0.9979 0.9665 0.7002 0.7260 0.7769 0.9957 1.0000 0.9467 0.4586 0.2044 0.2364 0.5734 0.6510 0.4416 0.2468	0.9520	−115.1	11.52	
	SA	0.1975 0.3803 0.4046 0.6823 0.4824 0.3973 0.9136 0.7014 0.9308 0.9736 0.8978 0.8516 0.9655 0.5969 0.9134 0.4990 0.3093 0.0245 0.0966 0	0.9989	−116.1	12.96	19.5717

(continued)

Table 3 (continued)

N	Algorithms	Optimized current excitation weights and $[I_1, I_2, I_3, I_4 \ldots I_N]$	Optimal inter-element spacing (λ_A)	SLL of hyper-beam with optimization (dB)	FNBW of hyper-beam with optimization (deg)	Directivity (dB)
	AIS	0 0.0250 0.1335 0.3851 0.6344	0.8042	−136.4	16.56	21.7815
		0.8001 0.9121 0.9432 0.8958 0.9064				
		0.9619 0.9205 0.7311 0.5706 0.5059				
		0.4439 0.4529 0.4836 0.3543 0.1851				
	ABC	0 0.1784 0.2709 0.3304 0.3437	0.9839	−156.6	16.56	21.1380
		0.5403 0.4797 0.6846 0.8459 0.8455				
		0.9433 0.8439 0.8557 0.8385 0.4966				
		0.4670 0.1755 0.0063 0 0				
	FPA	0 0 0.0064 0.2110 0.3786	0.9547	−170.8	16.56	19.3398
		0.5921 0.7978 0.8638 0.8769 0.9311				
		0.9475 0.9212 0.7570 0.5873 0.5320				
		0.4865 0.3984 0.3611 0.1813 0.0608				

NR* = Not Reported

$$t = \frac{\bar{\alpha}_{22} - \bar{\alpha}_{11}}{\sqrt{\left(\frac{\sigma_2^2}{n_A}\right) + \left(\frac{\sigma_1^2}{n_B}\right)}} \qquad (13)$$

$$Degree\ of\ freedom\ (\beta) = \frac{\left(\sigma_1^2/n_A + \sigma_2^2/n_B\right)^2}{\left\{\frac{\left[\sigma_1^2/n_A\right]^2}{(n_A - 1)} + \frac{\left[\sigma_1^2/n_B\right]^2}{(n_B - 1)}\right\}} \qquad (14)$$

where $\bar{\alpha}_{11}$ and $\bar{\alpha}_{22}$ are the mean values of the samples of the first and the second algorithms, respectively; σ_1 and σ_1 are the standard deviations of samples of the first and the second algorithms, respectively; n_A and n_B are the numbers of sample sizes in the first and the second algorithms, respectively. The degree of freedom for unequal variance and unequal sample size is given by (14). In this paper, $n_A = n_B$ is taken. Table 4 shows mean cost function value, standard deviation for the cost function, applied for the design of hyper-beamforming of linear antenna arrays, obtained by GA, PSO, DE, FFA, SA, AIS, ABC, and FPA. Table 5 shows t test values and p values obtained for FPA and RGA/PSO/DE/FFA/SA/AIS/ABC. The t test values obtained for different degrees of freedom (β) for all the approaches (FPA versus RGA, FPA versus

Table 4 Mean cost function value, standard deviation for the cost function, obtained by various algorithms

Algorithms	Mean cost function value	Standard deviation for the cost function
FPA	0.005189	0.007873
ABC	0.009545	0.01198
AIS	0.03533	0.02959
SA	0.03591	0.03021
FFA	0.09306	0.04149
DE	0.1177	0.04833
PSO	0.1545	0.03987
GA	0.1715	0.04456

Table 5 t test values and p values obtained for comparison of FPA with other algorithms

Algorithms	t test value	p value
PA/ABC	2.148632668041978	0.018315848805205
FPA/AIS	6.960571898179686	3.817690941154694e-009
FPA/SA	6.958263285059677	3.849258578547676e-009
FPA/FFA	14.713150417418458	0
FPA/DE	16.247103664596949	0
FPA/PSO	25.979107912539384	0
FPA /GA	25.988773161860340	0

PSO, and FPA versus DE, FPA versus FFA, FPA versus SA, FPA versus AIS, and FPA versus ABC) are larger than the corresponding critical value [66], meaning that there is a significant difference among FPA and RGA/PSO/DE/FFA/SA/AIS/ABC with 99.9% confidence. Thus, from statistical analysis, it is clear that FPA offers more robust and promising results.

The p value is computed based on the t test according to [66]. If the p value is lower than the implication point ($\alpha = 0.05$, by default), then null hypothesis can be rejected, which proves statistical presentation of the proposed algorithm is better than other algorithms. Table-V shows that all p values obtained for FPA with respect to the other algorithms are much smaller than 0.05. So, it is confirmed that FPA outperforms RGA, PSO, DE, FFA, SA, AIS, and ABC.

5 Convergence Characteristics of Different Algorithms

The algorithms can be compared in terms of the *CF* values, Fig. 9 shows the convergences of *CF* values obtained as RGA [18], PSO [18], DE [18], FFA [18], SA, AIS, ABC, and FPA are employed, respectively. As compared to other algorithms which yield suboptimal higher values of CF, FPA converges to the least minimum *CF* values in finding the near-global optimal current excitation weights and optimal inter-element spacing of hyper-beam of antenna arrays. With a view to the above fact, it may finally be inferred that the performance of FPA is the best among the other algorithms. The minimum *CF* values against number of iteration cycles are recorded

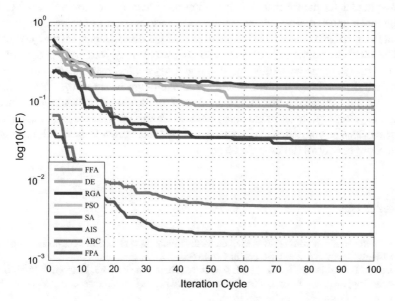

Fig. 9 Convergence profile of FPA and other algorithms for 10-element linear antenna array at $u = 1$

to get the convergence profile for each array set. The simulation programming was done in MATLAB language using MATLAB 7.5 on Dual Core(TM) Processor, 2.88 GHz with 2 GB RAM.

6 Conclusion

In this chapter, flower pollination algorithm (FPA) has been employed for the optimized hyper-beamforming and directivity of linear antenna array. To obtain the high reduction in side lobe level (SLL), the linear array is arranged in such a way that it will generate the sum and the difference patterns and with the help of the exponent parameter the hyper-beam pattern is obtained. Comparison of the results of various algorithms with FPA has been shown. The directivity is calculated by the Simpsons 1/3 method. The directivity obtained by the FPA is much better than those obtained by the conventional linear antenna. From the simulation results, it is clear that the SLL obtained by the FPA is the best among those of RGA, PSO, DE, FFA, SA, AIS, ABC, conventional linear antenna array, and the previous reported results [18]. Statistical results also prove the superiority and significance of the proposed algorithm.

7 Future Research Topics

In this chapter, the nature-inspired algorithms have been applied for the beamforming of the linear antenna array. However, there are various scopes of these algorithms in the field of optimization of pattern synthesis of the other antenna array geometries such as circular, planner, rectangular, and conformal array. Radiation pattern of any antenna array can be controlled by the three array control parameters such as excitation amplitude, inter-element spacing, and excitation phase of the each element. So, this nature-inspired algorithms can be used for the side-lobe-level reduction and radiation pattern synthesis of antenna arrays. Furthermore, the time modulation technique can also be applied in various array geometries to provide an additional degree of freedom to control the radiation pattern.

References

1. Schlieter H, Eigenbrod H (2000) Method for the formation of radiated beams in direction finder systems. February 1 2000. US Patent 6,021,096
2. Isernia T, Ares Pena FJ, Bucci OM, D'urso M, Fondevila Gomez J, Rodriguez JA (2004) A hybrid approach for the optimal synthesis of pencil beams through array antennas. IEEE Trans Antennas Propag 52(11):2912–2918

3. Walker R (1985) Bearing accuracy and resolution bounds of high-resolution beamformers. In: IEEE international conference on acoustics, speech, and signal processing, ICASSP'85, vol 10. IEEE, pp 1784–1787
4. Takao K, Fujita M, Nishi T (1976) An adaptive antenna array under directional constraint. IEEE Trans Antennas Propag 24(5):662–669
5. Schlieter H (2001) Method for three-dimensional beam forming in direction finding systems, January 23 2001. US Patent 6,178,140
6. Balanis CA (2005) Antenna theory analysis and design. Wiley, India
7. Kraus JD (1997) Antenna. TMH Publishing Co., Ltd., New Delhi
8. Anitha V, Lakshmi SSJ, Sreedevi I, Khan H, Ramakrishna KSKP (2012) An adaptive processing of linear array for target detection improvement. Int J Comput Appl (0975–8887) 42(4):33–36
9. Mailloux R (1986) Phased array architecture for millimeter wave active arrays. IEEE Antennas Propag Soc Newsl 28(1):4–7
10. Schrank H (1983) Low sidelobe phased array antennas. IEEE Antennas Propag Soc Newsl 25(2):4–9
11. Applebaum S, Chapman D (1976) Adaptive arrays with main beam constraints. IEEE Trans Antennas Propag 24(5):650–662
12. Chen S (2000) Iir model identification using batch-recursive adaptive simulated annealing algorithm
13. Černý V (1985) Thermodynamical approach to the traveling salesman problem: an efficient simulation algorithm. J Optim Theory Appl 45(1):41–51
14. Kirkpatrick S, Gelatt CD, Vecchi MP (1983) Optimization by simulated annealing. Science 220(4598):671–680
15. Randy L (1997) Haupt. Phase-only adaptive nulling with a genetic algorithm. IEEE Trans Antennas Propag 45(6):1009–1015
16. Haupt RL, Werner DH (2007) Genetic algorithms in electromagnetics. Wiley
17. Chung YC, Haupt RL (1999) Adaptive nulling with spherical arrays using a genetic algorithm. In: Antennas and propagation society international symposium, 1999. IEEE, vol 3. IEEE, pp 2000–2003
18. Ram G, Mandal D, Kar R, Ghoshal SP (2014) Optimized hyper beamforming of receiving linear antenna arrays using firefly algorithm. Int J Microwave Wirel Technol 6(2):181
19. Hardel GR, Yallaparagada NT, Mandal D, Bhattacharjee AK (2011) Introducing deeper nulls for time modulated linear symmetric antenna array using real coded genetic algorithm. In: 2011 IEEE symposium on computers informatics (ISCI), pp 249–254, March 2011
20. Eberhart RC, Shi Y (1998) Comparison between genetic algorithms and particle swarm optimization. In: International conference on evolutionary programming. Springer, pp 611–616
21. Yang X-S (2008) Nature-inspired metaheuristic algorithms. Luniver Press
22. Yang X-S (2009) Firefly algorithms for multimodal optimization. In: International Symposium on Stochastic Algorithms. Springer, pp 169–178
23. Yang X-S (2010) Firefly algorithm, stochastic test functions and design optimisation. Int J Bio-Inspired Comput 2(2):78–84
24. Yang X-S, Hosseini SSS, Gandomi AH (2012) Firefly algorithm for solving non-convex economic dispatch problems with valve loading effect. Appl Soft Comput 12(3):1180–1186
25. Fister I, Yang X-S, Brest J (2013) A comprehensive review of firefly algorithms. Swarm Evol Comput 13:34–46
26. Gandomi AH, Yang X-S, Talatahari S, Alavi AH (2013) Firefly algorithm with chaos. Commun Nonlinear Sci Numer Simul 18(1):89–98
27. Yang Xin-She (2013) Multiobjective firefly algorithm for continuous optimization. Eng Comput 29(2):175–184
28. Yang X-S (2010) Firefly algorithm. Engineering optimization, pp 221–230
29. Kennedy J, Eberhart R (1995) Particle swarm optimization. In: IEEE international conference on neural networks, 1995. Proceedings., vol 4. IEEE, pp 1942–1948
30. Mandal D, Yallaparagada NT, Ghoshal SP, Bhattacharjee AK (2010) Wide null control of linear antenna arrays using particle swarm optimization. In: 2010 Annual IEEE India conference (INDICON). IEEE, pp 1–4

31. Shi Y et al (2001) Particle swarm optimization: developments, applications and resources. In: Proceedings of the 2001 congress on evolutionary computation, 2001, vol 1. IEEE, pp 81–86
32. Durmuş B, Gün A (2011) Parameter identification using particle swarm optimization. In: International advanced technologies symposium (IATS 11), Elazığ, Turkey, pp 16–18
33. Hao Z-F, Guo G-H, Huang H (2007) A particle swarm optimization algorithm with differential evolution. In: 2007 international conference on machine learning and cybernetics, vol 2. IEEE, pp 1031–1035
34. Storn R, Price K (1995) Differential evolution-a simple and efficient adaptive scheme for global optimization over continuous spaces, vol 3. ICSI Berkeley
35. Storn R, Price KV (1996) Minimizing the real functions of the icec'96 contest by differential evolution. In: International conference on evolutionary computation, pp 842–844
36. Price KV, Storn RM, Lampinen JA (2005) Differential evolution a practical approach to global optimization
37. Lin C, Qing A, Feng Q (2009) Synthesis of unequally spaced antenna arrays by a new differential evolutionary algorithm. Int J Commun Netw Inf Secur 1(1):20–26
38. Lin C, Qing A, Feng Q (2010) Synthesis of unequally spaced antenna arrays by using differential evolution. IEEE Trans Antennas Propag 58(8):2553–2561
39. Rocca P, Oliveri G, Massa A (2011) Differential evolution as applied to electromagnetics. IEEE Antennas Propag Mag 53(1):38–49
40. Yap DFW, Koh SP, Tiong SK, Sim EYS, Yaw MW (2011) Artificial immune algorithm based gravimetric fluid dispensing machine. In: 2011 11th international conference on hybrid intelligent systems (HIS). IEEE, pp 406–410
41. Castro LN, Timmis JI (2003) Artificial immune systems as a novel soft computing paradigm. Soft Comput 7(8):526–544
42. Graaff AJ, Engelbrecht AP (2007) A local network neighbourhood artificial immune system for data clustering. In:2007 IEEE congress on evolutionary computation. IEEE, pp 260–267
43. Timmis J, Neal M (2001) A resource limited artificial immune system for data analysis. Knowl Based Syst 14(3):121–130
44. Dasgupta D, Ji Z, González FA et al (2003) Artificial immune system (ais) research in the last five years. IEEE Congr Evol Comput 1:123–130
45. Karaboga D, Basturk B (2008) On the performance of artificial bee colony (abc) algorithm. Appl Soft Comput 8(1):687–697
46. Karaboga D, Basturk B (2007) A powerful and efficient algorithm for numerical function optimization: artificial bee colony (abc) algorithm. J Global Optim 39(3):459–471
47. Rajo-Iglesias E, Quevedo-Teruel O (2007) Linear array synthesis using an ant-colony-optimization-based algorithm. IEEE Antennas Propag Mag 49(2):70–79
48. Mandelbrot BB (1982) The fractal geometry of nature
49. Kleinberg MJ (2000) Navigation in a small world. Nature 406(6798):845–845
50. Li G, Reis SD, Moreira AA, Havlin S, Stanley HE, Andrade JS Jr (2010) Towards design principles for optimal transport networks. Phys Rev Lett 104(1):018701–018701
51. Mantegna RN (1994) Fast, accurate algorithm for numerical simulation of levy stable stochastic processes. Phys Rev E Stat Phys Plasmas Fluids Relat Interdisc Top 49(5):4677–4683
52. Yang X-S (2010) Firefly algorithm, levy flights and global optimization. In: Research and development in intelligent systems XXVI. Springer, pp 209–218
53. Yang X-S, Deb S (2009) Cuckoo search via lévy flights. In: World congress on nature & biologically inspired computing, 2009. NaBIC 2009. IEEE, pp 210–214
54. Pavlyukevich Ilya (2007) Lévy flights, non-local search and simulated annealing. J Comput Phys 226(2):1830–1844
55. Yang X-S (2012) Flower pollination algorithm for global optimization. In: International conference on unconventional computing and natural computation. Springer, pp 240–249
56. Yang X-S, Karamanoglu M, He X (2014) Flower pollination algorithm: a novel approach for multiobjective optimization. Eng Optim 46(9):1222–1237
57. Yang X-S (2014) Swarm intelligence based algorithms: a critical analysis. Evol Intell 7(1):17–28

58. Yang X-S, Karamanoglu M, He X (2013) Multi-objective flower algorithm for optimization. Proc Comput Sci 18:861–868
59. Yang X-S (2014) Nature-inspired optimization algorithms. Elsevier
60. Yang X-S http://www.mathworks.com/matlabcentral/fileexchange/45112-flower-pollination-algorithm
61. Waser NM (1986) Flower constancy: definition, cause, and measurement. Am Nat 593–603
62. Alam DF, Yousri DA, Eteiba MB (2015) Flower pollination algorithm based solar pv parameter estimation. Energy Convers Manag 101:410–422
63. Łukasik S, Kowalski PA (2015) Study of flower pollination algorithm for continuous optimization. In: Intelligent systems' 2014. Springer, pp 451–459
64. Dubey HM, Pandit M, Panigrahi BK (2015) Hybrid flower pollination algorithm with time-varying fuzzy selection mechanism for wind integrated multi-objective dynamic economic dispatch. Renew Energy 83:188–202
65. Wang R, Zhou Y (2014) Flower pollination algorithm with dimension by dimension improvement. Math Prob Eng
66. Walpole RE, Myers RH, Myers SL, Ye K (1993) Probability and statistics for engineers and scientists, vol 5. Macmillan New York
67. Koç SNK, Köksal Ad (2011) Wire antennas optimized using genetic algorithm. Comput Electr Eng 37(6):875–885
68. Kułakowski P, Vales-Alonso J, Egea-López E, Ludwin W, García-Haro J (2010) Angle-of-arrival localization based on antenna arrays for wireless sensor networks. Comput Electr Eng 36(6):1181–1186
69. Zhang X, Feng G, Gao X, Dazhuan X (2010) Blind multiuser detection for mc-cdma with antenna array. Comput Electr Eng 36(1):160–168

Multi-Agent Optimization of Resource-Constrained Project Scheduling Problem Using Nature-Inspired Computing

Pragyan Nanda, Sritam Patnaik and Srikanta Patnaik

Abstract Multi-agent systems (MAS) are now being used for handling the complexity existing in the fast-changing business scenario. Scheduling of various activities and processes plays a major role in proper utilization of resources and timely completion of projects undertaken in various business management environments. Resource-constrained project scheduling problem (RCPSP) is one of the most challenging scheduling problems of business domain and MAS can capture well the complexities lying in the relationship between various activities and the resources that are scarcely available. Moreover, nature-inspired computing approaches can be integrated to multi-agent systems to enable the system to adapt to the dynamic and uncertain circumstances arising due to continuously changing business environments. This chapter presents RCPSP as a multi-agent optimization problem and studies how various nature-inspired approaches such as PSO, ACO and SFLA have been used for solving the RCPSP.

1 Introduction

Generally speaking, in the domain of e-commerce, supply chain management (SCM) and logistics, manufacturing, construction, telecommunications, insurance, leasing, health care, etc., involves dynamic and complex process to achieve their targets. In particular, the e-commerce has grown so fast in the past few years but at the

P. Nanda (✉) · S. Patnaik
Faculty of Engineering and Technology, Department of Computer Science
and Engineering, SOA University, Bhubaneswar 751030, Odisha, India
e-mail: n.pragyan@gmail.com

S. Patnaik
e-mail: patnaik_srikanta@yahoo.co.in

S. Patnaik
Department of Electrical and Computer Engineering, School of Computer Engineering,
National University of Singapore, Singapore 117583, Singapore
e-mail: sritampatnaik@u.nus.edu

© Springer International Publishing AG 2017   217
S. Patnaik et al. (eds.), *Nature-Inspired Computing and Optimization*,
Modeling and Optimization in Science and Technologies 10,
DOI 10.1007/978-3-319-50920-4_9

same time increase in consumer demands and interconnected markets has made the business scenario complicated. This needs development of a flexible and adaptive system for business process management. Traditional approaches such as system dynamics, mathematical and analytical models are not able to design the complex business scenarios as they need (i) exact specification of the requirements and (ii) centralised control over various operations. The traditional models fail to adapt to dynamic and uncertain circumstances as these models are problem specific and lack flexibility [3, 32, 34, 46]. Moreover, interpretations of these traditional models vary from individual to individual making it difficult for generalization.

To achieve success in any business environment, proper modelling of the system, planning the workflow process, scheduling and utilization of various activities and resources in timely manner is essential. Proper modelling of these complex systems is considered as the first step in achieving the goal. Although traditional approaches are able to capture the nonlinearity and dynamics of an organization, they model a system on the basis of certain assumptions which may not be applicable to real-time scenarios. On the other hand, in real-time scenarios the various participants in an organization such as the people working in the organization, the projects being carried on, the instruments or other assets required for the completion of the projects, the products obtained as the output of the completion of the projects possess their own individual characteristics that may or may not change according to dynamic situations, rules that may or may not vary for different goals, relationships with other participants and goals. These factors are mostly ignored in traditional modelling approaches but are considered as important while modelling the system using multi-agent systems [38].

1.1 Multi-agent System

Multi-agent systems (MAS) are well-suited for modelling various business environments as they are able to capture the complexity involved between the different states of individual processes as well as the relationships existing among various processes [26, 30, 34]. In addition to modelling, multi-agent systems can also be used for forecasting resource requirements to meet resource scarcity and compare various existing scenarios to guide the forecasting strategy, and finally, it can also be used for optimizing various resource scheduling strategies among competing processes for successful completions of the projects. In other words, multi-agent systems used for modelling of various business environments can be optimized for solving various decision-making problems such as (i) efficient usage of scarce and non-renewable resources, (ii) maintaining the relationships between the internal states of various processes and activities involved and (iii) finally cost and time optimization in the presence of various constraints such as time, precedence and resource constraints [9, 34]. Moreover, multi-agent systems suit best for applications with large-scale populations as each individual agent interacts with each other and tries to achieve its individual goal which in turn collectively results in the achievement of the ultimate goal of the system. For example, let us consider the case of supply chain system, in which the individual participants of the system consist of the various companies involved in buying, selling and supplying goods, the producers of the goods, the wholesalers selling the goods to the retailers and the retailers selling goods to the customers directly. Now, all these individual participants having their own individual goals and rules can be modelled as agents and the entire supply chain system as a multi-agent system. Similarly, the products or projects of a company can be modelled as a goal of a multi-agent system and the processes or individual activities involved in the successful completion of the project can be modelled as agents representing the dynamics and internal states of the company.

1.2 Scheduling

In business domain, the competition among various companies has exponentially increased in the markets due to the increasing growth in customer demands and expectations in terms of cost, quality and delivery time. The organizations have to handle the volatility existing in markets and adapt to the dynamic and uncertain situations efficiently by having flexible production systems. These production systems must be capable of meeting market needs and demands efficiently by properly organizing (i) production cycles, (ii) delivery dates and (iii) inventory. Scheduling plays a major role in solving the crucial problems in business production and organization domain. Scheduling properly arranges various activities and resources in an organization to successfully attain the production plan in order to meet the market trends and demands [21, 46]. Most of the scheduling problems are combinatorial problems but limited availability of resources and the precedence order of execution of the activities make them hard to solve. Failing in proper scheduling of activities and resources may lead to delay in the delivery of products leading to loss of revenue and reputation of the company. Various classical scheduling problems include job-shop scheduling problem, flow-shop scheduling problem, production scheduling problem, distribution scheduling problem, resource-constrained project scheduling problem, employee scheduling problem and transport scheduling.

One of the most widely addressed problems is the resource-constrained project scheduling problem which plans the scheduling of various activities involved in a project while optimizing the scarce availability of resources along the value chain. The resource-constrained project scheduling problem (RCPSP) is usually considered as a classical scheduling problem as it draws the attention of lots of researchers. The objective of the RCPSP is minimizing the total project make-span by scheduling various activities of a project in such a way that the precedence relationship between the activities and availability of limited resources are not hampered [1, 2, 42]. Traditional techniques such as branch and bound algorithms are capable of solving the RCPSP but are not suitable for large-scale applications as increase in number of activities increases computational time making it an NP-hard problem. Further, various heuristic and metaheuristic approaches have also been suggested in various literatures [7, 22, 23, 29, 33, 36], to find the near-optimum solutions but due to their problem specific nature, they cannot be applied to the new problems consisting of uncertainty or even to new instances of similar type of problems. However, the nature-inspired approaches have recently been used by many researchers for solving the classical RCPSP as well as the extended versions of RCPSP.

1.3 Nature-Inspired Computing

However, natural systems although complex by nature are capable of surviving in any kind of uncertain and extreme environments. This lead to the emergence of nature-inspired computing paradigm which draws inspiration from the adapting, self-organizing and collective behaviour of various natural processes and systems for solving many real-time complex problems [40, 50, 53]. Some of the widely used nature-based computing systems include the evolutionary algorithms that take inspiration from the Darwinian principles for survival of the fittest, the artificial neural networks based upon the nervous systems, ant colony optimization inspired by the foraging behaviour of ants, flocking behaviour of birds and fish schooling gave rise to the particle swarm optimization, foraging and mating behaviour of honey bees inspired artificial bee colony optimization and honeybee mating optimization, respectively. Some of the recently developed approaches include intelligent water drop optimization, invasive weed optimization, grey wolf optimization and gravitational search algorithm are inspired by the river formation method, hunting behaviour and how gravitational force works. In all the above-mentioned approaches, the system comprises of individual elements that perform the task assigned to them, communicate and interact with each other and the final outcome is the result of the collective behaviour of the individual elements. However, all these elements are autonomous, independent and capable of adapting to dynamic changes occurring in environment. These elements further self-organize themselves by interacting with other elements in order to tune their behaviour according to the environment. Moreover, due to the flexibility of the nature-inspired computing approaches they can be widely applied to various complex problems.

Thus, this chapter provides a detailed study of how the RCPSP has been solved by various nature-inspired approaches. The remainder of the chapter in organized as follows: Sect. 2 provides the background study of various approaches and the RCPSP in brief. Section 3 studies various existing approaches for solving RCPSP. Finally, Sect. 4 concludes the chapter.

2 Resource-Constrained Project Scheduling Problem

Resource-constrained project scheduling problem (RCPSP) is a classical problem that has attracted many researchers and academicians to develop various exact and heuristic scheduling procedures. The classical RCPSP consists of various activities which are supposed to be scheduled for the completion of the project in the presence

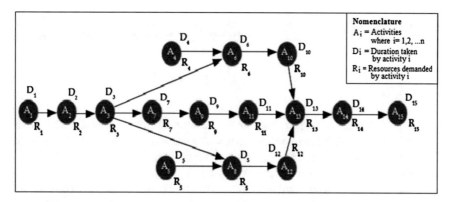

Fig. 1 Activity on node diagram

of precedence and resource constraints. It is usually characterized by the objective function of the problem, the features of the resources and the pre-emption conditions. Usually, minimization of the make-span of the project under consideration is considered as the objective of the resource-constrained project scheduling problem while minimizing total project cost and levelling of resource usage can be other possible objectives of the RCPSP [1, 27, 42]. However, the basic model is too restrictive for real-life applications leading to development of various extensions of the basic RCPSP. Cheng and Zhou [10] give an overview over these extensions in their paper and classify them according to the structure of the RCPSP. A schematic diagram of the resource-constrained project scheduling is given below in Fig. 1 showing the arrangement of activities of a project on activity network. Each node represents an activity and is associated with two values, the upper one representing the duration to be taken to complete the activity and the lower one representing the number of resources being demanded by the activity for successful completion of the activity. The arrows show the order of precedence the activities should follow and dependency on other activities.

Thus, the resource-constrained project scheduling problem (RCPSP) as discussed above can be viewed as a minimization problem where the make-span of the project is supposed to be minimized considering various resource constraints existing as well as scheduling the activities to be done in the required precedence order.

Mathematically, the RCPSP can be defined by a tuple (A, D, R, P, B), where A represents the set of activities required for the successful completion of the project and is given by $A = \{A_0, A_1, A_2, ..., A_n, A_{n+1}\}$, A_0 and A_{n+1} being two dummy activities representing the beginning and end of the schedule. The durations of the activities are given by a vector $D = \{D_0, D_1, D_2...D_n, D_{n+1}\}$ that stores the amount of time required for the completion of each activity. The time taken by the dummy activities d_0 and d_{n+1} for completion is considered as zero. The set of resources demanded by each activity is represented by $R = \{R_1, R_2...R_n\}$ and the number of units of each resource available is given by a vector $B = \{B_1, B_2...B_n\}$ such that B_i represents the number of units of R_i available for use. The precedence relation between two activities is represented as a set P, consisting of pairs of activities such that the pair $(A_i, A_j) \in P$ indicates that the former activity precedes the later. The set of activities constitute a schedule together. Further, the make-span of a schedule is given by

$$Make\text{-}span = \sum_{i=0}^{n+1} D_i \quad \forall A_i \in A \tag{1}$$

Traditional tools for project scheduling such as critical path method, critical chain analysis, resource levelling analysis, project evaluation and review technique and Gantt chart are usually used to analyse the minimum time required by various activities involved in a project in order to complete the project in hand. However, these tools work without considering the limited availability of resources in real-life due to cost or space factor, and moreover, the increasing personalization and customization of products due to diverse demands and other related constraints has made the structures of projects complicated. Therefore, finding a reasonable solution to the project scheduling problem which balances the resource availability and requirements along with project cost and duration to be completed is being considered as a NP-hard problem. Two types of approaches are usually considered to solve the RCPSP problem: (i) precise algorithms and (ii) heuristic-based approaches [10]. The precise algorithms such as linear programming and branch and bound approaches are suitable for small-scale projects. But for real-time projects are usually large-scale projects and precise approaches are quite expensive, both in terms of time and cost, whereas heuristic-based approaches are more suitable as they provide approximate solutions by using the heuristic information to search the optimal solutions rapidly. But in heuristic-based approaches, usually there are no systematic methods to select an appropriate heuristic rule and also heuristic-based approaches are usually trapped in local optima. Various metaheuristic approaches such as GA, PSO and ACO are also proposed as an alternative to solve the RCPSP.

MRCPSP is an extension of the classical RCPSP, where each of the activities can be executed in more than one mode and the objective is to find the mode in which the activity performs the best. The general MRCPSP problem addresses the time factor of a project as the objective; that is, it tries to minimize the make-span of a project, but it can further be transformed into a multi-objective problem by considering the cost and quality factors along with the time. This variant of the classical RCPSP is known as the TCQTP which stands for time–cost–quality trade-off problem; that is, it tries to maintain a balance among all the three factors by minimizing total time and cost while maximizing the quality [11, 17, 20, 44, 45, 49].

3 Various Nature-Inspired Computation Techniques for RCPSP

Since last few years, nature-inspired computing has been receiving huge popularity as they are capable of handling the complexity of large and dynamic systems. They provide efficient and acceptable solutions to the complex project scheduling problems which are otherwise known to be NP-hard problems. Some of the nature-inspired approaches adopted for solving the resource-constrained project scheduling problem are discussed below.

3.1 Particle Swarm Optimization (PSO)

PSO is an evolutionary optimization algorithm developed by Kennedy and Eberhart [16]. They defined a mathematical function as fitness function for each individual, and each individual element of the group is termed as particle and the group itself is termed as swarm and hence together represented swarm of particles. The particle swarm optimization algorithm is an evolutionary approach that draws inspiration from the social behaviour projected by individuals in a flock of birds or a school of fish. This social behaviour of the individual particles gets influenced by its own past experience as well as the past experience of its neighbourhood.

The procedure of PSO algorithm is given below:

  (i) First, a population of N number of particles is generated in the solution hyper-space.

 (ii) The initial positions x^0 and velocities v^0 of the particles are randomly assigned.

(iii) Each individual particle is usually represented using a vector of length d where d represents the dimension of the domain of the fitness function.

(iv) Once initialized, the fitness function of each particle is evaluated in each iteration and results in updating of the position and velocity of each particle.

 (v) Each particle remembers its own best position till the current iteration known as local best position and communicates the same with others.

(vi) The global best position in the entire swarm is then evaluated, and the position and velocity of each particle are updated with respect to the global best position of the swarm.

The equations for velocity and position are updated are as follows:

$$V_{i+1} = w_i V_i + c_1 r_1 (P_{best} - X_i) + c_2 r_2 (G_{best} - X_i) \tag{2}$$

$$X_{i+1} = V_{i+1} + X_i \tag{3}$$

where i = 1, 2 ... M.

V_i and V_{i+1} represents the velocity of the particle at current and new positions, respectively, while X_i and X_{i+1} represent the current position and the new position of the particle.

(vii) Next, P_{best} represents the local best position of the ith particle with respect to the fitness function evaluated whereas G_{best} represents the global best among the entire population.

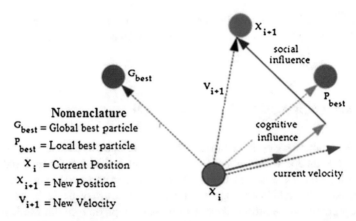

Fig. 2 Particle swarm optimization   .

(viii) The two positive constants c_1 and c_2 are the learning factors while r_1 and r_2 are the random numbers generated between $(0, 1)$ and finally, w_i is the inertia weight which controls the trade-off between exploration and exploitation.

(ix) Equation (1) computes the new velocity of the particle considering its previous velocity and its current distance from its local best and global best positions. And Eq. (2) computes the new position of the particle according to the new velocity and its current position. Figure 2, shows the progress mechanism of PSO where the new position of the particle is influenced by the local and global best particles and these influences are known as social and cognitive influence, respectively. The new position is first influenced by the current velocity and weight; next, it is influenced by the local best, and direction is changed accordingly, and finally, the direction of new position is influenced by the global best as shown below, and the new position is thus computed.

3.2 Particle Swarm Optimization (PSO) for RCPSP

A PSO-based approach was proposed by Zhang et al. to solve the resource-constrained project scheduling problem in [55]. They have considered two solution representations, namely the permutation-based and priority-based representations for developing the PSO-based framework. The objective of the proposed approach is to minimize the duration of the project, and the potential solutions to the resource-constrained project scheduling problem are represented by multidimensional particles in the search space. In an n-dimensional search space, Zhang et al. utilize priority-based representation by using an array consisting of n-elements to represent a PSO particle, and each particle represents a possible solution in the multidimensional space. Each element of the individual particle has been considered to represent an activity, and the position parameter of PSO represents the priority value assigned to the activity. Further, positioning of the parameters corresponds to the index numbers of an activity and the orders in which the elements are placed correspond to the sequence the activities are supposed to be accomplished. They initiate a permutation-based representation where the order of an activity indicates its priority in the sequence of the activities organized to start. Permutation-based representation is used satisfy precedence constraints. The velocity of the PSO particle corresponds to the gap between the current sequence and the local or global best; larger gaps indicate the increased probability of the particle representation to be updated. A hybrid particle-update mechanism has been used to update the parameters.

Next a serial generation method has been used to transform the solution to a feasible schedule, considering precedence and resource availability constraints. The serial generation method converts the solution into an active schedule by including the starting times with resource allocations for the activities. In the serial generation method, only a single activity is selected on the basis of the decreasing order of the priority values and availability of the required resources and is scheduled at each stage. However, if the amount of resources available is insufficient to finish the activity, the activity will be rescheduled to the earliest time its predecessors will be completed and resource feasible times. Zhang et al. have investigated the performance of the proposed approach while considering different solution approaches, through experimental analysis. Various parameters considered for the comparison purpose are (i) the inertia weight, (ii) the learning factors and (ii) the population size of particles (activities) in the project under study. They take advantage of the PSO-based features such as searching experience of a particle itself and from the experience all other particles. The computational results were studied in terms of (i) average deviation, (ii) maximum deviation and (iii) optimal project duration. While comparing the two solution representations, they have claimed that permutation-based representation outperforms priority-based representation considering permutation feasibility and precedence constraints. Again while comparing the PSO-based approach with other metaheuristic-based approaches, they have stated that the PSO-based approach shows better performance over tabu search, simulated annealing and genetic algorithm-based approaches for the RCPSP in searching feasible schedules.

Zhang et al. further extend the above approach by considering the priority-based representation in their paper [56] for solving the resource-constrained project scheduling problem. Instead of permutation of activity sequences, the proposed approach considers priority values of the activities as potential solutions to the scheduling problem. In the solution search space, PSO particles represented the potential solutions and each of the particles can be converted into a feasible schedule by a parallel technique which evaluates the potential solutions, considering both the resource and precedence constraints during the search process. The resource-constrained project scheduling problem considered by Zhang et al. assumes that (i) the duration of the activities of a construction project are known, (ii) precedence constraints exists, (iii) multiple resource constraints exists (varieties, amount available and renewability), (iv) non-pre-emptive nature of the activities and finally (v) minimizing project duration being the managerial objective.

According to the authors, different priority rules such as shortest activity duration (SAD), minimum late finish time (MILFT) or minimum total float (MITF) can be used for resolving conflicts arising between various activities due to competition for limited resources. But, since no systematic method exists for the selection of the priority rules, they adopt a local-search-based multiple-pass heuristic method [6] for selecting the best rule for solution. This multiple-pass heuristic method adjusts the starting times of some activities to improve the solution which in turn leads to a better schedule and hence forms the basis for PSO-based mechanism. The initial priorities were randomly initialized. Zhang et al. adopt the PSO-based representation of activities and parameters from their previous work [55]. The priority-based representation of the activities avoids infeasible sequences and transforms it into a feasible schedule considering precedence and resource constraints.

To begin with the proposed approach begins with the initialization part where the initial priorities, positions and velocities are randomly assigned. Then, the search process starts and the local best and global best particles are evaluated according to the fitness function. The next iteration executes the velocity and position updation processes according to Eqs. (1) and (2), respectively. Once the velocity and position of the priority-based particles are updated, the parallel transform scheme starts transforming each particle to a feasible schedule and duration of each schedule is listed. Again after the generation of the feasible schedules, the local best particles and the global best particle are evaluated. If the stopping criteria (e.g. maximum number of iteration or maximum iterations with stable global best) have been met, then the procedure will stop; otherwise, the procedure will be continued. For computational analysis, Zhang et al. considered a project with three types of renewable resources and 25 activities. They claimed PSO to outperforming GA and other heuristic-based approaches in obtaining the optimal schedule with shortest duration due to its past experience sharing mechanism. Moreover, with limited parameters PSO proves to be an efficient and easy to implement alternative for solving RCPSP.

Although various heuristic and metaheuristic approaches have been proposed in literature [7, 22, 23, 29, 33, 36], to find near-optimum solutions in considerable computational time, but their applicability to new problems or even new instances of similar type of problems is limited by their problem specific nature.

Hyper-heuristic methods are proposed to overcome the above-mentioned limitations by developing generic methods to produce acceptable solutions on the basis of good low-level heuristics. Burke et al. [8] proposed a definition for the term "hyper-heuristic" as a "learning mechanism for selecting or generating heuristics to solve computational search problems" in their paper [8]. They also classified the hyper-heuristic method into two major categories, namely (i) heuristic selection and (ii) heuristic generation. The heuristic selection deals with choosing existing heuristics whereas heuristic generation deals with generation of new heuristics from existing ones.

Koulinas et al. propose a particle swarm optimization (PSO)-based hyper-heuristic algorithm (PSO-HH) [24] that handles various solution methods which in turn controls several low-level heuristics and finally selects a suitable one or a sequence of low-level heuristics to solve the resource-constrained project scheduling problem

(RCPSP). In their proposed approach, random keys form the basis of the solution and a serial scheduling generation scheme is used to generate active schedules employing the activity priority which are again modified by the low-level heuristics. They adopt an acyclic activity-on-node (AoN) network topology for the project representation that consists of n activity nodes, n precedence relations between activities and R renewable resources. Koulinas et al. consider the minimization of the project duration maintaining the precedence relations as the objective of RCPSP. The start time of the project is assumed to be at zero while resource unit availability is considered to be constant throughout the progress of the project activity. The PSO particles representing the project activities are implemented using vectors consisting eight integer numbers where each of the number represents a low-level heuristic algorithm and the particles maintain the precedence relations of the low-level heuristic algorithms. As the fitness function is evaluated, the sequence in which the low-level heuristics are applied to find the best solution is stored for every individual particle. Similarly, the global best solution is evaluated and stored for further use. The vectors are then assigned priorities using the serial scheduling generation scheme (SSGS) that not only generates a feasible schedule but also calculates the make-span of the project, and theses sequences are further modified to improve local search. Also the subsequent velocities and inertia weights are computed and updated, further on the basis of these updated values the next position of each particle is generated. They have considered the maximum number of total iterations as the stopping criteria.

They have also considered three randomly chosen network instances with four resources from the problem instance library PSPLIB and compared the convergence history of the algorithm to validate the functionality and effectiveness of the proposed PSO-HH approach. They further observed that the parameter settings are acceptable as the comparison curves indicate that the algorithm converges relatively quickly. Moreover, the average CPU time cost of PSO-HH is acceptable and increases linearly with the increase in the number of schedules for each set. Also the adoption of hyper-heuristic approach makes the proposed technique more flexible and problem-independent since information regarding how low-level heuristics work is not required.

3.3 Ant Colony Optimization (ACO)

Marco Dorigo and his colleagues introduced ant colony optimization (ACO) in the early 1990s [12–15], as a nature-inspired metaheuristic for the solution of hard combinatorial optimization problems. Ant colony optimization (ACO) takes inspiration from the foraging behaviour of ants. The foraging behaviour of ants shows that ants wander randomly to explore the area surrounding their nest in search of food source. Once the food source is found, it evaluates the quantity and quality of the food and returns back to its colony leaving trails of pheromones on the path to mark the way to the food source. When other ants are wandering and searching for food, come across the pheromone trail, they decide probabilistically whether to follow the path or not.

If they follow the path, the pheromones generated by the new ants while bringing food back to the colony will make the trail stronger and hence will attract more ants. Since the ants drop pheromones each time they return with food, the shortest paths will be the stronger one. But as soon as the food source is emptied out, the path decays slowly due to the decrease in number of ants following the path and the evaporating nature of pheromone. The ant colony optimization is based upon this foraging behaviour of ants, and the goal is to find the optimal path between the food source and their nest. Further, the ACO works on a dynamic system and hence can be applied problems dealing with changing topology and having a requirement to find shortest path or optimal path such as computer networks and travelling salesman problem.

As discussed above, the ant colony optimization is a population-based metaheuristic technique used to find approximate solutions. ACO consists of a set of artificial ant agents that search for optimal solutions in a search space. To utilize the ACO, the targeted problem is usually transformed into a weighted graph problem and the objective to be achieved is searching the best path on the graph. The ant agents use a stochastic process to construct the solution by traversing the graph, and this solution building process is controlled by a pheromone model consisting of a set of parameters associated with each graph component.

Formal Model for a Combinatorial Optimization Problem

The formal model of a combinatorial optimization problem as defined by Dorigo et al. consists of a triplet (S, Ω, f), where

S is a search space defined over a finite set of discrete decision variables X_i, $i = 1, \ldots, n$;

Ω is a set of constraints among the variables; and

An objective function such as $f : S \rightarrow R_0^+$ *that* is to be minimized (since any maximization problem can be reduced to a minimization problem).

The search space is defined as follows:

The generic variable X_i takes values in $D_i = \{v_i^1 ..., v_i^{|D_i|}\}$. A feasible solution $s \in S$ is a complete assignment of values to variables that satisfies all constraints in Ω. Although several ACO algorithms have been proposed, the main characteristic of the original ACO is that the pheromone values are updated by all the solution ants in each of the iterations. The pheromone τ_{ij}, associated with the edge joining the adjacent nodes i and j, is updated as follows:

$$\tau_{ij} \leftarrow (1 - \rho).\tau_{ij} + \sum_{k=1}^{m} \tau_{ij}^k \tag{4}$$

where ρ is the evaporation rate, m is the number of ants, and $\Delta \tau_{ij}^k$ is the amount of pheromone laid on the edge connecting the nodes (i, j) by ant k and is given by:

$$\Delta \tau_{ij}^k = \begin{cases} Q/L_k & if\ the\ \text{edge is already visited} \\ 0 & \text{otherwise} \end{cases} \tag{5}$$

Fig. 3 Ant colony optimization

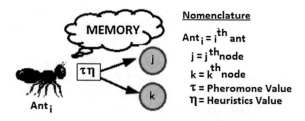

Nomenclature

$Ant_i = i^{th}$ ant

$j = j^{th}$ node

$k = k^{th}$ node

τ = Pheromone Value

η = Heuristics Value

where Q is a constant and L_k is the length of the tour constructed by ant k.

While constructing a solution, the next node to be visited by the ants is selected using a stochastic approach as shown in the Fig. 3. When ant k is at node i and has constructed a partial solution S^P, the probability of selecting node j as the next node is given by:

$$P_{ij}^k = \begin{cases} \frac{\tau_{ij}^{\alpha} \cdot \eta_{ij}^{\beta}}{\sum \tau_{ij}^{\alpha}} & \text{if } c_{ij} \in N(S^P) \\ 0 & \text{otherwise} \end{cases} \tag{6}$$

where $N(S^P)$ is the set of feasible components, i.e., the edges that are not visited by the ant k. The parameters α and β control the relative importance of the pheromone against the heuristic information η_{ij}, given by:

$$\eta_{ij} = \frac{1}{d_{ij}} \tag{7}$$

where d_{ij} is the distance between the nodes.

3.4 Ant Colony Optimization (ACO) for RCPSP

Zhang proposed an ant colony-based approach to solve the multi-mode resource-constrained project scheduling problem (MRCPSP) in his paper [57]. In their approach, Zhang has considered two different pheromone types for reflecting the memory of the ant, one for the activity sequence formation and the other is the selection of the mode. He provides the probabilistic and heuristic information for each corresponding pheromone and transforms the ACO solutions into feasible schedules employing the serial schedule generation scheme (SSGS). Previously, the ACO-based approach has been applied to solve the single-mode RCPSP, where each individual ant builds a solution for RCPSP, by finding an optimal sequence of activities required for the completion of the project by evaluating the probability of their feasibility with respect to the resource and precedence constraints. The probability of each selection evaluates how good the sequence of the activities selected is, and it is calculated using both the pheromone and heuristic information. The pheromone represents the past memory of feasible findings by individual ants while the heuristic value carries information related to the problem in hand. Once the activity has been identified in the activity list, the pheromone is deposited on the feasible selection by incrementing the pheromone value in each of the iteration. Again after the completion of each cycle, the pheromone values are decremented by certain amount to reflect the evaporation of pheromone with time. The declining of pheromone also prevents the stagnation in the searching operation, and the entire process continues until the stopping criteria are met.

However, in case of MRCPSP, the activities of a project can be executed in more than one mode. Therefore, along with the feasible sequence in which the activities can be optimally scheduled, the mode of execution is supposed to be considered for selecting a feasible schedule. First, the basic parameters such as the two types of pheromones, the initial solution and the feasible schedule, and the iteration count are initialized. Then, a set of eligible activities in the activity list are selected and their probability as well as the respective heuristic function are evaluated and on the basis of this evaluation the next feasible activity is selected. Next, a list of possible modes in which the selected activity can be executed is generated and both the second type of probabilistic and heuristic functions are evaluated and the executable mode is selected. In their approach for solving MRCPSP, two types of heuristic functions are adopted in order to address the project duration, the first one being minimum total slack which is responsible for the selection of the activities with smaller values for total slack. The second heuristics is the shortest feasible mode that gives priority to the mode having activities with short durations. After the feasible activity and the corresponding execution mode has been selected, a set of solution schedules are generated followed by evaluation and ranking of the best solutions by computing the project durations. This leads to the identification of the global best solution. Finally, the values of both the types of pheromones are updated by evaporation and increment equations. This entire process of searching, evaluating ranking and updation is carried out until the stopping criterion is met. The various parameters crucial for tuning the performance of their approach include the factors determining the influence of the pheromone and heuristic values on the probabilistic decision of an ant; the total amount of pheromone deposited by an individual ant in each search cycle; total number of ants and total number of elitist ants generated on the basis of the ranking index; maximum number of cycles defined for the search operation and maximum number of cycles for which the value of the global best is steady.

Thus, Zhang presents ACO as a metaheuristic-based approach for solving the MRCPSP by employing two separate levels of pheromones and heuristic information for the mode and sequence selection of the feasible activities. They reported their approach as an effective alternative to solve MRCPSP and can help practitioners to plan and utilize resources optimally for successful accomplishment of the project.

3.5   Shuffled Frog-Leaping Algorithm (SFLA)

Eusuff et al. [18] proposed a population-based metaheuristic algorithm inspired by natural memetics that performs a heuristic search to find a global optimal solution for solving discrete combinatorial optimization problems. SFLA is modelled by mimicking the foraging behaviour of frogs, where food is laid on randomly located stones in a pond. The SFLA combines the advantages of two metaheuristic approaches, namely the memetic algorithm (shuffled complex evolution algorithm (SCE)) and the particle swarm optimization (PSO), for solving hard combinatorial optimization problems. It considers the evolution of memes as a result of interaction between individuals as a base for the development of the algorithm. A meme in general means a contagious idea or behaviour or information pattern that spreads from one individual to another belonging to a particular culture by replicating itself. In other words, a meme acts as a carrier that carries cultural ideas or practices from one brain to another. A meme consists of memotypes in a way similar to the chromosomes of a gene.

Memetic algorithms are a class of algorithms gaining wide acceptance due to their ability to solve large combinatorial optimization problems where other metaheuristics have failed. The SFLA starts with a randomly generated population of potential solutions where each solution is represented by a frog. The SFLA divides the entire population of frogs into several partitions referred to as memeplexes. The memeplexes or partitions represent different cultures of frogs and allows them to evolve independently by performing local search (exploitation) in the search space. The local search is influenced by PSO in that the ideas of individual frogs in each memeplex can be improved by interacting and exchanging information with other frogs leading to memetic evolution. The frogs with worst fitness value can change improve their memes by using the information obtained from the memeplex best or best frog in the entire population. The frogs' new position corresponds to the new meme along with the incremental changes in memotypes corresponding to a leaping step size similar to PSO. After the exploitation process is over in each memeplex, the information in the memeplexes is shuffled among each other after some generations of each memeplexes for global evolution, i.e. extending the local search to global search process (exploration). The shuffling process is employed to enhance the quality of the memes. The local search within memeplexes and shuffling of memeplexes continue until convergence condition is met. Hence, it combines the goodness of social behaviour of PSO with evolutionary algorithm. SFLA consists of few parameters to be initialized as compared to other metaheuristics such as (i) number of frogs called population size (P) or initial solutions, (ii) number of memeplexes (m) (iii) number of frogs in each memeplex (n). Figure 4 shows the division of the entire frog population into memeplexes:

Basic steps of SFLA are as follows:

(i) The initial population of frogs (P solutions) is randomly generated in a D-dimensional search space. Each individual frog (solution) "i" is represented as $X_i = (X_{i1}, X_{i2}, \ldots, X_{is})$.

(ii) The fitness value for each frog is calculated and then the frogs are arranged in decreasing order of their fitness value. The position of the best frog P_x in the entire population is recorded.

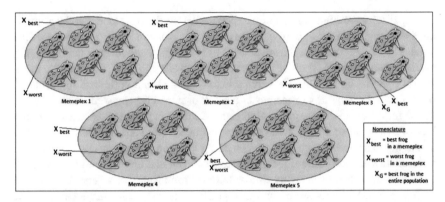

Fig. 4 Shuffled frog leap algorithm

(iii) The entire population is then divided into "m" number of memeplexes, each memeplex containing "n" frogs. Therefore, the entire population can be given by $P = m * n$.

(iv) The frogs are assigned to the memeplexes in a way such that the first frog is assigned to the first memeplex and the mth frog goes to the mth memeplex, and the $m + 1$th frog is assigned again to the first memeplex and the process continues up to the assignment of the nth frog.

(v) The best frog X_b and worst frog X_w are then identified in each memeplex.

(vi) Memetic evolution takes place in each memeplex. Memetic evolution in each memeplex Y_k, $k = 1 . . . m$ enhances the individual meme quality which in turn enhances the performance of the individual frog corresponding to its goal. Frogs with better memes are required to contribute more to the development of new memes as compared to others. Triangular probability distribution is used for ensuring competitiveness while selecting frogs for memetic evolution.

(vii) The memeplexes are shuffled after a predefined number of steps of memetic evolution are completed within each memeplex. Then, the frogs are again sorted in decreasing order of their performance value and the position of the best frog P_x is updated.

(viii) Finally, the convergence is checked; if the convergence criteria is satisfied, the execution will be stopped; otherwise, again the entire population will be partitioned into memeplexes and above-mentioned steps will be repeated.

3.6 Shuffled Frog-Leaping Algorithm (SFLA) for RCPSP

The shuffled frog-leaping algorithm has been applied to solve RCPSP by Fang and Wang [19]. Fang and Wang proposes a metaheuristic approach on the basis of SFLA which includes an encoding scheme and a decoding scheme on the basis of extended activity list (EAL) and SFLA-specific serial generation scheme (SSSGS), respectively. The EAL consists of three major lists: (i) a list containing all possible activities of the project, (ii) the second list consisting of the start time of all the activities present in the first list and (iii) the third list contains the completion times of the activities. Fang and Wang have proposed the SSSGS for transforming the EAL representation into potential schedules. Then for initializing the population, a regret-based biased sampling method has been adopted which is a priority rule-based heuristics and considers the latest finish time (LFT) of the activities of a project as the priority rule. This method arranges the activities in feasible sequences with precedence based upon the LFT priority rule. They applied the forward–backward improvement (FBI) approach to the activities to evaluate them where the first the backward pass processes the activities in the descending order of their finish time. The feasible sequence is built by scheduling the activities as late as possible whereas the forward pass re-sequences the activities in the schedule obtained from the backward pass as early as possible. Once the schedules are generated, the individuals are sorted in descending order of their make-span values as minimizing the make-span value is the objective of the RCPSP. The entire population is then partitioned into m memeplexes employing SFLA and each memeplex consisting of n frogs again performs the memetic search independently. As in SFLA, a sub-memeplex is generated for each memeplex and the best and the worst frogs are selected from the sub-memeplexes. Next, they perform crossover in each sub-memeplex in order to improve the positions of each of the frog as well as to proceed towards the global optimal one and utilize a resource-based crossover scheme (RBCO) for generating new child by applying it to the best and the worst frogs of the sub-memeplex. They have chosen RBCO here for the crossover operation because it satisfies precedence and resource constraints while conducting crossover. They further define a parameter named as the average resource utilization rate (ARUR), increasing which reduces the make-span of the project. Next, if the newly generated child is evaluated as worse than the worst frog, then RBCO is performed again between the worst frog and the best frog in the entire population. After that, if the latest obtained child is also worse than the worst frog of the sub-memeplex, a new frog is generated randomly to replace the newly generated child frog as a child, i.e. the new worst frog of the sub-memeplex. Once the worst frog position has been updated, a local search is carried out to improve the exploitation process.

The local search proposed by Fang and Wang combines the permutation-based local search (PBLS) with the FBI and applies to the new child generated above. The PBLS explores the neighbourhood of the new child maintaining the precedence feasibility present in the extended activity list (EAL). The FBI is then applied to the new frog for further improvement of exploitation while reducing the make-span of the project by scheduling and rescheduling the activities in the decreasing order of their finish times. The combination of PBLS along with FBI not only improves the quality of exploitation performed by the individuals but also guarantees precedence feasibility. The authors further propose speed-up evaluation methods to improve the efficiency of the SFLA and compare with existing approaches claiming that SFLA solves RCPSP effectively.

Further, Wang and Fang have also proposed a SFLA-based approach for solving the multi-mode RCPSP (MRCPSP), in their paper [47]. Firstly the authors have adopted a reduction procedure proposed by Sprecher et al. [41] for search space reduction as a preprocessing step of their approach. The purpose of their reduction procedure is to discard all the non-executable or inefficient modes along with the redundant resources from the initial search space to reduce it. They encode the list of activities with multiple execution modes (EMAL) as the virtual frogs of SFLA; that is, each encoded frog represents a potential solution in the actual search space. The previously discussed EAL consists of one additional list along with the previously discussed lists for the assignment of modes to the activities, hence known as extended multi-mode activity list (EMAL). They further have used the multi-mode SSGS (MSSGS) to decode the above-encoded list to schedules. This complete procedure is known as the transformation procedure of the activities list into solution schedules and the multiple execution modes of the activities make it more complicated. Initially, the modes are randomly assigned to the population and the regret-based sampling method along with the LFT heuristics is applied for probabilistically generating the activity lists while maintaining the precedence and resource constraints. Once the priority-based activity lists are generated, the individual frogs are evaluated and multi-mode FBI (MFBI) is applied in order to improve the quality of the initial population. The individual virtual frogs are then arranged in descending order of their make-span values and are divided into various memeplexes and sub-memeplexes to perform independent local search. Further, they propose a simplified two-point crossover (STPC) to be performed between the best and the worst frogs of each sub-memeplex in order to improve the position of the frogs towards global optima. First, the STPC is performed between the best and the worst frogs of each sub-memeplex to generate a new child that would replace the worst frog in the sub-memeplexes. Now if the newly generated child is worse than the worst frog of the sub-memeplex then again STPC is performed between the worst frog of the sub-memeplex and the best frog of the entire population. But if the new child is still worse, then a new frog will be randomly generated to replace the old worst frog of the sub-memeplex. Then, again the multi-mode PBLS and multi-mode FBI are applied to the new frog, respectively, to further improve the exploitation quality of the new frog. The MPBLS performs the neighbourhood search and swaps the adjacent activities without breaking the precedence relationship between the activities for better exploration. The MFBI further improves the fitness of exploration by repetitively applying the forward and backward serial generation scheme until saturation in make-span improvement is observed. Periodically, after the completion of certain number of iterations including both the local search and the crossovers, the virtual frogs of the entire population are shuffled for the evolution of next generation to maintain diversity in the memeplexes and the entire process is repeated by again sorting and dividing the entire population into new memeplexes. The authors claim this to be the first work on the SFLA-based approach for solving the MRCPSP. They have generalized the proposed approach by adopting various suitable methods and parameters for solving the multi-mode resource-constrained project scheduling problem. They also have compared the computational results against some existing benchmarks for validating the effectiveness of the proposed approach. Wang and fang further suggest exploration of network complexity, resource factors and resources strength as future directions.

3.7 Multi-objective Invasive Weed Optimization

The multi-objective invasive weeds optimization proposed by Kundu et al. [25] is based upon the soil occupying and new colony generation behaviour of the weeds. The seeds of the weeds are first scattered in suitable farmlands and grow into weeds. The newly generated colonies usually grow around the existing weeds and those grown in arable areas have higher chance of survival. This forms the base of the invasive weed-based multi-objective optimization algorithm. Firstly, in MOIWO, a set of weeds are generated randomly as the initial population and the objective function of each potential weed are evaluated. The entire population is then sorted using the fuzzy ranking and each member weed of the population is allowed to generate seeds with higher potential members. Generation of new weed colonies improves the search space exploration. The newly generated seeds are then scattered over the search space by using normal distribution and standard deviation functions while breeding is based upon competency between the members. As soon as the weed population crosses a threshold limit, fuzzy ranking is applied again and the best members are kept for further evolution whereas the weaker ones will be eliminated to maintain competitiveness between the members of the population. Consequently, new weeds come closer to the parent weeds, thus reducing the distance between them and transforming from diversification at the beginning to intensification by the end. This process is repeated until the stopping criterion is met.

3.8 Multi-objective Invasive Weed Optimization for MRCPSP

Maghsoudlou et al. have proposed a novel multi-objective model on the basis of discrete time–cost–quality trade-off (TCQT) for the multi-skill MRCPSP [31] and develop a multi-objective invasive weeds optimization-based approach (MOIWO) for solving the multi-skill MRCPSP with multiple objectives. The authors have integrated the MRCPSP with discrete TCQTP to transform it to a multi-objective problem by considering the time, cost and quality as the three objectives for the multi-skill MRCPSP. In other words, the three objectives can be stated in detail as (1) minimizing the make-span of the project, (2) minimizing the cost involved in worker allocation to different skills and (3) maximizing the quality of the activities being processed, i.e. the weighted sum of the individual activity qualities. Their model determines the best mode for each activity, the feasible start time and lastly assigns the best skill of the resources. Their model assumes manpower as the skill available along with other assumptions such as (i) each manpower is provided with predefined cost, duration and quality; (ii) a manpower can be allocated to one skill at a time; and (iii) all skills of an activity should perform simultaneously. Various constraints imposed on the model include that (i) each activity can be executed in one mode only at a time, (ii) the primary relationships between the activities such as precedence rules should be maintained, (iii) all work force assigned to an activity must start simultaneously and (iv) balance should be maintained between total workforces required and assigned. They developed the MOIWO for solving the above-mentioned multi-objective MRCPSP. While applying the MOIWO to the proposed model, the activities required for the project completion as presented as potential solutions and the objective values of each activity are evaluated considering the constraints. Now each activity can execute in different modes, and several skills are required by each activity, and moreover, these skills can be executed by certain workforces while each workforce is not capable of executing all the skills. So, along with the activities, the modes of executions, the skills required and the skills that each work force can execute are prioritized using fuzzy ranking and are assigned to the respective activities. All these are encoded into single chromosome structure and decoded afterwards into sequence of feasible executable activities. Now, the sequences of activities are sorted again in ascending order and the first feasible activity is identified as the other activities cannot start execution before the completion of the first activity. The sequence is then again searched for the next executable activity, and this process is continued iteratively until all the activities are identified with their best positions. The sequence thus generated by the above process is the most feasible sequence of the activity for successful completion of the project. They have reported that the proposed MOIWO-based approach has performed better than other multi-objective approaches in terms of diversification and better exploration.

3.9 Discrete Flower Pollination

The flower pollination algorithm was originally proposed by Yang based on the characteristics of flower pollination, to solve combinatorial problems [52]. The process of flower pollination usually consists of reproduction of flowers as a result of transfer of flower pollen by various pollinators as well as through abiotic pollination. Moreover, pollination can be divided into self- or cross-pollination while self-pollination maximizes reproduction of flowers of same species. Pollinators such as bees and flies exhibit Levy flights as step size while flying a distance which obeys Levy distribution. Optimal plant reproduction through pollination is considered as the objective function of the flower pollination algorithm (FPA). The primary rules forming the basis of their approach can be stated as (i) pollinators follow Levy flights and biotic and cross-pollination serves global optimization, (ii) abiotic and self-pollination serves local pollination, (iii) the similarity or difference between two flowers is considered as increment step and (iv) probability p controls the switching of local and global pollination.

3.10 Discrete Flower Pollination for RCPSP

Bibiks et al. [4, 5] have proposed a modified discrete version of the flower pollination algorithm for solving RCPSP in their paper. Bibiks et al. have stated in their approach that each individual flower is used to represent a candidate solution, i.e. a schedule. A schedule consists of a set of activities need to be executed in particular order to accomplish project completion. Thus, an individual flower represents a permutation vector consisting of elements representing the activities to be scheduled along with the precedence order in which the activities are supposed to be executed. Initial positioning of these vectors in the search space follows the component order and rearrangement of the order represents the movement of these vectors. Further, step size of the movement is derived from Levy flights with interval between [0, 1]. The make-span of the project in hand is mapped as the objective function which is evaluated to measure the quality of the solution. Again to attain global pollination, they have associated two types of mutations with the step sizes to accomplish the change in order of the activities, namely swap mutation and inverse mutation. Swap mutation is associated with small step size where two activities that are selected randomly switch position with each other. However, inverse mutation is similar to swap mutation but as it is associated with large step, all the activities positioned between the two randomly selected tasks are also swapped together but without affecting the precedence relationship existing between. Local pollination can be achieved by applying crossover method where two flowers are randomly selected and combined together to create a new one; that is, a subset of the tasks from one of the two flowers has been taken and combined with the remaining activities from the second flower to create the new one without hampering the precedence order. They have finally tested their DFPA using benchmark instances of RCPSP from PSPLIB and compared with other existing nature-based approaches such as PSO, genetic algorithm, ant colony optimization and simulated annealing showing that their approach outperforms the existing one in terms of deviation percentage.

3.11 Discrete Cuckoo Search

Cuckoo search technique was originally proposed by Yang and Deb [51] based upon the nest searching behaviour of cuckoo for laying eggs and is used for solving continuous problems. Further, Ouaarab et al. [35] developed the discrete cuckoo search algorithm by representing the solutions in the search space as permutations for solving combinatorial problems. The main goal of their approach is to replace the worse solution by new and better solutions. Primary steps of the cuckoo search algorithm are as follows:

(i) Each of the eggs laid by the cuckoo is dumped in randomly chosen nest;
(ii) Best nests containing high-quality eggs are carried to next generations;

(iii) Number of host nests is constant and if the host bird discovers the egg laid by cuckoo, the host bird will either abandon the nest or destroy the egg.

The CS algorithm maintains a balance between exploration and exploitation of the search space. Levy distribution is adopted by their approach for defining step size.

3.12 Discrete Cuckoo Search for RCPSP

Bibiks et al. modified the discrete version of the cuckoo search (CS) [5] in which, the egg and nest represent the candidate solution in the search space, i.e. the permutation vector consisting of the scheduled tasks and their order of execution. Initially, these permutations/solutions are positioned according to their component order, and afterwards, the order is changed by rearranging the components in the search space. This movement in the search space is accomplished by using Levy flights where the number of steps and their step size is determined by levy distribution within the interval [0, 1]. The objective function evaluates the quality of the solution that is tries to find the schedule with the shortest make-span. The step size of the movement in the solution space is determined by levy distribution where for small steps swap mutation is used and inverse mutation is applied for larger step size as discussed previously. The swap mutation and inverse mutations are employed without affecting the precedence order of the scheduled activities. The authors tested the efficiency of the proposed approach using PSPLIB benchmarks where it outperforms existing approaches and exhibits better computational time.

3.13 Multi-agent Optimization Algorithm (MAOA)

Since in multi-agent systems, multiple agents collaborate with potential agents to solve complex problems, MAS-based approaches [37] show better performance, where agents usually represent resources and their behaviours such as learning, collaborating and cooperating are used to handle the RCPSP. Zheng and Wang [58] proposed a new multi-agent optimization algorithm (MAOA) to solve the resource-constrained project scheduling problem (RCPSP). They combined and adopt the significant behaviours of both agent-based and population-based swarm intelligence techniques to solve RCPSP effectively. In their approach, each agent represents a feasible solution and collaborates with other agents via various elements (behaviours) to perform collaborative exploration and exploitation in the search space. The elements represent social behaviour, autonomous behaviour, environmental adjustment and self-learning. They created an environment, by dividing the agents into several groups and further considered the social behaviour at both global and local levels to perform better exploration and exploitation. Further, the autonomous behaviour corresponds to exploitation of its own neighbourhood while self-learning corresponds to thorough search of the potential region. Also some agents migrate from group to group to dynamically adjust to environment and share information. They adopted the extended activity list (EAL) concept proposed by Fang and Wang [48] to represent the agents as search operators with such an encoding scheme perform efficiently. A regret-based biased random sample method with the LFT priority rule [Hartmann, 1998] has been adopted by the authors to initialize the agents. The construction of the activity list of each agent is followed by the evaluation of the activity list and transformation into a schedule using serial scheduling generating scheme. The agents are randomly divided into groups after initialization.

Subsequently, Zheng and Wang employs a magnet-based crossover (MBCO) proposed by Zamani [54] with the advantage of the offsprings capable of inheriting partial sequences from both parents which in turn leads to better exploration. MBCO has been adopted as the global social behaviour so that leader agent in each group can interact with the global best leader agent to generate a set of offspring, and the best offspring is adopted to replace the corresponding leader agent. Further, a resource-based crossover (RBCO) mechanism [48] which states that average resource utilization is inversely proportional to the project duration has been adopted. RBCO can be used as a local social guidance for an agent in order to inherit the activity sub-sequence having higher resource utilization rate and can improve the agent itself. Further, Zheng [58] has employed a permutation-based swap operator to conduct the local search (autonomous behaviour) while maintaining the precedence constraint of an EAL. Again, they have adopted the forward–backward improvement (FBI) proposed in [28], which keeps adjusting a solution till the maximum reduction is attained, as a self-learning approach for the best leader agent to explore further. In order to share the information regarding the improvement of the social, autonomous and self-learning behaviour among the different groups, the environment is also adjusted in every 10 generations considering three major parameters, i.e. the number of groups, the size of the groups and the acceptance probability.

Finally, they investigate the effects of the above-mentioned major parameters on the basis of Taguchi method of design of experiment (DOE) and provide the test results using three sets of benchmark instances. They observe that small and large values of the number of groups and size of groups lead to weak and insufficient exploration ability, respectively, whereas medium values balance the exploration and exploitation but may trade off the population size and evolution of the population. Also the smaller values of the acceptance probability force the approach to accept inferior solutions. The authors claim that the experimental results show that the proposed approach shows competitive results for solving large-scale problems. MAOA thus provides better potential solutions with acceptable running time which increases linearly with respect to the total number of the generated schedules. They also claim MAOA to be an effective solver for the RCPSP in a dynamic environment. They identify development of the mathematical model, analysing the convergence property of MAOA and designing self-adaptive approaches for parameter settings as focus of further research.

4 Proposed Approach

4.1 RCPSP for Retail Industry

A huge number of literatures addressing the resource-constrained project scheduling problem exist, out of which few relevant ones are discussed in the above section. Most of the literatures discussed above have considered construction problems for studying the aspects of resource-constrained project scheduling problem. But in case of construction problems, most of the attributes are fixed or deterministic, so predicting and attaining objectives is not so difficult. But when real-time dynamic cases such as garment industry is being considered for optimal scheduling of activities and resources to deliver products timely during the various events such as various functions and seasons such as winter, summer and marriage. and maximize profit it is really hard. Some of the factors responsible for the increasing complexity of the garment industry include increasing variations in consumer need, fragmentation of the market and demand for quality products at low rates. One of the major challenges in retail industry is not only to predict the demand of the market but also to adapt to it dynamically. Figure 2 identifies some of the relevant activities in retail industry that are needed to be scheduled along with precedence and resource constraints for successful completions of projects and gain advantage over competitors by maximizing the profit.

In order to predict the demand of the continuously changing market, previous sales history needs to be reviewed and the demands and customizations are usually incorporated as product attributes for future production. Similarly, past purchase history and loan history need to be reviewed to identify reliable suppliers and working capital sources for obtaining resources before manufacturing activity is started. Once production of goods is over, warehousing of the finished goods, distribution through different channel according to priority, and collection of revenue and repayment of existing liabilities are some of the activities which requires optimal scheduling. But some of the common critical issues that cannot be controlled even through forecasting and which may lead to huge loss are early production, overproduction, underproduction and delay in production due to unavoidable circumstances. While early and overproduction may require extra storage facilities, underproduction may lead to loss of potential consumers and delay in production may lead to huge leftover products after certain events such as festivals or seasons such as winter or summer are over. Although existing scheduling approaches solve the scheduling problem optimally, but in most cases, the dynamic adaptation part in order to handle the complexity of the continuously changing market scenarios has been overlooked (Fig. 5).

4.2 Cooperative Hunting Behaviour of Lion Pride

Lions are one of the most ferocious yet social animals from the cat family that exhibits high level of cooperative behaviour while hunting in a group. Usually, a group of

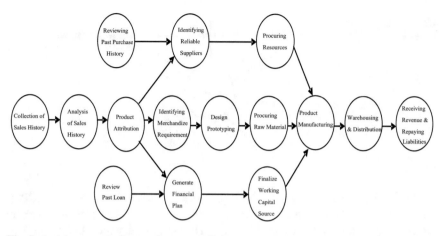

Fig. 5 Activity on node diagram for garment industry

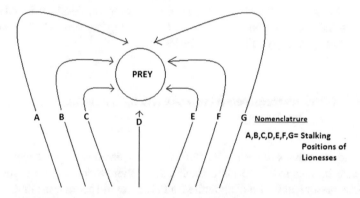

Fig. 6 The seven stalking positions of the lionesses while hunting

lions known as a pride consists of at least one adult male lion with 5–7 lionesses and their cubs [39]. In a pride, the lionesses cooperate with each other to successfully execute their hunting strategy, thus increasing the probability of killing a larger prey such as wildebeests and zebras, which would be sufficient to feed the whole pride. And capturing such a prey is not possible for individual lioness alone, so they follow a strategy where initially each of the lioness occupy one of the seven stalking positions as defined by Stander [43] as shown in Fig. 6.

After observing 486 coordinated hunts by lionesses, Stander has generalized the above stalking positions which are usually followed while hunting in a group. The stalking positions are divided into the left and right wings and the centre. During the hunting process, each of the lionesses updates its position while maintaining the strategy and tries to minimize its distance from the prey without coming into the sight of the prey. Finally, the lionesses encircle the prey into a death trap escaping which is rarely possible for the prey. The death trap along with the initial direction of approach is given in Fig. 7.

Fig. 7 Death trap by
lionesses

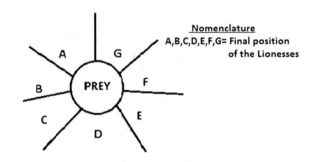

This hunting strategy of the lion pride provides inspiration for solving the RCPSP problem as most of the cooperative hunting behaviour of the lionesses is similar to the project scheduling problem of retail industry. The initial arrangement of the lionesses in different stalking positions can be related to the precedence order maintained among the various activities. Therefore, we are proposing a multi-agent-based system for solving the resource-constrained project scheduling problem which is inspired from the hunting strategy followed by the lion pride.

5 A Lion Pride-Inspired Multi-Agent System-Based Approach for RCPSP

The solution search space for the RCPSP is initially populated by n lioness agents, where each lion agent represents a candidate solution which in turn represents a schedule consisting of the set of activities for the successful completion of the project in hand. The initial position of the lioness agents follows the stalking positions according to the dominance hierarchy existing among them. In other words, the initial arrangement of the activities in each schedule maintains precedence relations. The individual candidate solution (schedule) is represented by permutation vector where the elements of the vector represent activities and the indexes represent the position of an activity in the schedule without violating the precedence relations between them. Once the initialization phase is over, the fitness value of each of the candidate solution is evaluated on the basis of minimum distance of each lion agent from the prey, that is, each of the schedules (solutions) are evaluated to find the one with minimum make-span which is given by the following equation:

$$\min \ dist = pos_{lion} - pos_{prey} \tag{8}$$

After the fitness value of each of the solution is evaluated, the schedule with the minimum make-span is considered as the global best solution of that iteration and the position of each of individual solution (lioness agent) is improved with respect to the global best solution to attain global optimization. But with each new position, the lioness agents consume some energy and the available energy level decreases consequently, that is, each activity in the schedule depends on the availability of certain resources without which execution of the activity is not possible. The remaining energy of the lion agent in each new position is given by Eq. (9) as follows:

$$E_{i+1} = E_i - 1/2m \times vel_{lion}^2 \qquad (9)$$

The new position of the ith lion agent is determined by Eq. (10) given below:

$$pos_{lion\,i+1} = (pos_{lion\,i} - pos_{prey} \times E_{i+1} + (vel_{lion} + \omega \times r) \qquad (10)$$

where ω is the relative direction of movement of the lion agent with respect to the prey and r is the distance between the prey and the lion agent with respect to relative angular velocity given by Eq. (11).

$$r = pos_{lion\,i} - pos_{prey} \qquad (11)$$

The local optimization among the activities can be obtained by rearranging the activities within the schedule while maintaining both the precedence and resource constraints. The above-mentioned process is repeated until an optimal schedule is generated; that is, no more changes in the value of make-span occur or maximum number of iterations is reached. Thus, both local optimization and global optimization are attained to generate the optimal schedule that minimizes the make-span of the project in hand. The pseudocode of the proposed approach is given below (Fig. 8):

1: Initially generate 'n' lion agent (lion$_i$). where i= 1,2,...,n
2: while(E$_i$=max_energy) do
3: Evaluate Dominance
4:Occupy position through dominance hierarchy
5: Objective function, f(x) = min (make-span) = min (dist) = Pos$_{lion}$ – Pos$_{prey}$.
6 : max_interations = Ei/stepsize.
       7: while (count < max_iteration) or (stopping criterion) do
      8: Evaluate distance ri between Poslion and Posprey
        9: if (ri > rj) then
        10: Update Pos$_{lion(i+1)}$
        11. Endif
    12: Endwhile
13: Endwhile
14: Return the optimal schedule

Fig. 8 Pseudocode

6 Conclusion

modelling and optimizing processes involved in business domain using traditional approaches such as system dynamics, mathematical, statistical and analytical models is quite complex since tuning and validating the parameters for new problems is not feasible. MAS can capture well the underlying complexity of various complex systems and processes of business domain. Scheduling is considered as one of the most important process for the successful completion of any project. Resource-constrained project scheduling is one of the most widely addressed problems of business domain. This chapter presents the study of resource-constrained project scheduling problem of business domain as multi-agent optimization problem. Various nature-inspired approaches used for solving the classical RCPSP and its variants have been studied. Nature-inspired computing can be integrated with multi-agent systems to solve this NP-hard problem since these approaches are based upon simple rules and are capable of handling the complexity and uncertainty existing in real-time environments. Therefore, a multi-agent-based approach inspired by the hunting strategy of lion prides has been proposed for solving the RCPSP. Various aspects of RCPSP will be considered for integration and validation of the model in further research work. A nature-inspired integrated multi-agent system-based approach has been proposed for solving the RCPSP which is inspired by the hunting strategy of lion pride. The mathematical model as well as pseudocode of the proposed approach has been discussed.

References

1. Artigues C, Demassey S, Neron E (eds) (2013a) Resource-constrained project scheduling: models, algorithms, extensions and applications. Wiley
2. Artigues C, Leus R, Nobibon FT (2013b) Robust optimization for resource-constrained project scheduling with uncertain activity durations. Flex Serv Manuf J 25(1–2):175–205
3. Becker J, Kugeler M, Rosemann M (eds) (2013) Process management: a guide for the design of business processes. Springer Science & Business Media
4. Bibiks K, Li JP, Hu F (2015a) Discrete flower pollination algorithm for resource constrained project scheduling problem. Int J Comput Sci Inf Secu 13(7):8
5. Bibiks K, Hu F, Li JP, Smith A (2015b) Discrete cuckoo search for resource constrained project scheduling problem. In: 2015 IEEE 18th international conference on computational science and engineering (CSE). IEEE, pp 240–245. (October)
6. Boctor FF (1990) Some efficient multi-heuristic procedures for resource-constrained project scheduling. Eur J Oper Res 49:3–13
7. Bouleimen K, Lecocq H (2003) A new efficient simulated annealing algorithm for the resource-constrained project scheduling problem and its multiple mode version. Eur J Oper Res 149(2):268–281
8. Burke EK, Hyde M, Kendall G, Ochoa G, Özcan E, Woodward JR (2010) A classification of hyper-heuristic approaches. In: Handbook of metaheuristics. Springer, US, pp 449–468
9. Chang JF (2016) Business process management systems: strategy and implementation. CRC Press
10. Cheng T, Zhou G (2013) Research on project scheduling problem with resource constraints. J Softw 8(8):2058–2063

11. De Reyck B, Herroelen W (1999) The multi-mode resource-constrained project scheduling problem with generalized precedence relations. Eur J Oper Res 119(2):538–556
12. Dorigo M, Maniezzo V, Colorni A (1991) The ant system: an autocatalytic optimizing process. Technical Report 91-016 Revised, Dipartimento di Elettronica, Politecnico di Milano, Italy
13. Dorigo M, Maniezzo V, Colorni A (1996) Ant system: optimization by a colony of cooperating agents. IEEE Trans Syst Man Cybern Part B (Cybernetics), 26(1):29–41
14. Dorigo M, Gambardella LM (1997) Ant colony system: a cooperative learning approach to the traveling salesman problem. IEEE Trans Evol Comput 1(1):53–66
15. Dorigo M, Di Caro G (1999) The ant colony optimization meta-heuristic. In: Corne D, Dorigo M, Glover F (eds) New ideas in optimization. McGraw Hill, London, UK, pp 11–32
16. Eberhart RC, Kennedy J (1995) A new optimizer using particle swarm theory. In: Proceedings of the sixth international symposium on micro machine and human science, vol 1, pp 39–43. (October)
17. El-Rayes K, Kandil A (2005) Time-cost-quality trade-off analysis for highway construction. J Constr Eng Manag 131(4):477–486
18. Eusuff M, Lansey K, Pasha F (2006) Shuffled frog-leaping algorithm: a memetic meta-heuristic for discrete optimization. Eng Optim 38(2):129–154
19. Fang C, Wang L (2012) An effective shuffled frog-leaping algorithm for resource-constrained project scheduling problem. Comput Oper Res 39(5):890–901
20. Hartmann S, Briskorn D (2010) A survey of variants and extensions of the resource-constrained project scheduling problem. Eur J Oper Res 207(1):1–14
21. Kerzner HR (2013) Project management: a systems approach to planning, scheduling, and controlling. Wiley
22. Klein R (2000) Project scheduling with time-varying resource constraints. Int J Prod Res 38(16):3937–3952
23. Kolisch R (1996) Efficient priority rules for the resource-constrained project scheduling problem. J Oper Manag 14(3):179–192
24. Koulinas G, Kotsikas L, Anagnostopoulos K (2014) A particle swarm optimization based hyper-heuristic algorithm for the classic resource constrained project scheduling problem. Inf Sci 277:680–693
25. Kundu D, Suresh K, Ghosh S, Das S, Panigrahi BK, Das S (2011) Multi-objective optimization with artificial weed colonies. Inf Sci 181(12):2441–2454
26. Küster T, Lützenberger M, Heßler A, Hirsch B (2012) Integrating process modelling into multi-agent system engineering. Multiagent Grid Syst 8(1):105–124
27. Lambrechts O, Demeulemeester E, Herroelen W (2008) Proactive and reactive strategies for resource-constrained project scheduling with uncertain resource availabilities. J Sched 11(2):121–136
28. Li KY, Willis RJ (1992) An iterative scheduling technique for resource-constrained project scheduling. Eur J Oper Res 56(3):370–379
29. Lova A, Tormos P, Cervantes M, Barber F (2009) An efficient hybrid genetic algorithm for scheduling projects with resource constraints and multiple execution modes. Int J Prod Econ 117(2):302–316
30. Macal C, North M (2014) Introductory tutorial: agent-based modeling and simulation. In: Proceedings of the 2014 winter simulation conference. IEEE Press, pp 6–20. (December)
31. Maghsoudlou H, Afshar-Nadjafi B, Niaki STA (2016) A multi-objective invasive weeds optimization algorithm for solving multi-skill multi-mode resource constrained project scheduling problem. Comput Chem Eng 88:157–169
32. Morecroft JD (2015) Strategic modelling and business dynamics: a feedback systems approach. Wiley
33. Nanobe K, Ibaraki T (2002) Formulation and tabu search algorithm for the resource constrained project scheduling problem. In: Essays and surveys in metaheuristics. Springer, US, pp 557–588
34. North MJ, Macal CM (2007) Managing business complexity: discovering strategic solutions with agent-based modeling and simulation. Oxford University Press

35. Ouaarab A, Ahiod B, Yang XS (2014) Discrete cuckoo search algorithm for the travelling salesman problem. Neural Comput Appl 24(7–8):1659–1669
36. Palpant M, Artigues C, Michelon P (2004) LSSPER: solving the resource-constrained project scheduling problem with large neighbourhood search. Ann Oper Res 131(1–4):237–257
37. Ren H, Wang Y (2011) A survey of multi-agent methods for solving resource constrained project scheduling problems. In: 2011 international conference on management and service science (MASS). IEEE, pp 1–4. (August)
38. Rosemann M, vomBrocke J (2015) The six core elements of business process management. In: Handbook on Business Process Management 1. Springer Berlin Heidelberg, pp 105–122
39. Scheel D, Packer C (1991) Group hunting behaviour of lions: a search for cooperation. Anim Behav 41(4):697–709
40. Siddique N, Adeli H (2015) Nature inspired computing: an overview and some future directions. Cogn Comput 7(6):706–714
41. Sprecher A, Hartmann S, Drexl A (1997) An exact algorithm for project scheduling with multiple modes. Oper Res Spektrum, 19(3):195–203
42. Sprecher A (2012) Resource-constrained project scheduling: exact methods for the multi-mode case, vol 409. Springer Science & Business Media
43. Stander PE (1992) Cooperative hunting in lions: the role of the individual. Behav Ecol Sociobiol 29(6):445–454
44. Tavana M, Abtahi AR, Khalili-Damghani K (2014) A new multi-objective multi-mode model for solving preemptive time-cost-quality trade-off project scheduling problems. Expert Syst Appl 41(4):1830–1846
45. Tiwari V, Patterson JH, Mabert VA (2009) Scheduling projects with heterogeneous resources to meet time and quality objectives. Eur J Oper Res 193(3):780–790
46. Vanhoucke M (2012) Project management with dynamic scheduling. Springer Berlin Heidelberg
47. Wang L, Fang C (2011) An effective shuffled frog-leaping algorithm for multi-mode resource-constrained project scheduling problem. Inf Sci 181(20):4804–4822
48. Wang L, Fang C (2012) A hybrid estimation of distribution algorithm for solving the resource-constrained project scheduling problem. Expert Syst Appl 39(3):2451–2460
49. Wuliang P, Chengen W (2009) A multi-mode resource-constrained discrete time-cost tradeoff problem and its genetic algorithm based solution. Int J Proj Manag 27(6):600–609
50. Xing B, Gao WJ (2014) Innovative computational intelligence: a rough guide to 134 clever algorithms. Springer, Cham, pp 250–260
51. Yang XS, Deb S (2009) Cuckoo search via Lévy flights. In: NaBIC 2009. World Congress on Nature & Biologically Inspired Computing, 2009. IEEE, pp 210–214. (December)
52. Yang XS (2012) Flower pollination algorithm for global optimization. In: International conference on unconventional computing and natural computation. Springer Berlin Heidelberg, pp 240–249
53. Yang XS (2014) Nature-inspired optimization algorithms. Elsevier
54. Zamani R (2013) A competitive magnet-based genetic algorithm for solving the resource-constrained project scheduling problem. Eur J Oper Res 229(2):552–559
55. Zhang H, Li X, Li H, Huang F (2005) Particle swarm optimization-based schemes for resource-constrained project scheduling. Autom Constr 14(3):393–404
56. Zhang H, Li H, Tam CM (2006) Particle swarm optimization for resource-constrained project scheduling. Int J Proj Manag 24(1):83–92
57. Zhang H (2011) Ant colony optimization for multimode resource-constrained project scheduling. J Manag Eng 28(2):150–159
58. Zheng X-L, Wang L (2015) A multi-agent optimization algorithm for resource constrained project scheduling problem. Expert Syst Appl 42(15):6039–6049

Application of Learning Classifier Systems to Gene Expression Analysis in Synthetic Biology

Changhee Han, Kenji Tsuge and Hitoshi Iba

Abstract Learning classifier systems (LCS) are algorithms that incorporate genetic algorithms with reinforcement learning to produce adaptive systems described by if-then rules. As a new interdisciplinary branch of biology, synthetic biology pursues the design and construction of complex artificial biological systems from the bottom-up. A trend is growing in designing artificial metabolic pathways that show previously undescribed reactions produced by the assembly of enzymes from different sources in a single host. However, few researchers have succeeded thus far because of the difficulty in analyzing gene expression. To tackle this problem, data mining and knowledge discovery are essential. In this context, nature-inspired LCS are well suited to extracting knowledge from complex systems and thus can be exploited to investigate and utilize natural biological phenomena. This chapter focuses on applying LCS to gene expression analysis in synthetic biology. Specifically, it describes the optimization of artificial operon structure for the biosynthesis of metabolic pathway products in *Escherichia coli*. This optimization is achieved by manipulating the order of multiple genes within the artificial operons.

Keywords Learning classifier systems · Gene expression analysis · Synthetic biology · Bioinformatics · Artificial operon · Consultation algorithm

C. Han (✉)
Graduate School of Information Science and Technology, The University of Tokyo,
122B1 Engineering Building 2, 7-3-1 Hongo, Bunkyo, Tokyo, Japan
e-mail: han@iba.t.u-tokyo.ac.jp

K. Tsuge
Institute for Advanced Biosciences, Keio University,
403-1 Nipponkoku, Daihoji, Tsuruoka, Yamagata, Japan
e-mail: ktsuge@ttck.keio.ac.jp

H. Iba
Graduate School of Information Science and Technology, The University of Tokyo,
101D4 Engineering Building 2, 7-3-1 Hongo, Bunkyo, Tokyo, Japan
e-mail: iba@iba.t.u-tokyo.ac.jp

© Springer International Publishing AG 2017
S. Patnaik et al. (eds.), *Nature-Inspired Computing and Optimization*,
Modeling and Optimization in Science and Technologies 10,
DOI 10.1007/978-3-319-50920-4_10

1   Introduction

How can gene expression be optimized? In synthetic biology, this remains a challenging and fruitful goal that would facilitate the mass production of biofuels, biomedicine and engineered genomes. Nature-inspired learning classifier systems (LCS) [28] were originally envisioned as *cognitive systems* with interactive components, permitting the classification of *massive and noisy datasets* in biological systems. Furthermore, LCS efficiently generate compact *rules describing complex systems* that enable knowledge extraction. Hence, LCS perfectly complement synthetic biology [6], which pursues the design and construction of *complex artificial biological systems* by combining ICT, biotechnology and nanotechnology.

LCS have performed well within many different biological domains, including gene expression analysis [2, 24, 68]. Implementation of LCS to deal with gene expression is poised to grow in the near future, as a number of sizable genomic datasets are made available by large-scale international projects, such as the 1000 Genomes Project, the 100,000 Genomes Project and the ENCODE Project.

This chapter presents a crucial first step of applying LCS to complex tasks in synthetic biology, especially in the context of gene expression analysis. This chapter consists of three parts. The first part introduces the framework of LCS and explains how common LCS work. LCS can be used to implement precise data mining methods with an *if-then rule structure* by adopting the strong points of both genetic algorithms (GA) and reinforcement learning (RL). Two major types of LCS exist: Michigan-style LCS, which assess the fitness of individual classifiers, and Pittsburgh-style LCS, which assess the fitness of whole populations. Whereas Pittsburgh-style LCS, such as GAssist and BioHEL, achieve good results in bioinformatics tasks because of their simple rules, Michigan-style LCS are more widely applicable to other practical tasks. For example, minimal classifier systems (MCS) [13] can serve as archetypes for more complex implementations, strength-based zeroth-level classifier systems (ZCS) [62] can solve simple problems precisely, and accuracy-based extended classifier systems (XCS) [63] show optimal solutions within many fields.

The second part of this chapter describes synthetic biology and its potential contributions to society [17]. Furthermore, it describes how LCS have tackled problems related to gene expression analysis [24] and how LCS can optimize an artificial operon structure. To fulfill the goals of synthetic biology, it is essential to optimize each stage in the *design cycle*, utilize *basic biological blocks* and assemble *DNA sequences*. LCS can be applied to fields ranging from renewable biofuels to biomedicine—for intractable diseases such as cancer, infectious diseases and autoimmune disorders—and engineered genomes of artificial organisms. LCS have outperformed gene expression analysis in many instances, especially in cancer classification. *Operons* are functioning units of genomic DNA that are transcribed as a unit and regulated together, thereby coordinating gene transcription. The functionality of bacteria, such as *Escherichia coli* (*E. coli*), can be maximized by modifying operon structure and thus altering gene expression.

The third part of this chapter focuses on the *first computational approach* to analyzing the rules describing the relationship between gene order within an operon and *E. coli* growth (i.e., production) by consultation algorithms, including LCS. Operon construction rules were extracted by machine learning, and those rules were verified by conducting the wet-lab experiments of designing novel *E. coli* strains predicted to have high growth rates. This research identified gene orders that significantly control the growth of *E. coli* and enable the design of new *E. coli* strains with high growth rates. These results indicate that LCS—followed by consultation with well-known algorithms—can efficiently extract knowledge from big and noisy biological system datasets as well as previously intractable dynamic systems.

2 Learning Classifier Systems: Creating Rules that Describe Systems

LCS are nature-inspired machine learning methods; accordingly, they can function as cognitive systems featuring complex interactive components with nonlinear collective activity [28, 29, 31]. For knowledge extraction, LCS generate new random if-then rules by GA and choose the best rule by RL. The domain hierarchy of LCS is described in Fig. 1. To give a general introduction to LCS, this section highlights the basic LCS mechanisms and the fundamental distinction between Michigan- and Pittsburgh-style LCS. More details about implementation optimization as well as a cohesive encapsulation of implementation alternatives are available in an outstanding LCS survey [58] (for a historical overview of LCS research, see [14]).

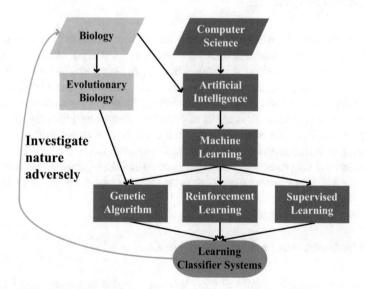

Fig. 1 Domain hierarchy of the LCS concept

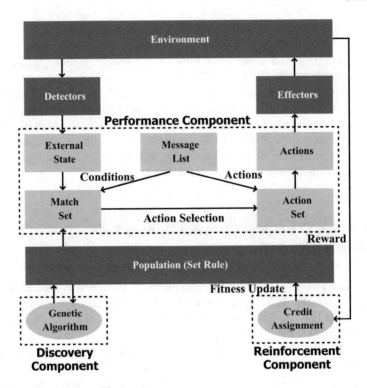

Fig. 2 Interactions among LCS components

2.1 Basic Components

It is not absolutely clear which components should be regarded as basic components of LCS because many different implementations of LCS exist. Holmes et al. [32] outline four practically minimal components: (1) A *finite population of classifiers* that describes the current knowledge of the system to solve distributed problems; (2) a *performance component*, which controls interactions with the environment; (3) a *reinforcement component* (or a credit assignment component [30]), which distributes the reward from the environment to the classifiers to obtain globally optimal solutions; and (4) a *discovery component*, which exploits GA to discover better rules and improve existing ones according to fitness estimates. Figure 2 demonstrates the highly interactive mechanisms and how these major components interact with each other.

2.2 Michigan- and Pittsburgh-style LCS

LCS research is fundamentally divided into two groups: Michigan-style LCS and Pittsburgh-style LCS. Holland [31], the father of LCS, proposed Michigan-style

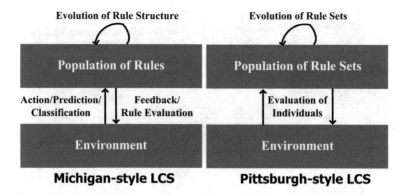

Fig. 3 Differences between Michigan- and Pittsburgh-style LCS

LCS originally, and Smith [54, 55] later introduced Pittsburgh-style LCS. While in Michigan-style LCS, individual classifiers cooperate to provide a collective solution for the entire population, in Pittsburgh-style LCS, individual complete and variable-length sets of classifiers compete to solve the problem. Figure 3 illustrates the main structural differences between these two systems. While Michigan-style LCS evolve at the level of individual rules, Pittsburgh-style LCS evolve at the level of multiple rule sets.

In general, Michigan-style LCS are more commonly used because of their simplicity, fewer evaluation periods and higher flexibility in dealing with a wide range of problems. Michigan-style LCS evaluate the fitness of individual classifiers, whereas Pittsburgh-style LCS assess the fitness of entire populations. Therefore, Michigan-style LCS are typically online learning systems that learn iteratively from sets of problem instances, and Pittsburgh-style LCS are more suitable for offline learning systems that learn iteratively from single problem instances. Moreover, Michigan-style LCS show more distributed solutions with a huge number of rules, while Pittsburgh-style LCS provide more compact solutions with few rules. Two examples of Pittsburgh-style LCS implementations are GAssist [7] and its successor BioHEL [8, 21], which is used widely for data mining, especially with large-scale bioinformatic datasets.

3 Examples of LCS

This section describes some widely used Michigan-style LCS.

3.1 Minimal Classifier Systems

MCS are simplified LCS implementations that were proposed by Bull [13] as platforms for advancing LCS theory. They provide helpful archetypes for more complicated implementations. Figure 4 illustrates how MCS work as a whole. Whereas

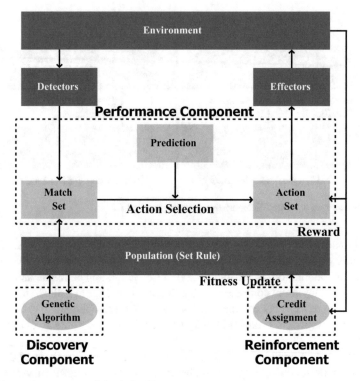

Fig. 4 Interactions among minimal classifier system components

typical LCS possess a message list in order to handle multiple messages both from the environment and from other components' feedback at previous time steps, MCS only learn iteratively from one data instance at a time. If the match set conflicts with a given input, a new rule—obtained randomly by GA to handle the input—replaces a previous one. Therefore, a population remains constant according to its fitness value. New rules usually inherit fitness from the previous rules [13].

MCS's RL begins with an reward from the environment according to the performed action. The following equation [13] updates the action set [A] at a learning rate β:

$$fitness([A]) \leftarrow fitness([A]) + \beta \left(\left(\frac{Reward}{|[A]|} \right) - fitness([A]) \right) \quad (1)$$

3.2 Zeroth-level Classifier Systems

To increase the clarity and performance of LCS, Wilson [62] proposed simple ZCS for practical use. Figure 5 shows the interactions among ZCS components. ZCS

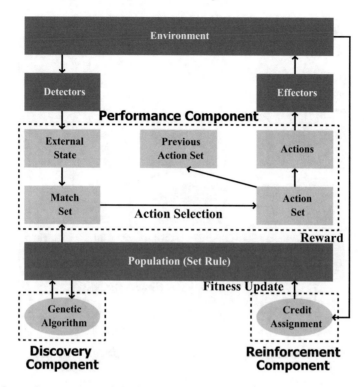

Fig. 5 Interactions among zeroth-level classifier system components

are often called strength-based LCS (or payoff-based LCS) since their rule fitness depends on the strength (i.e., payoff) defined by the rules. ZCS differ from the typical LCS in that they eliminate the message list and rule-binding, so inputs are straightforwardly matched with the population of rules and no prediction of each action exists. Furthermore, in ZCS, the credit assignment schema basically adopts an implicit bucket brigade [25], which distributes credit between all rules proposing a given action. ZCS perform well in a simple framework relative to original LCS. However, ZCS still find considerably suboptimal solutions in many applications because their simplicity rapidly increases the number of over-general classifiers [18].

ZCS's RL centers the previous and current action sets without predicting each action. The following equation [62] updates the action set [A], defining [A]' as the action set that follows [A] in time, β as a learning rate and γ as a predetermined discount factor:

$$\text{fitness}([A]) \leftarrow \text{fitness}([A]) + \beta(\text{Reward} + \gamma\, \text{fitness}([A]') - \text{fitness}([A])) \quad (2)$$

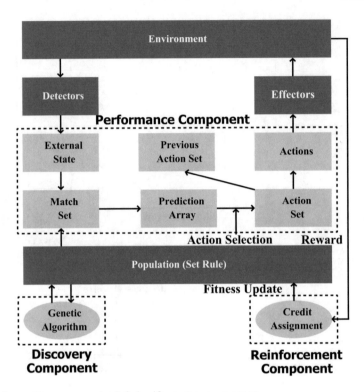

Fig. 6 Interactions among extended classifier system components

3.3 Extended Classifier Systems

One year after introducing ZCS, Wilson [63] proposed XCS to address the drawbacks
of ZCS, using accurate and general classifiers. As Fig. 6 illustrates, XCS inherited
many ZCS attributes. While ZCS are referred to as strength-based LCS, XCS are
called accuracy-based LCS because their rule fitness does not depend on the payoff
itself but on the accuracy of payoff predictions [37]. To prevent XCS rules from
consistently executing bad actions owing to accuracy-based fitness and to find an
optimal action-selection solution, XCS exploit Q-learning [59], while the previous
LCS often use TD(0) [56] instead. Q-learning is an off-policy RL technique that can
find an optimal action-selection policy for any Markov decision process; it learns
the optimal policy from its previous interactions with the environment. TD(0) is a
simple case of Q-learning that only updates the immediately preceding prediction.
In addition to using accuracy-based fitness, ZCS replaces standard GA with niche-
based GA to eliminate undesirable competition. This combination of accuracy-based
fitness, which promotes prediction accuracy, and the niche GA, which accelerates
the process of obtaining general rules, makes excellent predictions possible. As
a consequence, many researchers have adopted XCS extensively because of their

simplicity in generating optimal solutions across a wide range of research fields [1, 47].

XCS's Q-learning-based RL exploits predicted reward to choose the action. The following five steps [63] update the action set [A] with a relative accuracy:

1. Each rule's error ε_j is updated: $\varepsilon_j = \varepsilon_j + \beta(|P - p_j| - \varepsilon_j)$, where β is a learning rate and P is the resulting value.
2. Rule predictions p_j are updated: $p_j = p_j + \beta(P - p_j)$.
3. Each rule's accuracy κ_j is determined: $\kappa_j = \alpha \left(\frac{\varepsilon_0}{\varepsilon}\right)^\nu$ or $\kappa_j = 1$ where $\varepsilon < \varepsilon_0$ and ν, α and ε_0 control the shape of the accuracy function.
4. A relative accuracy κ'_j is determined by $\frac{\text{each rule's accuracy}}{\text{total accuracy in the action set}}$.
5. The relative accuracy modifies the classifier's fitness F_j: If the fitness is adjusted $1/\beta$ times, $F_j = F_j + \beta(\kappa'_j - F_j)$, otherwise $F_j = $ the average of κ'_j values.

4 Synthetic Biology: Designing Biological Systems

Engineering microorganisms requires a comprehensive understanding of natural biological systems, such as cryptic cellular behaviors and tools for controlling cells. To overcome these problems, synthetic biology aims to design and construct biological systems for practical applications. In order to avoid empirical studies without undertaking predictive modeling, synthetic biology is shifting from developing *proof-of-concept designs* to establishing *general core platforms* for efficient biological engineering based on *computer science* [17]. Optimizing gene expression computationally plays an essential role in mass producing useful materials through a synthetic biology approach. This section highlights the key progress and future challenges in synthetic biology and details some potential applications, namely biofuels, biomedicine and engineered genomes.

4.1 The Synthetic Biology Design Cycle

Figure 7 shows the design cycle for the core platforms in synthetic biology. To avoid laborious trial and error, this design cycle illustrates how to (1) design systems according to high-level concepts, (2) model these designs as circuits with efficient parts libraries, (3) simulate their functionality, (4) construct the design effortlessly, (5) probe the resulting circuits and (6) measure the results. In the design cycle, constant feedback between stages plays a key role in enhancing circuit functionality. Moreover, evolutionary strategies exist in the cycle to increase the performance of other steps, though these strategies remain underutilized. Phases such as circuit conceptualization, design and construction have advanced significantly, but many bottlenecks still exist at modeling, simulation, probing and measurement phases.

Fig. 7 Design cycle for
engineering circuits in
synthetic biology

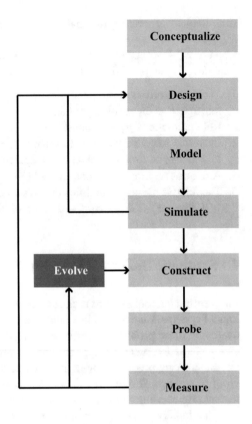

Despite the progress in this field, the design cycle for core platforms remains slow, costly and arduous; the goals of synthetic biology—understanding and manipulating biological systems—are still exceptionally challenging, as biological phenomena are complex and opaque. Though much progress has been made in several of these steps, others require more innovation. Modeling complex biological systems, in particular, requires novel computational analysis methods.

4.2 Basic Biological Parts

In synthetic biology, basic universal parts are commonly used to build higher-order synthetic circuits and pathways in order to simplify biological engineering (Fig. 8). DNA sequences form parts, parts form devices and devices form systems. These components fulfill particular functions, such as controlling transcription, regulating molecules and changing genetic material. Kushwaha and Salis [38] established regulatory genetic parts that can transfer a multienzyme circuit and pathway between *Bacillus subtilis*, *Pseudomonas putida* and *E. coli*. These basic parts usually come from natural systems. For example, natural promoters are exploited, such as P_L of

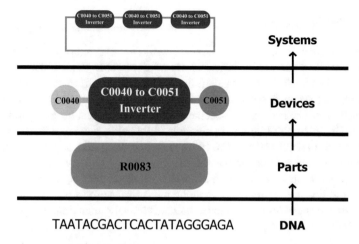

Fig. 8 Hierarchy of synthetic biology

phage lambda and the *lac* operon promoter, to regulate gene expression, both of which occur naturally in prokaryotes [20, 67].

4.3 DNA Construction

As synthetic biology designs complex biological systems from the bottom-up, standardized biological parts are vital. Parts can form devices, devices can be organized into systems and eventually systems can create large networks, synthetic chromosomes and genomes. To accomplish this goal, DNA must be assembled for the functional circuits.

Conventionally, the assembly of DNA sequences required laborious cloning steps, so constructing complex metabolic pathways and regulatory circuits was impossible; to deal with DNA sequences effortlessly, sequence-independent, one-pot and ordered assembly methods have been recently explored. As a result, many assembly methods have been proposed, such as the BioBrick and BglBrick standards [5], sequence- and ligase-independent cloning (SLIC) [41] and Gibson assembly [23]. Advancements in DNA synthesis technologies may lead to direct circuit design without multiple parts in the future, though this is currently impractical owing to the difficulty of constructing large combinatorial libraries of artificial circuits.

4.4 Future Applications

Standardized biological parts and circuit design can realize the goals of synthetic biology—designing and constructing biological systems for useful purposes.

4.4.1 Biofuels

Fossil fuels account for more than 80% of the world's energy supply, even though they harm the environment significantly and are nonrenewable. They must be replaced with renewable substitutes in the near future. Biofuels made from renewable and eco-friendly resources may become a great alternative, if they can compete economically.

Synthetic biology has pursued engineering many microbial hosts to produce various biofuels, including long-chain alcohols, fatty acids, alkanes and alkenes [52]. To produce biofuels, it is crucial to characterize the enzymes, regulation and thermodynamics of relevant native metabolic pathways. Until now, the most efficient production has been achieved with *E. coli* because of their significant resiliency, sufficient toolbox and peripheral metabolic pathways. Some advanced biofuel production methods have already achieved industrial-level production, though most are far from it.

4.4.2 Biomedicine

Functional devices—made by standardized biological blocks—can provide novel diagnostic and therapeutic tools for previously incurable or intractable diseases, such as cancer, infectious diseases and autoimmune disorders. To achieve this goal, it is necessary to identify the safety, side effects and *in vivo* performance of engineered circuits. In the context of therapy, it is especially important to determine how to deliver engineered cells and synthetic gene constructs to internal tissues.

Cancer

One factor that impedes curing cancer is the difficulty in differentiating between cancerous and healthy cells. The signature of hypoxia can help differentiate those cells. Therefore, synthetic biologists have engineered *E. coli* to attack mammalian cells selectively in hypoxic environments using heterologous sensors [4]. Other researchers have followed a similar logic to take advantage of the link between enzymatic activity and hypoxia [64].

Furthermore, Xie et al. [65] proposed a miRNA-detection strategy that can distinguish between cell types according to miRNA expression signatures. Separate circuits can identify a cell type by miRNA expression level; the combination of these circuits can perform as a cell-type classifier according to miRNA level. This approach has been shown to selectively kill HeLa cancer cells by regulating the expression of a proapoptosis protein by two high-expression and three low-expression miRNAs. In this way, different circuits can be blended to increase diagnosis and therapy performance.

Infectious Diseases

Antibiotic resistance and other properties, such as biofilm formation [44] and persistence [3], have intensified microbial infections. Treating severe antibiotic resistance in pathogens often requires the use of combinations of powerful antibiotics, which may ultimately promote further antibiotic resistance. Moreover, antibiotics can cause dreadful disorders of the human microbiome because they kill both pathogenic and nonpathogenic bacteria. As in the case of cancer, selective targeting of pathogens is essential in order to minimize this side effect. Other approaches are also possible through synthetic biology, such as targeting bacterial biofilms, reinforcing antibiotics [45] and engineering new treatment vehicles.

Autoimmune Disorders

Autoimmune disorders occur when the human immune system errantly attacks healthy body tissue covered in autoantigens. Autoimmune diseases include Hashimoto's thyroiditis, reactive arthritis and type I diabetes. Since precise causes are unknown, research programs should search for all potential targets. Larman et al. [39] utilized a synthetic human peptidome to find potential autoantigens and antibodies. This method enabled the investigation of all the coding regions within the human genome.

4.4.3　Engineered Genomes

Synthetic biology also aims to create artificial organisms and reveal the complexity of genetic variation with new enzymatic tools that enable precise genetic modifications. Recent advances in genetic assembly tools for synthetic biology have enabled directed mutagenesis to identify gene function, the correction of defective genes and the redesign genome structure. For example, Elowitz et al. [19] engineered genomes in an attempt to build a new organism and thus discover principles underlying life. In *E. coli*, Isaacs et al. [34] manipulated chromosomes by replacing all 314 TAG stop codons with TAA codons.

5　Gene Expression Analysis with LCS

Because LCS originated from biological systems and contain complex, interactive components that allow them to function as cognitive systems, they are well suited for classifying large-scale, noisy datasets that are often encountered in synthetic biology and systems biology. Furthermore, LCS generate sufficient rules describing systems with only a few interpretable key attributes, whereas state-of-the-art machine learning methods such as deep learning [26] cannot determine rules; LCS perform excellently for knowledge extraction to understand complex biological phenomena.

Accordingly, LCS and LCS-inspired systems have solved many bioinformatic and medical problems, such as automating alphabet reduction for protein datasets [9] and the classification of a primary breast cancer dataset [36]. LCS-inspired systems, such as BioHEL [8, 21], exist that are even designed for data mining large-scale bioinformatic datasets.

As explained in the previous section, gene expression analysis plays an essential role in solving a large number of problems in synthetic biology. Machine learning techniques have recently been attempted in a wide range of genetics and genomics fields related to the mechanisms of gene expression because of their ability to interpret large and complex genomic datasets [42]. For instance, machine learning can predict (1) gene expression from DNA sequences [10], (2) ChIP-seq profiles of histone modifications [35] and (3) transcription factors that bind to a gene promoter [49]. Researchers have also attempted to model all gene expression in a cell using a network model [22]. This trend will accelerate dramatically with the rapid advance of machine learning algorithms, tremendous increases in computational power and accessibility of massive datasets from large-scale international collaborations, such as the 1000 Genomes Project, the 100,000 Genomes Project and the ENCODE Project.

Glaab et al. [24] evaluated rule-based evolutionary machine learning systems inspired by Pittsburgh-style LCS, specifically BioHEL and GAssist, in order to increase the understandability of prediction models with high accuracy. Using three publicly available microarray cancer datasets, they inferred simple rule-based models that achieved an accuracy comparable with state-of-the-art methods, such as support vector machines, random forest and microarray prediction analysis. Furthermore, essential genes contained within BioHEL's rule sets beat gene rankings from a traditional ensemble feature selection. Specifically, BioHEL better predicted the relationships between relevant disease terms and top-ranked genes.

Zibakhsh et al. [68] suggested memetic algorithms, including LCS with a multiview fitness function approach. Fitness functions have two evaluation procedures: The evaluation of each individual fuzzy if-then rule in accord with the specified rule quality and the evaluation of each fuzzy rule quality according to the whole rule set performance. These algorithms significantly outperformed classic memetic algorithms in discovering rules. These new approaches have also extracted understandable rules differentiating different types of cancer tissues that were more accurate than other machine learning algorithms, such as C4.5, random forest and logistic regression model.

Abedini et al. [2] proposed two XCS-inspired evolutionary machine learning systems to investigate how to improve classification accuracy using feature quality information: FS-XCS, which utilize feature selection to reduce features, and GRD-XCS, which exploit feature ranking to modify the rule discovery process of XCS. The authors classified breast cancers, colon cancers and prostate cancers, each with thousands of features, from microarray gene expression data. They found that feature quality information can help the learning process and improve classification accuracy. However, approaches that restrict the learning process and rely exclusively on the feature quality information, such as feature reduction, may decrease classification accuracy.

6 Optimization of Artificial Operon Structure

A number of genes in genomes encode many proteins that modulate cellular activities or implement specific functionality. In bacterial genomes, such as *E. coli* genomes, such genes frequently act as an operon, that is, a functioning unit of genomic DNA that controls the transcription of multiple genes simultaneously with a single promoter. Figure 9 illustrates a typical operon with a promoter, an operator and structural genes. An operon is transcribed into a continuous mRNA strand and either translated in the cytoplasm, or trans-spliced to generate monocistronic mRNAs that are translated independently. As such, gene expression of elements within an operon decreases linearly with transcription distance [43]. To increase the productivity of synthetic metabolic pathways, the relative abundance of transcripts must be regulated accurately using an operon to achieve the balanced expression of multiple genes and avoid the accumulation of toxic intermediates or bottlenecks that inhibit the growth of microorganisms [61].

To approach this central challenge in synthetic biology, operon structure optimization has been pursued in recent years. There are several examples: The optimization and genetic implementation of a blueprint as an artificial operon, according to metabolic real-time analysis, tripled the production of dihydroxyacetone phosphate from glucose [12]; the amplification of genomic segments in artificial operons successfully controlled gene expression using selective RNA processing and stabilization (SRPS) [53], which transcribes primary mRNA into segments using nucleases and thus produces variation in stability among the segments [66]; libraries of tunable intergenic regions (TIGRs)—which recombine numerous post-transcriptional control elements and permit specifying the desired relative expression levels—have helped optimize gene expression in artificial operons [50].

Instead of modifying genes themselves, a completely different approach— reordering multiple genes into an operon structure with an appropriate promoter— may become a breakthrough in optimizing *E. coli* growth (i.e., production). A novel gene assembly method, the ordered gene assembly in *Bacillus subtilis* (OGAB) method [57], enables the changing of gene orders by the assembly of multiple DNA fragments in a fixed order and orientation. Newly designed operons, created by

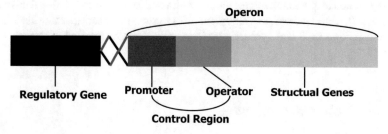

Fig. 9 Operon structure consisting of a promoter, an operator and structural genes, which is controlled by a regulatory gene

reordering genes from metabolic pathways, show significant differences in production because the abundance of specific mRNA sequences decreases as genes become separated from the promoter in *E. coli*.

The production of zeaxanthin increased considerably by rearranging the gene order of a five-gene operon [46, 48]. Similar results were also observed in the production of UHMW-P(3HB), which was achieved by reordering gene order within a three-gene operon [27].

However, optimizing more than five genes remains laborious as the number of gene orders increases factorially—a five-gene operon possesses 120 gene orders, but a ten-gene operon possesses 3,628,800 gene orders. Moreover, this research is typically carried out empirically without predictive modeling. As a result, a computational approach that optimizes operons accurately with respect to gene order is essential for comprehensive prediction. The next section presents a novel computational approach to analyzing the relationship between gene order within an operon and the growth rate of *E. coli* using a machine learning algorithm, such as LCS. The current work not only simulated those operon construction rules, but also verified the rules by conducting the wet-lab experiments of designing novel *E. coli* strains with expected high growth rates.

7   Optimization of Artificial Operon Construction by Machine Learning

7.1   Introduction

An aim was to investigate the influence of gene order among ten genes within an operon on the growth rate of *E. coli*. To do so, this research utilized consultation via machine learning algorithms that include LCS to analyze the relationship and then verified predicted gene orders with high growth rates using wet-lab experiments. This is the first computational approach for analyzing the relationship between operon gene order and *E. coli* growth rate. This research significantly contributes to the goal of designing efficient operons for the mass production of useful materials in synthetic biology—such as personalized medicine [60], medical diagnosis [16] and advanced biofuels [40]. Furthermore, this study provides a crucial step toward interdisciplinary research linking LCS and synthetic biology, which can be applied to various other tasks.

7.2   Artificial Operon Model

Machine learning was applied to investigate the construction principles relating gene order within an operon to the separation from a promoter involved in a metabolic pathway and thus the growth rate of *E. coli*. The operon contained ten genes (labeled A, B, C, D, E, F, G, H, I and J). The high growth-rate gene orders were then verified using wet-lab experiments.

The expression of multiple genes changes in accordance with changes in gene order [61]. Therefore, *E. coli* growth rates differ not only as a result of the presence and absence of genes, but also gene order within operons. Generally, if the expression of multiple genes is balanced, *E. coli* grows rapidly, but if it is not, *E. coli* grows poorly or is totally inhibited as toxic intermediates or bottlenecks accumulate. Yet, optimizing more than five genes has been impossible using conventional approaches because the number of gene orders increases factorially with the number of genes in an operon.

Ten genes were identified in this study. It was challenging to analyze them accurately for the following reasons: (1) The number of gene orders obtained from wet-lab experiments was only 0.004% of the total 3,628,800 gene orders (the dataset consists of 93 individual datasets from *E. coli* with gene orders similar to wild-type strains as well as 51 datasets with random gene orders); (2) even *E. coli* with identical gene orders exhibit a large standard deviation in growth rate (the maximum standard deviation of the dataset is around 0.05/h); (3) *E. coli* strains with high growth rates in the dataset possess similar gene orders—the highest growth rate was approximately 0.73/h.

Therefore, this research adopted algorithms of heuristic machine learning techniques that can resolve the trade-off between accuracy and smoothness to elucidate which gene orders significantly influence *E. coli* growth. As Fig. 10 shows, gene order refers to the arrangement of genes, and it significantly influences the growth rate of *E. coli*. Therefore, the current work investigated operon construction principles relating gene order to growth rate and tried to design new *E. coli* strains with high growth rates.

Figure 11 provides a flowchart describing the experimental procedure. The protocol consists of three steps: (1) Preprocessing, (2) supervised analysis and (3) postanalysis. Throughout the wet-lab experiment, *E. coli* was cultivated in duplicate. The OGAB method was exploited to reconstitute gene orders by assembling multiple DNA fragments with a desired order and orientation and thus generating an operon structure in a resultant plasmid. The raw data were normalized, and the gene orders were classified into several growth rates; a sampling of gene orders was selected and verified through wet-lab experiments.

7.3 Experimental Framework

Two experiments were conducted with consultation algorithms, one employing the results of two algorithms—Pittsburgh-style LCS [28, 29, 31] called GAssist [7] for compact solutions with few rules and a decision-tree induction technique, C5.0—and the other employing the results of four algorithms—other two well-known conventional machine learning techniques, random forest and multilayer perceptron, in

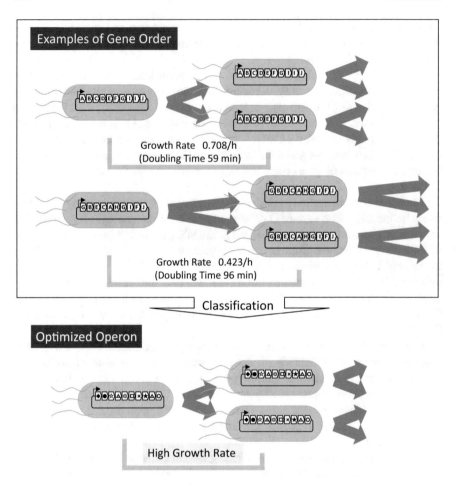

Fig. 10 Optimizing gene order of an artificial operon

addition to using LCS and C5.0. This was executed in order to analyze the relationship between the gene order within operons and *E. coli* growth rate. Then, six operons were designed per experiment according to the optimal predicted gene orders to assess the performance of the classifiers. Particularly, six operons per experiments were selected because of the difficulty in designing a large number of operons in terms of time and effort.

The term "consultation" refers to choosing gene orders with high growth rates in each algorithm by considering attributes of classifiers from C5.0 [51] in order to avoid over-fitting. The problem becomes a classification domain, and the parameters of the classifier model were selected to maximize ten-fold cross-validation accuracy.

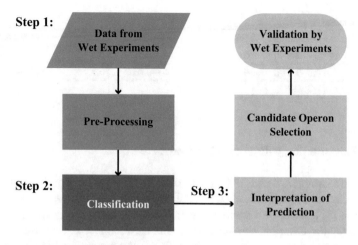

Fig. 11 Flowchart illustrating the experimental procedure

Table 1 Classes of growth rates

Classes	Growth rate	Classes	Growth rate	Classes	Growth rate
0	0–0.1/h	0.1	0.1–0.2/h	0.2	0.2–0.3/h
0.3	0.3–0.4/h	0.4	0.4–0.5/h	0.5	0.5–0.6/h
0.6	0.6–0.7/h	0.7	≥0.7/h		

7.3.1 Consultation with Two Algorithms

Using LCS and C5.0, 45 explanatory variables (describing the relative orders between two genes) were classified into eight growth rate groups (Table 1) defined by an equal interval—the growth rates take precise values, so they had to be converted into finite classes for the classification task. To test the classification performance in wet-lab experiments, 20 random datasets were examined out of the total 144 datasets as a test dataset.

Based on these classification results, six gene orders were selected within the operon that were predicted to increase *E. coli* growth, and then, strains were designed in order to test them using wet-lab experiments. First, four gene orders were identified that were classified as promoting growth rates exceeding 0.7/h in every algorithm considering C5.0 attributes. Furthermore, this work selected two gene orders predicted to have growth rates that were relatively high but significantly different from those of the original dataset to investigate the influence of modifying gene order remarkably.

Table 2 Classes of growth rates

Classes	Growth rate	Classes	Growth rate	Classes	Growth rate
0	0–0.4/h	0.4	0.4–0.5/h	0.5	0.5–0.6/h
0.6	0.6–0.65/h	0.65	0.65–0.7/h	0.7	0.7–0.72/h
0.72	≥0.72/h				

7.3.2 Consultation with Four Algorithms

In addition to using LCS and C5.0, random forest [11] was also exploited, which is an efficient ensemble learning method that employs many decision trees, and multilayer perceptron [33], which is a standard neural network model. Forty-five explanatory variables (again, describing the relative orders between two genes) were classified into seven growth rate groups (Table 2) with smaller ranges for high growth rates in order to predict gene orders with high growth rates more accurately. The results of the previous experiment (six datasets) were also used as the main dataset for this analysis, and 21 additional datasets from the total 150 datasets were used as the test dataset; three datasets per class were employed as test data.

From these classification results, six gene orders were selected for the operon structure, and they were designed for subsequent wet-lab experiments. First, this work identified two gene orders that were estimated to promote growth rates in excess of 0.72/h by LCS analysis and 0.7/h in the C5.0 analysis and random forest analysis, which considers the C5.0 attributes. In addition, this research selected two gene orders that were estimated to promote growth rates in excess of 0.7/h by LCS, C5.0 and random forest analyses and in excess of 0.65/h by multilayer perceptron analysis, which considers C5.0 attributes. Finally, two gene orders were identified that were classified as promoting growth rates in excess of 0.7/h by LCS and random forest analyses, 0.65/h by C5.0 analysis and 0.72/h by multilayer perceptron analysis, which considers C5.0 attributes.

7.4 Results

This subsection shows how these consultation algorithms utilizing LCS work in cases of consultation using two and four algorithms. The results include both computational simulations and their biological verification. This method performed effective data mining in this gene expression analysis owing to the high accuracy of LCS and its ability to determine definite understandable rules that describe complex systems efficiently.

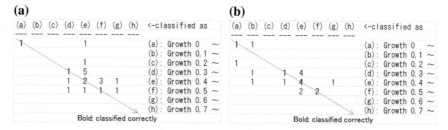

Fig. 12 **a** Classification of test data by learning classifier systems. **b** Classification of test data by C5.0

7.4.1 Consultation with Two Algorithms

Classification of Test Data by LCS

Figure 12 shows classification results, and the numbers of gene orders are presented with both the classified class and the actual class. Classification succeeded with 25% accuracy; if up to one class error is allowed, it achieved 80% classification accuracy. Most gene orders were classified as promoting growth rates between 0.4/h and 0.5/h.

According to a rule set of eight simple and interpretable rules, a gene order was determined that was given the highest class assignment ($\geq$0.7/h; the $\rightarrow$ operator represents the rule stating that the gene preceding the operator is located in front of the gene following the operator; e.g., A$\rightarrow$B means gene A is located before gene B). According to this rule set, gene A tends to be assigned to the front of the operon, while gene J tends to be assigned to the back. The rule set inferred to describe the highest growth rate classification using LCS is as follows.

- A$\rightarrow$B, A$\rightarrow$G, B$\rightarrow$H, C$\rightarrow$I, D$\rightarrow$F, E$\rightarrow$I, E$\rightarrow$J, H$\rightarrow$J

Classification of Test Data by C5.0

C5.0 produced classifications with 40% accuracy (Fig. 12); permitting one error class, this method obtained 80% of classification accuracy. Six gene orders promoting growth rates between 0.3/h and 0.4/h or between 0.5/h and 0.6/h were incorrectly classified as promoting growth rates between 0.4/h and 0.5/h.

Growth Rates of Newly Designed Operons

This work selected six gene orders within the operon that were classified by two-algorithm consultation to promote high growth rates, and then, they were designed for wet-lab verification experiments (Fig. 13). The novel strain was cultured in duplicate, and the error was estimated to be less than 6% per sample. Strains (Gene

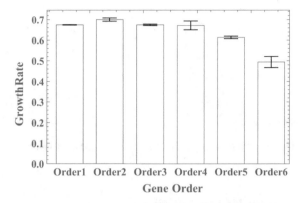

Fig. 13 Growth rates of *E. coli* with novel gene orders shown as means ± SD

orders): Order1 (ABEICHDFGJ); Order2 (ABDCEGFIHJ); Order3 (ABCDEIH-FGJ); Order4 (ABCDFGEIHJ); Order5 (AGBIEJCHDF); and Order6 (AEICHDF-BGJ). Novel operons exhibited high growth rates. In particular, Order2 gave a growth rate comparable to the highest in the dataset (around 0.73/h), considering that the maximum standard deviation of the dataset is around 0.05/h. Order5 and Order6, which each have completely different gene orders from the dataset, demonstrated low growth rates compared with the other orders.

7.4.2 Consultation with Four Algorithms

Classification of Test Data by LCS

Figure 14 shows classification results, and the numbers of gene orders are presented with both the classified class and the actual class. LCS yielded classifications that succeeded with around 38% accuracy (Fig. 14); if up to one class error is allowed, it achieved around 95% classification accuracy. Six gene orders promoting growth

(a)   **(b)**

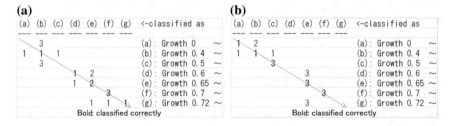

Fig. 14 a Classification of test data by learning classifier systems. **b** Classification of test data by C5.0

rates between 0/h and 0.4/h or between 0.5/h and 0.6/h were classified incorrectly as promoting growth rates between 0.4/h and 0.5/h.

This analysis produced a restrictive rule set of 13 rules that determined a gene order that was assigned to the highest class ($\geq$0.72/h). Genes A, B, C and D each tended to be assigned to the front of the operon, while gene J tended to be assigned to the back. The rank order of gene A has an especially strong influence on the growth rate. The rule set inferred to describe the highest growth rate classification using LCS is as follows.

- A→B, A→C, B→D, B→G, C→G, C→I, D→H, E→I, F→J, G→J, H→E, H→F, I→J

Classification of Test Data by C5.0

C5.0 produced a classification with approximately 52% accuracy (Fig. 14); within one class error, 86% classification accuracy was obtained. Six gene orders promoting growth rates between 0.6/h and 0.65/h or exceeding 0.72/h were incorrectly classified as promoting growth rates between 0.65/h and 0.7/h.

Classification of Test Data by Random Forest

The random forest analysis performed classification with around 48% accuracy (Fig. 15); if up to one class error is allowed, this method reached 100% classification accuracy. Three gene orders promoting growth rates between 0.6/h and 0.65/h were incorrectly classified as promoting growth rates between 0.65/h and 0.7/h.

Classification of Test Data by Multilayer Perceptron

The multilayer perceptron classification yielded 57% accuracy (Fig. 15); within one class error, this method achieved 90% classification accuracy. Five gene orders pro-

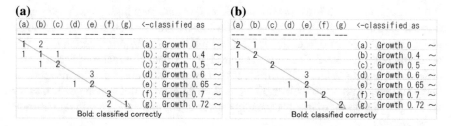

Fig. 15 a Classification of test data by random forest. **b** Classification of test data by multilayer perceptron

Fig. 16 Growth rates of *E. coli* with novel gene orders presented as means ± SD

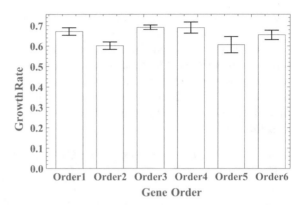

moting growth rates between 0.6/h and 0.65/h or exceeding 0.7/h were incorrectly classified as promoting growth rates between 0.65/h and 0.7/h.

Growth Rates of Newly Designed Operons

Six novel operons were designed with gene orders predicted to have high growth rates according to four-algorithm consultation (Fig. 16). The empirical error was less than 7% per sample. Strains (Gene orders): Order1 (ABCDFEHIGJ); Order2 (ACBD-FGEHIJ); Order3 (ABDCEFIGHJ); Order4 (ABDCGEHFIJ); Order5 (ABCDFEI GHJ); and Order6 (ABCDFIEGHJ). All of the newly designed operons showed high growth rates (>0.6/h). However, no operon promoted a higher growth rate than that which was obtained by the two-algorithm consultation. The gene orders are more similar to each other compared with those in two-algorithm analysis.

7.5   Conclusion

This research found that certain two-gene order rules can be used to design operons that significantly effect *E. coli* growth (i.e., production) using consultation algorithms that include LCS. Moreover, new *E. coli* strains with high growth rates were also successfully created using these operon construction rules. Genes that are closer to the promoter in an operon exhibit higher mRNA expression in general, as supported by real-time RT-PCR results. However, the explanation for severe growth rate differences among strains with different novel operons is unclear. The interactions between genes may differ substantially as their sequential order change. Yet, other potential explanations of the relationship between gene order and growth rate warrant consideration.

Most operon rearrangement studies without computational predictive modeling suggest that the gene orders that resemble those of wild-type strains tend to have high growth rates [46, 48]. The computationally optimized operons are similar to wild-type strains in gene order to some extent, as expected. Gene A, which exhibits much higher mRNA expression levels than other genes, was consistently located in the front of the operon in order to obtain a high growth rate. However, except for several genes strongly linked to their positions, reordering genes for growth optimization is possible. LCS rule sets, such as the optimal spatial relationship between gene E and gene I, provide further details on how gene orders can be optimized. While this is difficult to predict experimentally, the results are easily understood by biologists. However, if understandable rules are ignored, random forest might be the most suitable method as it achieves 100% accuracy if up to one-class incorrect classification is allowed.

This study also reveals that surpassing the highest growth rate of the dataset is challenging for the following reasons: (1) Although the classification of test data went well, no operon promoted growth rates exceeding those previously found; (2) the dataset is small and noisy; and (3) the dataset is biased, and efficient operons share similar gene orders.

This research aimed to optimize operon structure for the largest number of genes thus far—ten genes, which can be rearranged into 3,628,800 orders—by a novel computational approach. Current research focused on relative orders between two genes as the explanatory variables. However, more comprehensive variables, such as those used in natural language processing, may enhance classification accuracy; the structure involved in assessing sentences through word order is similar to the operon structure involved in predicting growth rates. These findings should also be confirmed using more randomized data to avoid over-fitting, especially when the number of genes within an operon is more than ten. Furthermore, this research focused on operons that promote efficient growth; however, future studies should also explore operons that inhibit *E. coli* growth.

Taken together, these findings illustrate that machine learning—especially the use of consultation algorithms utilizing LCS to avoid over-fitting—can help identify the most efficiently structured operons even when the number of genes within an operon is large. Changes in mRNA expression of genes and gene interactions altered by gene order may cause these results. Computational results must be interpreted with caution, but newly designed operons tested in wet-lab experiments support this approach. This first computational study proves that pair-wise order relationships between genes produce significant differences in operon efficiency; given the difficulty in understanding all interactions between genes, future studies with more comprehensive explanatory variables are needed. Furthermore, this study suggests that LCS can play a significant role in data mining from large and noisy datasets extracted from biological systems, especially gene expression analysis for the mass production of useful materials in synthetic biology.

8 Summary

This chapter has described what LCS and synthetic biology are and how they are closely related in the context of gene expression analysis. Specifically, it has illustrated the first computational approach to optimizing operon structure by altering gene order to match optimization rules produced by consultation with machine learning algorithms that use LCS. This research included both computational simulation and biological verification. Overall, these results indicate that LCS and LCS-inspired systems can perform effective data mining and knowledge discovery in gene expression analysis and synthetic biology broadly because LCS can extract definite understandable rules describing complex systems efficiently while retaining high accuracy.

This chapter both extends LCS and connects LCS with synthetic biology. This study confirms that LCS—followed by consultation with well-known algorithms to avoid over-fitting and obtain better solutions—can provide excellent knowledge discovery from huge and noisy datasets in biological systems or previously intractable dynamic systems. Moreover, this chapter provided a crucial first step of interdisciplinary research linking LCS and synthetic biology by illustrating both the core concepts as well as a clear relationship between these domains that can be applied to various other tasks.

Acknowledgements Dr. Yoshihiko Hasegawa and Dr. Hiroshi Dohi of The University of Tokyo give insightful discussions that were significant in this analysis of the operon structure. The authors would also like to especially thank UTDS members Marishi Mochida, Akito Misawa, Takahiro Hashimoto, Kazuki Taniyoshi, Yuki Inoue and Mari Sasaki for making a writing environment so pleasant.

References

1. Abedini M, Kirley M (2011) Guided rule discovery in XCS for high-dimensional classification problems. In: Wang D, Reynolds M (eds) AI 2011: advances in artificial intelligence. Springer, Heidelberg, pp 1–10
2. Abedini M, Kirley M, Chiong R (2013) Incorporating feature ranking and evolutionary methods for the classification of high-dimensional DNA microarray gene expression data. Australas Med J. doi:10.4066/AMJ.2013.1641
3. Allison KR, Brynildsen MP, Collins JJ (2011) Metabolite-enabled eradication of bacterial persisters by aminoglycosides. Nature. doi:10.1038/nature10069
4. Anderson JC, Clarke EJ, Arkin AP, Voigt CA (2006) Environmentally controlled invasion of cancer cells by engineered bacteria. J Mol Biol. doi:10.1016/j.jmb.2005.10.076
5. Anderson JC, Dueber JE, Leguia M, Wu GC, Goler JA et al (2010) BglBricks: A flexible standard for biological part assembly. J Biol Eng. doi:10.1186/1754-1611-4-1
6. Andrianantoandro E, Basu S, Karig DK, Weiss R (2006) Synthetic biology: new engineering rules for an emerging discipline. Mol Syst Biol. doi:10.1038/msb4100073
7. Bacardit J (2004) Pittsburgh genetics-based machine learning in the data mining era: representations, generalization, and run-time. PhD thesis, Ramon Llull University, Barcelona
8. Bacardit J, Burke EK, Krasnogor N (2009) Improving the scalability of rule-based evolutionary learning. Memetic Comput 1:55–67

9. Bacardit J, Stout M, Hirst JD, Valencia A, Smith RE et al (2009) Automated alphabet reduction for protein datasets. BMC Bioinformatics. doi:10.1186/1471-2105-10-6

10. Beer MA, Tavazoie S (2004) Predicting gene expression from sequence. Cell 117:185–198

11. Breiman L (2001) Random Forests. Mach Learn 45:5–32

12. Bujara M, Schmperli M, Pellaux R, Heinemann M, Panke S (2011) Optimization of a blueprint for in vitro glycolysis by metabolic real-time analysis. Nat Chem Biol 7:271–277

13. Bull L (2005) Two simple learning classifier systems. In: Bull L, Kovacs T (eds) Foundations of learning classifier systems. Springer, Heidelberg, pp 63–89

14. Bull L (2015) A brief history of learning classifier systems: from CS-1 to XCS and its variants. Evol Intell 8:55–70

15. Bull L, Bernado-Mansilla E, Holmes J (2008) Learning classifier systems in data mining: an introduction. In: Bull L, Bernado-Mansilla E, Holmes J (eds) Learning classifier systems in data mining. Springer, Heidelberg, pp 1–15

16. Chen YY, Smolke CD (2011) From DNA to targeted therapeutics: bringing synthetic biology to the clinic. Sci Transl Med. doi:10.1126/scitranslmed.3002944

17. Cheng AA, Lu TK (2012) Synthetic biology: an emerging engineering discipline. Annu Rev Biomed Eng. doi:10.1146/annurev-bioeng-071811150118

18. Cliff D, Ross S (1994) Adding temporary memory to ZCS. Adapt Behav 3:101–150

19. Elowitz M, Lim WA (2010) Build life to understand it. Nature 468:889–890

20. Elvin CM, Thompson PR, Argall ME, Hendr NP, Stamford PJ et al (1990) Modified bacteriophage lambda promoter vectors for overproduction of proteins in *Escherichia coli*. Gene 87:123–126

21. Franco MA, Krasnogor N, Bacardit J (2012) Analysing BioHEL using challenging boolean functions. Evol Intell 5:87–102

22. Friedman N (2004) Inferring cellular networks using probabilistic graphical models. Science 303:799–805

23. Gibson DG, Young L, Chuang RY, Venter JC, Hutchison CA et al (2009) Enzymatic assembly of DNA molecules up to several hundred kilobases. Nat Methods 6:343–345

24. Glaab E, Bacardit J, Garibaldi JM, Krasnogor N (2012) Using rule-based machine learning for candidate disease gene prioritization and sample classification of cancer gene expression data. PLoS ONE. doi:10.1371/journal.pone.0039932

25. Golberg DE (1989) Genetic algorithms in search, optimization, and machine learning. Addison-Wesley, Boston

26. Hinton GE, Salakhutdinov RR (2006) Reducing the dimensionality of data with neural networks. Science 313:504–507

27. Hiroe A, Tsuge K, Nomura CT, Itaya M, Tsuge T (2012) Rearrangement of gene order in the *phaCAB* operon leads to effective production of ultrahigh-molecular-weight poly[(R)-3-Hydroxybutyrate] in genetically engineered *Escherichia coli*. Appl Environ Microbiol. doi:10.1128/AEM.07715-11

28. Holland JH (1975) Adaptation in natural and artificial system: an introduction with application to biology, control and artificial intelligence. University of Michigan Press, Ann Arbor

29. Holland JH (1980) Adaptive algorithms for discovering and using general patterns in growing knowledge bases. Int J Policy Anal Inf Syst 4:245–268

30. Holland JH (1986) Escaping brittleness: the possibilities of general-purpose learning algorithms applied to parallel rule-based systems. In: Michalski RS, Carbonell JG, Mitchell TM (eds) Machine learning: an artificial intelligence approach. Morgan Kaufmann, Los Altos, pp 593–623

31. Holland JH, Reitman JS (1978) Cognitive systems based on adaptive algorithms. In: Waterman DA, Hayes-Roth F (eds) Pattern directed inference systems. Academic Press, New York, pp 313–329

32. Holmes JH, Lanzi PL, Stolzmann W, Wilson SW (2002) Learning classifier systems: new models, successful applications. Inf Process Lett. doi:10.1016/S0020-0190(01)00283-6

33. Hornik K, Stinchcombe M, White H (1989) Multilayer feedforward networks are universal approximators. Neural Netw 2:359–366

34. Isaacs FJ, Carr PA, Wang HH, Lajoie MJ, Sterling B et al (2011) Precise manipulation of chromosomes in vivo enables genome-wide codon replacement. Science 333:348–353

35. Karlic R, Chung HR, Lasserre J, Vlahovicek K, Vingron M (2010) Histone modification levels are predictive for gene expression. Proc Natl Acad Sci USA. doi:10.1073/pnas.0909344107

36. Kharbat F, Odeh M, Bull L (2008) Knowledge discovery from medical data: an empirical study with XCS. In: Bull L, Bernado-Mansilla E, Holmes J (eds) Learning classifier systems in data mining. Springer, Heidelberg, pp 93–121

37. Kovacs T (2004) Strength or accuracy: credit assignment in learning classifier systems. Springer, London

38. Kushwaha M, Salis H (2015) A portable expression resource for engineering cross-species genetic circuits and pathways. Nat Commun. doi:10.1038/ncomms8832

39. Larman HB, Zhao Z, Laserson U, Li MZ, Ciccia A et al (2011) Autoantigen discovery with a synthetic human peptidome. Nat Biotechnol 29:535–541

40. Lee SK, Chou H, Ham TS, Lee TS, Keasling JD (2008) Metabolic engineering of microorganisms for biofuels production: from bugs to synthetic biology to fuels. Curr Opin Biotechnol 19:556–563

41. Li MZ, Elledge SJ (2007) Harnessing homologous recombination in vitro to generate recombinant DNA via SLIC. Nat Methods 4:251–256

42. Libbrecht MW, Noble WS (2015) Machine learning applications in genetics and genomics. Nat Rev Genet. doi:10.1038/nrg3920

43. Lim HN, Lee Y, Hussein R (2011) Fundamental relationship between operon organization and gene expression. Proc Natl Acad Sci USA. doi:10.1073/pnas.1105692108

44. Lu TK, Collins JJ (2007) Dispersing biofilms with engineered enzymatic bacteriophage. Proc Natl Acad Sci USA. doi:10.1073/pnas.0704624104

45. Lu TK, Collins JJ (2009) Engineered bacteriophage targeting gene networks as adjuvants for antibiotic therapy. Proc Natl Acad Sci USA. doi:10.1073/pnas.0800442106

46. Nakagawa Y, Yugi K, Tsuge K, Itaya M, Yanagawa H et al (2010) Operon structure optimization by random self-assembly. Nat Comput. doi:10.1007/s11047-009-9141-0

47. Nakata M, Kovacs T, Takadama K (2014) A modified XCS classifier system for sequence labeling. In: Proceedings of the 2014 conference on Genetic and evolutionary computation. ACM Press, New York, pp 565–572

48. Nishizaki T, Tsuge K, Itaya M, Doi N, Yanagawa H (2007) Metabolic engineering of carotenoid biosynthesis in Escherichia coli by ordered gene assembly in Bacillus subtilis. Appl Environ Microbiol. doi:10.1128/AEM.02268-06

49. Ouyang Z, Zhou Q, Wong WH (2009) ChIP-Seq of transcription factors predicts absolute and differential gene expression in embryonic stem cells. Proc Natl Acad Sci USA. doi:10.1073/pnas.0904863106

50. Pfleger BF, Pitera DJ, Smolke CD, Keasling JD (2006) Combinatorial engineering of intergenic regions in operons tunes expression of multiple genes. Nat Biotechnol. doi:10.1038/nbt1226

51. Quinlan JR (1993) C4.5: programs for machine learning. Morgan Kaufmann, San Mateo

52. Rabinovitch-Deere CA, Oliver JW, Rodriguez GM, Atsumi S (2013) Synthetic biology and metabolic engineering approaches to produce biofuels. Chem Rev. doi:10.1021/cr300361t

53. Rochat T, Bouloc P, Repoila F (2013) Gene expression control by selective RNA processing and stabilization in bacteria. FEMS Microbiol Lett. doi:10.1111/1574-6968.12162

54. Smith SF (1980) A learning system based on genetic adaptive algorithms. PhD thesis, University of Pittsburgh

55. Smith SF (1983) Flexible learning of problem solving heuristics through adaptive search. In: Proceedings of the eighth international joint conference on artificial intelligence. Morgan Kaufmann, San Francisco, pp 421–425

56. Sutton RS, Barto AG (1998) Reinforcement learning. MIT Press, Cambridge

57. Tsuge K, Matsui K, Itaya M (2003) One step assembly of multiple DNA fragments with a designed order and orientation in Bacillus subtilis plasmid. Nucleic Acids Res. doi:10.1093/nar/gng133

58. Urbanowicz RJ, Moore JH (2009) Learning classifier systems: a complete introduction, review, and roadmap. J Artif Evol Appl. doi:10.1155/2009/736398
59. Watkins C (1989) Learning from delayed rewards. PhD thesis, University of Cambridge
60. Weber W, Fussenegger M (2012) Emerging biomedical applications of synthetic biology. Nat Rev Genet. doi:10.1038/nrg3094
61. White MM (2006) Pretty subunits all in a row: using concatenated subunit constructs to force the expression of receptors with defined subunit stoichiometry and spatial arrangement. Mol Pharmacol 69:407–410
62. Wilson SW (1994) ZCS: a zeroth level classifier system. Evol Comput 2:1–18
63. Wilson SW (1995) Classifier fitness based on accuracy. Evol Comput 3:149–175
64. Wright CM, Wright RC, Eshleman JR, Ostermeier M (2011) A protein therapeutic modality founded on molecular regulation. Proc Natl Acad Sci USA. doi:10.1073/pnas.1102803108
65. Xie Z, Wroblewska L, Prochazka L, Weiss R, Benenson Y (2011) Multi-input RNAi-based logic circuit for identification of specific cancer cells. Science 333:1307–1311
66. Xu C, Huang R, Teng L, Jing X, Hu J et al (2015) Cellulosome stoichiometry in Clostridium cellulolyticum is regulated by selective RNA processing and stabilization. Nat Commun. doi:10.1038/ncomms7900
67. Yanisch-Perron C, Vieira J, Messing J (1985) Improved M13 phage cloning vectors and host strains: nucleotide sequences of the M13mp18 and pUC19 vectors. Gene 33:103–119
68. Zibakhsh A, Abadeh MS (2013) Gene selection for cancer tumor detection using a novel memetic algorithm with a multi-view fitness function. Eng Appl Artif Intell 26:1274–1281

Ant Colony Optimization for Semantic Searching of Distributed Dynamic Multiclass Resources

Kamil Krynicki and Javier Jaen

Abstract In this chapter, we discuss the issues related to the problem of semantic resource querying in dynamic p2p environments and present the current approximations and successful solutions, with a special emphasis on their scalability. We focus on the use of nature-inspired metaheuristics, especially the ant colony optimization, and describe in detail the fundamental challenges an efficient p2p resources querying algorithm must overcome. The outlined approaches are evaluated in terms of the quality and completeness of the searches, as well as the algorithmic overhead. We introduce the notions of information diffusion in a p2p network as a means of combating the resource redistribution, and multipheromone approaches, that are often used to enable efficient semantic queries.

Keywords Ant colony optimization · Semantic searching · Distributed queries

1 Introduction

The principal use of computer systems, both in their origins as well as nowadays, is to aid humans with time-consuming tasks, such as storage, retrieval, and processing of large amounts of data. Initially, the considered scope of tasks put before computers was sufficiently small to be resolved on the spot, but, with time, the size and complexity of data outgrew the capacity of a singular, centralized processing entity. This motivated the formulation of the concepts of distributed computation: a number of independent, even heterogeneous, computation units, working in parallel and coordinating their efforts in order to achieve one, globally defined task; too complex of each one to solve on its own in any reasonable amount of time.

K. Krynicki (✉) · J. Jaen
DSIC, Universitat Politècnica de València, Cami de Vera S/N, 46022 Valencia, Spain
e-mail: kkrynicki@dsic.upv.es

J. Jaen
e-mail: fjaen@upv.es

© Springer International Publishing AG 2017
S. Patnaik et al. (eds.), *Nature-Inspired Computing and Optimization*,
Modeling and Optimization in Science and Technologies 10,
DOI 10.1007/978-3-319-50920-4_11

Fig. 1 P2P network query.
Note the self-check of the
node *A*. *Blue* continuous
arches represent the query
routing, and the *dotted red
arches* represent the response
routing

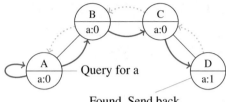

A specific subset of distributed computation in general is the concept of distributed resources. The idea behind it is simple: the substitution of a large centralized resource repository with an interconnected network of smaller ones. Examples of such systems include the following: distributed databases and p2p file exchange as data-centered networks, and processing grids as repositories of computational resources. Such resource repositories, be it resources of computational or informational nature, boast much higher availability, efficiency, extendibility, robustness, and load balancing than their centralized counterparts. Distributed approach to data storage is not without its shortcomings, however, as the cost of coordination and data transmission might outweigh the benefits, if handled trivially or incorrectly.

One of the main challenges of such systems is the efficient implementation of a resource query. A resource query, or simply a *query*, is an act of requesting resources of given parameters from the network, issued by either one of the entities within it or by an external source. The first step in tackling this issue is to model the concrete elements of computer networks with an abstract mathematical model of a symmetric directed graph $G = (V, E)$. The vertices of this graph, often referred to as nodes or peers, represent computation units, while the edges of the graphs are the connections of a network. In addition, every node $n \in V$ is a container of abstract resources that can be added, used, or depleted over time. It simplifies the problem conceptually and allows the transition of various graph search techniques (k-depth flood, random walks, etc.) alongside their well-established mathematical apparatus to the query execution and evaluation in graphs. In the field of computation, an equivalent of the graph of nodes is the network of peers, so called a peer-to-peer network, or *p2p*. See Fig. 1 for a graphical example of a very simple query in a 4-node network.

The essence of efficient p2p searches can be boiled down to a trade-off. On the one hand, in the resolution of a query, we must aim to analyze as many nodes as possible, in order to guarantee the completeness of the response. On the other hand, in order to guarantee algorithmic scalability, we would like to put as little strain on the network as possible. This means analyzing the minimum number of nodes necessary and aiming toward the $O(1)$ asymptotic complexity with a small constant factor. The global challenge is to enable efficient, scalable, robust semantic query routing through a p2p network for every peer, while taking into consideration potential problems, such as network and resource dynamism, failures, redundancy, and external attacks.

2   P2p Search Strategies

The simplest p2p search algorithm is a *full lookup* of every node in the network. The obvious limitation of the full lookup is that the scalability of the system which uses it is severely impaired, and it is made completely impractical for any large-scale solutions, where the only feasible approach is partial lookup. The upside, however, is that in this approach we are certain that if a resource exists in the network, it will be always retrieved successfully. Guaranteeing completeness of the response is a highly desirable factor in some industrial environments, where every resource must be accounted for. The full lookup strategies can be drawn directly from the graph theory and grouped into two broad classes: floods, based on breadth-first search (*BFS*), and in-depth floods, related to depth-first search (*DFS*). See Fig. 2 for an example of a BFS flood. The computational complexity of this class of solutions tends to be $O(E + V)$, where E is the number of edges and V is the number of nodes in the graph.

The most natural first step in increasing the scalability of full lookup is limiting its width or depth, trading off the completeness for scalability. There is a score of partial flood-like solutions, such as *k-depth flood* or *k-width flood*, where, rather

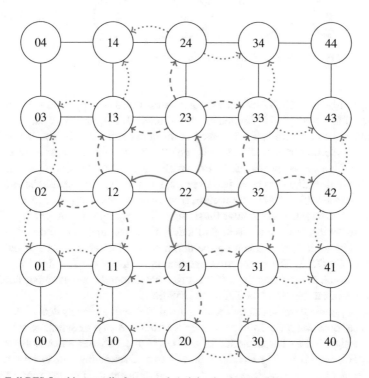

Fig. 2 Full BFS flood in a small p2p network (originating in node 22)

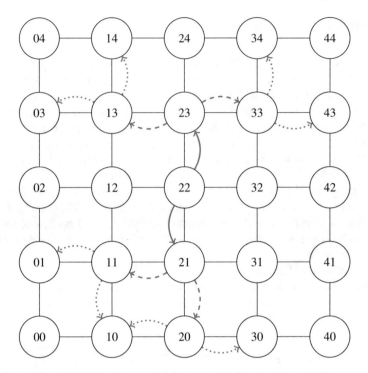

Fig. 3 Pruned ($k = 2$) DFS flood in a small p2p network (originating in node 22)

than exploring the entire network, we put an arbitrary, a priori pruning. More recent versions of Gnutella [3, 15] implemented a pruned BFS approach, with an iterative broadcast flood of the immediate vicinity of the query emitter node. Another solution of this kind was employed in Freenet [2], which, rather than broadcasting the query to all the neighbor nodes, propagates it to a small, random subset. This idea was based on the fact that similar data tends to cluster up, so a rough search should be enough to reveal it. However, this assumption holds true for only certain types of resources and specific networks, such as popular music files. The asymptotic complexity of these algorithms varies, but it tends to be a function of the parameters of the partial flood, usually polynomially dependent on the depth of flood and the degree of partiality, written as $O(f(k))$. The conceptual extreme opposite of the full flood is *random walks*, in which at each step the query is propagated to only one neighbor node. See Fig. 3 for an example of a $k = 2$ limited BFS flood.

A more practical implementation of a partial lookup is a hierarchical lookup, only applicable to structured p2p networks. In a structured p2p network, certain nodes can be designated as parent nodes of subsets of the network and resolve queries locally. This idea is much more scalable in terms of searching, but suffers in terms of maintainability, as any changes to the network, departure or apparition of nodes or resources, must be accounted for in the central node. The extreme variant of this approach was implemented in a number of early p2p networks, which relied on a centralized, server-like node. A good example of this approach is one of the first large-scale distributed file-sharing networks, Napster [1]. The Napster network consisted of an all-knowing server that resolved queries by providing response lists of nodes that contained queried resources. Naturally, this organization made Napster vulnerable to attacks as well as lawsuits. Another industrial example of this lookup is the early version of Gnutella, which implemented a central peer lookup coupled with a flood model. The unquestionable upside of the central server approach is that the query response is a constant time $O(1)$ problem, rather than $O(E + V)$ of a distributed full lookup, albeit it does suffer from a high constant factor. Additionally, this class of solutions relies on the existence of a predefined structure in the network, which is often not desirable or possible.

Rather than using a single super-peer, one can implement a full super-hierarchy. A simple idea, based on this approach, is to divide the network into a number of *buckets* of size b, each managed by its own dedicated super-peer, which can also be a member of the bucket. The search here boils down to two floods: first, between the super-peers, in order to locate the appropriate section of the network, and a second one, inside the bucket. The asymptotic time of this approach is $O(n/b + b)$, where b is the size of the bucket and n is the number of peers.

The simple approaches can be summarized in a compact pseudocode as shown in Fig. 4. Two code lines require an additional comment. In the code line 6 (A), the algorithm checks if the search has come to an end. Depending on the actual implementation, we can opt to explore the full network (full lookup) and reach the k-th depth, ttl distance, or T time (pruned lookups) from the emitter. In the code line 8 (B), a flood-like broadcast is performed. A full flood is achieved if we broadcast

```
1: let n₀ ∈ N be the emitter of query q

2: create agent a₀
3: PROCESS_AGENT(a₀, q, n₀)

4: function PROCESS_AGENT(agent, query, node)
5:     execute query in node
6:     if not stop_condition then                          ▷ (A)
7:         // in parallel
8:         for all nᵢ ∈ (N ⊆ N'_G(node))) do               ▷ (B)
9:             create agent aᵢ
10:            PROCESS_AGENT(aᵢ, query, nᵢ)
11:        end for
12:    else
13:        return to n₀
14:    end if
15: end function
```

Fig. 4 Simple query propagation pseudocode

to all the nodes $N = N'_G(node)$, where $N'_G(node)$ is the set of all the neighbors of the *node*, partial floods are achieved for $N \subset N'_G(node)$, and an in-depth random walks for $|N| = 1$.

Naturally, these solutions can only be considered adequate for small networks. In order to achieve a truly scalable approach, we must take the first step toward the *content analysis*. Thus far, the routing was detached from the actual content of the network and queries. It was based on the network's topology only, with a high degree of randomness and are pruned on arbitrary grounds. Many authors propose variants of the classical approaches with content analysis [4, 6, 7, 13, 14, 26]. In general, the content analysis has the objective of reducing the randomness of lookups by giving the nodes the capacity to choose routing paths that have the highest potential to yield best results. The content analysis can be divided into *active*, in which nodes analyze their neighbors and locally store an index based on this analysis, and passive, where nodes analyze the flows of queries and resources through them. On average, content-based approaches report an order of magnitude improvement over simple methods, but introduce new challenges and increase overall costs, in terms of, both, memory and computation.

In parallel, a new promising technique for distributed query propagation was conceived, called *distributed hash table* (DHT). DHT attempts to draw best aspects of the aforementioned approaches, combining the completeness of full lookups with the decentralization and scalability of partial lookup, with limited use of content analysis. It introduces a structured overlay onto unstructured p2p networks. Each query is converted into a 160 bit SHA hash value, which labels uniquely the section of the overlay where the queried resource must be located. Next, the algorithm performs an exhaustive check of all the indicated peers, with the guarantee that at least one of them contains the requested resource. The storage of resources is the reverse. SHA hash is used to choose the section of the network, within which the resources can be placed randomly. Albeit very promising, DHT presents a critical drawback. It only allows an exact match for name-based searches, which is sufficient for file searches, but cannot be considered a true content analysis or semantic search. DHT becomes increasingly inefficient if we attempt to request resources by indicating a set of metadata values, a semantic definition, or a collection of loose constraints. There exists a number of modern DHT implementations such as CAN, Chord, Pastry, and Tapestry. All use simple topologies (mainly ring) with complex overlays. None, despite being subject to intensive research, is fully free of the aforementioned limitation on the query structure. DHT can resolve queries in $O(\log n)$ or even $O(\log(\log n))$ time.

Further improvements to the problem can be achieved exploring the passive approaches to content analysis, coupled with search heuristics and the notion of evolutionary computation. A well-known and accepted family of solutions of this type is the nature-inspired ant colony optimization [9], which we explain in detail in Sect. 3. In Table 1, we summarize the so far discussed classes' approaches.

Table 1 Summary of approaches

Name	Complexity	Scalable	Complete	Bound	Random	Structured	Vulnerable	Content	Evolutive
BFS/DFS	$O(n+l)$	✗	✓	✗	✗	✗	✗	✗	✗
Pruned BFS/DFS	$O(f(k))$	✗[a]	✗	✓	✓	✗	✗	✗	✗
Central lookup	$O(1)$	✓	✓	✓	✗	✓	✓	✓[a]	✗
Super-peers	$O(n/b + f(b))$	✓[a]	✓	✓	✗	✓	✗[a]	✓[a]	✗
DHT	$O(\log(n))$	✓	✓	✓	✗	✓[a]	✗	✓[a]	✗
Content-Based	$O(TTL)$	✓	✗	✓	✓[a]	✗	✗	✓	✓[a]
Evolutionary	$O(TTL)$	✓	✗[a]	✓	✓[a]	✗	✗	✓	✓

[a] Potentially/partially

3 Nature-Inspired Ant Colony Optimization

The notion of transferring the computational effort to the peers is at the very core of p2p networks. One can take another step and enable p2p search algorithm to deposit algorithm-bound data in the peers as well. This way subsequent executions of the search algorithm over the same network would have an easier task, due to being able to draw conclusions from past findings. These algorithm-generated traces are referred to as *stigmergic medium,* and the high-level concept of depositing and reading it is called *stigmergy.*

Stigmergy can be found well outside the field of computation, in the nature itself. Many kinds of social insects secrete chemical stigmergic substance, called pheromone, that is used with the idea of allowing the transfer of knowledge to future entities. From a biologists' perspective, the behavior of a social insect, for example an ant, can be summarized as follows. Insects seek ways to benefit the colony. They can set out on a random sweep of the nearby territory in search for goods or they can participate in the transportation of goods from the sources already located. These two broad types of behavior are known as exploration and exploitation, respectively. An insect performing exploration might stumble upon a valuable object. In such an event, it carries out an evaluation and returns to the nest, secreting a trail of pheromone on its way back. The higher the estimated value, or *goodness,* of the newly found source, the stronger the pheromone trail will be. This, in turn, can entice more ants to choose to exploit the trail over performing exploration. A self-reinforcing process takes place, where more and more ants deposit pheromone and the trail gains appeal and eventually may become densely used. In computation, these entities are instances or iterations of algorithm execution.

The arguably most prominent example of a nature-inspired algorithm is the ant colony optimization (ACO). ACO is a metaheuristic for solving optimization problems, often reduced to path searches in graphs. As we established, this property makes it an interesting candidate for carrying out resource searches in graphs [19].

Real-life ants have a tendency to build and maintain the shortest paths between their points of interest, using the aforementioned stigmergic approach. This process was illustrated in the double bridge experiment [8] and modeled mathematically by Goss et al. [12]. Goss' model led Marco Dorigo to formulate the first computational model of their behavior, named ant system (AS) [10]. AS models the notions of pheromone deposition and evaporation as well as the detailed behavior of ants. More exact studies by Dorigo himself and others soon followed, leading to many improved variants of AS. The one, which is to this day highly regarded as the one of the most accurate is called ant colony system (ACS) [11].

In ACO, every edge of the search graph is assigned a numerical value, called *pheromone.* The totality of the pheromone values of all the edges is called the pheromone state of the network. Rather than fully randomly, as in previous methods, the query tends to be propagated to the edges with the highest pheromone, which are likely to yield resources efficiently. As mentioned, there exist two global modalities of query propagation: exploration (stochastic) and exploitation (deterministic). At

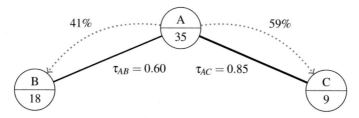

Fig. 5 Ant colony optimization query routing. *Blue dotted arches* symbolize the potential routes of exploration, with a percentual probability of choosing a particular edge. Parameter $\beta = 1$

each step, the query progresses randomly either via one or the other. The exploration from the node r to the next node s is chosen with weighted probability p_{rs}, where the weights are calculated as $\tau_{rs} \times \eta_{rs}^{\beta}$ (Eq. 1), while exploitation is a deterministic best pheromone approach (Eq. 2)

$$p_{rs} = \begin{cases} \frac{\tau_{rs} \times \eta_{rs}^{\beta}}{\sum_{z \in N_G(r)} \tau_{rz} \times \eta_{rz}^{\beta}} & \text{if } s \in N_G(r) \\ 0 & \text{otherwise} \end{cases} \tag{1}$$

$$s = \text{argmax}_{s \in N_G(r)} \, \tau_{rs} \times \eta_{rs}^{\beta} \tag{2}$$

where

- $N_G(r)$ is the set of neighbors of r
- $\tau_{rs} \in [0, 1]$ is the pheromone value on the edge rs
- η_{rs} is the cost value on the edge rs
- $\beta \in [0, 1]$ is a parameter

An example of a routing decision is displayed in Fig. 5. The percentual values over the edges AB and AC express the probability of the query being broadcasted to either B or C from A in case of exploration. If the exploitation is performed, the query is always forwarded to C.

Upon resolution, each query backtracks the exact route to its emitting node. The pheromone on the traversed edges is modified according to the quality of the results obtained, using Eqs. 3 and 4. Two opposite processes take place: on the one hand, the edge that participated in obtaining high-quality solutions gets reinforced, on the other hand, all the edges get their pheromone decreased (or *evaporated*). See Fig. 6 for an example of a pheromone update process by a resolved query.

The classically used goodness measure is *Hop per Hit*, which is a simple ratio of the node-to-node transitions the agent performed (Hop) to resources it retrieved (Hit). A special case is reserved for $Hit = 0$, which can be penalized with $HpH = constant factor$. There exist other quality measures, for instance, in clock-based systems, a useful one is Time per Hit. Other may include the following: the inverse Hit per Hop, pure Hit, pure Hop, and more. We, however, chose to use Hop per Hit in all our studies, as it is the most prevalent goodness measure in the literature.

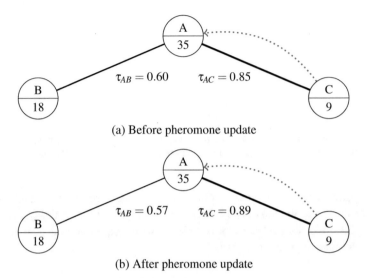

(a) Before pheromone update

(b) After pheromone update

Fig. 6 Ant colony optimization pheromone update. The *red arch* symbolizes a resolved query that provided resources of goodness $g = \frac{5}{2}$. Parameters $\rho = 0.05$, $\alpha = 0.05$

$$\tau_{rs} \leftarrow (1 - \rho) \cdot \tau_{rs} + \rho \cdot \gamma \cdot \max_{z \in s} \tau_{rz} \tag{3}$$

where

- $\rho \in [0, 1]$ and $\gamma \in [0, 1]$ are parameters

$$\tau_{rs} \leftarrow (1 - \alpha) \cdot \tau_{rs} + \alpha \times g \tag{4}$$

where

- g is the goodness of the solution
- $\alpha \in [0, 1]$ is a parameter

ACO-powered resource search in p2p networks can be reduced to the pseudocode shown in Fig. 7. The major differences with respect to the simple strategies are step 7 (A), where the routing decision is taken, based on the pheromone state, and step 12 (B), where the pheromone state is updated, based on the quality of the findings. ACS, in its state-of-the-art formulation, requires only minor changes in order to become fully p2p compatible. Our modified version of ACS, called RC-ACS [20], is a robust base for further study of the use of ACO in p2p.

The key change of RC-ACS, with respect to the pure ACS, is the way RC-ACS handles so called solution candidates. The traditional ACS is normally executed in a centralized environment, in which the cost of node-to-node transition is negligible. Therefore, in ACS, it is commonly accepted to generate a number of t candidate solutions, with the use of t agents. From this set, only the best solution is used for

1: **let** $n_0 \in N$ be the emitter of query q

2: create agent a_0
3: PROCESS_AGENT(a_0, q, n_0)

4: **function** PROCESS_AGENT(*agent*, *query*, *node*)
5:     execute *query* in *node*
6:     **if not** stop_condition **then**
7:         read from *pheromone state* $\triangleright (A)$
8:         **let** n_i be the outcome of exploration/exploitation
9:         PROCESS_AGENT(a_0, q, n_i)
10:     **else**
11:         calculate solution quality
12:         contribute to *pheromone state* $\triangleright (B)$
13:         return to n_0
14:     **end if**
15: **end function**

Fig. 7 Nature-inspired query propagation pseudocode

pheromone deposition, while the remainder is discarded. In a distributed environment, this kind of approach would be very wasteful. The cost of transitioning between nodes is a large portion of the overall cost of any distributed algorithm execution, so any found resources should be aggregated in the result, and nothing should be discarded.

Even though RC-ACS is a potentially good mechanism for p2p networks and it improved significantly the usability limits of distributed resource queries, it does present some challenges, mainly regarding the query's complexity and semantics. The problems can be classified into two major categories. First and foremost, ACO suffers so called *Network Dynamism Inefficiency*, which we analyze in detail in [18] and in Sect. 4. Second, and most importantly, ACO is inherently incompatible with semantic routing. In an environment, in which resources are categorized and organized in taxonomies, an important idea to consider is the concept of a *multiclass query*, that is, queries constrained to a subset of the resource taxa. Despite the unquestionable success of ACO, multiclass resource queries are often indicated as an example of a problem that is not satisfactorily solved with it. It is well known that resource querying involving multiclass resources has a less natural representation within ACO's metaheuristic, which manifests itself in reduced efficiency. We explore the second subject, known as *semantic queries inefficiency*, in detail in Sect. 5 as well as in [16, 17].

4 Nature-Inspired Strategies in Dynamic Networks

As mentioned, the ACO process of establishing the pheromone state, good paths, and nodes can be lengthy. Typically, the time period between the uninitialized pheromone state and a stable, efficient pheromone state is called pheromone convergence or

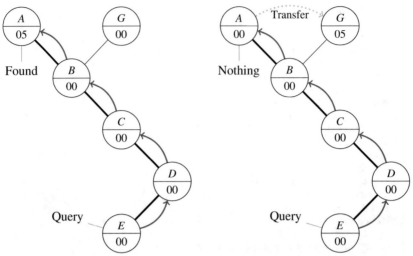

(a) Correct pheromone state before re-source transfer. Successfully resolved query (blue continuous arches)

(b) Invalidated pheromone state after re-source transfer (red dotted arch). Unsuc-cessfully resolved query (blue continuous arches)

Fig. 8 Example of network dynamism inefficiency

simply *convergence*. Once the convergence is achieved in a graph, the routing of queries tends to remain unaltered with very high probability.

4.1 Network Dynamism Inefficiency

In dynamic networks, ACO, or RC-ACS specifically, struggles. Any changes to the network, be it the content of the nodes or the structure of the network itself, invalidate the existing pheromone state and force the algorithm into a brief period of a secondary convergence, called *reconvergence*. See Fig. 8 for an illustration of the problem. The node A contains 5 abstract resources, which are requested by the node E (subfigure a). A strong pheromone path, visually marked as thicker edges, leads directly from the node E, which makes the query resolution quick and optimal. In subfigure b, there occurs a transfer of resources from the node A to the node G. Note, however that the pheromone state remains unchanged, yet it is not valid anymore. A new query emitted from the node E would likely reach the node A and be resolved unsuccessfully, even though that the number of resources present in the network is constant. It would take several failed queries for the pheromone state to reconverge.

The situation in the Fig. 8 is a very extreme example of inefficiency caused by the dynamism of the network. In our work [18], we examine a number of more realistic setups. We create a 1024-node network and subject it to a number of dynamism

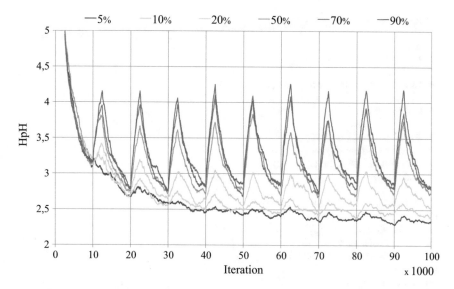

Fig. 9 The effects of node migration in a p2p network on Hop per Hit (HpH) query quality measure (less is better). No information diffusion

types that a real-life p2p network might be subjected to, which include the following: resources appearance, node migration, and edge appearance. Each of the examined network dynamisms corresponds to a possible real-life p2p problem, such as node failures, resource depletion, and more.

In Fig. 9, we present an example of findings we report in the mentioned study. The network impact is performed at every 10^4 iterations. The percentual values correspond to the magnitude of the network impact, which in this case was a node migration. We can very clearly observe a saw-like behavior of the quality measure HpH, the higher, the stronger the impact was. Another evident conclusion is that the stronger magnitudes of the network migration impede the full convergence to take place. While the mildly affected network (5%-level migration) improves throughout the experiment and reaches values below 2.4 Hop per Hit, the strongest impacts (90%-level migration) barely achieve minimums of 2.7 HpH and an average of 3.4 HpH. Similar conclusions are reported in [22].

4.2 Solution Framework

In the article [18], we establish a possible algorithmic solution that could potentially mitigate or eliminate the said inefficiency of ACO in p2p. We hoped to solve it fully within the ACO metaheuristic, so that it does not forfeit any of the benefits of ACO. We called the resulting complementary algorithm *information diffusion*.

```
 1: let n_0 ∈ N be the emitter of query q

 2: create agent a_0
 3: PROCESS_AGENT(a_0, q, n_0)

 4: function PROCESS_AGENT(agent, query, node)
 5:     (...)
 6: end function

 7: function NODE_CHANGE(node)                                    ▷ (A)
 8:     create diffusion agent d_0
 9:     DIFFUSE_CHANGES(d_0, node)
10: end function

11: function DIFFUSE_CHANGES(node)                                ▷ (B)
12:     contribute to pheromone state in node
13:     for all n_i ∈ (N ⊆ N'_G(node))) do
14:         create agent d_i
15:         DIFFUSE_CHANGES(d_i, n_i)
16:     end for
17: end function
```

Fig. 10 Nature-inspired query propagation pseudocode with dynamism

See pseudocode in Fig. 10 for a conceptual-level explanation of the ideas behind the information diffusion. Note code line 7 (A), where the effect of a node change creates, what we call a *diffusion agent*, which is then released from their emitting nodes and performs a pheromone state update corresponding to the change in question, see code line 11. The diffusion agents are highly similar to the classical ACO's query agents, with the difference that, rather than performing both the search and pheromone update, they only perform the latter.

The idea of information diffusion is rather straightforward. If a change in the node occurs, rather than waiting for regular queries to discover it and accommodate for it, we force a pheromone update in the direct vicinity of the changed node. For a visual illustration, let us return to the simplified dynamism example from Fig. 8. In Fig. 11, we present how the information diffusion works. First node A, which loses resources, communicates this fact to its neighbor (node B). Later, node G, which in turn gains resources, communicates this fact as well (subfigure a). This way node B has a complete knowledge of the modification of the network and can immediately start routing queries efficiently (subfigure b).

4.3 Experimental Evaluation

We design four simple information diffusion strategies based on the classical graph theory approaches: k-pruned breadth-first search, which we call *in-width k-depth*, and k-pruned depth-first search, called *in-depth k-depth*. Both strategies come in two

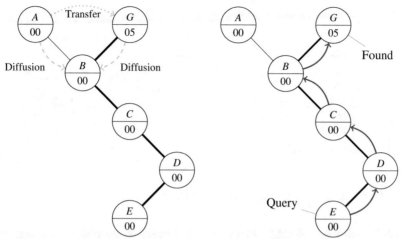

(a) Resource transfer (red dotted arch). Pheromone correction with the information diffusion (red dashed arches)

(b) Correct pheromone state after resource transfer. Successfully resolved query (blue continuous arches)

Fig. 11 Network dynamism problem solved with information diffusion. E is the query emitter

Table 2 Information diffusion types

	BFS (k-depth)	DFS (k-depth)
Online	In-width strain	In-depth strain
Offline	In-width non-strain	In-depth non-strain

distinct variants: online, meaning that the information is performed in parallel to the query issuing, and offline, in which we assume that the network has a downtime period, when to perform the diffusion. Note that the offline approach might not be suitable for a network that experiences a constant, uninterrupted flow of queries. See Table 2 for the summary of information diffusion approaches.

In our experiments, we confirm several facts about the information diffusion. First, unsurprisingly, we establish that offline diffusion is unconditionally beneficial and yields about a 30% improvement in terms of average HpH goodness measure. Naturally, the offline approach is rather idealistic and might not be suitable for every situation. However, even these trivial strategies (BFS, DFS) had a positive impact on the quality and speed of reconvergence in certain online variants. The in-width approach, the more resource-consuming of the two, has reported a statistically significant improvement for $k = 2$ in our experiments, while the lightweight in-depth approach is statistically beneficial for any $k > 0$. See Table 3, for a comparison of normalized results.

Table 3 Normalized HpH goodness for various executions of information diffusion algorithms

In-width			In-depth		
k	Offline	Online	k	Offline	Online
0	1	1	0	1	1
1	0.91[a]	0.93	5	0.872[a]	0.892
2	0.83[a]	0.892[a]	10	0.782[a]	0.832[a]
3	0.82[a]	1.02	15	0.726[a]	0.778[a]
4	0.798[a]	1.352	20	0.704[a]	0.768[a]

[a] Statically significant improvement

5 Nature-Inspired Strategies of Semantic Nature

The more profound and subtle of the two challenges mentioned in Sect. 3 is the semantic query inefficiency. The main reason behind the difficulties of resolving semantic queries with ACO is that ACO entails strict limitations, which are consequences of the constrained nature of real-life ants' communication capabilities [5]. As mentioned, ants leave trails of pheromone, connecting the colony to locations of interest. The pheromone level can be seen only as a general goodness of what is at the end of the trail, without specifying directly what it is, nor the type of resource it might be. Therefore, ACO search agents possess only a rough indication of a good direction to follow. A problem arises if the model in question includes various types of objectives.

5.1 Semantic Query Inefficiency

Consider the example given in the Fig. 12. In subfigure a, the node E releases a query for resources of class a and it is resolved successfully, note that thicker edges symbolize higher pheromone values. However, as subfigure b displays, if the same node queries for resources of class b, the query would end up in a a-rich node (A in our example) and yield no results, despite the fact that there are class b resources in the network. Clearly, a single pheromone value is not sufficient to guarantee efficient routing in models with ontology- or taxonomy-based resource classifications.

In a more general experiment in [16], we demonstrate that even in a small p2p network of 1024 nodes, the HpH-measured goodness decreases drastically in function of number of query classes C. See Table 4. With $C = 100$, the behavior of the ACS algorithm is asymptotically tending toward random walks.

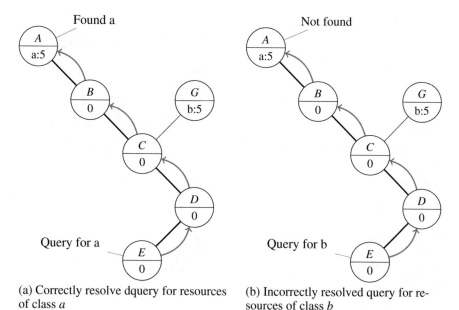

(a) Correctly resolve dquery for resources of class *a*

(b) Incorrectly resolved query for resources of class *b*

Fig. 12 Example of semantic query routing inefficiency

Table 4 HpH goodness decrease in function of number of query classes C, with ACS for $C = 1$, $C = 10$, $C = 100$, and random walks (RW)

C	Iterations			
	Initialized	2×10^6	4×10^6	Improvement (%)
1	1.46	0.63	0.61	−60.2
10	2.34	1.77	1.70	−27.3
100	3.08	2.89	2.88	−6.5
RW	4.16	4.18	4.15	−0

5.2   Solution Framework

One way to avoid this problem is by preventing ants of different tasks from interacting with each other. It is commonly achieved by separating pheromone for distinct objectives the search agents may have, see Fig. 13 for a visual example of this notion. This approach was explored by many authors [21, 23–25], to a different extent. Using different pheromone types (*multipheromone*, also referred to as *pheromone levels*) to mark trails leading toward distinct resources is widely accepted as a good idea.

The equations governing the interaction with the pheromone state must be extended to include the notion of multipheromone. The most straightforward approach to this problem is limiting the routing to only one pheromone level. Each query is fully executed only within one pheromone level ℓ, every query routing deci-

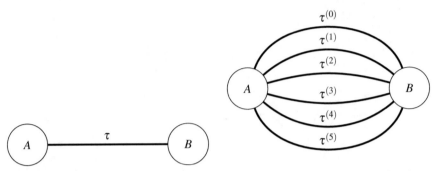

(a) Single pheromone connected nodes   (b) Multi pheromone connected nodes

Fig. 13 Single τ- and Multi 5-pheromone $\tau^{(0)} - \tau^{(5)}$ node connections

Fig. 14 Multipheromone
solution to semantic query
routing inefficiency. Queries
for resources *a* should only
interact with *dashed
pheromone*, while queries for
resources *b* only with *dotted
pheromone*

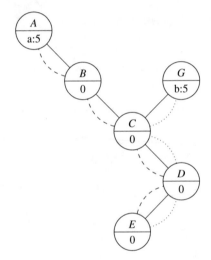

sion, as well as pheromone updates, use only the pheromone within the selected
ℓ level. The pheromone level selection, which happens at query launch in a node,
is a static and deterministic process. For our initial example, we propose the static
multipheromone solution in Fig. 14. The interaction between the queries searching
for resources *a* and *b* is fully prohibited: *a*-queries use the dashed pheromone, while
b-queries only the dotted pheromone.

There is a number of benefits coming from this simplistic approach. First and
foremost, the classic equations are easily rewritten with the multipheromone in mind,
by simply substituting the single τ pheromone value with a $\tau^{(\ell)}$ multipheromone:

$$s = \mathrm{argmax}_{u \in N_G(r)} \{\tau_{ru}^{(\ell)} \times \eta_{ru}^{\beta}\} \tag{5}$$

$$p_{rs} = \frac{\tau_{rs}^{(\ell)} \times \eta_{rs}^{\beta}}{\sum_{z \in N_G(r)} \tau_{rz}^{(\ell)} \times \eta_{rz}^{\beta}} \tag{6}$$

$$\tau_{ru}^{(\ell)} \leftarrow (1 - \alpha) \cdot \tau_{ru}^{(\ell)} + \alpha \cdot \delta\tau \tag{7}$$

$$\tau_{ru}^{(\ell)} \leftarrow (1 - \rho) \cdot \tau_{ru}^{(\ell)} + \rho \cdot \gamma \cdot \max_{z \in N_G(r)} \tau_{rz}^{(\ell)} \tag{8}$$

Second, especially for low numbers of pheromone levels, below 10, this approach is sufficiently robust and efficient. We shall call this idea *static multipheromone*.

Unsurprisingly, the static multilevel approach, in which one pheromone level is dedicated to queries of just one resource class, is suboptimal and redundant, and it suffers from several serious drawbacks, especially as the number of levels gets large. First, the entire taxonomy of resources must be known in advance. Second, it fails if the resource classes are added or removed during the algorithm's execution. Third, the evolution of the system is slowed down by a factor of the number of levels, due to the fact that, on average, only $1/|L|$ of queries contribute to each pheromone level. In addition, there is no intra-level knowledge transfer, even if certain types of resources always coincide. From a purely computational point of view, maintaining data structures for all the pheromone values is memory demanding.

In [16, 17], we proposed a dynamic multipheromone ACO variant called Angry Ant Framework (*AAF*). Rather than having one, fixed and a priori established mapping of queries to pheromone levels, in AAF the mapping happens spontaneously. Moreover, the number of pheromone levels is dynamic as well. At first, the graph is initialized with just one pheromone level $\tau^{(0)}$ and the totality of the query traffic is routed through it. Under certain circumstances, pheromone levels can be added or even abandoned. See Fig. 15 for the pseudocode of our solution. Note the code line 8 (*A*), where the pheromone level is evaluated, and the code line 9 (*B*), where new pheromone levels are potentially created.

Concretely, in order to implement any dynamic multipheromone ACO, one must solve two algorithmic questions:

1. The assigning of queries to pheromone levels. In a static multipheromone model, this step is performed manually, prior to the execution of the algorithm and it is inflexible.
2. The condition under which new pheromone levels are created and changed.

In AAF, the first problem is solved with, so called, level assignment matrices $\mathscr{L}$, which are maintained in every node. The columns of $\mathscr{L}(r)$ correspond to the types of queries that have originated at the node r, while the rows to the pheromone levels that are present in the r

```
1: let n_0 ∈ N be the emitter of query q

2: create agent a_0
3: PROCESS_AGENT(a_0, q, n_0)

4: function PROCESS_AGENT(agent, query, node)
5:     execute query in node
6:     if not stop_condition then
7:         read from pheromone state
8:         if pheromone state not satisfactory then        ▷ (A)
9:             expand/reassign pheromone state levels       ▷ (B)
10:        end if
11:        let n_i be the outcome of exploration/exploitation
12:        PROCESS_AGENT(a_0, q, n_i)
13:    else
14:        calculate solution quality
15:        contribute to pheromone state
16:        return to n_0
17:    end if
18: end function
```

Fig. 15 Nature-inspired query propagation pseudocode with semantic queries

$$\mathscr{L}(r) = \begin{array}{c} \\ \tau_0 \\ \tau_1 \\ \vdots \\ \tau_L \end{array} \begin{pmatrix} q_0 & q_1 & \cdots & q_{max} \\ 0.928 & 0.233 & \cdots & \mathscr{L}_{0\ max} \\ 0.623 & 0.853 & \cdots & \mathscr{L}_{1\ max} \\ \vdots & \vdots & \ddots & \vdots \\ 0.544 & 0.630 & \cdots & \mathscr{L}_{L\ max} \end{pmatrix} \tag{9}$$

where

τ_ℓ is the pheromone level
q_i is the query type

The probability for the query q to be assigned to ℓ-th level is given by Eq. 10, which is executed independently for each row in $\mathscr{L}(r)$. Two outcomes must be considered:

1. If multiple levels are chosen, multiple agents are sent out
2. If no levels are chosen, the agent is sent to the best level (Eq. 11)

$$p[a \to a^{(\ell)}] = \frac{\mathscr{L}_{\ell q}(r)}{\sum_{\ell' \in \{1..L\}} \mathscr{L}_{\ell' q}(r)} \tag{10}$$

$$\ell_{best} = \mathrm{argmax}_{\ell' \in \{1..L\}}\{\mathscr{L}_{\ell' q}(r)\} \tag{11}$$

The values in $\mathscr{L}$ are updated by resolved queries, upon returning to the emitter node:

$$\mathscr{L}_{\ell q}(r) \leftarrow (1 - \alpha_*) \cdot \mathscr{L}_{\ell q}(r) + \alpha_* \cdot \delta\tau \tag{12}$$

$$\mathcal{L}_q(r) \leftarrow (1 - \rho_*) \cdot \mathcal{L}_q(r) \tag{13}$$

The second aforementioned concept, the pheromone level creation, is much more subtle. As we indicate in Fig. 15, any currently used pheromone level can be abandoned when it is deemed *not satisfactory*. The satisfaction evaluation must be performed in an informed way, one that reflects the true nature of the resources present in the system. The probability of abandoning the current pheromone level ℓ is given by Eq. 14:

$$p_{irr}^{(\ell)}(r) = \iota \times \underbrace{\frac{1}{\tau_{max}}}_{(a)} \times \underbrace{\frac{\tau_{sum}^{(\ell)}(r)}{deg(r)}}_{(b)} \times \underbrace{\sum_{z \in N_G(r)} \frac{\tau_{rz}^{(\ell)}}{\tau_{sum}^{(\ell)}(r)} \ln \frac{\tau_{sum}^{(\ell)}(r)}{\tau_{rz}^{(\ell)}}}_{(c)} \tag{14}$$

where

(a)   is the normalization factor, guarantees $p_{irr}^{(\ell)}(r) \in \mathbb{R}_0^1$
(b)   is the average pheromone per edge in the node r
(c)   is the entropy of the pheromone distribution in the node r on the level ℓ
$\tau_{sum}^{(\ell)}(r)$ is the sum of all the pheromone values in a given node, on a given level (Eq. 15)
$deg(r)$ is the degree of the node r

$$\tau_{sum}^{(\ell)}(r) = \sum_{s \in N_G(r)} \tau_{rs}^{(\ell)} \tag{15}$$

If the decision of abandoning the current pheromone level ℓ is taken, the query $q^{(\ell)}$ has three distinct possibilities of reassignment:

$$q^{(\ell)} \rightarrow \begin{cases} q^{(\hat{\ell})} & \text{if } P > p_{irr}^{(\ell)}(r) \text{ and } (R \geq \frac{H(\mathcal{L}_q)}{H_{max}(|Q|)} \text{ and } q \in Q), \\ q^{(\ell+1)} & \text{if } P > p_{irr}^{(\ell)}(r) \text{ and } (R < \frac{H(\mathcal{L}_q)}{H_{max}(|Q|)} \text{ or } q \notin Q), \\ q^{(\ell)} & \text{otherwise,} \end{cases} \tag{16}$$

where

$\hat{\ell}$ is the best possible level (Eq. 11)
$H(\mathcal{L}_q)$ is the Shannon entropy of $\mathcal{L}_q$ (Eq. 17)
$H_{max}(|Q|)$ is the maximum Shannon entropy of $\mathcal{L}_q$ (Eq. 18)
P, R are uniform random variables in the $[0, 1]$ range

$$H(\mathcal{L}_q) = \frac{ln(\sum_{\ell' \in \mathcal{L}_q} \mathcal{L}_{\ell'q})}{\sum_{\ell' \in \mathcal{L}_q} \mathcal{L}_{\ell'q}} \times \sum_{\ell' \in \mathcal{L}_q} \mathcal{L}_{\ell'q} ln(\mathcal{L}_{\ell'q}) \tag{17}$$

$$H_{max}(|Q|) = ln(|Q|) \tag{18}$$

Equation 16 can be understood as follows:

1. If $\mathscr{L}$ in the current node has a column corresponding to the current query q ($q \in Q$) and the routing knowledge accumulated is sufficient ($R \geq \frac{H(\mathscr{L}_q)}{H_{max}(|Q|)}$), choose the best pheromone level ($\hat{\ell}$)
2. If $\mathscr{L}$ in the current node does not have a column corresponding to the current query q ($q \in Q$) or the routing knowledge accumulated is not sufficient ($R < \frac{H(\mathscr{L}_q)}{H_{max}(|Q|)}$), choose the next higher pheromone level ($\ell + 1$)
3. Otherwise, keep using the current pheromone level

If at any moment the reassignment algorithm chooses a level that does not exist, the level is created. This is referred to as *pheromone split*. Note that the pheromone split occurs only locally and does not affect the totality of the graph, which makes it a very scalable procedure. This incremental process takes place without any prior knowledge about the ontology of resources or their distribution, and will eventually stabilize, as shown in Fig. 16. Depending on the execution detail, the final number of pheromone levels can be strongly sublinear, possibly logarithmic, with respect to the number of query classes, which is much more scalable than the static multipheromone approach presented earlier.

5.3 Experimental Evaluation

The most basic evaluation of AAF must begin with analyzing the behavior of the system when the full model is not needed, when the number of query classes is 1,

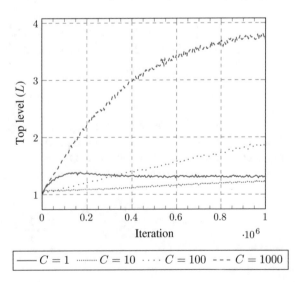

Fig. 16 Top used pheromone level for C different resource and query types

or the resource distribution is completely disjoint. This way we can establish what degree of overhead, if any, this model introduces.

The experiment to examine this fact was originally presented in [17]. Assume the existence of only two resource classes: c_1 and c_2 ($C = 2$). Next, we divide the p2p network G twofold. First, we select two continuous subnetworks of nodes G_{c1} and G_{c2}, each of which only contains resources of class c_1 and c_2, respectively. Independently, we select two other continuous subnetworks: G_{q1} and G_{q2} of nodes that will be allowed to query for either c_1 or c_2 only. The divisions are not exclusive; a node can find itself belonging to both subgraphs of each division.

The division into G_{c1} and G_{c2} is constant. They are both the precise halves of the initial network G, spanning across 50% of G. For simplicity, we can assume that G_{c1} is the left side of the network and G_{c2} is the right side of the network. The division into G_{q1} and G_{q2} comes in three variants:

$v0.5$ G_{q1} and G_{q2} are identical to G_{c1} and G_{c2}. There must be no query traffic between G_{c1} and G_{c2} as all the queries should be solved in their corresponding section.

$v0.7$ G_{q1} and G_{q2} are of 70% of N, which means that there is a 40% overlap between G_{q1} and G_{q2} in the center of the network. There should be moderate query traffic between G_{c1} and G_{c2}.

$v1.0$ $G_{q1} = G_{q2} = G$. All the nodes can query for both types of resources.

Intuitively, this experiment can be understood as two domes of query-generating nodes progressively overlapping each other, with complete disjoint at $v0.5$ and full overlap at $v1.0$. Naturally, the bigger the overlap section, the more need there is to distinguish between both types of queries c_1 and c_2 and direct them toward the appropriate part of the network and the more necessary AAF becomes.

In Fig. 17, we can compare AAF ($\iota = 1$) and ACS ($\iota = 0$) in the three aforementioned execution variants. In the $v0.5$ variant, there is no mismatch between the distribution of resources and queries and, in consequence, no need for pheromone splits to take place. We observe a very similar evolution for both versions of the algorithm. Just as in the $C = 1$ experiment, Angry Ant scores slightly better. However, as soon as the overlap is introduced, the AAF ($\iota = 1$) starts showing its superiority, achieving a 13% HpH improvement in the $v0.7$ case and 22% HpH improvement in the $v1.0$ case. We conclude that in this particular non-random distribution, we achieve better results than ACS ($\iota = 0$).

This way we have demonstrated that the dynamic multipheromone approach is beneficial, when required, and it does not contribute any unnecessary or uncompensated load. Additional analysis performed in [16] demonstrated that AAF improves, rather than simply matching, the performance of ACS when $C = 1$. The suggestion stemming from this analysis is that ACS could be substituted by AAF without penalties.

Having shown that the overhead introduced by AAF with respect to ACS is fully compensated by the increase in query efficiency, we proceeded to perform analysis in a multiclass environment, $C = 10$ and $C = 100$. Conclusions of the full analysis, available in [18], state that AAF improves the quality of p2p network searches in

Fig. 17 Base comparison
experiment of AAF ($\iota = 1$)
with ACS ($\iota = 0$). Less is
better

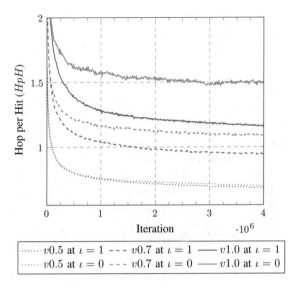

Table 5 Multiclass queries for $C = 10$ and $C = 100$ classes. *Hit* is resources found, while *Hop* is the length of the path. HpH is a goodness metric, Hop per Hit, which is the ratio of the two

Algorithm	Hit			Hop			HpH		
	t_0	t_m		t_0	t_m		t_0	t_m	
ACS ($C = 10$)	3.17	4.81	+52%	7.96	7.41	−7%	2.52	1.54	−39%
AAF ($C = 10$)	4.08	8.63	+112%	10.34	6.46	−38%	2.54	0.75	−70%
ACS ($C = 100$)	2.74	2.91	+6%	8.19	8.11	−1%	2.99	2.79	−6%
AAF ($C = 100$)	4.24	7.35	+73%	12.90	7.30	−43%	3.04	0.99	−67%

t_0: Average of the initial 10^4 iterations
t_m: Average of the final 10^4 iterations

a significant manner. See Table 5 for a summary of results. We can observe there that AAF finds both, shorter paths and more resource-rich paths than ACS. It also achieves better relative improvement over the course of the experiment in every tracked metric.

Finally, we performed an analysis of the memory consumption of the three aforementioned approaches: single pheromone (represented by ACS), static multipheromone (represented by Semant [23]), and dynamic multipheromone (AAF), in function of resource and query classes. We initialized a small p2p network of 4096 nodes with $C = 1$, $C = 10$, $C = 100$, and $C = 1000$ resource classes and executed 10^6 random resources queries. The conclusions are presented in Fig. 18. We can report that, due to the reduced pheromone level creation, AAF has introduced only a very limited overhead in comparison with the single pheromone approach and it is constant, for all practical purposes. Meanwhile, the memory consumption of the static multipheromone approach grows linearly with C. Note that the experiment failed to complete for $C = 1000$ in the case of static multipheromone.

Fig. 18 Memory consumption, less is better

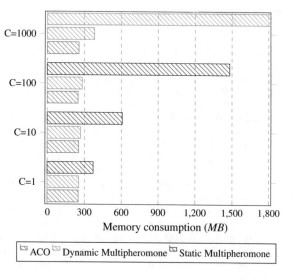

Memory consumption (*MB*)

ACO Dynamic Multipheromone Static Multipheromone

6 Conclusions and Future Developments

We have presented here an abbreviated summary of the recent developments in the field of p2p resource searches, with a special emphasis on the ACO-based approaches in dynamic networks and semantic queries. We laid out the most challenging aspects of this problem, alongside some successful solutions. The two major issues that needed to be addressed, the network dynamism inefficiency and the semantic query inefficiency, have sparked a series of empirical studies and were solved satisfactorily. It needs to be pointed out, however, that the solutions outlined in this chapter can be only considered partial, as there is yet to be proposed an algorithm capable of coping with both, a dynamic network and semantic queries simultaneously.

It can be claimed that a trivial merger of AAF with information diffusion is an appropriate initial proposal for such an algorithm. Nevertheless, to the best of our knowledge, no study to examine the validity of this statement has ever been performed.

Even though a simple composition of the two partial solutions has the potential to resolve the global problem, it is not guaranteed to do so, as the interaction between the partial solutions might generate unpredictable results. Therefore, another way is to propose a uniform algorithmic approach, which encompasses both of these problems in a single mathematical apparatus.

All the aforementioned problems and their potential solutions are of broad and current interest, as they can provide fundamental improvements in p2p networks and data-grid efficiencies.

References

1. Napster (1999). http://www.napster.com/, [Currently unavailable]
2. Freenet (2000). https://freenetproject.org/. Accessed 23 Dec 2015
3. Guntella (2000). http://www.gnutella.com/. [Offline]
4. Cohen E, Fiat A, Kaplan H (2003) A case for associative peer to peer overlays. SIGCOMM Comput Commun Rev 33(1):95–100
5. Colorni A, Dorigo M, Maniezzo V (1991) Distributed optimization by ant colonies. In: Proceedings of the european conference on artificial life, pp 134–142
6. Crespo A, Garcia-Molina H (2002) Routing indices for peer-to-peer systems. In: Proceedings 22nd international conference on distributed computing systems, pp 23–32
7. Crespo A, Garcia-Molina H (2005) Semantic overlay networks for P2P systems. In: Lecture notes in computer science (including subseries lecture notes in artificial intelligence and lecture notes in bioinformatics), vol 3601. LNAI, pp 1–13
8. Deneubourg JL, Aron S, Goss S, Pasteels JM (1990) The self-organizing exploratory pattern of the argentine ant. J Insect Behav 3(2):159–168. (March)
9. Dorigo M, Di Caro G (1999) Ant colony optimization: a new meta-heuristic. In: Proceedings of the 1999 congress on evolutionary computation-CEC99 (Cat. No. 99TH8406), 2
10. Dorigo M (1992) Optimization, learning and natural algorithms. Ph.D thesis, Politecnico di Milano
11. Dorigo M, Gambardella LM (1997) Ant colony system: a cooperative learning approach to the traveling salesman problem. IEEE Trans Evol Comput 1:53–66
12. Goss S, Aron S, Deneubourg JL, Pasteels JM (1989) Self-organized shortcuts in the Argentine ant. Naturwissenschaften 76:579–581
13. Joseph S, Hoshiai T, Member R (2003) Decentralized meta-data strategies: effective peer-to-peer search. Strategies, E 86B(6):1740–1753
14. Kalogeraki V, Gunopulos D, Zeinalipour-Yazti D (2002) A local search mechanism for peer-to-peer networks. In: Proceedings of the eleventh international conference on information and knowledge management, pp 300–307
15. Klingberg T (2002) Gnutella 6
16. Krynicki K, Houle ME, Jaen J (2015) A non-hybrid ant colony optimization heuristic for convergence quality. In: IEEE international conference on systems, man, and cybernetics. Accepted for presentation
17. Krynicki K, Houle ME, Jaen J (2015) An efficient aco strategy for the resolution of multi-class queries. Under review
18. Krynicki K, Jaen J, Catala A (2013) A diffusion-based ACO resource discovery framework for dynamic p2p networks. In: 2013 IEEE congress on evolutionary computation. IEEE, pp 860–867. (June 2013)
19. Krynicki K, Jaen J, Mocholí JA (2013) On the performance of ACO-based methods in p2p resource discovery. Appl Soft Comput J 13:4813–4831
20. Krynicki K, Jaen J, Mocholí JA (2014) Ant colony optimisation for resource searching in dynamic peer-to-peer grids. Int J Bio-Inspired Comput 6(3):153–165
21. Sim KM, Sun WH (2003) Ant colony optimization for routing and load-balancing: survey and new directions. IEEE Trans Syst Man Cybern Part A Syst Hum 33:560–572
22. Mavrovouniotis M, Müller FM, Yang S (2015) An ant colony optimization based memetic algorithm for the dynamic travelling salesman problem. In: Proceedings of the 2015 on genetic and evolutionary computation conference, GECCO'15. ACM, New York, NY, USA, pp 49–56
23. Michlmayr E (2007) Self-organization for search in peer-to-peer networks. Stud Comput Intell 69:247–266
24. Salama KM, Abdelbar AM, Freitas AA (2011) Multiple pheromone types and other extensions to the Ant-Miner classification rule discovery algorithm. Swarm Intell 5:149–182

25. Salama KM, Abdelbar AM, Otero FEB, Freitas AA (2013) Utilizing multiple pheromones in an ant-based algorithm for continuous-attribute classification rule discovery. Appl Soft Comput J 13:667–675
26. Tsoumakos D, Roussopoulos N (2003) Adaptive probabilistic search for peer-to-peer networks. In: Proceedings third international conference on peer-to-peer computing (P2P2003), pp 1–18

Adaptive Virtual Topology Control Based on Attractor Selection

Yuki Koizumi, Shin'ichi Arakawa and Masayuki Murata

Abstract One approach for accommodating traffic on a wavelength-routed optical network is to construct a virtual topology by establishing a set of lightpaths between nodes. To adapt to various changes in network environments, we propose an adaptive virtual topology control method, which reconfigures virtual topologies according to changing network environments, in IP over wavelength-routed wavelength division multiplexing networks. To achieve adaptability in the virtual topology control method, we focus on *attractor selection*, which models behaviors where biological systems adapt to unknown changes in their surrounding environments. The biological system driven by attractor selection adapts to environmental changes by selecting *attractors* of which the system condition is preferable. Our virtual topology control method uses deterministic and stochastic behaviors and controls these two appropriately by simple feedback of IP network conditions. Unlike current heuristic virtual topology control methods developed in the area of engineering, our method does not rely on pre-defined algorithms and uses stochastic behaviors for adapting to changes in network environments. The simulation results indicate that our virtual topology control method based on attractor selection adaptively responds to changes in network environments caused by node failure and constructs operational virtual topologies in more than 95% of simulation trials when 20% of nodes in the physical network fail simultaneously.

Keywords Wavelength-routed optical network · Virtual topology control · Adaptability · Biologically inspired networking · Attractor selection

Y. Koizumi (✉) · S. Arakawa · M. Murata
Osaka University, Osaka, Japan
e-mail: ykoizumi@ist.osaka-u.ac.jp

S. Arakawa
e-mail: arakawa@ist.osaka-u.ac.jp

M. Murata
e-mail: murata@ist.osaka-u.ac.jp

© Springer International Publishing AG 2017
S. Patnaik et al. (eds.), *Nature-Inspired Computing and Optimization*,
Modeling and Optimization in Science and Technologies 10,
DOI 10.1007/978-3-319-50920-4_12

1   Introduction

With the growth of the Internet, new application layer services, such as peer-to-peer networks, voice over IP, and IP television, have emerged. Such applications cause large fluctuations in network environments.

Wavelength division multiplexing (WDM) networks offer a flexible network infrastructure by using wavelength-routing capabilities. In such wavelength-routed WDM networks, a set of optical transport channels, called lightpaths, are established between nodes via optical cross-connects (OXCs). Much research has been devoted to developing methods of carrying IP traffic, which is the majority of Internet traffic, over wavelength-routed WDM networks [1, 7, 11, 17, 20, 22, 28, 34, 35]. One approach for accommodating IP traffic on a wavelength-routed WDM network is to configure a *virtual topology*, which consists of lightpaths and IP routers. To efficiently transport traffic, *virtual topology control*, which configures a virtual topology on the basis of given traffic demand matrices, has been investigated [25, 29]. By reconfiguring virtual topologies, wavelength-routed WDM networks offer the means of adapting to changing network environments. As one of changes in network environments, the huge fluctuations in traffic demand caused by emerging application layer services have been revealed [18, 24]. Furthermore, as networks play an increasingly important role as a social infrastructure, the ability to withstand or recover from various changes in network environments, such as network failures or changes in traffic demand, becomes a crucial requirement of networks.

Current virtual topology control methods, which are based on the control paradigm developed in the area of engineering, mainly take into account a certain set of scenarios for environmental changes and prepare countermeasures to those changes as algorithms. For these environmental changes, these virtual topology control methods may guarantee optimal performance, but they cannot achieve expected performance if unexpected changes occur.

One of the best examples of adapting to various environmental changes is a biological system, which is discussed in the life sciences [15]. We adopt mechanisms found in biological systems to achieve adaptability and robustness against various environmental changes in networks, such as network failures and changes in traffic demand. To achieve adaptability to various environmental changes, an approach that keeps adapting to any change in a network is indispensable. Such behaviors are often found in the area of life sciences, especially in biology [15]. In contrast to engineering systems, biological systems do not rely on pre-defined algorithms. Instead they mainly exploit stochastic behaviors for adapting to environmental changes. Therefore, they do not guarantee optimal performance but can adapt to unexpected environmental changes. Unlike most other engineering-based virtual topology control methods, we develop a virtual topology control method that is adaptive to various environmental changes by exploiting mechanisms found in biological systems.

We focus on *attractor selection*, which models the behaviors of organisms when they adapt to unknown changes in their surrounding environments and recover their conditions. Kashiwagi et al. [16] showed an attractor selection model for Escherichia coli cells to adapt to changes in nutrient availability. Furusawa and Kaneko [9] showed another attractor selection model for explaining the adaptability of a cell, which consists of a gene regulatory network and a metabolic network. The fundamental concept underlying attractor selection is that a system adapts to environmental changes by selecting a suitable *attractor* for the current surrounding environment. This selection mechanism is based on deterministic and stochastic behaviors, which are controlled by a simple feedback of the current system conditions, as we will describe in Sect. 3. This characteristic is one of the most important differences between attractor selection and other heuristic or optimization approaches developed in the area of engineering. While current engineering approaches cannot handle unexpected changes in the environment, attractor selection can adapt to unknown changes since a system is driven by stochastic behavior and simple feedback of the current system conditions. Therefore, we adopt attractor selection as the key mechanism in our virtual topology control method to achieve adaptability to various environmental changes.

In [19], we developed a virtual topology control method based on attractor selection that is adaptive to changes in traffic demand. This method constantly keeps adapting to changes in traffic demand by only using information on link utilization. The quantity of link utilization information is less than that obtained from traffic demand matrices, which most engineering-based virtual topology control methods use, but link utilization information is retrieved quickly and directly using the simple network management protocol (SNMP) [5]. Therefore, our method reacts to changes in traffic demand as quickly as possible. To achieve adaptability to environmental changes, we need to appropriately design a virtual topology control method based on attractor selection, even though the control paradigm of our proposed method is based on biological systems, which have inherent adaptability. Our previously proposed method [19] requires elaborative design of its internal structure in advance of its operation. More precisely, it must have appropriate attractors to adapt to changes in network environments. We designed the internal structure using information on an underlying physical topology, focusing on changes in traffic demand as one of changes in network environments. Though it is adaptable to changes in traffic demand, it is difficult to keep its adaptability if its topological structure changes due to changes in network environments, such as node failure. To overcome this problem, we need a system design approach without a priori knowledge of environmental changes. For the present research, we extend our previously proposed virtual topology control method. The method in this chapter dynamically reconfigures its internal structure, i.e., attractors, according to changing network environments and achieves adaptability to network failures and to changes in traffic demand, as with our previous virtual topology control method [19].

The rest of this chapter is organized as follows. In Sect. 2, we briefly discuss related approaches. We introduce the concept of attractor selection [9] in Sect. 3, and then, we propose our adaptive virtual topology control method based on attractor selection in Sect. 4. We evaluate the adaptability of our method to network failures in Sect. 5. Finally, we conclude this chapter in Sect. 6.

2 Related Work

The approaches for recovering from network failures in wavelength-routed optical networks can be classified into two categories: protection and restoration [38].

With protection approaches, a dedicated backup lightpath for each working light-path is reserved for recovery from network failures at network design time. Protection approaches generally enable fast recovery from expected network failures [23, 31, 32]. However, it is obvious that such approaches cannot handle unexpected network failures since they exploit several assumptions or priori knowledge about network failures and pre-design those backup lightpaths at the network design time. For instance, most protection approaches do not take into account the situation in which both working and backup lightpaths fail simultaneously. Though several protection approaches that enable recovery from dual-link failures are proposed [33], they also exploit several assumptions or priori knowledge about network failures. Therefore, protection approaches cannot handle unexpected network failures.

In contrast, when a network failure occurs, restoration approaches dynamically reconfigure an alternative virtual topology for the lightpath affected by the failure [6]. To establish an alternative virtual topology, restoration approaches must discover spare resources in the network by collecting network information, such as surviving optical fibers and OXCs. If restoration approaches collect the network information, they maintain connectivity of virtual topologys. However, restoration approaches do not enable recovery from failure if they cannot collect the network information due to that failure. For instance, failure in a node, including both an IP router and an OXC, makes the collection of the network information difficult, as we will discuss in Sect. 5.

Since protection and restoration approaches, which are based on the control paradigm developed in the area of engineering, take into account a certain class of failures and are optimized for achieving fast and efficient recovery from the assumed failures, they may not enable recovery from unexpected network failures, such as multiple and series of network failures caused by a disaster. It is difficult for engineering approaches to adapt to various environmental changes as long as they use pre-designed algorithms. Therefore, the development of a virtual topology control method that adapts to various environmental changes in networks including network failures is important.

3 Attractor Selection

We briefly describe attractor selection, which is the key mechanism in our adaptive virtual topology control method. The original model for attractor selection was introduced in [9].

3.1 Concept of Attractor Selection

A dynamic system driven by *attractor selection* uses noise to adapt to environmental changes. In attractor selection, *attractors* are a part of the equilibrium points in the phase space in which the system conditions are preferable. The basic mechanism consists of two behaviors, i.e., deterministic and stochastic behaviors. When the current system conditions are suitable for the environment, i.e., the system state is close to one of the attractors, deterministic behavior drives the system to the attractor. When the current system conditions are poor, stochastic behavior dominates deterministic behavior. While stochastic behavior is dominant in controlling the system, the system state fluctuates randomly due to noise, and the system searches for a new attractor. When the system conditions have recovered and the system state comes close to an attractor, deterministic behavior again controls the system. These two behaviors are controlled by simple feedback of the system conditions. Therefore, attractor selection adapts to environmental changes by selecting attractors using stochastic behavior, deterministic behavior, and simple feedback. In the following section, we introduce attractor selection that models the behavior of the gene regulatory and metabolic reaction networks in a cell.

3.2 Cell Model

Figure 1 is a schematic of the cell model used in [9]. It consists of two networks, i.e., the gene regulatory network in the dotted box at the top of the figure and the metabolic reaction network in the box at the bottom.

Each gene in the gene regulatory network has an expression level of proteins, and deterministic and stochastic behaviors in each gene control the expression level. Deterministic behavior controls the expression level due to the effects of activation and inhibition from the other genes. In Fig. 1, the effects of activation are indicated by the triangular-headed arrows, and those of inhibition are indicated by the circular-headed arrows. In stochastic behavior, inherent noise randomly changes the expression level.

In the metabolic reaction network, metabolic reactions consume various substrates and produce new substrates. These metabolic reactions are catalyzed by proteins on corresponding genes. In the figure, metabolic reactions are illustrated as fluxes of substrates, and catalyses of proteins are indicated by the dashed arrows. The changes in the concentrations of metabolic substrates are given by metabolic reactions and the transportation of substrates from outside the cell. Some nutrient substrates are supplied from the environment by diffusion through the cell membrane.

The growth rate is determined by dynamics in the metabolic reactions. Some metabolic substrates are necessary for cellular growth; thus, the growth rate is determined as an increasing function of their concentrations. The gene regulatory network

Fig. 1 A schematic of gene
regulatory and metabolic
reaction networks

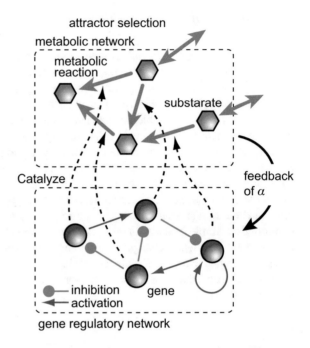

uses the growth rate as feedback of the conditions on the metabolic reaction network
and controls deterministic and stochastic behaviors by using the growth rate. If the
metabolic reaction network is in poor condition and the growth rate is small, the effect
of stochastic behavior dominates that of deterministic behavior, triggering a search
for a new attractor. During this phase, the expression levels are randomly changed
by noise, and the gene regulatory network searches for a state that is suitable for the
current environment. After the conditions of the metabolic reaction network have
been recovered and the growth rate increases, deterministic behavior again drives
the gene regulatory network to stable states.

The following section describes the mathematical model of attractor selection in
more detail.

3.3 Mathematical Model of Attractor Selection

The internal state of a cell is represented by a set of expression levels of proteins on n genes, $(x_1, x_2, \ldots, x_n)$, and concentrations of m metabolic substrates,
$(y_1, y_2, \ldots, y_m)$. The dynamics of the expression level of the protein on the ith gene,
x_i, is described as

$$\frac{dx_i}{dt} = \alpha \left(\varsigma \left(\sum_j W_{ij} x_j - \theta \right) - x_i \right) + \eta. \tag{1}$$

The first and second terms on the right-hand side represent the deterministic behavior of gene i, and the third term represents stochastic behavior. In the first term, the regulation of protein expression levels on gene i by other genes are indicated by regulatory matrix W_{ij}, which takes 1, 0, or -1, which correspond to activation, no regulatory interaction, or inhibition of the ith gene by the jth gene, respectively. The rate of increase in the expression level is given by the sigmoidal regulation function, $\varsigma(z) = 1/(1 + e^{-\mu z})$, where $z = \sum W_{ij} x_j - \theta$ is the total regulatory input with threshold θ for increasing x_i, and μ indicates the gain parameter of the sigmoid function. The second term represents the rate of decrease in the expression level on gene i. This term means that the expression level decreases depending on the current expression level. The last term at the right-hand side in Eq. (1), η, represents molecular fluctuations, which is Gaussian white noise. Noise η is independent of production and consumption terms, and its amplitude is constant. The change in expression level x_i is determined by deterministic behavior, the first and second terms in Eq. (1), and stochastic behavior η. The deterministic and stochastic behaviors are controlled by the growth rate α, which represents the conditions of the metabolic reaction network.

In the metabolic reaction network, metabolic reactions, which are internal influences, and the transportation of substrates from the outside of the cell, which is an external influence, determine the changes in the concentrations of metabolic substrates y_i. The metabolic reactions are catalyzed by proteins on corresponding genes. The expression level x_i determines the strength of the catalysis. A large x_i accelerates the metabolic reaction, and a small one suppresses it. In other words, the gene regulatory network controls the metabolic reaction network through catalyses.

Some metabolic substrates are necessary for cellular growth. Growth rate α is determined as an increasing function of the concentrations of these vital substrates. The gene regulatory network uses α as the feedback of the conditions on the metabolic reaction network and controls deterministic and stochastic behaviors. If the concentrations of the required substrates decrease due to changes in the concentrations of nutrient substrates outside the cell, α also decreases. By decreasing α, the effects that the first and second terms in Eq. (1) have on the dynamics of x_i decrease, and the effects of η increase relatively. Thus, x_i fluctuates randomly and the gene regulatory network searches for a new attractor. The fluctuations in x_i lead to changes in the rate of metabolic reactions via the catalyses of proteins. When the concentrations of the required substrates again increase, α also increases. Then, the first and second terms in Eq. (1) again dominate the dynamics of x_i stochastic behavior, and the system converges to the state of the attractor. Since we mainly use the gene regulatory network and only mentioned the concept of the metabolic reaction network, we omitted the description of the metabolic reaction network from this chapter. Readers can refer to [9] for detailed description of the metabolic reaction network. The next section explains our virtual topology control method based on this attractor selection model.

4   Virtual Topology Control Based on Attractor Selection

We propose our virtual topology control method based on the attractor selection model. We first introduce the network model that we use. Then, we describe our method.

4.1   Virtual Topology Control

Our network consists of nodes having IP routes overlaying OXCs, with the nodes interconnected by optical fibers, as shown in Fig. 2a. This constitutes the physical topology of the network. Optical de-multiplexers allow each optical signal to be dropped to IP routers, or OXCs enable to pass through those signals. In such wavelength-routed networks, nodes are connected with dedicated virtual circuits called lightpaths. Virtual topology control configures lightpaths between IP routers via OXCs on the WDM network, and these lightpaths and IP routers form a virtual topology, as shown in Fig. 2b. When lightpaths are configured in the WDM network, as illustrated in Fig. 2a, the virtual topology in Fig. 2b is constructed. The IP network uses a virtual topology as its network infrastructure and transports IP traffic on the virtual topology. By reconfiguring virtual topologies, that is, where to establish lightpaths, wavelength-routed optical networks offer the means of adapting to changing network environments. It is indispensable to consider how to reconfigure virtual topologies and where to establish lightpaths in order to develop an adaptive virtual topology control method.

4.2   Overview of Virtual Topology Control Based on Attractor Selection

In attractor selection, the gene regulatory network controls the metabolic reaction network, and the growth rate, which is the status of the metabolic reaction network, is recovered when the growth rate is degraded due to changes in the environment. In our virtual topology control method, the main objective is to recover the performance of the IP network by appropriately constructing virtual topologies when performance is degraded due to changes in network environments. Therefore, we interpret the gene regulatory network as a WDM network and the metabolic reaction network as an IP network, as shown in Fig. 3. By using stochastic behavior, our virtual topology control method adapts to various changes in network environments by selecting suitable attractors, which correspond to virtual topologies in our method, for the current network environment, and the performance of the IP network recovers after it has degraded due to network failures.

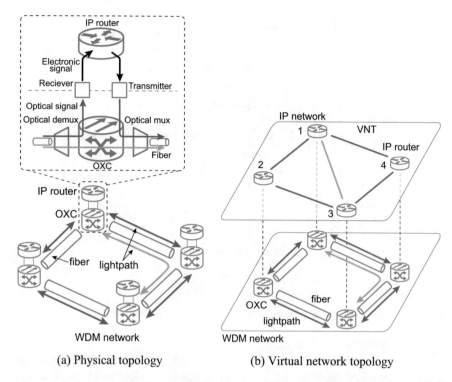

Fig. 2 Example of wavelength-routed WDM networks; physical and virtual network topologies. In (**b**), *lower layer* represents WDM network, which consists of OXCs and fibers, and *upper layer* represents IP network, which uses a virtual topology constructed due to wavelength-routing capability as network infrastructure

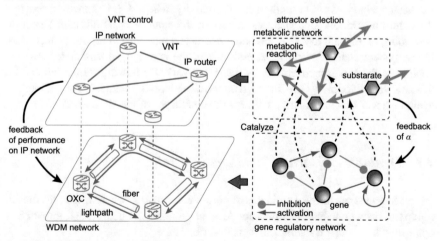

Fig. 3 Interpretation of attractor selection into virtual topology control

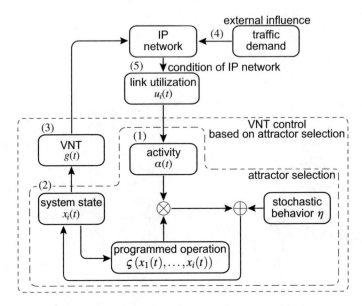

Fig. 4 Control flow of virtual topology control based on attractor selection

A flowchart for our virtual topology control method is illustrated in Fig. 4. Our proposed approach works on the basis of periodic measurements of conditions of IP networks. As one example of the conditions, we use the link load, which is the volume of traffic on links. This information is converted to activity, which is the value to control deterministic and stochastic behaviors. We describe the deterministic and stochastic behaviors of our method and the activity in Sect. 4.3. Our method controls the deterministic and stochastic behaviors in the same way as attractor selection depending on the activity. Our method constructs a new virtual topology according to the system state of attractor selection, and the constructed virtual topology is applied as the new infrastructure for the IP network. By flowing traffic demand on this new virtual topology, the link load on the IP network is changed, and our method again retrieves this information to know the conditions of the IP network.

4.3 Dynamics of Virtual Topology Control

We place genes on all the candidates of possible lightpaths, where l_i denotes the ith lightpath. Each gene has an expression level of proteins x_i, and the ith lightpath l_i is controlled by x_i. The dynamics of x_i is described as

$$\frac{dx_i}{dt} = \alpha \left(\varsigma \left(\sum_j W_{ij} x_j \right) - x_i \right) + \eta, \tag{2}$$

where η represents white Gaussian noise, $\varsigma(z) = 1/(1 + \exp(-z))$ is the sigmoidal regulation function, and the activity, α, which is equivalent to the growth rate in cells introduced in Sect. 3, indicates the condition of the IP network. We give the definition of α below. We use the same formula as in Eq. (1) to determine x_i. We determine whether the ith lightpath is established or not with x_i. In our method, we establish lightpath l_i if $x_i > 0.5$; otherwise, we do not establish l_i. Therefore, our method interprets x_i as the virtual topology, and x_i converges to the state of attractors.

In a cell, α indicates the conditions of the metabolic reaction network, and the gene regulatory network seeks to optimize this α. In our virtual topology control method, we use the maximum link utilization on the IP network as a metric indicating the conditions of the IP network. To retrieve the maximum link utilization, we collect the traffic volume on all links and select their maximum values. This information is easily and directly retrieved using SNMP. The activity must be an increasing function for the goodness of the conditions of the target system, i.e., the IP network in our case, as mentioned in Sect. 3. Note that any metric that indicates the condition of an IP network, such as average end-to-end delay, average link utilization, or throughput, can be used for defining α. We use the maximum link utilization, which is one of the major performance metrics for virtual topology control and used in many studies [10, 29] as the IP network condition. Therefore, we convert the maximum link utilization on the IP network, $u_{\max}$, into α as

$$\alpha = \frac{1}{1 + \exp\left(\delta\left(u_{\max} - \zeta\right)\right)}, \tag{3}$$

where δ represents the gradient of this function, and the constant number, ζ, is the threshold for α. If the maximum link utilization is more than ζ, α rapidly approaches 0 due to the poor conditions of the IP network. Therefore, the dynamics of our virtual topology control method is governed by noise and the search for a new attractor. When the maximum link utilization is less than ζ, we rapidly increase α to improve the maximum link utilization.

The smooth transition between a current virtual topology and a newly calculated virtual topology is also one of important issues for virtual topology control as discussed in [8, 36]. Our virtual topology control method constructs a new virtual topology on the basis of the current virtual topology, and the difference between these two virtual topologies is given by Eq. (2). The high degree of activity means that the current system state, x_i, is near the attractor, which is one of the equilibrium points in Eq. (2), and therefore, the difference given by this equation is close to zero. Consequently, our virtual topology control method makes small changes to virtual topology enabling adaptation to changes in network environments. Where there is a low degree of activity due to poor conditions in the IP network, stochastic behavior dominates deterministic behavior. Thus, x_i, fluctuate randomly due to noise η to search for a new virtual topology that has the higher activity, i.e., the lower maximum link utilization. To discover a suitable virtual topology efficiently from the huge number of possible virtual topologies, our method makes large changes to the virtual topology. In this way, our proposed scheme modifies virtual topologies depending on the maximum link utilization on IP networks and adapts to changes in traffic demand. We have already shown that our method achieves the smooth transition between virtual topologies in [19].

Since attractors are a part of the equilibrium points in the phase space, as described in Sect. 3, the definition of attractors is a challenging and important aspect of our proposed method. In Sect. 4.4, we describe how to define attractors in the phase space.

4.4 Attractor Structure

The regulatory matrix W_{ij} in Eq. (2) is an important parameter since it determines the locations of attractors in the phase space. Since our method selects one of the attractors with Eq. (2) and constructs the virtual topology corresponding to the selected attractor, defining W_{ij} is a challenge. To define arbitrary attractors in the phase space, we use the knowledge about the Hopfield neural network, which has a similar structure to gene regulatory networks.

The dynamics of our proposed method is expressed by Eq. (2). From the perspective of dynamical systems, α is regarded as a constant value that determines the convergence speed. The noise η is Gaussian white noise with mean 0. These values do not affect equilibrium points, that is, attractors in our method, in the phase space. Therefore, the equilibrium points are determined by the following differential equation,

$$\frac{d}{dt}x_i = \varsigma \left(\sum_j W_{ij} x_j \right) - x_i.$$

This is the same formula as a continuous Hopfield network [14]. Therefore, we use the knowledge of associative memory to store arbitrary attractors in the phase space [3, 30].

Suppose, we store a set of virtual topologies $g_k \in G$ in the phase space defined by Eq. (2). Let $\boldsymbol{x}^{(k)} = (x_1^{(k)}, x_2^{(k)}, \ldots, x_i^{(k)})$ be the vector of the expression levels corresponding to virtual topology g_k. To store $\boldsymbol{x}^{(k)}$ in the phase space, we adopt the method introduced in [3], which stores patterns in the phase space by orthogonalizing them. Due to space limitations, we omitted detailed description of this method. Readers can refer to [3] for detailed description. We store m virtual topologies, $\boldsymbol{x}^{(1)}, \boldsymbol{x}^{(2)}, \ldots, \boldsymbol{x}^{(m)}$, in the phase space. Let X be the matrix whose rows are $\boldsymbol{x}^{(1)}, \boldsymbol{x}^{(2)}, \ldots, \boldsymbol{x}^{(m)}$. The regulatory matrix $W = \{W_{ij}\}$, whose attractors are $\boldsymbol{x}^{(1)}, \boldsymbol{x}^{(2)}, \ldots, \boldsymbol{x}^{(m)}$, is defined as

$$W = X^+ X, \tag{4}$$

where X^+ is a pseudo inverse matrix of X.

Though pattern orthogonalization results in high stability of stored patterns [3], our method can also use simpler memorization approaches, such as Hebbian learning [13]. By using Hebbian learning, W_{ij} is defined as follows,

$$W_{ij} = \begin{cases} \displaystyle\sum_{g_s \in G} (2x_i^{(s)} - 1)(2x_j^{(s)} - 1) & \text{if } i \neq j \\ 0 & \text{if } i = j. \end{cases} \tag{5}$$

Our method can use either Eq. (4) or Eq. (5) to define the attractor structure. We mainly use Eq. (4) for defining the regulatory matrix, since the authors in [3] found that pattern orthogonalization results in higher memory capacity than Hebbian learning. In the next section, we describe the method of dynamically reconfiguring the attractor structure to adapt to dynamically changing network environments.

4.5 Dynamic Reconfiguration of Attractor Structure

If we use Eq. (4) to define the attractor structure, expression levels converge to any of the stored attractors. In other words, our method constructs any of the stored virtual topologies $g_s \in G$. One approach of adapting to various changes in network environments is to store all possible virtual topologies. However, this approach is almost impossible due to the memory capacity limitation of the Hopfield network [3, 13]. Therefore, it is important to develop a method to increase the number of attractors suitable for the current network environment in the phase space. By incorporating new attractors suitable for the current network environment into the attractor structure, we attain adaptability to various environmental changes when only a limited number of attractors are held in the attractor structure.

To achieve this, we update the regulatory matrix W_{ij} in the case that our proposed method finds a virtual topology that is suitable for the current network environment and not a member of G when α is low. Before describing this approach in detail, we show the control flow of our proposed method in Fig. 4. For simplicity, we use the terms used in Eqs. (2)–(4) with time t, e.g., the expression level at time t is expressed as $x_i(t)$, in this section. In step (1) of Fig. 4, we calculate α from the maximum link utilization at time t, $\alpha(t)$. In step (2), we update $x_i(t)$ with Eq. (2). Next, we convert $x_i(t)$ to the virtual topology $g(t)$ and provide it as the network infrastructure of the IP network in step (3). In step (4), traffic flows on this virtual topology $g(t)$. In step (5), the resulting flow of traffic determines link utilization $u_i(t)$ on the lightpath l_i. Then, we again calculate α at time $t + 1$, $\alpha(t + 1)$, from $u_i(t)$. Therefore, $\alpha(t + 1)$ indicates the goodness of virtual topology $g(t)$, which is converted from x_i. Therefore, we can determine the goodness of virtual topology $g(t)$ for the current network environment by observing $\alpha(t + 1)$.

We can determine that the performance on virtual topology $g(t)$ is high enough if $\alpha(t + 1) > A$, where A is a threshold value. We add $g(t)$ to the set of attractors G if $\alpha(t + 1) > A$, $\alpha(t) \leq A$, and $g(t) \notin G$. Then, we update W_{ij} with Eq. (4). We add another condition $\alpha(t) \leq A$ to prevent unnecessary updates of W_{ij}. The expression levels $x_i(t)$ always fluctuate due to the constant noise term, η, in Eq. (2). This results in the generation of many attractors close to one of the stored attractors, resulting in the lack of the diversity of attractors. In our method, x_i always fluctuate due to η in Eq. (2). Even if α is high and the deterministic behavior dominates the dynamics of x_i over stochastic behavior, x_i change slightly; thus, our method sometimes constructs a slightly different virtual topology from the already constructed virtual topology. Without the condition $\alpha(t) \leq A$, this behavior results in the generation of many attractors that are close to one of the stored attractors, resulting in the lack of the diversity of attractors. For this reason, we add $\alpha(t) \leq A$ and add virtual topology $g(t)$ as a new attractor to the attractor structure only if the stochastic behavior dominates the dynamics of x_i.

Finally, we describe the deletion of attractors. Since the memory capacity of the Hopfield network is limited, as mentioned above, we cannot keep all of the added virtual topologies in the attractor structure. We have to delete some of the stored attractors when we add virtual topology $g(t)$ as a new attractor. We simply use a first-in first-out (FIFO) policy for managing attractors when a new attractor is added. Though we can use a more sophisticated policy for managing attractors, our method achieves sufficient adaptability, as discussed in the next section.

5   Performance Evaluation

In this section, we evaluate the adaptability of our proposed method through simulation experiments.

5.1   Simulation Conditions

5.1.1   Parameters

The physical topologies used in this evaluation are generated randomly according to Erdös-Rényi model [26]. The physical topology has 100 nodes and 496 optical fibers, one optical fiber for each direction. Each node has 16 transmitters and 16 receivers. Since the number of transmitters and receivers is a more strict constraint for constructing virtual topologies than the number of wavelengths, we set the number of wavelengths on each optical fiber to 1024, that is, we mainly consider the number of transmitters and receivers as a constraint for constructing virtual topologies. We assume that the capacity of each lightpath is 100 Gbps. We use randomly generated traffic demand matrices having elements that follow a log-normal distribution with a mean of 1 Gbps and with variance in the variable's logarithm, 1, according to the observation in [27]. In this evaluation, we store 30 attractors in the attractor structure and use 30 randomly generated virtual topologies for defining the initial W_{ij}. Though we use randomly generated virtual topologies for defining the initial regulatory matrix, the matrix is dynamically reconfigured depending on the network environments, as described in Sect. 4.5; therefore, the setting of the initial W_{ij} does not have much impact on the functionality of our proposed method. We set the constant values in the definition of α in Eqs. (2) and (3) as follows, $\mu = 20, \delta = 50$, and $\zeta = 0.5$, respectively. For each evaluation, we conduct 1000 simulation trials by using 10 randomly generated traffic demand matrices and 100 randomly selected node failure patterns.

5.1.2 Node Failure

We focus on node failures as environmental changes in a network. In this section, we describe how we handle node failure in a simulation.

When a node fails, incoming and outgoing optical fibers to the node also fail at the same time. Figure 5b shows an example of a node failure when nodes 0 and 10 fail. In this case, seven incoming optical fibers to node 0 and seven outgoing optical fibers from node 0 fail. Three incoming and outgoing optical fibers at node 10 also become unavailable. In this sense, node failures are tougher situations for virtual topology control methods to adapt to it than link failures, since one node failure also results in several link failures. Next, all lightpaths that are routed on those failed nodes or optical fibers become unavailable. In our simulation evaluation, we do not reroute the failed lightpaths when node failure occurs. The failed lightpaths are always unavailable during node failure. We evaluate our method under the situation in which virtual topology control methods cannot retrieve the information of failed nodes. More precisely, simultaneous node failures often cause the issue that arise from synchronization of Link State Advertisements (LSAs), and the synchronization makes the collection of the network information difficult as discussed in [4]. Without the information, a virtual topology control method cannot reroute the failed lightpaths. Note that the situation in which virtual topology control methods can retrieve the information is equivalent to that evaluated in [19], and we have shown that our method is adaptable to changes in traffic demand in such situations. In this paper, therefore, we investigate the adaptability of our method in the different situation from our previous work to prove the adaptability of our proposed method.

Our proposed method constructs virtual topologies by only using the feedback of the IP network condition, α, without relying on any failure detection mechanism. That is, it does not exploit the information about node failure and the availability of lightpaths. Our method does not change its algorithm to construct virtual topologies during node failure and constructs virtual topologies with α and the deterministic and stochastic behaviors regardless of lightpath availability.

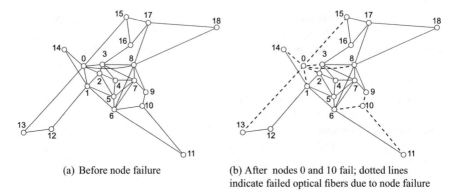

(a) Before node failure   (b) After nodes 0 and 10 fail; dotted lines
                                 indicate failed optical fibers due to node failure

Fig. 5 Example of node failure

5.1.3 Comparisons

For purposes of comparison, we use two existing heuristic virtual topology control methods. A heuristic virtual topology design algorithm called the minimum delay logical topology design algorithm (MLDA) is proposed in [29]. The MLDA constructs virtual topologies on the basis of a given traffic demand matrix. The main objective with MLDA is to minimize the maximum link utilization. The basic idea behind MLDA is to place lightpaths between nodes in descending order of traffic demand. We use another virtual topology control method, the increasing multi-hop logical topology design algorithm (I-MLTDA) [2, 21]. The main objective with I-MLTDA is also to minimize the maximum link utilization. To achieve this objective, I-MLTDA maximizes multi-hop traffic, that is, maximizes accommodated traffic on the virtual topology by minimizing traffic on the multi-hop lightpaths. To determine where to establish lightpaths, I-MLTDA uses hop counts on the constructing virtual topology and traffic demand matrix information. Let H_{sd} and Δ_{sd} denote the minimum number of hops to send traffic from source s to destination d and traffic from s to d, respectively. Note that I-MLTDA re-computes H_{sd} each time to place a lightpath. I-MLTDA places lightpaths in descending order of weights $\Delta_{sd} \times (H_{sd} - 1)$. Since the largest $\Delta_{sd} \times (H_{sd} - 1)$ would correspond to the largest amount of multi-hop traffic that would be carried in one hop as opposed to using multiple hops, this would result in the maximization of the accommodated traffic on the resulting virtual topology.

While most current heuristic virtual topology control methods, including MLDA and I-MLTDA, use the information of the traffic demand matrix, our proposed method uses only the link utilization information. It is generally difficult to retrieve the information of the traffic demand matrix [5, 37]. Therefore, several methods for estimating traffic matrix information from link load information have been proposed [37]. However, the estimation accuracy of the methods depends on the type of traffic, as discussed in [5]. Therefore, to evaluate the potential adaptability of the current heuristic virtual topology control methods, MLDA and I-MLTDA, we do not use traffic estimation methods, and they are given real traffic matrix information.

5.2 Dynamics of Virtual Topology Control Based on Attractor Selection

This section explains the basic behaviors of our virtual topology control method. We evaluate our virtual topology control method with the maximum link utilization, as shown in Fig. 6. The horizontal axis is the time, that is, the number of virtual topology reconfigurations in our method, and the vertical axis is the maximum link utilization. In this simulation, 20 randomly selected nodes fail simultaneously at time 200. Figure 6a shows that the maximum link utilization degrades drastically at time 200 due to node failures, but our proposed method recovers the maximum link utilization shortly after this degradation. In contrast, both heuristic methods cannot construct virtual topologies that have low maximum link utilization. Since our proposed method constructs virtual topologies by using stochastic behavior, it cannot guarantee fast recovery, as shown in Fig. 6b. While our proposed method converges to suitable virtual topologies for the current network environment within 30 virtual topology reconfigurations, as shown in Fig. 6a, it reconfigures virtual topologies 190 times, as shown in Fig. 6b.

To visualize the overall performance of our proposed method, Figs. 7 and 8 are histograms of the maximum link utilization and convergence time of our method, which is defined as the period until the maximum link utilization is recovered. The histograms summarize the results of 1000 simulation trials by using 10 randomly generated traffic demand matrices and 100 randomly selected node failure patterns.

As shown in Fig. 7, our method constructs virtual topologies, having the maximum link utilization less than the target maximum link utilization $\zeta = 0.5$ in Eq. (3), in 955 out of 1000 simulation trials, whereas the maximum link utilization on the virtual topologies constructed using MLDA and I-MLTDA are widely distributed. This result indicates that the performance of MLDA and I-MLTDA depends on traffic or node failure patterns, but our proposed method achieves almost the same level of

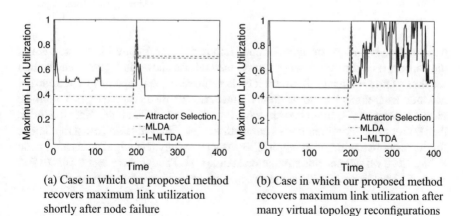

(a) Case in which our proposed method recovers maximum link utilization shortly after node failure

(b) Case in which our proposed method recovers maximum link utilization after many virtual topology reconfigurations

Fig. 6 Dynamics of virtual network topology control based on attractor selection

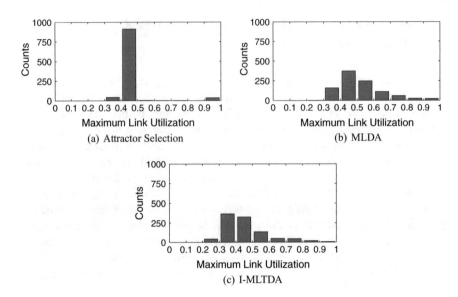

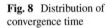

Fig. 7 Distribution of maximum link utilization

Fig. 8 Distribution of
convergence time

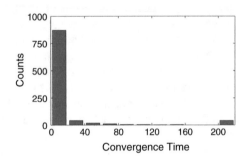

performance regardless of traffic or node failure patterns. Figure 8 is a histogram of
the convergence time with our proposed method. As shown in Fig. 6b, our method
requires a large amount of time until it finds a suitable virtual topology for the current
network environment in some cases. However, our method converges within 20
virtual topology reconfigurations in more than 87% simulation trials. Note that more
than 200 convergence times means our method cannot find suitable virtual topologies
for the network environment within 200 virtual topology reconfigurations after node
failure since we stopped this simulation at time 400. Though our method fails to find
suitable virtual topologies in some cases, it mostly achieves fast convergence.

5.3 Adaptability to Node Failures

In this section, we investigate the adaptability of our proposed method to node failure. We evaluate the number of simulation trials where virtual topology control methods constructs a virtual topology with a maximum link utilization of less than the target link utilization ζ after node failure occurs. We refer to this number normalized by the total amount of simulation trials as the operational ratio. In this evaluation, we set ζ to 0.5, assuming that the network is operated in the domain where the maximum link utilization is below 0.5.

We evaluate the operational ratio by changing the number of failed nodes. Figure 9 shows the operational ratio of each virtual topology control method over the number of failed nodes. The horizontal axis is the number of failed nodes, and the vertical axis is the operational ratio. The operational ratio of both MLDA and I-MLTDA decreases linearly as the number of failed nodes increases. In contrast, our proposed method maintains high operational ratio regardless of the number of failed nodes. Note that we do not omit the results of the simulation trial, in which no feasible virtual topologies for the failed network exist, from this evaluation. Though the operational ratio of our proposed method decreases in the case that the number of failed nodes is more than 10, i.e., more than 10% of nodes fail simultaneously, our method constructs operational virtual topologies in more than 95% of cases.

Next, we investigate the adaptability of our proposed method under a tougher situation. We use a physical topology having 100 nodes and 400 optical fibers. Since this physical topology has less optical fibers than the physical topology in Fig. 9, the effects of node failures are more critical. For simplicity, we refer to the physical topologies with 496 optical fibers and 400 optical fibers as physical topologies A and B, respectively. Figure 10 shows the operational ratio versus the number of failed nodes in the case that physical topology B is used. The operational ratio with physical topology B is smaller than that with physical topology A, as shown in Fig. 9, since the impacts of node failure are more critical in physical topology B. However, when

Fig. 9 Operational ratio versus number of failed nodes in physical topology with 100 nodes and 496 optical fibers

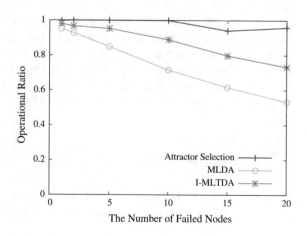

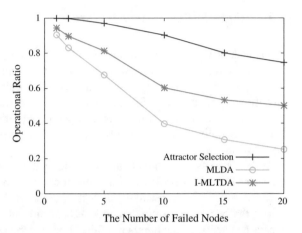

Fig. 10 Operational ratio versus number of failed nodes in physical topology with 100 nodes and 400 optical fibers

either physical topology A or B is used, our method achieves a higher operational ratio than MLDA and I-MLTDA, and the differences of the operational ratio between our method and the two heuristic methods are almost the same.

The results discussed in Sects. 5.2 and 5.3 indicate that our proposed method is highly adaptable to node failure. In the following section, we show the effects of parameters of our method on the adaptability.

5.4 Effects of Noise Strength

Since stochastic behavior, i.e., noise, plays an important role in achieving adaptability to changes in network environments, we evaluate the effects of noise strength on the adaptability of our proposed method. Figure 11 shows the operational ratio versus the noise strength, i.e., the deviation of Gaussian white noise, σ. The horizontal axis is the noise deviation. When $\sigma = 0$, i.e., there is no noise in the dynamics of our method, our method always constructs the same virtual topology; thus, it does not recover from changes in the network caused by node failure, since the dynamics of our method are driven by η. The operational ratio increases as noise becomes strong, and our method shows the best operational ratio when $\sigma > 0.15$. Furthermore, the effects of noise strength do not depend on the number of failed nodes. The effect of variance parameter has been analyzed in detail [12].

5.5 Effects of Activity

The activity α has two parameters, ζ and δ, in Eq. (3). The parameter ζ is the target link utilization, and we set $\zeta = 0.5$, as mentioned above. In this section, we investigate

Fig. 11 The effects of the
noise strength; operational
ratio over the noise strength

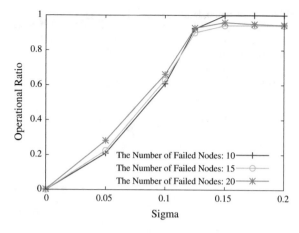

Fig. 12 Effect of the
activity; operational ratio
over the parameter in the
activity, δ

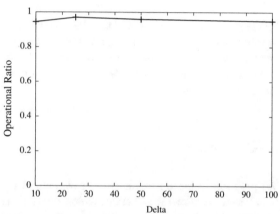

the effects of δ on the adaptability of our method. Parameter δ represents the gradient of α. Since the gradient of α is steep with large δ, our method becomes sensitive to the changes in the maximum link utilization. Figure 12 shows the operational ratio depending on δ, where the horizontal axis is δ. The operational ratio does not change depending on the setting of δ. This result indicates that δ does not greatly affect the adaptability of our method in terms of operational ratio.

5.6 Effects of Reconfiguration Methods of Attractor Structure

As mentioned in Sect. 4.5, our method can use either Eq. (4) or Eq. (5) to define the attractor structure. This section compares orthogonalization and Hebbian learning. Note that we use orthogonalization for reconfiguring the attractor structure, except

Fig. 13 Operational ratio versus number of failed nodes in the case that Hebbian learning is used for reconfiguring attractor structure

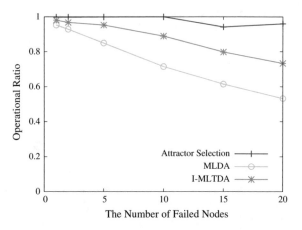

for the results in this section. Figure 13 shows the operational ratio versus the number of failed nodes. There are no observable differences between the results in Figs. 9 and 13. These results indicate that we can use Hebbian learning, which is simpler and less computationally complex than pattern orthogonalization, for reconfiguring the attractor structure.

6 Conclusion

We proposed a virtual topology control method that is adaptive to node failure. It is based on attractor selection, which models the behaviors of biological systems that adapt to environmental changes and recover their conditions. Unlike most engineering-based virtual topology control methods, our method does not rely on predefined algorithms and uses stochastic behaviors for adapting to changes in network environments. Our method dynamically reconfigures its attractor structure according to changing network environments and becomes adaptable to network failure in addition to changes in traffic demand.

We applied the attractor selection mechanism found in biological systems to virtual topology control in wavelength-routed WDM networks. The proposed approach can be applied to network virtualization, which has attracted significant attention recently. Extending our approach for virtualization-based networks, such as network virtualization or OpenFlow, and achieving adaptability in such networks is one of our future research directions.

References

1. Arakawa S, Murata M, Miyahara H (2000) Functional partitioning for multi-layer survivability in IP over WDM networks. IEICE Trans Commun E83-B(10):2224–2233
2. Banerjee D, Mukherjee B (2000) Wavelength-routed optical networks: Linear formulation, resource budgeting tradeoffs, and a reconfiguration study. IEEE/ACM Trans Network 8(5)
3. Baram Y (1988) Orthogonal patterns in binary neural networks. Technical memorandum 100060, NASA
4. Basu A, Riecke JG (2001) Stability issues in OSPF routing. In: Proceedings of ACM SIG-COMM, pp 225–236
5. Callado A, Kamienski C, Szabó G, Gerő BP, Kelner J, Fernandes S, Sadok D (2009) A survey on internet traffic identification. IEEE Commun Surv Tutorials 11(3):37–52
6. Cheng X, Shao X, Wang Y (2007) Multiple link failure recovery in survivable optical networks. Photonic Netw Commun 14(2):159–164
7. Comellas J, Martinez R, Prat J, Sales V, Junyent G (2003) Integrated IP/WDM routing in GMPLS-based optical networks. IEEE Netw Mag 17(2):22–27
8. Durán RJ, Lorenzo RM, Merayo N, de Miguel I, Fernández P, Aguado JC, Abril EJ (2008) Efficient reconfiguration of logical topologies: multiobjective design algorithm and adaptation policy. In: Proceedings of BROADNETS, pp 544–551
9. Furusawa C, Kaneko K (2008) A generic mechanism for adaptive growth rate regulation. PLoS Comput Biol 4(1):e3
10. Gençata A, Mukherjee B (2003) Virtual-topology adaptation for WDM mesh networks under dynamic traffic. IEEE/ACM Trans Netw 11(2):236–247
11. Ghani N, Dixit S, Wang TS (2000) On IP-over-WDM integration. IEEE Commun Mag 38(3):72–84
12. Hanay YS, Arakawa S, Murata M (2015) Network topology selection with multistate neural memories. Exp Syst Appl 42:3219–3226
13. Hopfield JJ (1982) Neural networks and physical systems with emergent collective computational abilities. Proc Natl Acad Sci USA 79(8):2554–2558
14. Hopfield JJ (1984) Neurons with graded response have collective computational properties like those of two-state neurons. Proc Natl Acad Sci USA 81:3088–3092
15. Kaneko K (2006) Life: an introduction to complex systems biology. Understanding complex systems. Springer, New York
16. Kashiwagi A, Urabe I, Kaneko K, Yomo T (2006) Adaptive response of a gene network to environmental changes by fitness-induced attractor selection. PLoS ONE 1(1):e49
17. Kodialam M, Lakshman TV (2001) Integrated dynamic IP and wavelength routing in IP over WDM networks. In: Proceedings of IEEE INFOCOM, pp 358–366
18. Koizumi Y, Miyamura T, Arakawa S, Oki E, Shiomoto K, Murata M (2007) On the stability of virtual network topology control for overlay routing services. In: Proceedings of IEEE fourth international conference on broadband communications, networks, and systems (IEEE Broadnets 2007), Raleigh, NC, USA, pp 810–819
19. Koizumi Y, Miyamura T, Arakawa S, Oki E, Shiomoto K, Murata M (2010) Adaptive virtual network topology control based on attractor selection. IEEE/OSA J Lightw Technol 28(11):1720–1731
20. Lee K, Shayman MA (2005) Rollout algorithms for logical topology design and traffic grooming in multihop WDM networks. In: Proceedings of IEEE global telecommunications conference 2005 (GLOBECOM '05), vol 4
21. Leonardi E, Mellia M, Marsan MA (2000) Algorithms for the logical topology design in WDM all-optical networks. Opt Netw 1:35–46
22. Li J, Mohan G, Tien EC, Chua KC (2004) Dynamic routing with inaccurate link state information in integrated IP over WDM networks. Comput Netw 46:829–851
23. Lin T, Zhou Z, Thulasiraman K (2011) Logical topology survivability in IP-over-WDM networks: Survivable lightpath routing for maximum logical topology capacity and minimum

spare capacity requirements. In: Proceedings of the international workshop on the design of reliable communication networks

24. Liu Y, Zhang H, Gong W, Towsley D (2005) On the interaction between overlay routing and underlay routing. In: Proceedings of IEEE INFOCOM, pp 2543–2553

25. Mukherjee B, Banerjee D, Ramamurthy S, Mukherjee A (1996) Some principles for designing a wide-area WDM optical network. IEEE/ACM Trans Netw 4(5):684–696

26. Newman MEJ, Strogatz SH, Watts DJ (2001) Random graphs with arbitrary degree distributions and their applications. Phys Rev E 64:026,118

27. Nucci A, Sridharan A, Taft N (2005) The problem of synthetically generating IP traffic matrices: initial recommendations. Comput Commun Rev 35(3):19–32

28. Rahman Q, Sood A, Aneja Y, Bandyopadhyay S, Jaekel A (2012) Logical topology design for WDM networks using tabu search. Distributed computing and networking. Lecture notes in computer science, vol 7129. Springer, Berlin, pp 424–427

29. Ramaswami R, Sivarajan KN (1996) Design of logical topologies for wavelength-routed optical networks. IEEE J Sel Areas Commun 14:840–851

30. Rojas R (1996) Neural networks: a systematic introduction. Springer

31. Sahasrabuddhe L, Ramamurthy S, Mukherjee B (2002) Fault management in IP-over-WDM networks: WDM protection versus IP restoration. IEEE J Sel Areas Commun 20(1):21–33

32. Shen G, Grover WD (2003) Extending the p-cycle concept to path segment protection for span and node failure recovery. IEEE J Sel Areas Commun 21(8):1306–1319

33. Sivakumar M, Sivalingam KM (2006) On surviving dual-link failures in path protected optical wdm mesh networks. Opt Switch Network 3(2):71–88

34. Xin Y, Rouskas GN, Perros HG (2003) On the physical and logical topology design of large-scale optical networks. IEEE/OSA J Lightw Technol 21(4):904–915

35. Ye T, Zeng Q, Su Y, Leng L, Wei W, Zhang Z, Guo W, Jin Y (2004) On-line integrated routing in dynamic multifiber IP/WDM networks. IEEE J Sel Areas Commun 22(9):1681–1691

36. Zhang Y, Murata M, Takagi H, Ji Y (2005) Traffic-based reconfiguration for logical topologies in large-scale wdm optical networks. J Lightw Technol 23:1991–2000

37. Zhang Y, Roughan M, Duffield N, Greenberg A (2003) Fast accurate computation of large-scale IP traffic matrices from link loads. Proc ACM Sigmetrics 31:206–217

38. Zhou D, Subramaniam S (2000) Survivability in optical networks. IEEE Netw 14(6):16–23

CBO-Based TDR Approach for Wiring Network Diagnosis

Hamza Boudjefdjouf, Francesco de Paulis, Houssem Bouchekara,
Antonio Orlandi and Mostafa K. Smail

Abstract Wiring networks are vital connections in which power and signals can be transmitted. Defects in these networks can have dramatic consequences, and it is therefore of paramount importance to quickly detect and accurately locate and characterize defects in these networks. In one side, the time-domain reflectometry (TDR) is a measurement concept that exploits reflected waveforms in order to identify the characteristics of wiring networks. In the other side, the colliding bodies optimization (CBO) algorithm has proven to be efficient and robust for solving optimization problems. The aim of this chapter was to combine both TDR and CBO in one approach for the diagnosis of wiring networks (DWN). In this approach, the DWN is formulated as an optimization problem, where the aim was to minimize the difference between the measured TDR response (of the network under test) and a generated one in order to get information about the status of this network. The proposed approach is validated using six experiments with two different configurations

H. Boudjefdjouf · H. Bouchekara
Constantine Electrical Engineering Laboratory, LEC, Department of Electrical
Engineering, University of Constantine 1, 25000 Constantine, Algeria
e-mail: hamza.boudjefdjouf@lec-umc.org

H. Bouchekara
e-mail: bouchekara.houssem@gmail.com

F. de Paulis · A. Orlandi (✉)
UAq EMC Laboratory, Department of Industrial and Information Engineering
and Economics, University of L'Aquila, L'Aquila, Italy
e-mail: antonio.orlandi@univaq.it

F. de Paulis
e-mail: francesco.depaulis@univaq.it

M.K. Smail
Institut Polytechnique des Sciences Avancées (IPSA),
7-9 Rue Maurice Grandcoing, 94200 Ivry-sur-Seine, France
e-mail: mostafa-kamel.smail@ipsa.fr

© Springer International Publishing AG 2017
S. Patnaik et al. (eds.), *Nature-Inspired Computing and Optimization*,
Modeling and Optimization in Science and Technologies 10,
DOI 10.1007/978-3-319-50920-4_13

of wiring networks. The results presented in this chapter show that the proposed approach can be used for a reliable DWN.

Keywords Time-domain reflectometry · Colliding bodies optimization · Diagnosis of wiring network · Fault detection · Inverse problem

1 Introduction

The increasing development of electric and electronic systems in industrial and consumer products makes electrical interconnects of vital importance to guarantee the performance of a system. The information transfer among silicon chips in a digital electronic system as well as in a sensor network, as simple examples in a large variety of applications, is achieved by passive printed circuit board and package microstrips or striplines, and by cables or wires, respectively. Whatever is the application domain, but mainly in electric systems subject to aggressive environmental conditions, the interconnects (wire, cables, connectors) can experience performance degradation up to the interconnect damage and interruption. The defects occurring in the wire/cable can lead to unavailability of safety equipment and compromise public health and security. It is therefore important to quickly and accurately locate the defects in order to repair the faulty section, to restore the power delivery and the information exchange, and to reduce the outage period as much as possible as Fantoni did in [1].

In the recent years, inspection methods have been developed to identify the fault location and level, in particular reflectometry methods as proposed by Abboud and Furse in [2, 3], respectively. They are the most commonly used methods for the diagnosis of wiring networks referred to as DWN from here on. The basic idea relies on the transmission line concept, where an electrically short signal launched down the line is reflected back to the source from any impedance discontinuity along its path as described by Furse in [4].

Two method families are currently available based on the reflectometry principle: frequency-domain reflectometry (FDR) and time-domain reflectometry (TDR). The FDR method injects into the line a set of sine waves and analyzes the standing wave due to the superposition of the injected and reflected signals. This analysis is quite easy for a simple point-to-point connection cable, but it becomes too complicated for more complex networks as highlighted by Auzanneau in [5].

The TDR-based methods have greatly improved the network monitoring due to their ability to detect and locate faults on simple wiring circuits by looking at the reflected signal reaching back the source point. In fact, one of the advantages of TDR is that it requires the network connection to only one location as Paul suggests in [6]. The technique is based on injecting a fast rising and narrow pulse into the line under test, and on evaluating and interpreting the resulting reflections. Ideally, a perfectly matched line will generate no reflections, whereas a mismatched line, either due to defects, will provide a reflection whose pulse shape and amplitude contain information on the impedance profile or on the fault condition of the line under test. The easiest faults to be detected are hard faults, namely short circuit or open circuit, since they generate a total pulse reflection with negative or positive polarity, respectively (the reflection coefficient is -1 for the short circuit and $+1$ for open circuit). Intermediate faults are characterized by any value of the reflection coefficient between -1 and $+1$ (0-valued reflection coefficient stands for matched line, thus no reflection). The distance of the fault with respect to the injected/detection point is determined by the evaluation of the delay of the reflected pulse, taking into account the propagation velocity in the line, as in (1).

$$D_{fault} = \left(V_{propagation} \cdot \Delta t\right)/2 \qquad (1)$$

where D_{fault}: distance to fault, $V_{propagation}$: propagation velocity, and Δt: time delay from the initial pulse to the reflected pulse.

Although TDR is a very well-established method firstly developed for point-to-point line characterization, the reflection pattern of complex networks is very difficult to interpret. Such complex networks are often branched, challenging the ability of traditional TDR-based procedure (which only measure the time of flight to the defect) to unambiguously extract all the faults from the time-domain responses.

The key point is to initially characterize the healthy network (both by measurements and by simulation models) such that any changes in reflection pattern allow to quickly identify the faulty condition. Then, based on the knowledge of the network topology, the fault location and type (open or short fault is considered here in) can be extracted. This problem was analyzed from the perspective of big data by Sharma and Sugam in [7–10], in which the system is made to learn from data which can be used to make fault predictions; the only disadvantage of using big data analysis is that we have to build a database that contains all scenarios of faults that can affect the network. This can increase the computational times and affect the result accuracy.

In this chapter, a "forward" model is firstly developed to generate the TDR response based on the knowledge of the cable network topology and it is validated in comparison with the vector network analyzer (VNA) experimental response. In order to perform the model to experimental comparison, the VNA response (namely the frequency-domain S-parameter S_{11}) an inverse fast Fourier transform (FFT) needs to be applied to obtain the time-domain data. Once the model of the healthy network is refined, it is employed in combination with the colliding bodies optimization (CBO) method, used to solve the inverse problem, in order to identify all the fault information from the TDR response of a faulty network. The proposed method consists on the minimization of the differences between the signature pattern of the measured faulty network and the simulated one. During the optimization process, the design variables of the electrical network are modified until its pulse response matches the measured response. The final optimized model is characterized by the actual faults on the network located at the right position.

The main advantage of the suggested method is that several hard faults which can affect the electrical network can be correctly identified simultaneously. The faulty network reconstructed following the proposed method allows considering the performance of the method as effective for hard fault detection under various fault conditions, with an average error in the fault localization being less than 4% among all considered fault conditions and networks.

The present chapter is organized as follows. In Sect. 2, the proposed approach is detailed. In Sect. 3, the experiments used to validate the proposed approach are given and the obtained results presented and analyzed. Finally, the conclusions are drawn in Sect. 4.

2 The Proposed TDR-CBO-Based Approach

As previously mentioned, the proposed approach combines TDR and CBO in order to detect, locate, and characterize defects in wiring networks. Each wiring network is composed of lines or branches and some branches are connected to junctions while others are terminated with loads. The length of a branch is noted as L_i and termination load of a branch is noted as R_i. For shorthand, we assume that if a branch ends by an open circuit its termination load is 1, whereas if it is terminated by a short circuit its termination load is 0. Furthermore, for simplicity, a branch can be identified by its order in the network topology, i.e., the third branch, for example, can be referred to as L_3.

The TDR generated using the developed forward model using the lengths and termination loads generated using the CBO algorithm is compared to the one collected from measurements. Once the two responses fit with each other, the optimal lengths and termination loads found represent the parameters of the faulty network and thus allow detecting, localizing, and characterizing the defects. The flowchart of the proposed approach is given in Fig. 1.

2.1 Problem Formulation

2.1.1 Objective Function

As aforesaid, in this chapter the DWN is formulated as an optimization problem. Mathematically, this can be expressed as follows:

$$\min F(x) = |TDR_{MEASURED} - TDR_{GENERATED}| \tag{2}$$

where F is the objective function, x is the vector of design variables, and $TDR_{MEASURED}$ and $TDR_{GENERATED}$ are the TDR responses measured using the experimental setup and generated using the forward model.

Fig. 1 Flowchart of the proposed TDR-CBO-based approach for the DWN

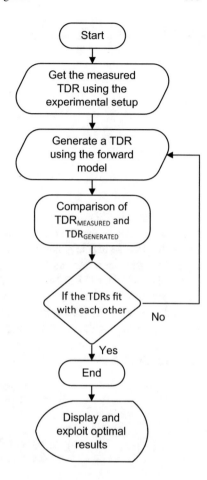

2.1.2 Design Variables

The design variables here are the branch lengths L_i and the branch termination loads R_i. L_i are continuous variables and they are bounded by upper and lower bounds, whereas R_i are discrete variables that can take the values 0 or 1 only. Recall that $R_i = 1$ means that branch ith is terminated with an open circuit, whereas $R_i = 0$ means that branch ith is terminated by a short circuit.

2.2 The Forward Model

2.2.1 Characteristic Impedance and Measurement of the Coaxial Cable

The RG-58 C/U is a type of coaxial cable often used in low-power signal transmission and RF connections; the cable has a characteristic impedance of 50 Ω. Most two-way

Fig. 2 RG-58 C/U coaxial
cable

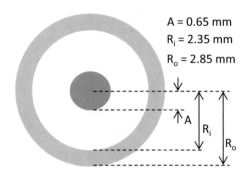

A = 0.65 mm

R$_i$ = 2.35 mm

R$_o$ = 2.85 mm

radio communication systems, such as marine, citizen band, amateur radio, police, firemen, WLAN antennas, are designed to work with this type of cable. However, the method presented is general and can be applied without any restriction to other type of transmission lines and cable types.

In this analysis, the RG-58 C/U coaxial cable shown in Fig. 2 has been used. The cable-per-unit-length parameters, such as capacitance C, inductance L, conductance G, and resistance R, are functions of cable geometry and are evaluated in (3)–(6) taking into account the typical values for the material properties such as dielectric permittivity $\varepsilon_r = 2.2$, dielectric loss tangent $tg\delta = 0.002$, metal conductivity $\sigma = 5.8 \times 10^7$ S/m.

$$C = \frac{2\pi \varepsilon_r \varepsilon_0}{\ln\left(\frac{R_i}{A}\right)} = 95.2 \cdot 10^{-12} \frac{F}{m} \tag{3}$$

$$L = \frac{\mu_0}{2\pi} \ln\left(\frac{R_i}{A}\right) = 257 \cdot 10^{-9} \frac{H}{m} \tag{4}$$

$$G = C \cdot \omega \cdot tg\delta = 2 \cdot \omega \cdot 10^{-13} \frac{S}{m} \tag{5}$$

$$R = \frac{1}{\pi A^2 \sigma} + \frac{1}{\pi \sigma (R_o + R_i)(R_o - R_i)} = 0.015 \frac{\Omega}{m} \tag{6}$$

where A, R_i and R_0 are the radii of the inner, outer, and external conductors, and μ_0 is the free-space magnetic permeability.

2.2.2 Wave Propagation Model in a Transmission Line

This section briefly reviews the forward model implemented using the finite-difference time-domain (FDTD) algorithm. The mathematics for modeling propagating waveforms on uniform transmission lines is well established and easily derived from the Maxwell's equations.

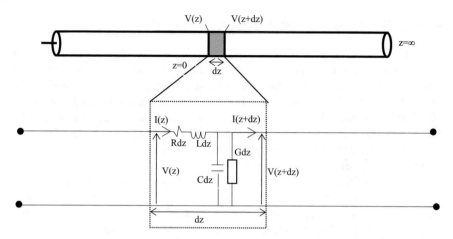

Fig. 3 RLCG model propagation of a multiconductor transmission line

Most actual TDR applications are intended for detection of discrete discontinuities such as open and short circuits. For such situations, each abrupt change of the line impedance causes a reflected wave. Conventional techniques fail when there are multiple or continuously varying discontinuities. Therefore, the main motivation behind this work is to overcome this limitation providing an effective method to identify multiple faults on complex network.

The propagation in a two-conductor transmission line TL (such as a coaxial cable) can be modeled by a RLCG circuit model as described by Halliday and Sullivan in [11, 12], respectively, and as shown in Fig. 3.

Writing Kirchhoff's law leads to the following differential equations widely known as telegrapher's equations:

$$\frac{\partial V(z, t)}{\partial z} = -R \cdot I(z, t) - L \cdot \frac{\partial I(z, t)}{\partial t} \tag{7}$$

$$\frac{\partial I(z, t)}{\partial z} = -G \cdot V(z, t) - C \cdot \frac{\partial V(z, t)}{\partial t} \tag{8}$$

where V and I are line voltage and line current, respectively. The position along the line is denoted as z, and the time is denoted as t. The R (resistance), L (inductance), C (capacitance), and G (conductance) are the per-unit-length parameters, and their values are computed analytically as in (3)–(6).

The method used for obtaining the time-domain solution of (7) and (8) and thus for simulating the healthy as well as the faulty networks is based on the FDTD formulation which converts the differential equations into recursive finite-difference equations [12].

2.2.3   Finite-Difference Time-Domain (FDTD) Method

The one-dimensional FDTD method is very efficient and easy to implement for modeling complex networks and to include inhomogeneous configurations as well as impedance discontinuities; the method requires that the entire transmission line network is divided into a mesh discretization in which the current and voltage are calculated. Normally, for a quick algorithm implementation, this mesh can be uniform; in this work, the line axis is discretized in Δz increments, the time variable t is discretized in Δt increments, and the derivatives in the telegrapher's equations are approximated by the finite differences.

The length of the spatial cell size Δz and sampling interval Δt are chosen to be $\Delta z = \lambda/60$ and $\Delta t = (2 \cdot v)$, respectively, where λ is the wavelength of the source signal, and v is the velocity of the propagation along the line. This choice insures the stability on the time-stepping algorithm as pointed out by Smail in [13].

Replacing the derivatives with centered finite differences in (7) and (8), we obtain the following recurrence equations:

$$\left[\frac{L}{\Delta t} + \frac{R}{2} \right] I_k^{n+1} = \left[\frac{L}{\Delta t} - \frac{R}{2} \right] I_k^n - \frac{1}{\Delta z} \left[V_{k+1}^n - V_k^n \right] \tag{9}$$

$$\left[\frac{C}{\Delta t} + \frac{G}{2} \right] V_k^{n+1} = \left[\frac{C}{\Delta t} - \frac{G}{2} \right] V_k^n - \frac{1}{\Delta z} \left[I_{k+1}^n - I_k^n \right] \tag{10}$$

2.2.4   Time-Domain Reflectometry Measurements

In order to obtain the TDR measurement-based response, the inverse fast Fourier transform (IFFT) of the S_{11} parameter obtained by a vector network analyzer (VNA) is applied.

The typical measurement setup for implementing this methodology is based on the TDR instrument; it is a large bandwidth oscilloscope (>20 GHz as typical value) that implements a step source characterized by a fast ($\approx$ps) rising edge. The step reflected from the network discontinuity is detected and displayed on the scope.

An alternative setup is proposed in this paper based on frequency-domain measurements using a vector network analyzer (10 MHz–9 GHz Anritsu MS4624B) to measure the network reflection properties.

The one-port VNA-measured data, the S_{11}, represents the impulse response of the network; thus, this parameter can be simply multiplied by the spectrum of the same input pulse used in the FDTD simulation. The IFFT can them be applied to obtain the time-domain response to the input pulse, to be compared to the corresponding impulse response of the forward model simulated using the FDTD algorithm.

The measurement setup employs the VNA to measure twice the same network, the first one in the 10 MHz–1 GHz band and the second in the 1–2 GHz band; then, the two sets of data are combined together to achieve a 2-GHz bandwidth data with a doubled frequency resolution ($\Delta f = 618$ kHz based on 1601 samples per

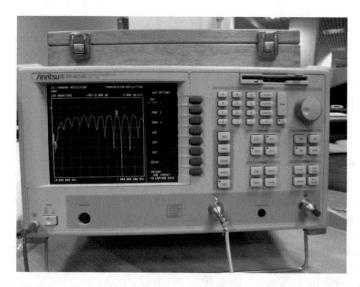

Fig. 4 The VNA (Anritsu MS4624B 10 MHz–9 GHz) used for experiments to measure the S_{11} and thus, after the post-processing, the TDR responses

measurement). The DC to 10 MHz missing measured data are extrapolated linearly interpolating the low-frequency trend of the measured S_{11}, both in magnitude and phase. The reconstructed S_{11} is multiplied by the spectrum of the input pulse. The latter is the same used as input for the FDTD code, and it is a raised cosine pulse, with a rising time of 0.4 ns and amplitude of 1 V as proposed by Boudjefdjouf in [14, 15] (Fig. 4).

2.3   Colliding Bodies Optimization (CBO)

Optimization is widely used in a wide variety of fields, from engineering design to business planning. In these activities, certain objectives need to be achieved, or quantities such as profit, quality, and time have to be optimized. In the case of optimization objectives related to variables that are subject to constrains in real-world applications, the optimization process comes up to help searching for solutions to optimally employ these valuable resources. Optimization is the study of such planning strategies and design problems using mathematic tools, as discussed by Yang in [16] and Binitha in [17].

The CBO is a new optimization algorithm that is inspired from the law of collision between the two bodies as discussed by Kaveh in [18, 19]. The CBO uses a population of colliding bodies (CBs) where each CB has a mass and a velocity; this last one varies before and after the collision.

The CBO main steps are as follows:

- Step 1: initialize the population of CBs randomly in the search space.
- Step 2: calculate the mass of each CB using the following expression:

$$m_k = \frac{\frac{1}{F(k)}}{\sum_{i=1}^{n} \frac{1}{F(i)}}, k = 1, 2, ..., n \tag{11}$$

where F is the objective function and n is the population size.

- Step 3: rank the population where the best CB is ranked first. After that, the population is divided into two groups (with equal size). The first group starts from the first (best) CB to the one of the middle and the second group starts from the middle to the end of the population. The CBs of the first group (i.e., the best ones) are considered as stationary while the CBs of the second group (the worst ones) move toward the first group. Therefore, the velocities before collision of the first group (stationary CBs) are given by

$$v_i = 0, i = 1, ..., \frac{n}{2} \tag{12}$$

And for the second group (moving CBs) they are given by

$$v_i = x_i - x_{i-\frac{n}{2}}, i = \frac{n}{2} + 1, ..., n \tag{13}$$

where v_i and x_i are the velocity and the position of the ith CB, respectively.

- Step 4: calculate the new velocities after the pairwise collision between the members of the first group and those of the second group. Therefore, the velocities after collision of the first group are given by

$$v_i' = \frac{\left(m_{i+n/2} + \varepsilon m_{i+n/2}\right) v_{i+n/2}}{m_i + m_{i+n/2}}, i = 1, ..., \frac{n}{2} \tag{14}$$

and for the second group they are given by

$$v_i' = \frac{\left(m_i - \varepsilon m_{i-n/2}\right) v_i}{m_i + m_{i-n/2}}, i = \frac{n}{2} + 1, ..., n \tag{15}$$

where v_i' is the velocity after collision and ε is the coefficient of restitution (COR).

- Step 5: evaluate the new positions of CBs after collision using the following expression:

$$x_i^{new} = x_i + rand\,ov_i', \ i = 1, ..., n \tag{16}$$

where x_i^{new} is the new position of the ith CB after collision.

It is worth mentioning here that the COR is defined in order to have a good balance between the exploration and the exploitation phases of the CBO and it is given by

$$\varepsilon = \frac{1 - iter}{iter_{max}} \tag{17}$$

where $iter$ and $iter_{max}$ are the current iteration number and the maximum number of iterations, respectively.

Steps from 2 to 5 are repeated until a termination criterion is met.

2.3.1 Implementation of the CBO Algorithm for the DWN

The CBO has been implemented for the DWN as follows. First, the CBO parameters such as the number of CBs and the maximum number of iterations $iter_{max}$ are initialized. Then, the CBO algorithm generates a population of CBs where the dimensions correspond to the vector of design variables of the DWN problem that are lengths and termination loads. Each CB corresponds to a different topology of the considered network. The generated CBs collide and consequently move inside the search space (following the different steps of the CBO algorithm described before) looking for better topologies of the considered network. The best CBs are the ones that have topologies closer to the network under test. The similitude between the topologies is computed using the objective function given by (2). In other words, each CB has a TDR (generated using the direct model); this TDR is compared with the measured one, and the closer the TDRs (generated and measured) are, the better the CB is and consequently the closer the topology is compared with the one of the networks under test. The CBs continue to move until the number of iterations reaches its maximum value. Finally, the best CB is printed and the results are analyzed by the specialized user in order to make the final diagnosis of the network under test.

3 Applications and Results

In order to validate the proposed TDR-CBO-based approach, six practical experiments have been carried out. In these experiments, we have used two different configurations based on the Y-shaped wiring network and the YY-shaped wiring network, as described in the following paragraphs. Figure 5 shows the experimental setup used for measuring TDR responses of the tested networks using the VNA.

In order to assess the performance of the proposed approach, the error is calculated using the following expression:

$$Error = \frac{\|L_{REAL} - L_{CALCULATED}\|_\infty}{\|L_{REAL}\|_\infty} + 100 \times \sum |R_{iREAL} - R_{iCALCULATED}| \tag{18}$$

Fig. 5 Sketch of the
experimental setup showing
a coaxial wiring network
connected to a VNA

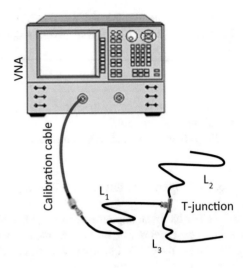

The first term in (17) is used to measure the error when calculating different lengths, i.e., for the localization phase while the second term is used to measure the error for the characterization phase. It is worth to mention that there is no room for making a mistake in the characterization phase. In other words, it is not acceptable to confuse a short circuit fault with an open circuit one; this is why the second term is multiplied by 100. In the other side, a small mistake in the calculation of lengths can be acceptable; however, it reflects the accuracy of the proposed approach.

3.1 The Y-Shaped Wiring Network

The Y-shaped wiring network is composed of three lines (or branches) L_1, L_2, and L_3 and a single junction as shown in Fig. 6. Schematic representations of the investigated experiments for this network are shown in Fig. 7a. The main line is L_1, while L_2 and L_3 are open terminated or, more rigorously, are terminated by loads R_2 and R_3 of infinite value.

Experiment 1:
The first experiment conducted in this paper considers the healthy Y-shaped network given in Fig. 7a. Lines L_1, L_2 and L_3 are 1m, 4m, and 1 m long, respectively. Furthermore, L_2 and L_3 are terminated with open circuits loads, thus $R_2 = 1$ and $R_3 = 1$.

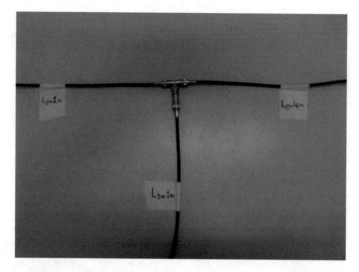

Fig. 6 The Y-shaped wiring network with BNC T junction

Experiment 2:
In this experiment, we consider a faulty Y-shaped network affected by a short circuit in L2 at a distance of 2 m from the junction and 3 m from the origin as shown in Fig. 7b.

Experiment 3:
In this third experiment, we consider a faulty Y-shaped network affected by two hard faults; the first fault is an open circuit in L_2 at 3 m from the origin, the second fault is a short circuit in L_3 at 1.53 m from the origin as shown in Fig. 7c.

The proposed approach has been run for these experiments and the obtained results are summarized in Table 1. From this table, we can make the following remarks:

- The errors are very low for the three experiments.
- The error for Experiment 3 is higher than the error for Experiment 2.
- The proposed approach was able to identify the network of Experiment 1 as a healthy network and to detect the presence of faults in the networks of Experiment 2 and Experiment 3. Therefore, the detection phase is very effective.
- The localization phase is very accurate.
- There is no mistake while calculating the termination loads. In other words, the characterization phase is successful.
- Among the three experiments, Experiment 3 is the most difficult one to treat because the network in this experiment is affected by two hard faults.
- The diagnosis of the investigated configurations using the proposed approach is reliable.

Fig. 7 Different
experiments investigated for
the Y-shaped wiring
network: **a** healthy case, **b**
case with one fault and **c**
case with multiple faults

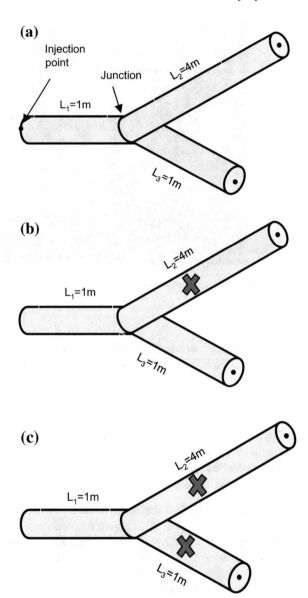

Table 1 Optimal results found for Experiment 1, Experiment 2 and Experiment 3

Experiment	Design variables					Error
	L_1[m]	L_2[m]	L_3[m]	R_2	R_3	
Experiment 1	1.00	4.00	1.00	1	1	0.0000
Experiment 2	1.00	2.05	1.00	0	1	0.0244
Experiment 3	1.00	2.08	0.53	1	0	0.0385

Fig. 8 TDR responses of the three experiments investigated using the Y-shaped network. **a** Experiment 1, **b** Experiment 2 and **c** Experiment 3

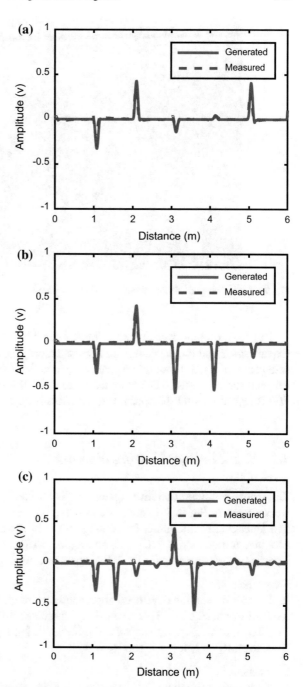

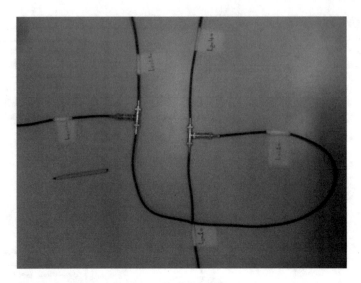

Fig. 9 The YY-shaped wiring network with BNC T junctions

The TDR responses generated using the developed approach for Experiment 1, Experiment 2 and Experiment 3 are sketched along with those obtained using measurements in Fig. 8. From this figure, we can notice that there is a good matching between the measured TDR responses and the TDR pattern generated using the FDTD algorithm after the optimization process is applied.

3.2 The YY-shaped Wiring Network

The second configuration investigated in this chapter is the YY-shaped wiring network shown in Fig. 9. It is composed of five branches L_1, L_2, L_3, L_4 and L_5 and two junctions. Furthermore, this configuration has two main branches L_1 and L_2 and three termination loads R_2, R_4 and R_5. A schematic representation of the three investigated experiments for the YY-shaped network is given in Fig. 10.

Experiment 4:
In this experiment, the developed approach is tested on the healthy YY-shaped network shown in Fig. 11a. This network has five lines that have a length of 1 m, 4 m, 1 m, 0.5 m, and 1.5 m, respectively. Furthermore, L_2, L_4, and L_5 have open circuit ends; thus, $R_2 = 1$, $R_4 = 1$, and $R_5 = 1$.

Experiment 5:
In this experiment, a faulty YY-shaped network is under test. It is affected by an open circuit fault in L_2 at 2 m from the first junction and 3 m from the origin as shown in Fig. 11b.

Fig. 10 Different experiments investigated for the YY-shaped wiring network: **a** healthy case, **b** case with one fault, and **c** case with multiple faults

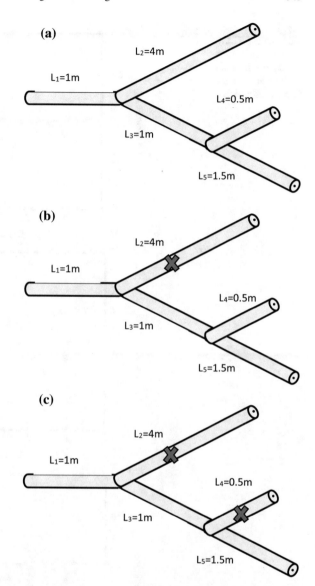

Fig. 11 TDR responses of the three experiments investigated using the YY-shaped network **a** Experiment 4, **b** Experiment 5, and **c** Experiment 6

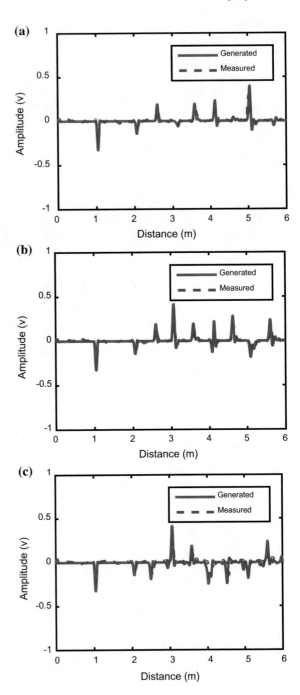

Table 2 Optimal results found for Experiment 4, Experiment 5, and Experiment 6

Experiment	Design variables								Error
	L_1[m]	L_2[m]	L_3[m]	L_4[m]	L_5[m]	R_2	R_4	R_5	
Experiment 4	1.00	3.99	1.00	0.49	1.50	1	1	1	0.0000
Experiment 5	1.00	2.03	1.00	0.48	1.50	1	0	1	0.0147
Experiment 6	1.00	2.02	1.00	0.44	1.50	1	0	1	0.0198

Experiment 6:

The last experiment carried out in this chapter considers an YY-shaped network affected by two hard faults; the first is an open circuit fault in L_2, at 2 m from the first junction and 3 m from the origin, the second fault is a short circuit in L_4, at 0.4 m from the second junction and 2.4 m from the origin.

The proposed TDR-CBO-based approach has been run for these experiments and the obtained results are given in Table 2. This table provides the following comments:

- The errors are very low for the three experiments investigated for the second configuration.
- The error for Experiment 6 is higher than the error for Experiment 5.
- The proposed approach was able to detect the presence or absence of faults in the investigated network. Therefore, the identification phase is still very effective.
- Once again, the localization phase is very accurate even for a more complicated configuration.
- There is no mistake while calculating the termination loads. Therefore, the characterization phase is successful.
- Experiment 6 is the most difficult case to treat because the network here is affected by two faults.
- The diagnosis of complicated configurations using the proposed approach is reliable.

Figure 11 shows a comparison between the TDR responses generated using the proposed approach and those measured experimentally using the VNA. Therefore, Fig. 11 shows that the proposed approach achieves high-quality results compared to measurements.

4 Conclusion

The purpose of the current study was to propose a new approach for the diagnosis of hard faults in wiring networks. This approach is based on the time-domain reflectometry and colliding bodies optimization algorithm. The diagnosis of wiring networks has been formulated as an optimization problem where the objective function consists of the comparison between the TDR responses measured and generated using the forward model constructed from the design variables such as branch lengths and termination loads (either open or short). Six experiments have been carried out using two different network configurations, the Y-shaped network and the YY-shaped network. Furthermore, healthy networks and faulty networks affected with one or more faults have been tested. The obtained results show that the proposed approach is highly efficient and reliable for the diagnosis of wiring networks. This can be justified by the low values of errors found. Finally, the most suitable extension for this work is to investigate the validity of the proposed approach for soft faults.

References

1. Fantoni PF (2006) Wire system aging assessment and condition monitoring using Line Resonance Analysis (LIRA). In: Owems, 20–22 April. Citavecchia, Italy
2. Abboud L, Cozza A, Pichon L (2012) A matched-pulse approach for soft-fault detection in complex wire networks. In: IEEE Transactions on Instrumentation and Measurement, vol. 61, no. 6, pp 1719–1732, June 2012
3. Furse C, Chung YC, Dangol R, Mabey G, Woodward R (2003) Frequency-domain reflectometry for on-board testing of aging aircraft wiring. IEEE Trans Electromagn Compat 45(2):306–315
4. Furse C, Chung YC, Lo C, Pendayala P (2006) A critical comparison of reflectometry method for location of wiring faults. Smart Struct Syst 2(1):25–46
5. Auzanneau F (2013) Wire troubleshooting and diagnosis: review and perspectives. Progr Electromagn Res B, 49:253–279
6. Paul CR (1994) Analysis of multiconductor transmission lines. Wiley, New York
7. Sharma S et al (2014) A brief review on leading big data models. Data Sci J 13:138–157
8. Sharma S, Tim US, Gadia S, Shandilya R, Sateesh P (2014) Classification and comparison of leading NoSQL big data models. Int J Big Data Intell (IJBDI). Inderscience
9. Sharma S (2015) Evolution of as-a-service era in cloud. Cornell University Library. (http://arxiv.org/ftp/arxiv/papers/1507/1507.00939.pdf)
10. Sharma S (2015) Expanded cloud plumes hiding Big Data ecosystem. Future Gener Comput Syst
11. Halliday D, Resnick R (1962) Physics, Part II, 2nd edn. John Wiley & Sons, New York
12. Sullivan DM (2000) Electromagnetic simulation using the FDTD method. IEEE Press, New York, Piscatway
13. Smail MK, Hacib T, Pichon L, Loete F (2011) Detection and location of defects in wiring networks using time-domain reflectometry and neural networks. IEEE Trans Magn 47(5)
14. Boudjefdjouf H, Mehasni R, Orlandi A, Bouchekara HREH, de Paulis F, Smail MK (2014) Diagnosis of multiple wiring faults using time-domain reflectometry and teaching-learning-based optimization. Electromagnetics 35(1):10–24
15. Boudjefdjouf H, Bouchekara HREH, de Paulis F, Smail MK, Orlandi A, Mehasni. R (2016) Wire fault diagnosis based on time-domain reflectometry and backtracking search optimization algorithm. ACES J 31(4)
16. Yang X-S (2010) Nature inspired metaheuristic algorithms, 2nd edn. Luniver press, University of Cambridge, UK
17. Binitha S, Sathya SS (2012) A survey of bio inspired optimization algorithms. IJSCE 2(2)
18. Kaveh A, Mahdavi VR (2014) Colliding bodies optimization: a novel meta-heuristic method. Comput Struct 139:18–27
19. Kaveh A, Mahdavi VR (2014) Colliding bodies optimization method for optimum design of truss structures with continuous variables. Adv Eng Softw 70:1–12

Morphological Filters: An Inspiration from Natural Geometrical Erosion and Dilation

Mahdi Khosravy, Neeraj Gupta, Ninoslav Marina, Ishwar K. Sethi and Mohammad Reza Asharif

Abstract Morphological filters (MFs) are composed of two basic operators: dilation and erosion, inspired by natural geometrical dilation and erosion. MFs locally modify geometrical features of the signal/image using a probe resembling a segment of a function/image that is called structuring element. This chapter analytically explains MFs and their inspirational features from natural geometry. The basic theory of MFs in the binary domain is illustrated, and at the sequence, it has been shown how it is extended to the domain of multivalued functions. Each morphological operator is clarified by intuitive geometrical interpretations. Creative natural inspired analogies are deployed to give a clear intuition to readers about the process of each of them. In this regard, binary and grayscale morphological operators and their properties are well defined and depicted via many examples.

M. Khosravy (✉)
Faculty of Computers Networks and Security (CNS), University of Information Science and Technology, Ohrid, Republic of Macedonia
e-mail: mahdi.khosravy@uist.edu.mk; mahdikhosravy@yahoo.com

N. Gupta
Faculty of Machine Intelligence and Robotics (MIR), University of Information Science and Technology, Ohrid, Republic of Macedonia
e-mail: neeraj.gupta@uist.edu.mk

N. Marina
University of Information Science and Technology, Ohrid, Republic of Macedonia
e-mail: rector@uist.edu.mk

I.K. Sethi
Department of Computer Science and Engineering, Oakland University, Rochester, Oakland, MI, USA
e-mail: isethi@oakland.edu

M.R. Asharif
Faculty of Engineering, University of the Ryukyus, Nishihara, Okinawa, Japan
e-mail: asharif@ie.u-ryukyu.ac.jp

© Springer International Publishing AG 2017
S. Patnaik et al. (eds.), *Nature-Inspired Computing and Optimization*,
Modeling and Optimization in Science and Technologies 10,
DOI 10.1007/978-3-319-50920-4_14

1 Natural Geometrical Inspired Operators

When we are talking about the operators and filters, normally our mind focuses on number sequences, matrices, or tensors. The conventional operators are applied to numerical operands as in classical mathematics; the basic operators addition, subtraction, and their extensions, such as multiplication and division, are implemented to numbers as operands. Apart from the numbers and algebraic relations, we have the geometry that works on shape and size of the objects, which are easier to visualize and analyze in comparison with numbers. By bare eyes, they can be seen easily while going through the natural environment. To formulate this theory inspired by nature, we consider some operations over shapes, where shapes and operation are operand and operator, respectively.

Let us have a shape and see the simplest possible operation on it. When we go through nature, the main stable shapes in our view are mountains, cliffs, rocks, valleys, etc. Also, a satellite view of the earth shows different colored shapes and geometries of the lands, islands, and lakes, etc. The main shape of operations are inspired from the nature based on natural enlargement and abridgment. Figure 1 shows two natural geometrical abridgments.

Let us just simply consider a rock as a natural shape in our vision. Two changes in the shape of the rock can be seen by the nature: (i) deposition of soil and snow on the rock and (ii) retrogression and degradation of rock. In the first case, deposited rock by snow/soil has same expanded shape from the original shape. In the second case, the degraded rock by weather change has the similar phenomenon but depleted from original shape (see Fig. 2). These natural geometrical expansion and abridgment inspire the basic morphological addition and subtraction operators to the shapes and establishing mathematics on shapes known as "*mathematical morphology.*"

Fig. 1 *Left* Wind erosion of solid rocks, leaving rocks looking like skeletons in south of Iran (courtesy: the-village.ru). *Right* The wave erosion of a rock in Oroumieh Lake, Iran (courtesy: thewatchers.adorraeli.com)

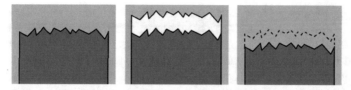

Fig. 2 Natural geometrical expansion and abridgment inspire the basic morphological addition and subtraction operators to shapes and establishing mathematics on shapes known as "mathematical morphology." The side view of rock (*bottom left*), when the rock is snow-covered (*bottom middle*) and if the rock surface is washed out by rain (*bottom right*)

Fig. 3 Example of a morphological operator applied on two shapes operands

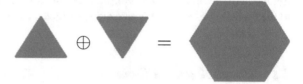

Normally, we are more familiar with conventional mathematics, where numerical operators have been applied to the numbers. However, the morphological operators are applied on shapes in mathematical morphology, and the resulting output is a shape. For instance, see Fig. 3, where the morphological addition has been applied on a triangle and its reverse, and the result is a hexagon. How this result is obtained is explained in the following sections of this chapter.

Section 2 talks about mathematical morphology and their resultant operators and filters. Section 3 explains how morphological operators were born in binary domain and how their theory has been established based on set theory. After that, the geometrical features of binary morphological operators and filters are intuitively illustrated in Sect. 4. Sections 5 and 6 describe the evolution of morphological operators to grayscale domain. The authors give an intuitive understanding of geometrical features of morphological filters on functions in Sect. 7. Finally, section MFvalley concludes the chapter.

2 Mathematical Morphology

Mathematical morphology deploys set theory, lattice theory, topology, and random functions for the analysis and processing of geometrical structures. The history of mathematical morphology dates back to 1964 in the result of collaborative work of Georges Matheron [1, 2] and Jean Serra [3, 4], at Mines Paris Tech, France, wherein Matheron supervised the PhD thesis of Serra. Later in 1968, a Center of mathematical morphology was established in Mines Paris Tech, and Matheron and Serra led it. At first, mathematical morphology was essentially applied in binary domain until the mid-1970s. Thereafter, it was generalized to grayscale functions and images. The generalization of the main operators to multivalued functions yielded operators such as morphological gradients, top-hat transform, and the Watershed, too. In 1986, Serra brought more flexibility to the theory of mathematical morphology by its further generalization based on complete lattices. As well, together with Matheron, they formulated a theory for morphological filtering. Since the first International Symposium on Mathematical Morphology (ISMM) in 1993 in Barcelona, Spain, up to the moment, ISMMs are organized in different countries for twelve times.

2.1   Morphological Filters

Morphological filters (MFs) are based on mathematical morphology. MFs as a non-linear processing methodology presents an approach different from classical linear methods in signal/image processing. In the linear approach of signal and image processing, the processing is through operations of addition, subtractions, multiplications, and divisions on the sample/pixel value and its neighboring values, e.g., nth derivative or averaging of the values. The fundamental principle of linear signal/image processing is the closeness of the samples/pixel values in a neighborhood, that is, the continuity assumption. In this type of processing, the determining characteristic of a neighborhood for deploying its values is exclusively the certainty level of continuity. Therefore, it can be concluded that linear signal/image processing is not matched with the edges and noncontinuities inside the signal or image. Indeed, it erodes and blurs the edges and borders within signal and image, respectively. However, mathematical morphology performs on the edges and borders in a different manner. It does not blur the edges and borders, but modifies it for better analysis. This is because of the theoretical base of mathematical morphology, where comparison operators, neighboring maxima and minima are in use. Therefore, edge and noncontinuity points are playing the key roles in operation in each neighborhood and have a direct impact on them. As a result, new edges and noncontinuity will appear in the neighborhood of same edge and noncontinuity.

The main function undergoing during each of morphological operators is mutual interaction between signal/image and structuring element(SE). Here, SE is a set or a function in the finite domain. This interaction is performed along the signal/image at each point, between the neighborhood of the point, and all the points of the SE. Although SEs in MFs is similar to a mask in linear signal/image processing, their way of being chosen and their practical interpretation is different from the masks. SE is often chosen in a way to be matched with the subjects of detection or removal by MFs. The key role of the SE will be explained more in subsequent sections of this chapter.

Nowadays, the main role of mathematical morphology can be seen in the grayscale domain, which works on multivalued functions. However, it was born based on set theory and grew up in binary domain [1]; later its operators were extended to multi-valued function domain [5–8]. Morphological operations are easier to understand in the binary domain rather than the grayscale. Thus, it is advised to learn about them in binary domain first, and then it would be much easier to understand the choice of mathematical equation for each operator in the domain of grayscale values. Consider a binary image with pixel values of 0 or 1. The connection between the binary image and set theory is as follows: By choosing an origin point, each pixel of the image represents a location with respect to the origin. The value of the pixel 0 or 1 indicates whether the pixel belongs to background or foreground sets, respectively. In other words, all the vectors belong to either the foreground or background sets.

3 Morphological Operators and Set Theory

Morphological operators were first defined and introduced based on set theory. Mathematical morphology observes signal/image as a set, and its nonlinear operators are applied based on set theory. Here, each signal/image is a set, and each nonlinear operator is a set transform. Let X be a set of space vectors in continuous/discontinuous two-dimensional space $\mathbb{Z}^2$. By considering a coordination origin, the connection between a binary image and set theory can be clarified; Each pixel of an image is representing a vector, pointing to the pixel from the origin. The value of the pixel such as 1 and 0 indicates that vector belongs to foreground set and background set, respectively (See Fig. 4).

Fig. 4 By considering a coordination origin, a binary image can be depicted as two foreground and background sets of vectors

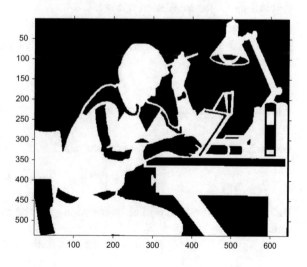

Fig. 5 Set translation in
binary domain

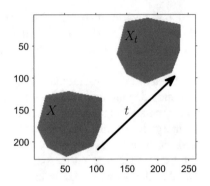

In mathematical language, a binary image can be represented as a function $f(x)$
in $\mathbb{Z}^2$ space (or $\mathbb{R}$) with codomain of $\{0, 1\}$. The foreground set is $S = \{x | f(x) = 1\}$,
where background set is its complement set $S^c = \{x | f(x) = 0\}$.

3.1 Sets and Corresponding Operators

Similar to conventional mathematics, the theory of mathematical morphology is
based on basic operators; here, we illustrate their definitions and properties.

Definition 1 (*Set translation*) Translation of the set X by vector t is denoted as X_t
and defined as follows:

$$X_t = \{x | x - t \in X\}. \tag{1}$$

Figure 5 shows the translation of a set.

Definition 2 (*Set complement*) The complement of set X is denoted as X_c and
defined as follows:

$$X_c = \{x | x \notin X\}. \tag{2}$$

Figure 6 shows the complement of a set.

Definition 3 (*Union*) The union of two sets X and Y is defined as follows:

$$X \cup Y = \{x \mid x \in X \text{ or } x \in Y\}. \tag{3}$$

Definition 4 (*Intersection*) The intersection of X and Y, viz., two sets, is defined as
follows:

$$X \cap Y = \{x \mid x \in X \text{ and } x \in Y\}. \tag{4}$$

Figure 7 shows the union and intersection of two sets.

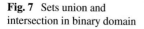

Fig. 6 Set complement in binary domain

Fig. 7 Sets union and
intersection in binary domain

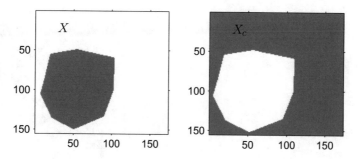

Fig. 8 Binary set and its
transpose

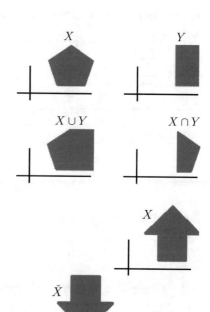

Definition 5 (*Set transpose*) The transpose of a set X is denoted as $\check{X}$ and defined
as follows:

$$\check{X} = \{x \mid -x \in X\}. \tag{5}$$

Figure 8 shows the transpose of set. The five preceding transforms suffice all the
morphological operators requirements. In this regard, if a system is programable for
applying these transforms, the system is capable of processing binary images based
on mathematical morphology.

Definition 6 (*Set scaling*) Let X be a set in $\mathbb{R}^n$ space, and $\alpha x \in \mathbb{R}$, then the scaled
set of X by the scale of α is denoted as αX and defined as follows:

Fig. 9 Scaling X by the scale of $\alpha = 2$

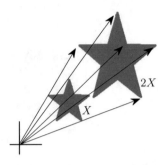

Table 1 Notations and definitions of basic operators in set theory

Operator	Notation	Definition
Translation	X_t	$X_t = \{x \mid x - t \in X\}$
Scaling	αX	$\alpha X = \{\alpha x \mid x \in X\}$
Complement	X_c	$X_c = \{x \mid x \notin X\}$
Transpose	$\check{X}$	$\check{X} = \{x \mid -x \in X\}$
Union	$X \cup Y$	$X \cup Y = \{x \mid x \in X \text{ or } x \in Y\}$
Intersection	$X \cap Y$	$X \cap Y = \{x \mid x \in X \text{ and } x \in Y\}$

$$\alpha X = \{\alpha x \mid x \in X\}. \tag{6}$$

Figure 9 the illustrates set scaling, Basic operations in set theory are summarized in Table 1.

3.2 Basic Properties for Morphological Operators

In a view, mathematical morphology has been built on physical foundations. They perform on images with an emphasis on the interaction between mathematical operators and physical characteristics of the image. Morphological operators have three properties: (i) translation invariance, (ii) translation compatibility, and (iii) scaling compatibility, which is explained here.

Definition 7 (*Translation Invariance*) The transform φ is invariant to translation, if

$$\forall t : \varphi(X_t) = [\varphi(X)]_t \tag{7}$$

This property is fulfilled for the transforms, independent of origin. Otherwise, an arbitrary origin cannot be chosen, if the transform is origin-dependent. An example of an origin-dependent transform is $\varphi^o = X \cap Z$, wherein Z is a set of vectors surrounded by a rectangle with a center o, which has been shown in the upper index of φ^o. Clearly, such a transform does not have the property of translation invariance; $X_h \cap Z \neq (X \cap Z)_h$.

Fig. 10 Translation
invariance property in set
domain

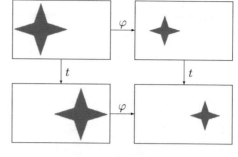

Fig. 11 Scaling
compatibility of a set
transform

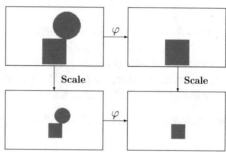

Figure 10 clarifies the translation invariance property.

Definition 8 (*Translation compatibility*) The transform φ is translation compatible,
if

$$\varphi^o(X_h) = [\varphi^{-h}(X)]_h \tag{8}$$

Although the transforms with the property of translation invariance have ownership
of translation compatibility, not all translation compatible transforms have translation invariance property. Indeed, translation compatibility covers a wider domain of
transforms compared to translation invariance.

Definition 9 (*Scaling compatibility*) Transforms φ_λ are compatible to translation if

$$\varphi_\lambda(X) = \lambda\varphi(\frac{X}{\lambda}) \tag{9}$$

See the Fig. 11.

3.3   Set Dilation and Erosion

As we have discussed in Sect. 1, mathematical morphology has two main nature-
inspired operators called "dilation" and "erosion"; in simple terms, they can be

Fig. 12 Dilation in binary
domain

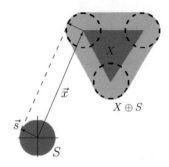

described as expansion and contraction, respectively. There is a taste of diversity in
definition and notation of dilation and erosion in the literature. Here, the authors have
utilized the definitions and notations described by Sternberg [7–9]. These definitions
are deployed by Maragos [10–13] also, wherein dilation and erosion are based on
Minkowski [14] addition and subtraction, respectively.

Definition 10 (*Dilation*) The dilation of set X by S is defined as follows:

$$X \oplus S = \{ x + s \mid x \in X \ and \ s \in S \} \tag{10}$$

The $X \oplus S$ result is a set of all translated vectors of X by the vectors in S. Figure 12
shows a dilation operation with a circular symmetric structuring element.

Corollary 1 expresses the dilation in a simpler way.

Corollary 1 *Alternative expression of dilation is in the following form:*

$$X \oplus S = \bigcup_{t \in S} X_t \tag{11}$$

$\square$

Definition 11 (*Erosion*) The erosion of set X by S is defined as follows:

$$X \ominus S = \bigcap_{t \in \check{S}} X_t \tag{12}$$

or;

$$X \ominus S = \bigcap_{t \in S} X_{-t} \tag{13}$$

In signal/image processing by morphological operators, the foreground is acquired
as set X, and the set S is called "structuring element." Figure 13 shows an example
of dilation and erosion in binary domain.

Fig. 13 Example of dilation and erosion in binary domain

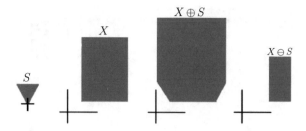

3.4 A Geometrical Interpretation of Dilation and Erosion Process

In geometrical view, dilation process can be expressed in two steps: First, structuring element (SE) S is transposed and translated by each pixel (vector t) of the image foreground. Second, if the translated transposed SE $\check{S}_t$ hits the foreground of the image, ($\exists x, x \in \check{S}_t$ and $x \in X$), that pixel t belongs to $X \oplus S$. Considering this view to the dilation process, it can be said in the author words:

The border of the dilation set $X \oplus S$ is the trace of the origin of structuring element in the movement of its transposed version $\check{S}$ exterior and tangent to the border of set X.

Corollary 2 is the mathematical expression of the above interpretation of dilation.

Corollary 2 *Dilation can alternatively be expressed as:*

$$X \oplus S = \left\{ t \mid \check{S}_t \cap X \neq \emptyset \right\} \tag{14}$$

In the same way, a similar interpretation is presented for erosion. This time, the structuring element itself slides on the foreground of the image. Each pixel t-corresponding vector t from origin-that the translated structuring element S_t is entirely in coverage of X as $S_t \subset X$, it belongs to the erosion set $X \ominus S$. In the words of the authors:

The border of the erosion set is the trace of origin of structuring element in movement interior and tangent to the boundary of set X.

Corollary 3 *The alternative expression of erosion is as follows:*

$$X \ominus S = \{ t \mid S_t \subset X \} \tag{15}$$

Figure 14 depicts the above interpretation of the dilation and erosion process in binary domain.

Fig. 14 Dilation and erosion process using Corollaries 2 and 3

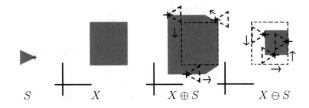

3.5 Direct Effect of Edges and Borders on the Erosion and Dilation

In preceding parts, it was observed that dilation $X \oplus S$ is obtained by the union of all translations of the points in X by vectors in S. Due to the translation of foreground (set X) by the vector s where $s \in S$, the same shape of the foreground is obtained for the location difference induced. As an evident amble, the translated points of the border figure the boundaries of the translated shape. Therefore, it can be concluded that the translated set is the set of translated border points together with the point surrounded by them. Considering the above point of view and dilation, as the result of union operation, it can be said:

> If the origin of the structuring element S slides on the border of X, the exterior border of the structuring element trace is the border of the dilation set $X \oplus S$.

Typically, the points inside the image borders do not change due to dilation, and the borders are the subject of change, dependent on the size and shape of the structuring element. In other words, in dilation operation, the boundary points have the key role.

It is similar in the case of erosion operation. With this difference that the transpose of the structuring element is used, and since erosion is by intersection operation, the interior border of the movement trace of the transposed structuring element is considered. Therefore,

> If the origin of the transposed structuring element S slides on the border of X, the interior boundary of the structuring element trace is the boundary of the erosion set $X \ominus S$.

Considering the above-mentioned impact of dilation and erosion through the shape boundaries, a simple, handy way to drawing the dilation and erosion result can be expressed. To obtain the dilation/erosion set quickly and directly, we consider just the borders, and the origin of the structuring element (its transpose for erosion) is slid on the borders; the exterior (interior for erosion) border of the trace is the border of dilation/erosion set. The structuring element resembles a tool in our hand as it

Fig. 15 Dilation and erosion
in one step by considering
just the effect of structuring
element on the set borders

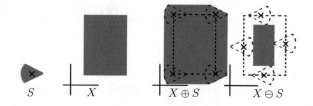

is applied to the original set to reform it and makes a new shape. This shape that
we obtain—in one step—is indeed the morphological operation output. Figure 15
clarifies obtaining dilation and erosion in one step by considering just the effect of
structuring element on the borders.

One of the advantages of dilation and erosion is their increasing property. Increasing property is crucial for an operator in signal/image processing that guarantees the
preservation of size ratios during the transform.

Definition 12 (*Increasing property*) In set domain, transform φ is increasing if

$$X \subset Y \implies \varphi(X) \subset \varphi(Y) \tag{16}$$

Corollary 4 (Translation invariance) *Both dilation and erosion are translation
invariant as explained here:*

$$X_t \oplus S = (X \oplus S)_t \tag{17}$$
$$X_t \ominus S = (X \ominus S)_t \tag{18}$$

Dilation is known as an extensive operation that extends the set. Corollary 5 clarifies
this property.

Corollary 5 (Extensivity) *If S is a structuring element containing its origin $O \in S$,
then,*

$$X \subset X \oplus S \tag{19}$$

The condition $O \in S$ makes the guarantee that the dilation does not remove any
point of the set. Hence, it adds if structuring element includes that one point. The
extensivity property can be even more generalized to involve the case that $O \notin S$ as
explained in Corollary 6.

Corollary 6 (Extensivity) *Let S an arbitrary structuring element, then extensivity
property can be extended as follows:*

$$\exists t, \ \forall X, \ X \subset (X \oplus S)_t \tag{20}$$

Corollary 7 (Anti-extensivity) *Let S an nonempty structuring element including its origin (O ∈ S), then:*

$$X \ominus S \subset X \tag{21}$$

For the case O ∉ S:

$$\exists t, \forall X, (X \ominus S)_t \subset X \tag{22}$$

This property brings the "set shrinking" effect for erosion. The properties of extensivity and anti-extensivity of the dilation and erosion, respectively, can be seen in the Figs. 13, 14, and 15.

Here, more properties of dilation and erosion are presented in order to increase the reader's knowledge about them.

Corollary 8 (Commutativity)

$$X \oplus S = S \oplus X \tag{23}$$

Despite dilation, erosion does not fulfill this property.

Corollary 9 (Associativity)

$$(X \cup Y) \oplus S = (X \oplus S) \cup (Y \oplus S) \tag{24}$$

$$(X \cap Y) \oplus S = (X \ominus S) \cap (Y \ominus S) \tag{25}$$

Corollary 10 (Distributivity) *Dilation is distributive:*

$$(X \oplus Y) \oplus T = X \oplus (Y \oplus T) \tag{26}$$

Figure 16 clarifies the distributivity of dilation.

Fig. 16 Distributivity of dilation in set domain

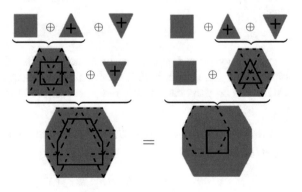

3.6 Closing and Opening

In this part, the operators closing and opening in the binary domain are presented. Opening and closing operators are special algebraic operators. In algebraic view, every operator φ with the following properties is an opening:

$\forall X$

$\varphi(X) \subset X$ .... Anti-Extensivity

$X \subset \acute{X} \implies \varphi(X) \subset \varphi(\acute{X})$ .... Increasing

$\varphi(\varphi(X)) = \varphi(X)$ .... Idempotent

In short, an opening is an operator with properties of anti-extensivity, increasing, and idempotence.

As well, in algebraic view, every operator φ with the following properties is a closing:

$\forall X$

$\varphi(X) \supset X$ .... Extensivity

$X \subset \acute{X} \implies \varphi(X) \subset \varphi(\acute{X})$ .... Increasing

$\varphi(\varphi(X)) = \varphi(X)$ .... Idempotent

In short, a closing is an operator with properties of extensivity, increasing, and idempotence.

Morphological opening and closing are defined by the combination of dilation and erosion.

Definition 13 (*Opening*) The opening of set X by structuring element S is defined as first erosion of X by S and then dilation of the result by S:

$$X \circ Y = (X \ominus S) \oplus S \tag{27}$$

Definition 14 (*Closing*) The closing of set X by structuring element S is defined as first dilation of X by transposed structuring element $\check{S}$ and then erosion of the result by the transposed structuring element $\check{S}$:

$$X \bullet Y = (X \oplus \check{S}) \ominus \check{S} \tag{28}$$

Figure 17 shows an example of the opening in binary domain.

A crucial theorem pioneered by Matheron [1] shows that every translation invariant set transform can be expressed by the union of some morphological operators (in some cases with an infinite number of them).

Corollary 11 *If φ is an algebraic opening, then*

$$\varphi(x) = \bigcup_{S_i \in K} X \circ S_i, \tag{29}$$

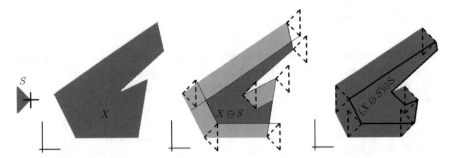

Fig. 17 Example of opening operation in binary domain by erosion and dilation cascade

and for every algebraic closing

$$\varphi^*(x) = \bigcap_{S_i \in K} X \circ S_i, \tag{30}$$

where K is set space.

Corollary 11 conveys the presented morphological opening and the closing as the construction breaks of any arbitrary opening and closing. In this chapter, hereafter whenever we talk about opening and closing, the morphological opening and closing are the point.

Corollary 12 (Morphological orientation of opening and closing) *Opening and closing are independent of translation of structuring element:*

$$\forall t$$

$$X \circ S_t = X \circ S \tag{31}$$
$$X \bullet S_t = X \bullet S \tag{32}$$

and for every algebraic closing;

$$\varphi^*(x) = \bigcap_{S_i \in K} X \circ S_i, \tag{33}$$

where K is set space.

Corollary 12 brings this advantage for opening and closing in their dependency to just size and shape of the structuring and element but not the location of its origin. This property is due to the cascading deployment of dilation and erosion operators where in one of them, despite the other one, the transposed structuring element is used, and they cancel the general translation effect of each other.

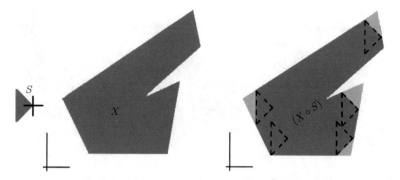

Fig. 18 Opening in one step by a direct handy drawing technique

In former sections, it has been observed that the dilation with a structuring element including just one point t is the translation of the set by vector measure of $+t$. Considering this point, the opening and closing of a set with an structuring element of just one point is the same set. This will be clear to the reader after next corollaries and the interpretations given by the authors. The structuring element with just one point is the simplest structuring element in the binary domain.

Corollary 13 *The opening of X by S is the union set of all structuring element translations S_t which are completely covered by X:*

$$X \circ S = \bigcup_{S_y \subset X} S \tag{34}$$

Corollary 13 guides us to a handy direct drawing technique to obtain opening.

If structuring element slides interior to the borders of the set in the way to be completely covered by the set, the exterior border of the movement trace of the structuring element is the border of the opening set. See Fig. 18.

In Figure 17, the opening is obtained by cascade operations of erosion and dilation in two steps, while in Figure 18 shows its obtaining one step.

Corollary 14 (Duality of opening and closing) *Opening and closing are dual of each other:*

$$[X^c \circ S]^c = X \bullet S \tag{35}$$

Corollary 15 *Considering the duality of opening and closing, we can express the closing in the following form:*

$$X \bullet S = \Big\{ \bigcup_{S_y \subset X^c} S_y \Big\}^c = \Big\{ \bigcup_{(S_y \cap X) \neq \emptyset} S_y \Big\}^c \tag{36}$$

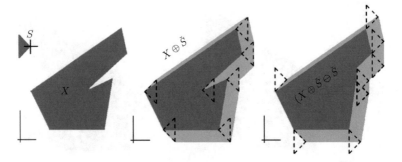

Fig. 19 Closing in two steps by cascade of the dilation and erosion operators

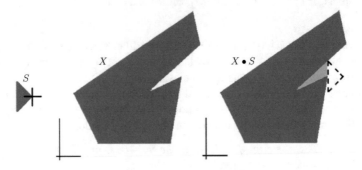

Fig. 20 Closing in one step by a direct handy drawing technique

Corollary 14 guides us to a handy direct drawing technique for obtaining binary closing:

> If structuring element slides exterior to the borders of the set in a way to be completely covered by the background set, the interior boundary of the movement trace of the structuring element is the border of the closing set. See Fig. 20.

In Fig. 19, the closing is obtained in two steps via the cascade of dilation and erosion operations, while in Fig. 20 it is achieved in one step using Corollary 15 and its interpretation of closing.

Definition 15 (*To be open for*) The set S is open for set T, if

$$S \circ T = S \tag{37}$$

Definition 16 (*To be closed for*) The set S is closed for set T, if

$$S \bullet T = S \tag{38}$$

Table 2 Notations and definitions of basic morphological operators

Operator	Notation	Definition
Dilation	$X \oplus S$	$\{x + s \mid x \in X \text{ and } s \in S\}$
Erosion	$X \ominus S$	$\bigcap_{t \in \check{S}} X_t$
Opening	$X \circ S$	$\bigcup_{S_t \subset X} S_t$
Closing	$X \bullet S$	$\left\{ \bigcup_{S_t \cap X \neq \emptyset} S_t \right\}^c$

Definition 17 (*Sieving*) If X is open for T, then (Table 2):

$$X \circ S \subset X \circ T \subset X \subset X \bullet T \subset X \bullet S \qquad (39)$$

3.7 A Historical Review to Definitions and Notations

Despite the long history of mathematical morphology, morphological operators suffer from diversity in definitions and notations. Maragos in a part of Ref. [10] has comprehensively reviewed this variety and tried to make the notation and definitions as much as possible close to each other. He emphasizes that the differences are not fundamental, and they are just in the way of presentation of the same concept.

The diversity at the first level is in notations used for dilation and erosion operators. The notation diversity does not matter, and a more important thing is the difference in definitions and equations for dilation and erosion. The definitions presented for dilation and erosion in this chapter are the same as the one by Hadwiger [15] for Minkowski addition and subtraction, respectively.

Matheron [1, 2] and Serra [3, 4] have used a different way by a deployment of transposed structuring element. They used the same definition for Minkowski addition as Hadwiger and used the notation $\oplus$ to express it. Their definition of Minkowski subtraction is as follows:

$$X \ominus S = \bigcap_{s \in S} X_s \qquad (40)$$

$$= \left\{ x \mid \check{S}_x \subseteq X \right\}, \qquad (41)$$

which is the same as Hadwiger's definition except the structuring element is transposed. Matheron and Serra have defined dilation and erosion by using Minkowski addition and subtraction. They have defined the dilation as follows:

$$X \oplus \check{S} = \{z \mid S_z \cap X \neq \emptyset\} \qquad (42)$$

Table 3 Definitions and notations in the literature for dilation, erosion, and Minkowski addition and subtraction

Reference	Minkowski addition	Minkowski subtraction
Hadwiger	$\{x+s \mid x \in X; \ s \in S\}$	$\{z \mid z+s \in X; \ s \in S\}$
Matheron	$X \oplus S = \{x+s \mid x \in X; \ s \in S\}$	$X \ominus S = (X^c \oplus S)^c = \{z \mid S_z \subset X\}$
Sternberg	$X \oplus S = \{x-s \mid x \in X; \ s \in S\}$	$X \ominus S = \{z \mid S_z \subset X\} = \{z \mid z+s \in X; \ s \in S\}$
Haralick	–	–
Maragos	$X \oplus S = \bigcup_{s \in S} X_s = \{x+s \mid x \in X; \ s \in S\}$	$X \ominus S = \bigcup_{s \in S} X_{-s} = \{z \mid S_z \subset X\}$
This chapter	$X \oplus S = \{x+s \mid x \in X; \ s \in S\} = \bigcup_{s \in S} X_s$	$X \ominus S = \{z \mid S_z \subset X\} = \bigcap_{s \in S} X_s$

Reference	Binary dilation	Binary erosion
Hadwiger	–	–
Matheron	$X \ominus \check{S} = \{z \mid X \cap S_z \neq \emptyset\} \{x-s \mid x \in X, \ s \in S\}$	$X \ominus \check{S} = \{z \mid S_z \subset X\} = \{z \mid z+s \in X, \ s \in S\}$
Sternberg	$X \oplus S = \{x-s \mid x \in X, \ s \in S\}$	$X \ominus S = \{z \mid S_z \subset X\} = \{z \mid z+s \in X, \ s \in S\}$
Haralick	$X \oplus S = \bigcup_{s \in S} X_s = \{x+s \mid x \in X, \ s \in S\}$	$X \ominus S = \bigcup_{s \in S} X_{-s} = \{z \mid S_z \subset X\}$
Maragos	$X \oplus S = \bigcup_{s \in S} X_s = \{x+s \mid x \in X; \ s \in S\}$	$X \ominus S = \bigcup_{s \in S} X_{-s} = \{z \mid S_z \subset X\}$
This chapter	$X \oplus S = \{x+s \mid x \in X; \ s \in S\} = \bigcup_{s \in S} X_s$	$X \ominus S = \{z \mid \check{S}_z \subset X\} = \bigcap_{s \in S} X_{\check{s}}$

The operation $\oplus$ is the same as Minkowski addition for sets. As well they define the erosion as follows:

$$X \ominus \check{X} = \{z \mid S_z \subseteq X\} \tag{43}$$

where it is the same as Minkowski erosion.

Heijmans and his colleagues [16, 17] used the same definitions as Matheron and Serra. Matheron and Serra define the opening and closing, respectively, as $X_S = (X \ominus \check{S}) \oplus S$ and $X^S = (X \oplus \check{S}) \ominus S$. Their opening has the same result as the opening mentioned in this chapter. However, their closing is different; if transposed structuring element were in use, it would be the same as the closing defined in this chapter.

The standard point of the all the above definitions is an identical definition of dilation and erosion as the same presented in this chapter. This definition is the same as one used by Sternberg [7, 18, 19] that was used later by Maragos [10]. Table 3 indicates different definitions and notations in the literature for dilation and erosion as well for Minkowski addition and subtraction.

4 Practical Interpretation of Binary Opening and Closing

Here, we present a practical interpretation of opening and closing in binary domain. As we know, a binary image is composed of foreground and background. As an analogy, we consider the foreground as an island surrounded by background sea, and we are going to analyze the opening effect and closing effect on the area, edges, cliffs, gulfs, straights, land necks, and the lakes inside the island. Figure 21 shows the binary image of the island as mentioned above together with a circular structuring element, and the resultant dilation, erosion, opening and closing islands.

Considering all the preceding definitions and corollaries, opening can be sufficiently expressed as follows:

> The opening is the set of all the points of the foreground, which in structuring element movement inside the foreground are hit by the structuring element.

In other words, in the movement of structuring element inside the set, every point that can be reached by structuring element belongs to opening set. Now, it can be understood why the name of the operation is opening.

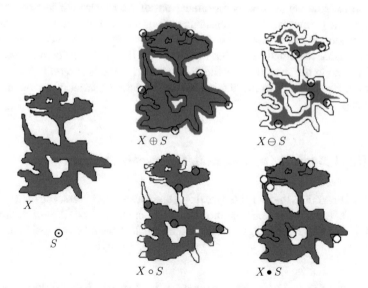

Fig. 21 Interpretation of binary morphological dilation, erosion, and opening and closing in island analogy

Indeed, opening result is the set of points open to (reachable by) the structuring element movement inside the set.

As a result of opening, the small islands will disappear since they do not have any space inside open enough for structuring element movement. Land necks will be cut and removed in the parts narrower than structuring element corresponding dimension. In general, every point of the set not open to the structuring element because of tightness, narrowness, or being cornered is removed. Therefore, opening results in removing the corners, necks, and small particles and decreased area of the islands (foreground) and increased sea area (background).

Based on the corollaries and definitions mentioned above, the closing can be expressed as follows:

The closing set is the set of all the points of foreground and background which are not hit by the structuring element in its movement out of the foreground.

Indeed, the closing set points are guarded by the foreground set not be reachable by the structuring element in its movement out of the foreground set. These points include all the points inside the foreground and some point out of its border in support of the borders. In other words, the points inside the tight gulfs and straits, as well the point inside the small lakes that together with the points inside the foreground set are closed to the structuring element, are added to the island area (foreground). Analyzing Fig. 21 clarifies the above geometrical interpretation of the opening and the closing.

5 Morphological Operators in Grayscale Domain

In the former section, the morphological operators were defined and clarified in binary domain. This section extends the morphological operators and filter to multivalued functions and grayscale images, wherein the signal/image takes all real values.

5.1 Basic Morphological Operators in Multivalued Function Domain

We aim to apply morphological operators on signal/image through a nonlinear system by using the set theory. To catch this goal, we should transfer the signal in the form of a set. The transfer of the signal to the form of a set results in the extension of morphological operators from binary domain to multivalued function domain. This transfer of the binary domain to multivalued function domain is done by the bless of "umbra" concept and thresholding [7, 19].

Definition 18 (*Umbra*) Consider the function $f : x \mapsto f(x)$: If $f(x)$ is defined in a n dimension space $x \in \mathbb{Z}^n$, its umbra $U(f)$ is defined in $n + 1$ dimensions space as follows:

$$U(f) = \{ (x; a) \in F^{n+1} \mid a \leq f(x) \} \tag{44}$$

Indeed, the umbra of each function is the set of the points under the function. Interestingly, there is a one to one correspondence between the functions and umbra sets. Every function in $\mathbb{Z}^n$ space corresponds to a set in $\mathbb{Z}^{n+1}$ space as its umbra. However, not every set is umbra of a function. A function can be obtained by its umbra;

$$f(x) = max\{ a \mid (x; a) \in U(f) \} \tag{45}$$

With the benefit of umbra definition, every signal corresponds to the umbra set of its function. The morphological operators can be done in binary domain over the umbra set of the signal and using the umbra concept; the resultant set corresponds a new signal in multivalued function domain as the result of morphological operation.

Umbra is defined not only for functions, but also it is defined for each set in $n + 1$ space as a set in the same space. The umbra of a set is defined as follows:

Definition 19 (*Umbra of a set*) If X is a set in $n + 1$ vector space, its umbra is defined as follows:

$$U(X) = X \oplus T \tag{46}$$
$$i.e. \quad T = \{(x, t)|x = 0, t \geq 0\} \tag{47}$$

As it is mentioned before, not every set can be umbra of a function. Serra in [3] proves a set is umbra of a function if $U(X) = X$. Figure 22 shows the umbra of a set and the umbra of a function.

With a practical tool of umbra, we can have a mapping from multivalued function domain to binary domain and can open the way of extension of the operators from binary domain to multivalued function domain. It might be a reader thinks for every morphological operation in the grayscale domain, we need to use the umbra as a return ticket to the binary domain. It is not the case; the umbra helps us to understand the morphological operation in the grayscale domain in a better way by directly observing the effect in binary domain. Here, by deploying umbra, the equivalent of each basic operation in the binary domain is defined for multivalued function domain to define dilation and erosion directly in the grayscale domain.

Definition 20 (*Grayscale function translation*) Translation of function f by vector $(x; t)$ is denoted as $f_{(x;t)}(.)$ and defined as follows:

$$f_{(x;t)}(y) = f(y - x) + t \tag{48}$$

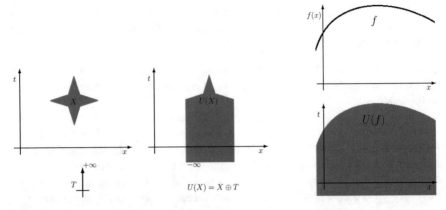

Fig. 22 Umbra of a set X, and umbra of function f

Equivalent to the union operation of two sets in the set domain is the maximum operation between two functions in multivalued function domain. The maximum operation is denoted by "$\vee$."

Definition 21 (*Maximum of two grayscale functions*) The maximum of two functions f and g is denoted as $f \vee g$ and is defined as follows:

$$(f \vee g)(x) = max_x[f(x), g(x)] \tag{49}$$

$$= \begin{cases} f(x) & \text{if} \quad f(x) \geq g(x), \\ g(x) & \text{if} \quad g(x) > f(x). \end{cases} \tag{50}$$

Equivalent to the intersection operation of two sets in set domain is the minimum operation between two functions in multivalued function domain. The minimum operation is denoted by "$\wedge$."

Definition 22 (*Minimum of two grayscale functions*) The minimum of two functions f and g is denoted as $f \wedge g$ and is defined as follows:

$$(f \wedge g)(x) = min_x[f(x), g(x)] \tag{51}$$

$$= \begin{cases} f(x) & \text{if} \quad f(x) \leq g(x), \\ g(x) & \text{if} \quad g(x) < f(x). \end{cases} \tag{52}$$

Corollary 16 *There is a one-to-one correspondence between maximum of functions and union of sets based on umbra concept:*

$$U(f \vee g) = U(f) \cup U(g) \tag{53}$$

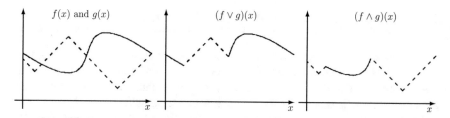

Fig. 23 Maximum and minimum operations between two functions

Corollary 17 *There is a one-to-one correspondence between minimum of functions and intersection of sets based on umbra concept:*

$$U(f \wedge g) = U(f) \cap U(g) \tag{54}$$

Figure 23 shows two functions as well the maximum and minimum operations between them.

Definition 23 (*Transpose of a function*) The transpose of a function is defined as follows:

$$\overset{\vee}{f}(x) = f(-x) \tag{55}$$

In analyzing the transposing effect on the umbra of a function, it can be seen that the umbra of the transposed function is mirrored with respect to the axis $x = 0$. If the umbra of the function is to be transposed with respect to the origin, that is, the same as inversion concept in binary domain, the transposed function should be inversed with respect to $t = 0$ too. This operation is shown in binary domain as

$$\hat{X} = \{ (x; -t) \mid (x; t) \in X \} \tag{56}$$

Using the above concept,

$$U(\overset{\vee}{f}) = \widehat{U(\overset{\vee}{f})} \tag{57}$$

Definition 24 (*Complement of a function*) The complement of a function is defined as follows:

$$f^c(x) = -f(x) \tag{58}$$

As the umbra of the function and umbra of its transposed are not transpose of each other, similarly, the umbra of the function and the umbra of the complement of the function are not the complement of each other, they fulfill the following equation:

$$[U(f)]^c = \widehat{U(f^c)} \tag{59}$$

5.2 Dilation and Erosion of Multivalued Functions

In the former part, we defined the required operators for the extension of morphological operators to function domains from binary domain. Although here, independent from umbra, we define dilation and erosion for multivalued functions; umbra, as a geometrical tool, is used for better inquisitional understanding.

Dilation and erosion are both increasing operations and independent from the shape and size of structuring element S, for every function f, always the functions g and h exist which fulfill the following equations:

$$U(g) = U(f) \oplus S \tag{60}$$
$$U(h) = U(f) \ominus S \tag{61}$$

Since an umbra set corresponds just to one function, dilation is simply expressed as $f \oplus g$ and erosion as $f \ominus g$.

Definition 25 (*Dilation for multivalued functions*) Dilation of function f by structuring element function g is defined as follows:

$$(f \oplus g)(n) = \max_v \{ f(v) + g(n - v) \} \tag{62}$$

Definition 26 (*Erosion for multivalued functions*) Erosion of function f by structuring element function g is defined as follows:

$$(f \ominus g)(n) = \min_v \{ f(v) - g(v - n) \} \tag{63}$$

Figure 24 dilation and erosion of a quantized signal by a flat and symmetric structuring element.

Corollary 18 (Dilation and erosion are increasing in function domain)

$$f \leq g \rightarrow f \oplus h \leq g \oplus h \tag{64}$$
$$f \leq g \rightarrow f \ominus h \leq g \ominus h \tag{65}$$

Corollary 19 (Extensivity of dilation)

$$\forall g(n), \ f(n) \leq (f \oplus g)(n) \tag{66}$$

Corollary 20 (Anti-extensivity of erosion)

$$\forall g(n), \ (f \ominus g)(n) \leq f(n) \tag{67}$$

Extensivity of dilation and anti-extensivity of erosion can be seen in Fig. 24.

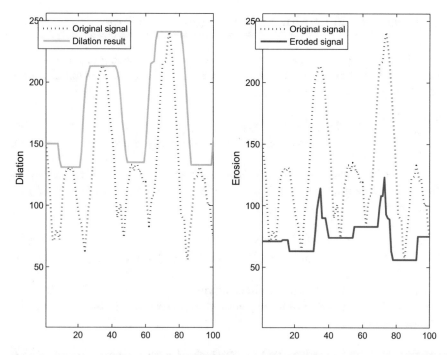

Fig. 24 Dilation and erosion of a quantized signal by a flat and symmetric structuring element

Table 4 Two forms of formulas for dilation and erosion in function domain

	$(f \oplus g)(n) =$	$(f \ominus g)(n) =$
Form 1	$\max_v\{f(n-v)+g(v)\}$	$\min_v\{f(n+v)-g(v)\}$
Form 2	$\max_v\{f(v)+g(n-v)\}$	$\min_v\{f(v)-g(v-n)\}$

5.3 Two Forms of Presentation for Dilation and Erosion Formula

Dilation and erosion in function domain are formulated in two forms. Table 4 shows these two forms of formulas for dilation and erosion in function domain.

Simply, it can be said that both forms lead to the same results. In erosion, it is the same either shifting the function to the center and obtaining its difference from structuring element function in structuring element domain (first form), or shifting structuring element function from the center to point n and obtaining the functions differences in neighborhood of n (second form). For dilation, similarly, both forms conclude the same result.

6   Opening and Closing of Multivalued Functions

Definition 27 (*Opening of a multivalued function*) The opening of function f by function g (structuring element) is the cascading process of the erosion of f by g and the dilation of the result by g:

$$f \circ g = (f \ominus g) \oplus g \tag{68}$$
$$= \max_w \{ \min_v \{ f(v) - g(v - w) \} + g(n - w) \} \tag{69}$$

Definition 28 (*Closing of a multivalued function*) The closing of function f by function g (structuring element) is the cascading process of the first dilation of f by g and then erosion of the result by g:

$$f \bullet g = (f \oplus g) \ominus g \tag{70}$$
$$= \min_w \{ \max_v \{ f(v) + g(v - w) \} - g(n - w) \} \tag{71}$$

Corollary 21 (Extensivity and anti-extensivity of opening and closing in function domain) *Opening and closing on functions are, respectively, anti-extensive and extensive operations:*

$$(f \circ g)(n) \leq f(n) \leq (f \bullet g)(n) \tag{72}$$

Figure 25 shows opening and closing of a quantized signal $f(n)$ by a flat structuring element.

Corollary 22 (Shape orientation of opening and closing of multivalued functions) *The translation of structuring element does not affect the result of opening and closing multivalued functions:*

$$f \circ g_{(x, t)} = f \circ g \tag{73}$$
$$f \bullet g_{(x, t)} = f \bullet g \tag{74}$$

Corollary 23 (Function open to function) *Function f is open to function g, if*

$$f \circ g = f \tag{75}$$

Corollary 24 (Function closed to function) *Function f is closed to function g, if*

$$f \bullet g = f \tag{76}$$

Corollary 25 (Sieving) *If function g is open to function h,*

$$f \circ g \leq f \circ h \leq f \leq f \bullet g \leq f \bullet g \tag{77}$$

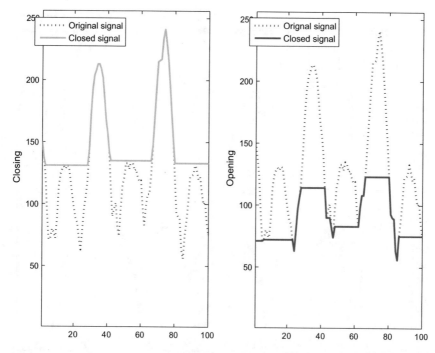

Fig. 25 Opening and closing of a quantized signal by a flat and symmetric structuring element

7 Interpretation and Intuitive Understanding of Morphological Filters in Multivalued Function Domain

In Sect. 4, an intuitive understanding and interpretation or binary morphological filters were presented using the island and sea analogy for binary images. Here, we try to give a similar intuitive analogy for morphological operators in multivalued function domain. A reader by having the background from binary domain interpretations can have a better understanding and intuition in this part.

Figure 26 shows the morphological operations over a signal by a flat structuring element. As it can be observed in Fig. 26, dilation has an expanding effect on the signal; it reduces the effect of valleys and flattens the peaks as it makes them wider with the same height. Dilation process can be interpreted as follows:

> The dilation of a function by a structuring element function segment can be drawn by tracing the origin of transposed structuring element in its movement tangent to signal surface from up.

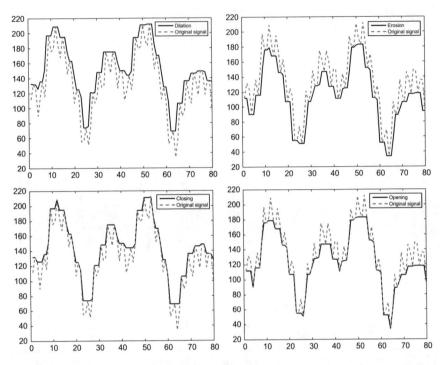

Fig. 26 Dilation, erosion, opening and closing of a quantized signal of 80 samples with a flat and symmetric structuring element of length 3 samples

Despite dilation, erosion has abridgment effect on the signal as it reduces the number of peaks and widens the valleys. Erosion process can be interpreted as follows:

> The erosion of a function by a structuring element function segment can be drawn by tracing the origin of structuring element in its movement tangent to the signal surface from below.

The morphological effect of opening and closing is smoothing. Opening smooths the signal by cutting the peaks, and so closing does by filling the valleys. The peaks and valleys are, respectively, cut and filled exactly from the point with the width of the same as structuring element width. Therefore, acquiring the wider structuring element leads to the stronger smoothing.

8 Conclusion

The authors aimed to give a simple, practical intuition and understanding of morphological filters by showing their inspiration from the natural geometry through a lot of illustrative, visual examples. This chapter makes the readers intuitively grasp the simplicity and power of morphological filters. The resultant picturesque view to the theory of mathematical morphology and the resultant operators help readers in practical block diagram design of morphological algorithms for different applications. As well, the researchers who are looking for an extension of a new version of morphological operators can achieve their invention in an easier way.

References

1. Matheron G (1975) Random sets and integral geometry. Wily, New York
2. Matheron G (1979) Iterative image transformations for an automatic screening of cervical. J Histochem Cytochem 27:128–135
3. Serra J (1982) Introduction to mathematical morphology. Academic press, London
4. Serra J (1982) Image analysis and mathematical morphology. Academic press, London
5. Geotcherian V (1980) From binary to graytone image processing using fuzzy logic concepts. Pattern Recogn 12:7–15
6. Rodenacker K, Gais P, Jutting U, Burger G (1983) Mathematical morphology in grey images. In: Proceedings of the European signal processing conference
7. Sternberg SR (1986) Gray-scale morphology. Comput Vision Graph Image Proc 35:333–355
8. Sternberg SR (1983) Biomedical image processing. IEEE Comput Mag 22–34
9. Haralick RM, Sternberg SR, Zhuang X (1987) Image analysis using mathematical morphology. IEEE Trans Pattern Anal Mach Intell P AMI-9(1987), no 4:532–550
10. Maragos P, Schafer RW (1990) Morphological systems for multi-dimensional signal processing. Proc IEEE 78(4):690–710
11. Maragos P (1989) A representation theory for morphological image and signal processing. IEEE Trans Pattern Anal Match Intell P AMI-11, 586–599
12. Maragos P, Schafer RW (1987) Morphological filters-part I: their set theoretic analysis and relations to linear shift-in variant filters. IEEE Trans Acoust Speech Signal Process ASSP-35, 153–1169
13. Maragos P, Schafer RW (1987) Morphological filters-part I: their set theoretic analysis and relations to linear shift-in variant filters. IEEE Trans Acoust Speech Signal Process ASSP-35, 1170–1184
14. Minkowski H (1903) Volumen Und Obertlach. Math Ann 57:447–495
15. Hadwiger H (1957) Vorlesungen uber Inhalt, Oberflach, und Isoperimetrie. Springer, Berlin
16. Heijmans H (1995) Morphological filters. In: Proceeding of summer school on morphological image and signal processing, Zakopane, Poland
17. Heijmans H, Goutsias J, Sivakumar K (1995) Morphological operators for image sequences, vol 62(3). pp 326–346
18. Sternberg SR (1979) Parallel architectures for image processing. In: Proceedings of the IEEE conference on computers, software, and applications. Chicago
19. Sternberg SR (1980) Cellular computers and biomedical image processing. In: Sklanskv J, Bisconte Jc (eds) Biomedical images and computers. Springer, Berlin

Brain Action Inspired Morphological Image Enhancement

Mahdi Khosravy, Neeraj Gupta, Ninoslav Marina, Ishwar K. Sethi
and Mohammad Reza Asharif

Abstract The image perception by human brain through the eyes is not exactly what the eyes receive. In order to have an enhanced view of the received image and more clarity in detail, the brain naturally modifies the color tones in adjacent neighborhoods of colors. A very famous example of this human sight natural modification to the view is the famous Chevreul–Mach bands. In this phenomenon, every bar is filled with one solid level of gray, but human brain perceives narrow bands at the edges with increased contrast which does not reflect the physical reality of solid gray bars. This human visual system action in illusion, highlighting the edges, is inspired here in visual illusory image enhancement (VIIE). An algorithm for the newly introduced VIIE by deploying morphological filters is presented as morphological VIIE (MVIIE). It deploys morphological filters for boosting the same effect on the image edges and aiding human sight by increasing the contrast of the sight. MVIIE algorithm is explained in this chapter. Significant image enhancement, by

M. Khosravy (✉)
Faculty of Computers Networks and Security (CNS), University of Information Science
and Technology, Ohrid, Republic of Macedonia
e-mail: mahdi.khosravy@uist.edu.mk; dr.mahdi.khosravy@ieee.org

N. Gupta
Faculty of Machine Intelligence and Robotics (MIR), University of Information Science
and Technology, Ohrid, Republic of Macedonia
e-mail: neeraj.gupta@uist.edu.mk

N. Marina
University of Information Science and Technology, Ohrid, Republic of Macedonia
e-mail: rector@uist.edu.mk

I.K. Sethi
Department of Computer Science and Engineering, Oakland University, Rochester,
Michigan, USA
e-mail: isethi@oakland.edu

M.R. Asharif
Faculty of Engineering, Information Department, University of the Ryukyus,
Okinawa, Japan
e-mail: asharif@ie.u-ryukyu.ac.jp

© Springer International Publishing AG 2017
S. Patnaik et al. (eds.), *Nature-Inspired Computing and Optimization*,
Modeling and Optimization in Science and Technologies 10,
DOI 10.1007/978-3-319-50920-4_15

MVIEE, is approved through the experiments in terms of image quality metrics and visual perception.

1 Introduction

Image enhancement is to protrude the visibility of the subjective details of the image, and it is a required preprocessing step before any image analysis technique. The lucidity of an image can be affected by conditions such as atmosphere degradation, imaging system quality, environmental noise, moving impulses, and abnormal light condition, and therefore the quality of the image is decreased and eventually desirable information and data of the image are lost.

Image enhancement is done either in the spatial domain, frequency domain, or special frequency domain. There are plenty of image enhancement techniques in the literature. The simpler spatial domain methods for enhancing images are by point processing operations as they are reviewed in the literature by Anil K. Jain in 1989 [1], Shapiro in 1992 [2], Morishita in 1992 [3], Ramesh Jain in 1995 [4], and Bovik in 1995 [5]. The neighborhood of a pixel in point processing operations involves just itself such as negative operation, thresholding transformation, intensity transformation, logarithmic transformation, powers-law transformations, piecewise linear transformation, and gray-level slicing. Aghagolzadeh performed enhancement of the image in frequency domain in 1992 [6]. Image enhancement in the frequency domain can be done through the image transforms by linear or nonlinear modification of spectral coefficients resulting in the spectral decomposition of the image. A category of transform-based methods deploys some differential operators to extract geometric features. Gaussian derivative is one among them; it estimates the gradient of the smoothed image by a Gaussian function. Another successfully employed approach is Laplacian pyramid image representation; Greenspan enhanced the perceptual sharpness of an image in 2000 [7] as deployed frequency augmentation in shape-invariant properties of edges generated with phase-coherent higher harmonics.

Besides, histogram equalization (HE) methods are very popular for image enhancement. It remaps the intensity levels of an image based on their CDF, in which the dynamic range of histogram is expanded; as a result, the contrast of the image is improved. However, it brings the disadvantage of significant change in the image brightness and consequence artifacts. Cheng in 2004 [8] introduced a simple and effective approach for HE in image enhancement. Contrast entropy was deployed by Agaian in 2007 [9] for HE-based image enhancement using transform coefficient. Ibrahim and Kong in 2007 [10] invented a dynamic HE for contrast enhancement which preserves the image brightness. They introduced the extension of their methodology for color images in 2008 [11].

Along with the methods mentioned above, wavelet domain filters have been widely used for image enhancement as Xu deployed it in 1994 [12]. Huang et. al. in 2004 and 2006 [13, 14] employed the responding function of the cones in the human visual system (HVS) to perceive contrast in the mapping function to process the coefficients of wavelet for color image enhancement. Demirel and Anbarjafari in 2011 [15] presented a significant enhancement to resolution and edges of an image by discrete and stationary wavelet methods.

Here, illusory visual aid of human visual system is inspired for a novel image enhancement technique as we call it *visual illusory image enhancement (VIIE)*.

2 Human Visual Perception

In order to enhance an image, studying and understanding the human visual system is an essence. It is due to subjectivity of image enhancement and human vision judgement as criteria of image quality. However, human vision is a big topic, and our primary focus is on the aspects which inspired us to invent the image enhancement presented, especially the human vision adaptation concerning the changes in illuminations.

When human eye focuses properly, the eye view image is projected in upside-down direction on the retina. The retina is a layer which is inner surface of eyeball back. It includes a dense texture of light-sensitive photoreceptors. The photoreceptors are of two types: cones and rodes. The central area of the retina is called *fovea*, wherein the cones are mainly located. Fovea is very much sensitive to the colors. Rodes are 10–20 times more than cones in number and are distributed in the retina.

Digital images are composed of different levels of intensities and illuminations. Since image enhancement involves modification of intensity levels, it is important to study eye's perception in the domain of intensity and its adaptation to intensity variation. Although the human eye adaptation to light intensities covers a large dynamic range, it cannot be simultaneously operated over the whole range. What makes the visual system able of perception of the light intensities over a range about $10e^{16}$ from scotopic threshold to the glare limit is its *brightness adaptation*. This is the capability of changing its overall sensitivity. Brightness adaptation besides brightness discrimination is an action of the human eye system.

Based on rodes and cones, the human eye has two distinct visual systems: (i) scotopic vision and (ii) photopic vision. Scotopic vision is the eye vision under the conditions of low light, and it is produced by rod cells. The wavelengths around 498 nm—the green-blue light—is the peak point of rod cells sensitivity. Rod cells are not sensitive to wavelengths longer than 640 nm (red color). We can observe that scotopic vision occurs at luminance levels from $10^{-3.5}$ to 10^{-6} cd/m^2 and is not color sensitive. Since several rodes together are connected to a nerve, rode area of the retina makes a vision with fewer details. Scotopic vision is colorless as what we can see in the moonlight, and it is called dim-light vision too.

The vision of the eye under well-lighted conditions of luminance level from 10 to 10^8 cd/m^2 is the photopic vision. It is color-sensitive but less sensitive to light. Cone cells in the retina are mediators of photopic vision. Since each cone is connected exclusively to its nerve, photopic vision provides much higher visual insight as well higher temporal resolution comparative to scotopic vision. The perception of three

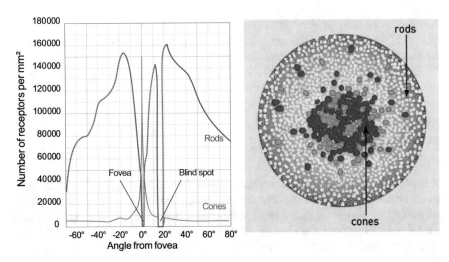

Fig. 1 The change in the number of rodes and cones, as well their geometrical location on the surface of the retina (courtesies: https://en.wikipedia.org/wiki/Retina and http://www.webexhibits. org)

bands of colors in the human eye is performed by sensing the light via three types of cones with biological peak color absorption values at wavelengths of about 420 nm (blue), 534 nm (bluish-green), and 564 nm (yellowish-green) [16]. Figure 1 shows the change in the number of rodes and cones over the surface of the retina, as well their geometrical location.

The human eye perceives the intensity in subjective brightness manner. Experimental evaluations have revealed that subjective brightness is a logarithmic function of the intensity of the exposed light to the eye. Figure 2 shows the overall range of subjective brightness as well the brightness adaptation under a particular condition, where the latter is much shorter. Although the range of subjective brightness is extra wide, an observer has the ability to discriminate a limited number of different intensities—around twenty—at every point of a grayscale image. However, the eye adaptation by roaming the eye focus around the image helps in increasing the above-described number up to the aforementioned wide range of intensity. This phenomenon is called *brightness adaptation*. When the eye is adapted to a certain level of light intensity, the simultaneous brightness discrimination range is shorter as is shown in Fig. 2 by shorter intersecting logarithmic curve B_b.

3 Visual Illusions

It may come to mind that whether studying visual illusions helps for scientists? Professor Michael Bach in 2006 [17] firstly asks the following question and then he answers with some reasons:

Fig. 2 Subjective brightness range for human eye shows an exceedingly large dynamic range

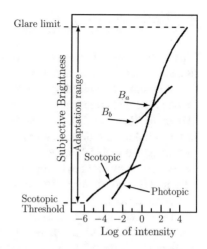

Is it only the playful child in scientists that drives them to study optical illusions?

His answer to the question is *no*. As a reason, he mentions the illusion effect in sports judgment that how a referee's decision can be affected by "flash-lag effect" as studied by Baldo et. al. in 2002 [18]. The second reason of him, which makes studying optical illusions much more than child play, is that illusions help us to understand the perception mechanisms. Besides, he lists clinical conditions wherein the optical illusion has the key role, such as a migraine, organic psychoses, epileptic aura, and Charles Bonnet syndrome [19]. Finally, he reminds us that Rorschach test [20] is partially performed by optical illusions. Professor Bach has presented a comprehensive list of optical illusions and visual phenomena in 2006 [17, 21]. However, his emphasis is on the beauty, bilateral experiments, and an attempt to explain the mechanisms of human vision. In most of the literature, visual illusion—in other words, visual phenomena—has mentioned the limitations of human vision in the perception of reality.

Here, we have a novel view on this issue; these phenomena are arranged by the human visual system in order to aid the human perception. In the next section, we clarify our view to optical illusions by answering the following questions:

How can illusions be employed by human visual system to improve the vision?

How can human visual illusions be inspired for image enhancement mechanism?

4 Visual Illusions

This section presents a brief review of visual phenomena. Remarkably, the optical illusion initiates a thought that all human vision is in illusion, that is true in a sense that our brain interprets photoreceptors signal, what we call it here *brain illusion aid*. Some of the most famous optical illusions are as follows:

Waterfall Illusion It is a motion aftereffect visual illusion. Aristotle has mentioned this effect: After looking at a waterfall for a minute then looking at the stationary rock just beside the waterfall, it seems that rock slightly moves upward. Motion aftereffect can be caused by looking at any moving visual stimulus. While one focuses his eyes on a moving object for a time and immediately shifting to a stationary object, a deceptive movement appears in the direction opposite to the initial moving stimulus. The motion adaptation of human vision explains this illusion.

Hermann Grid This illusion was reported by Ludimar Hermann in 1870 [22]. During reading, a book from John Tyndall, he noticed gray blobs at intersections of the empty spaces of matrix arranged figures. The blobs did not exist; when directly gazed, they disappeared. Similarly, the illusion can be observed in the scintillating grid that is a variation in Hermann grid illusion as is shown in Fig. 3. It happens in viewing at a structure orthogonal gray bars with white circles in their intersections and black background. Dark spots give the impression of appearing at the intersections of gray bars while not looking directly at them, while each of them disappear by directly looking on.

Fig. 3 Scintillating grid, an example of Hermann illusion

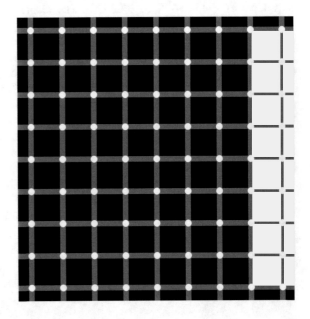

© Wikipedia, the courtesy of images

Fig. 4 *Left* Fraser spiral illusion [24]. The arc segments located on the concentric circles appear spiral. *Right*, *up* and *right bottom*, respectively, Zollner illusion [26] and cafe wall illusion [27, 28], wherein both of them parallel straight lines appear as converging and diverging lines

The neural process called *natural inhibition* [23] has been used to explain the Hermann illusion. In the human visual system, the intensity of each spot is not only the result of one receptor but a group of receptors, which together make the receptive field. At the intersection of white bars, most of the photoreceptors of the receptive field which inhibit the ganglion cell are in lower density, and in the absence of dominant central cells, due to indirect view, the intersections seem as gray.

In indirect view of the intersections of white bars, the central cells are absent in making the vision, while the surrounding neighbor cells are more dominant. Because the neighbor cells are affected by a black background, the illusory gray spots appear. **Spiral Illusion** The British psychologist named James Fraser had first studied the visual illusions in 1908, and described a spiral illusion, which is called after him as *Fraser spiral illusion* [24, 25]. Also, it is known as the *false spiral*. A black rope and a white rope are twisted to make a rope that is used for making concentric circles. Since the rope is wrapped, the arc segments of each black and white appear to form a spiral. However, the series of the arcs are located on the concentric circles. The spiral form of the arcs is illusory as shown in Fig. 4.

Moreover, two other well-known illusions—Zollner's illusion [26] and the cafe wall illusion [27, 28]—are based on the same principle. In both of them, parallel straight lines appear as converging and diverging lines. Similar to Fraser spiral illusion, illusory deviations are perceived because of the series of tilted colored segments. **Pinna Illusion** Pinna illusion was the first illusory rotating motion reported by Baingio Pinna in 1990 [29, 30]. Squares which each of them delineated by two black

Fig. 5 Pinna Illusion: Upon looming the center of concentric rings of elements with opposing polarity, they appear to rotate against each other. Upon physical rotation of the rings, they appear to expand or contract [30]

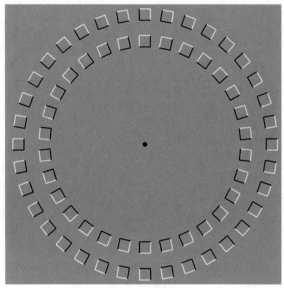

© Wikipedia, the courtesy of the image

and two white sides are arranged in two concentric rings of different diameters. All the squares have same dimensions and orientation with respect to the center of the rings. In squares of both rings, the sides interior and exterior to the ring are, respectively, black and white. The other two sides of the squares are black and white in the clockwise direction and counter-clockwise direction in the exterior ring and interior ring, respectively (See Fig. 5). By fixing the gaze precisely on the center point of the rings and looming toward the paper, two rings appear to rotate in opposite direction to each other. The direction of rotation changes upon looming away from the paper.

Another motion illusion appears while the same arrangement physically rotates clockwise, wherein the inner ring of squares seems to contract and the outer ring appears to expand. Upon rotating counter-clockwise, the inner ring appears to expand and the outer ring appear to contract. For more details, read Refs. [29, 30].

4.1 Rotating Snakes

In general, the illusory motion is a phenomenon that stationary images of arranged patterns and luminance make human vision percept motion inside the image. Chi et. al. in 2008 introduced self-animating images [31] based on this phenomenon with potential application for entertainment and advertisement designs.

Kitaoka in 2003 created an illusory motion called *Rotating Snakes* [32]. He created a still image of concentric circles of semi-snake patterns and luminance. Although the

© Akiyoshi Kitaoka, the courtesy of the image

Fig. 6 Rotating snakes illusion by Akiyoshi Kitaoka [32]

image was static, the snakes appeared to rotate around the center (Fig. 6). Kitaoka in 2006 found that the most illusive combination of colors for rotating snakes is blue and yellow together [33]. In Fig. 6, the motion seems from black → blue → white → yellow → black. Conway and Kitaoka have shown in 2005 [34] that the illusion works even if the image is achromatic: Yellow is replaced by light gray, and blue is replaced by dark gray. The reason behind this illusion is small spontaneous eye movements when looking at the image and fast and slow changes over time of neuronal representation of contrast and luminance as explained by Backus in 2005 [35].

Wave line Illusion In 2006, Seiyu Sohmiya reported a color spreading illusion [36]. He aligned three unconnected sinusoidal curves on a white background parallel to each other, while the middle one is of different intensity or of different color than the others.

In the case of the brighter middle curve than the side ones, its background seems darker while it is the same white color as the others. This is called *wave-line brightness illusion* as shown in left side of Fig. 7. For the colored curves, dependent on the color combination, illusory background colors appear not only around the middle curve but also around the side ones. This is called *wave-line color illusion* shown in right side of Fig. 7.

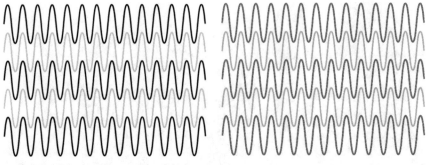

Fig. 7 *Left*, Wave-line brightness illusion; The white background of the gray waves is illusory darker than the rest. *Right*, Wave-line color illusion; there is the illusory appearance of *tinted orange* in the *white background* of the *orange waves*. Seiyu Sohmiya [36] discovered the wave illusion

Fig. 8 Shaded diamonds illusion

Shaded Diamonds Another optical illusion firstly presented by Watanabe is *shaded diamonds* [37]. In a tiling arrangement of the same shaded diamonds, as it can be seen in Fig. 8, the diamonds on the left appear lighter than the ones on the right side. This luminance gradient from right darker-appearance diamonds to the left lighter-appearance diamonds is an illusion as a result of human vision system function. While the neural processing in the retina suppresses slight changes in lightness, the contrast step from darker corner of each diamond to the lighter corner of the next diamond is transmitted from the human eye to the visual centers in the brain. Brain, by receiving the information regarding changing gradient of light from each layer of diamonds, perceives that each left layer is darker than the next right.

Tusi Motion The thirteenth-century Persian astronomer and mathematician Nasir al-Din Tusi showed how to create a straight line from circular motion. The same effect is used by Arthur Shapiro and Axel Rose-Henig in 2013 [38] in motion illusion as they named it after Nasir al-Din Tusi as Tusi illusion, wherein a circular motion is appeared by moving objects on straight lines. Tusi indicates a principle that *"if a circle rotates on the inner boundary of a circle with diameter of twice its diameter,*

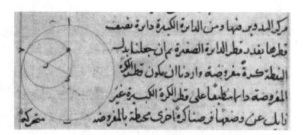

Fig. 9 A part of Tusi manuscript from thirteenth century in the Vatican exhibition [39]

Tusi Movement

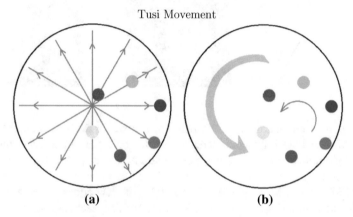

(a)   **(b)**

Fig. 10 Tusi motion illusion: **a** originalmovement of balls on straight lines, **b** the perception of rotational movement of the balls

then every point of the inner circle circumference traces a straight line on a diameter of the big circle." Fig. 9 shows Tusi manuscript from thirteenth century [39] in the Vatican exhibition which indicates the same principle. In Tusi illusion, a number of balls appear rotating around a circle. However, by following each ball, it can be observed that the ball swings on a line but does not rotate. Tusi motion was entered into the 2013 Illusion Contest and won the prize of the best illusion of the year [40]. Figure 10 shows Tusi motion illusion.

Stepping Feet It is a motion optical illusion that was firstly described by Stuart Anstis in 2003 [41]. A red rectangle and a blue one move horizontally passing black and white bars (see Fig. 11). Dependent on the width of the black and white bars in the background, their movement look like antiphase; however, it is completely synchronized. This effect has been shown clearly in the following video link: https://vimeo.com/34381164. Stepping feet illusion can be observed if the two moving rectangles are with different luminance (black and white or for the colored case like blue and red). Also, the background bars should have different luminance.

The reason of this illusion can be explained by the brightness contrast between the rectangle and the background around its head and tail. Let us see when the head and tail edges of rectangles are on light bars, like as in (b) in the Fig. 11. As it can

Fig. 11 "Stepping feet," a motion illusion

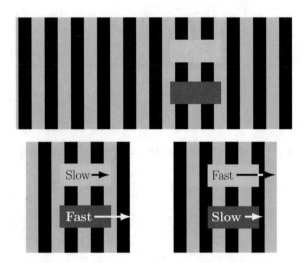

be seen in this condition, the contrast mentioned above for the blue rectangle is little because of their both bright luminance (blue and white). The perceived motion is affected by contrast, and hence, the blue rectangle seems slower than the red one. The red one has more contrast compared to blue one in head and tail with respect to the background. Conversely, in (c) in Fig. 11, the low contrast between the red rectangle and dark background bars makes its movement seem slower than the blue rectangle. This contrast effect and its opposite change for two rectangles create the visual illusion in seeing their movement as they appear antiphase with respect to each other.

Freezing Rotation Illusion Max R Dürsteler created a motion illusion that won the first prize of the best illusion of 2006 [42]. The term "Freezing rotation illusion" was used by him to designate illusory perception of a strong decrease in angular velocity of a rotating object. Two objects rotate around the same center. The front object has uniform continuous angular speed, while the background object has pendulum motion. Interestingly, while the back object swings in the direction opposite to the front one, the rotation of front one appears faster. Upon the same direction swing of the back object, the front object seems to reduce its speed and seems almost frozen. This illusory angular speed oscillation is perceived while the front object has constant angular speed.

Optical illusions are much more than the cases as mentioned above and out of the capacity of this book chapter. On the other hand, the text ability in showing the movement optical illusions is limited. Here, we have provided few significant and famous of them. Among the optical illusions, this chapter focuses on Mach Band illusion which inspires us for the introduced image enhancement technique.

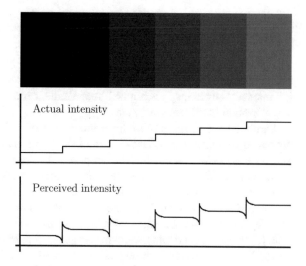

Fig. 12 Mach bands illusion: On the edges of intensity, human eye perceived the luminance with undershoot and overshoot

5 Mach Bands Illusion

An important characteristic of human vision is that the brightness perceived by eye is not a simple linear function of intensity. As a matter of fact, when human eye roams around the edges of two different intensities, it overshoots or undershoots by passing through the edges. This phenomenon has been demonstrated very well in *Mach bands* [43] that is named after Ernst Mach, who firstly discovered this illusion by human vision in 1865. Mach bands are the set of uniform luminance gray strips of different intensity levels, which are arranged sequentially beside each other. While each stripe has uniform flat illumination, it looks like as gradient curved in their luminance. At each strip, the area adjacent to the darker strip appears lighter and the area adjacent to the lighter strip appears darker. Figure 12 demonstrates Mach bands illusion, wherein the human eye perceives the intensity on the edges with an undershoot and overshoot in color intensity. An explanation for Mach bands illusion has been mentioned in the literature by the effect of light reflection from bands in striking origin-surround receptive fields in the eye. However, it is challenged by Purves team in 1999 arguing that it is the result of the characteristics of reflected light and its frequency [44].

6 Image Enhancement Inspiration from Human Visual Illusion

As observed in Match Bands phenomenon in preceding section, human visual system centers of brain highlight the intensity edges by an illusory overshoot and undershoot effect on them. It is by the mean of modification of real intensity values during passing from one intensity level to another intensity level. This modification leads to the illusion of intensity shades in uniform intensity levels around the edges in a way in which the darker side appears darker and the lighter side appears brighter. This illusion leads to an enhanced vision with a more clear perception of luminance changes. The action as mentioned above of the brain visual centers in highlighting the intensity edges inspires a novel image enhancement technique by assisting the visual system in the same manner.

Figure 12 shows the real uniform intensity around the Mach bands' edges and enhanced perception by visual centers of the brain via the addition of undershoots and overshoots to the edges. The proposed inspired *visual illusion-based image enhancement (VIIE)* artificially introduces similar overshoot and undershoot effect around the edges of an image. The additive overshoot and undershoot assists human visual system in its natural accentuation of the edges, and therefore resulting in a visually enhanced image. Practically, VIIE is performed in three steps: (i) edge detection, (ii) overshoot and undershoot masks generation around the detected edges, and (iii) addition of the overshoot/undershoot mask to the image. Figure 13 demonstrates VIIE application on a simple image composed of three bands of gray. As it can be seen, first the two vertical edges located between gray bands are detected. Thereafter, the overshoot and undershoot additive masks are made on the detected edges. The mask is added to the image to highlight the edges and conclude the enhanced image.

7   Morphological Image Enhancement Based on Visual Illusion

The preceding section presented the novel idea of visual illusion-based image enhancement (VIIE) inspired from brain visual centers action. Here, morphological filters (MFs) are deployed to be applied to the image through an algorithm to obtain the VIIE mask. To accomplish this goal, first, the edges of the images are detected, then the VIIE mask is made of the detected edges and it is to the image. As a result, the image is enhanced. Figure 13 illustrates the general block diagram of VIIE. It is remarkable to remind that the effect of MFs on the images is through the edge-resembled corners and valleys of the image surface. Thereof, any deployed morphological algorithm for VIIE can be performed with skipping the first step of edge detection.

In order to make the morphological VIIE mask, the overshoot/undershoot (O/U) shape modification should be applied on edge-esembled surface points of the image. For the sake of simplicity, let us consider the local cross section of the image around an edge. The image cross section should be in any direction different from the edge direction to cross the edge. As explained in the preceding section, VIIE effect on an edge is an overshoot and an undershoot, respectively, on the upper and lower sides of the edge as is shown in Fig. 14. Because, the introduced O/U modifications have a particular morphological shape and are independent of the local structure

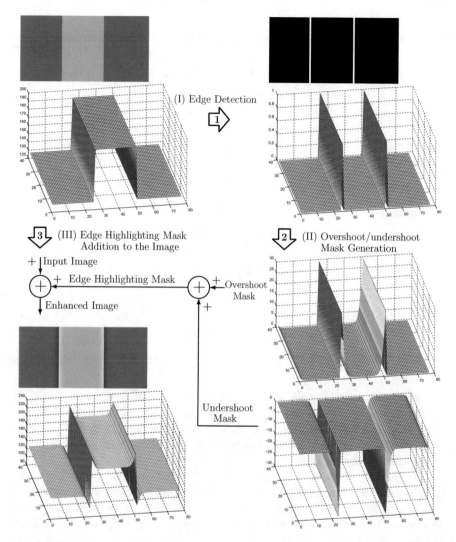

Fig. 13 Visual illusion based image enhancement general block diagram

of the image; the desired shape effect is applied through the structuring element in an algorithm of morphological opening and closing filters. The reason for the deployment of opening and closing is in their characteristic effect on the edge-resembled points of the image/signal surface. In morphological opening and closing, such points that are morphologically tight and not accessible by structuring element are replaced by the corresponding surface of the structuring element.

Figure 15 by using the aforementioned cross section of the image shows the direct appearance of the structuring element surface around the edge points in the morphological opening and closing operations.

Fig. 14 VIIE effect on an edge of an image has been shown through the cross section of the image

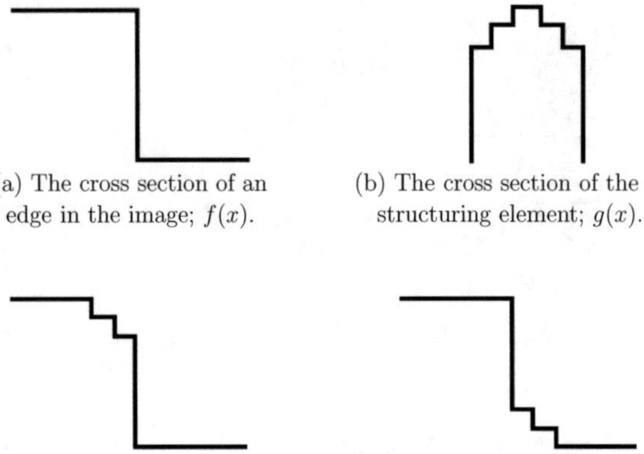

Fig. 15 The direct appearance of the structuring element surface around the edge-resembled points, in the opening (*up*) and closing (*bottom*) filtering

(a) The cross section of an edge in the image; $f(x)$.

(b) The cross section of the structuring element; $g(x)$.

(c) Opening; $(f \circ g)(x)$.

(d) Closing; $(f \bullet g)(x)$.

Therefore, opening and closing are constructing breaks of the algorithm for introducing the O/U modifications on the upper and lower sides of the edges. However, as it can be seen in the Fig. 15, what appears on the edge is the corresponding surface of the structuring element, not in the direction of sharpening the edge but oppositely in the direction of softening.

Therefore, firstly the structuring element $g(x, y)$ is provided, the one with a proper surface which its cross section in all directions matches the required overshoot and undershoot by the proposed VIIE. Then, the additive overshoot and undershoot images are, respectively, obtained by the subtraction of opening and closing from the image as follows:

$$o(x, y) = f(x, y) - (f \circ g)(x, y) \tag{1}$$

$$u(x, y) = f(x, y) - (f \bullet g)(x, y) \tag{2}$$

where $\circ$ and $\bullet$ are, respectively, morphological opening and closing. The VIEE corrective O/U image, $V_{ou}(x, y)$, is obtained by summation of the obtained overshoot and undershoot images:

$$v_{OU}(x, y) = o(x, y) + u(x, y) \tag{3}$$

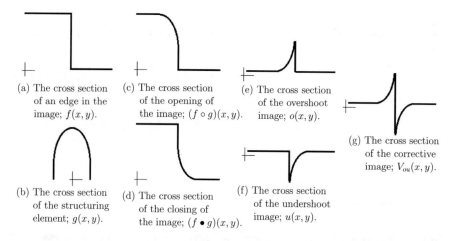

(a) The cross section of an edge in the image; $f(x, y)$.

(b) The cross section of the structuring element; $g(x, y)$.

(c) The cross section of the opening of the image; $(f \circ g)(x, y)$.

(d) The cross section of the closing of the image; $(f \bullet g)(x, y)$.

(e) The cross section of the overshoot image; $o(x, y)$.

(f) The cross section of the undershoot image; $u(x, y)$.

(g) The cross section of the corrective image; $V_{ou}(x, y)$.

Fig. 16 The cross section of an image around the edge in different stages of the morphological algorithm for obtaining the corrective VIIE O/U image

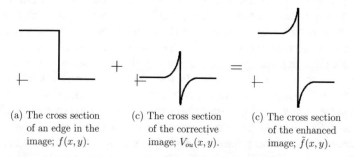

(a) The cross section of an edge in the image; $f(x, y)$.

(c) The cross section of the corrective image; $V_{ou}(x, y)$.

(c) The cross section of the enhanced image; $\tilde{f}(x, y)$.

Fig. 17 The cross section of the image around an edge in morphological visual illusory image enhancement (MVIIE)

Figure 16 depicts the cross section of the image around an edge in different stages of the morphological algorithm for obtaining the corrective O/U image.

Finally, the *morphological visual illusory image enhancement (MVIIE)* is concluded by the addition of the visual illusory O/U image to the image:

$$f_{MVIIE}(x, y) = f(x, y) + v_{OU}(x, y), \tag{4}$$

as is shown in the cross-sectional view in Fig. 17. The formulation of MVIIE can be simplified as follows:

$$f_{MVIIE}(x, y) = 3f(x, y) - \left((f \circ g)(x, y) + (f \bullet g)(x, y)\right) \tag{5}$$

Figure 18 shows the simplified block diagram of MVIIE.

Fig. 18 The simplified
block diagram of
*morphological visual
illusory image enhancement*

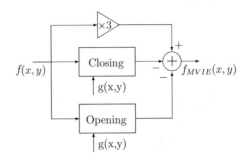

8   Results and Discussion

In preceding two sections, first, the visual illusory image enhancement based on
Mach bands illusion phenomenon was clarified. Then, MVIIE, the morphological
technique for image enhancement based on visual illusion, has been presented. Here,
the newly introduced MVIIE is analyzed over a range of images of different types
such as flower vector, Barbara, bridge, building, animated girl (Elsa in Disney's
Frozen), fighter above snowcapped mountains, woman image, text image, and Lenna.
However, measurement of image enhancement without any ideal reference for comparison is a difficult task. Apart from measuring the enhancement, just determining
whether any enhancement is done is a task dependent on the observer's perception
of desirable information in the image after and before enhancement process. Here,
the human eye perception-based measurement of image enhancement (EME) [9] has
been used for the evaluation of the proposed image enhancement technique. EME is
a measure of image contrast based on Weber–Fechner law in human perception. It
was first practically developed as a contrast measure by Agayan in [45]. Later, the
measure was modified and developed by [46] as follows:

$$EME = \frac{1}{k_1 k_2} \sum_{l=1}^{k_1} \sum_{l=1}^{k_2} 20 log \frac{I_{max;k,l}^w}{I_{min;k,l}^w} \tag{6}$$

wherein, the under work image $x(n, m)$ is split into $k_1 k_2$ subimages $w_{k,l}(i, j)$ of sizes
$l_1 \times l_2$. $I_{min;k,l}^w$ and $I_{max;k,l}^w$ are, respectively, minimum and maximum of the subimage
$w_{k,l}$.

The first evaluation of the proposed image enhancement is done over a vector
image of a rose flower. It has been shown before and after enhancement in Fig. 19.
The image size is 1024×768. The deployed morphological structuring element is an
elliptic paraboloid surface with domain range of 30×30 size as is shown in Fig. 20.

The enhancement of the image can be clearly observed in terms of visual effect in
Fig. 19. The image after MVIIE not only has higher contrast, more clear parts, and
better visual appearance, but also it seems more realistic like a three-dimensional
visualization of the image before MVIIE, what we call it a 3D effect. We have applied

The Original Image and the Morphological Visual Illusory Enhanced Image

Fig. 19 A vector image of a rose flower; *left* and *right*, before and after morphological visual illusory image enhancement, respectively

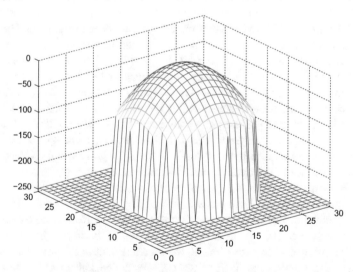

Fig. 20 The 30 × 30 pixels domain elliptic paraboloid structuring element for morphological visual illusory enhancement of the rose image shown in Fig. 19

the EME image quality measure with different sizes of acquired sub-blocks; 10 × 10, 20 × 20, ..., 100 × 100, over the image before and after enhancement. The result has been shown in the Fig. 21. As it can be seen, EME measure indicates the higher quality of the enhanced image for all the evaluation of different subimage blocks.

Fig. 21 EME the rose image before and after MVIIE, respectively, versus the subimage blocks size, l

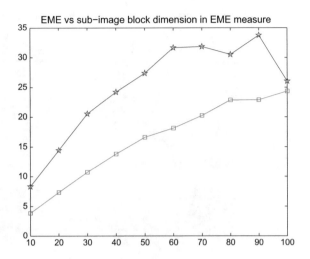

In particular for the $l = 60$ and $l = 90$ which are, respectively, double and triple the structuring element dimensions, EME shows a comparable high-quality effect.

We have acquired Barbara image of 256×256 as the 2nd image for the evaluation of MVIIE. It is a grayscale image with natural face parts, hairs, cloth tissues, and a detailed background. Figure 22 shows the comparison of the original and enhanced images by MVIIE. An epileptic paraboloid structuring element of a base surface size of 7×7 has been deployed (Fig. 22, bottom, left). The numerical evaluation of enhancement by EME for subimage block sizes of $l = 5, 10, \ldots, 40$ has been obtained and shown in Fig. 22, bottom-right. Apart from clear visible enhancement, the EME values approve the enhancement by superior difference of 5–10 EME value.

As a different case of evaluation, we have applied MVIIE on 256×256 bridge image. Bridge image is a grayscale image with a lot of edge and spacial details and fast variations in luminance in trees' leafs in the back, grasses at the riverside, and on the water surface of the river. Figure 23 shows the original and enhanced images by MVIIE. Similar to the enhancement of Barbara image, an epileptic paraboloid structuring element of the base surface size of 7×7 has been deployed (Fig. 23, bottom-left corner). EME evaluation of enhancement has been done for subimage block sizes of $l = 5, 10, \ldots, 45$ as is shown in Fig. 22, bottom-right corner. Here, despite Barbara image and rose image, the enhancement is slightly clear in terms of visual effect. As well, the EME values approve the enhancement by the superior difference of EME values around 5. The enhancement can be seen more clearly on the water surface and white guard of the bridge.

The proposed MVIIE has been evaluated on 256×256 gray-scaled building image too. The building image, before and after the enhancement, has been shown in Fig. 24. Here, an epileptic paraboloid structuring element of the base surface size of 9×9 has been used as shown in left-bottom of Fig. 24. Building image has more areas with uniform luminance such as window glasses and the walls, and parts with details of fast variations in luminance as the leafs of the tree in front of the building.

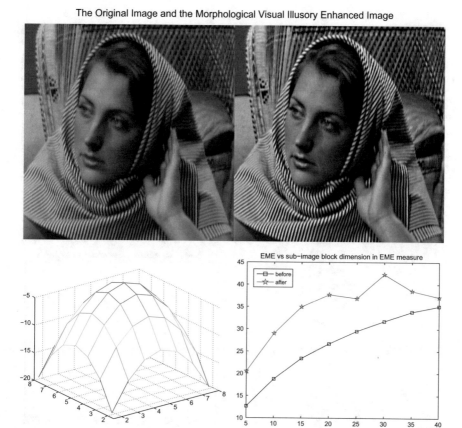

The Original Image and the Morphological Visual Illusory Enhanced Image

Fig. 22 Barbara image; *top-left* and *top-right*, before and after morphological visual illusory enhancement, respectively. *Bottom-left*; the 7×7 pixels domain elliptic paraboloid structuring element deployed for MVIIE. *Bottom-right*; The EME versus the subimage blocks sizes, l used in EME formula

As expected, because of more areas with uniform luminance, the enhancement by MVIIE is more efficient on the building image. It can be seen in Fig. 24 an apparent enhancement effect, while the luminance of image preserved, wall, windows, and even the tree leaves are visually improved. In order to have numerically approved and quantified enhancement, EME has been applied for subimage block sizes of $l = 5, 10, \ldots, 100$ and depicted in Fig. 24, bottom-right corner. In average, 30 EME values are observed in improvement of the image quality.

Another image that we have used for the evaluation of MVIIE is Elsa image from Disney Frozen cartoon movie as shown in Fig. 25. The image size is 400×400. We have used similar elliptic paraboloid structuring element but of size 15×15. The structuring element has been shown in bottom-left corner of the Fig. 25. The EME evaluation of the MVIIE over the Elsa image has been shown in the bottom-right

The Original Image and the Morphological Visual Illusory Enhanced Image

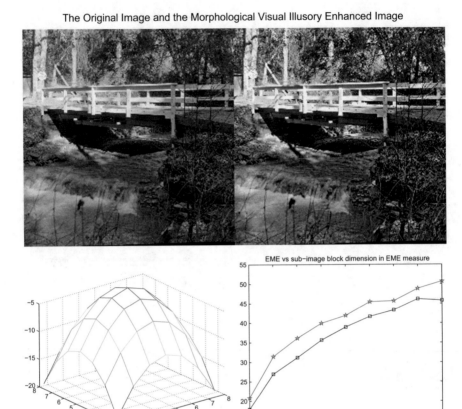

Fig. 23 Bridge image; *top-left* and *top-right*, before and after morphological visual illusory enhancement, respectively. *Bottom-left*; the 7×7 pixels domain elliptic paraboloid structuring element deployed for MVIIE. *Bottom-right*; The EME versus the subimage blocks sizes, l used in EME formula (*square-curve* before, *pentagram-curve* after)

corner of the Fig. 25. Although the chosen image of Elsa has a high quality, visually we can observe further improvement of the image quality after MVIIE. The effect can be observed better in Elsa's eyes and hairs. As well, EME indicates an average increase of 3 in image quality as is shown for different sizes of subimage blocks from 5 to 100. Apart from the above-described images, the images of airplane, woman, text, and Lenna have been analyzed by the proposed method, and the effect before and after the enhancement with EME evaluation curve is shown all together in Fig. 26. Similarly, a 5×5 elliptic paraboloid structuring element has been employed for all the images in Fig. 26. EME evaluation has been done for square subimage blocks of 5, 10, 15, and 20 pixels at each dimension that is up to four times the dimension of the structuring element. Visually and numerically, further enhancement of the images is verified.

The Original Image and the Morphological Visual Illusory Enhanced Image

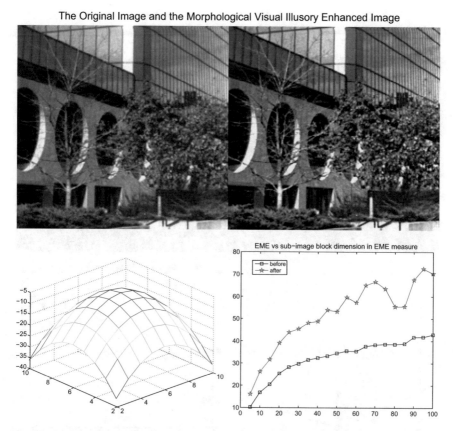

Fig. 24 Building image: *top-left* and *top-right*, before and after morphological visual illusory enhancement, respectively. *Bottom-left* the 9×9 domain elliptic paraboloid structuring element deployed for MVIIE. *Bottom-right* the EME versus the subimage blocks sizes, l is used in EME formula (*square-curve* before, *pentagram-curve* after)

9 Summary

This chapter presents a novel image enhancement technique based on the human visual system illusory perception of luminance changes around the intensity edges. In Mach bands illusion, the human visual system percepts a uniform luminance in the form of intensity overshoot and undershoot by passing through edges from low- to the high-intensity gradient. The observed intensity overshoot and undershoot are due to the illusory brain action to aid visual perception of the edges. Since, in an image the edges include the main data contents, this illusory brain action leads to highlight the image edges and resulting image with higher visual quality. Thus, we proposed an image enhancement technique inspiring from this illusion for having increasing image sharpening and contrast. The visual illusory image enhancement is performed here by deploying morphological filters for modifying the morphological features of

The Original Image and the Morphological Visual Illusory Enhanced Image

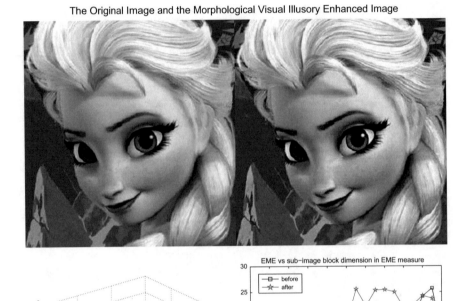

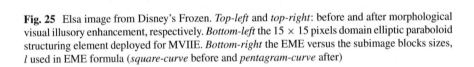

Fig. 25 Elsa image from Disney's Frozen. *Top-left* and *top-right*: before and after morphological visual illusory enhancement, respectively. *Bottom-left* the 15 × 15 pixels domain elliptic paraboloid structuring element deployed for MVIIE. *Bottom-right* the EME versus the subimage blocks sizes, *l* used in EME formula (*square-curve* before and *pentagram-curve* after)

the image and introducing the overshoot and undershoot effect around the edges in the same manner as human visual system behaves in Mach bands illusion. The newly introduced technique is called here *morphological visual illusory image enhancement (MVIIE)*. It has been analytically evaluated in the enhancement of a range of standard images. The enhancement can be clearly observed by visual comparison of the images before and after MVIIE. As well, the EME measure of image quality approves the performed enhancement, while its quantified metric shows a considerable increment. The employment of human visual illusion for image enhancement as introduced in this chapter draws a novel research orientation in brain action inspiration for artificial intelligence applications.

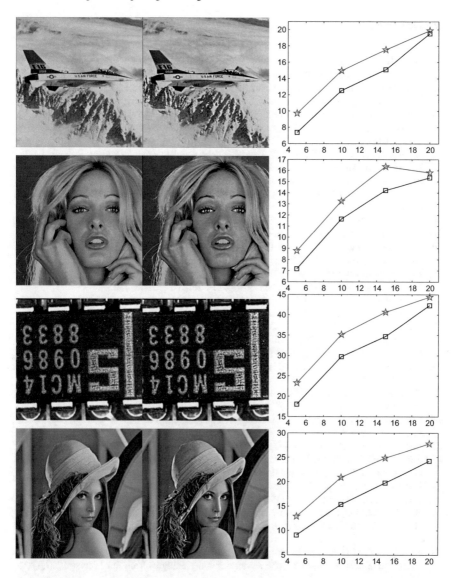

Fig. 26 MVIIE has been applied and evaluated on the *left column images*. The *middle column* shows the images after enhancement. Same 5×5 elliptic paraboloid structuring element has been employed for all the images. The *right column* shows the EME evaluation of the enhancement for square subimage blocks of 5, 10, 15 and 20 pixels at each dimension

References

1. Jain AK (1989) Fundamentals of digital image processing. Prentice-Hall, Inc
2. Shapiro L, Haralick R (1992) Computer and robot vision. Addison-Wesley, Reading 8
3. Morishita K, Yokoyama T, Sato K (1992) U.S. Patent No. 5,150,421. U.S. Patent and Trademark Office, Washington, DC
4. Jain R, Kasturi R, Schunck BG (1995) Machine vision, vol 5. McGraw-Hill, New York
5. Bovik AC (1995) Digital image processing course notes. Dept. of electrical engineering, U. of Texas, Austin
6. Aghagolzadeh S, Ersoy OK (1992) Transform image enhancement. Opt Eng 31(3):614–626
7. Greenspan H, Anderson CH, Akber S (2000) Image enhancement by nonlinear extrapolation in frequency space. IEEE Trans Image Process 9(6):1035–1048
8. Cheng HD, Shi XJ (2004) A simple and effective histogram equalization approach to image enhancement. Digit Sig Process 14(2):158–170
9. Agaian SS, Silver B, Panetta KA (2007) Transform coefficient histogram-based image enhancement algorithms using contrast entropy. IEEE Trans Image Process 16(3):741–758
10. Ibrahim H, Kong NSP (2007) Brightness preserving dynamic histogram equalization for image contrast enhancement. IEEE Trans Consum Electron 53(4):1752–1758
11. Kong NSP, Ibrahim H (2008) Color image enhancement using brightness preserving dynamic histogram equalization. IEEE Trans Consum Electron 54(4):1962–1968
12. Xu Y, Weaver JB, Healy DM Jr, Lu J (1994) Wavelet transform domain filters: a spatially selective noise filtration technique. IEEE Trans Image Process 3(6):747–758
13. Huang K, Wang Q, Wu Z (2004) Color image enhancement and evaluation algorithm based on human visual system. In: Proceedings of IEEE International Conference on Acoustics, Speech, and Signal Processing, May 2004 (ICASSP'04), vol 3. IEEE, pp III–721
14. Huang KQ, Wang Q, Wu ZY (2006) Natural color image enhancement and evaluation algorithm based on human visual system. Comput Vis Image Underst 103(1):52–63
15. Demirel H, Anbarjafari G (2011) Image resolution enhancement by using discrete and stationary wavelet decomposition. IEEE Trans Image Process 20(5):1458–1460
16. Pelz J (1993), In: Stroebel L, Zakia RD (eds) The focal encyclopedia of photography, 3E edn. Focal Press, p 467
17. Bach M, Poloschek CM (2006) Optical illusions. Adv Clin Neurosci Rehabil 6(2):20–21
18. Baldo MVC, Ranvaud RD, Morya E (2002) Flag errors in soccer games: the flash-lag effect brought to real life. Perception 31(10):1205–1210
19. Bonnet C (1775) Essai analytique sur les facultés de l'âme, vol 1. Cl. Philibert
20. Rorschach H, Morgenthaler W (1921) Rorschach psychodiagnostik. Ernst Bircher, Bern
21. Bach M (2006) Optical illusions and visual phenomena
22. Hermann L (1870) Eine erscheinung simultanen contrastes. Pflügers Arch Eur J Physiol 3(1):13–15
23. Baumgartner G (1960) Indirekte grössenbestimmung der rezeptiven felder der retina beim menschen mittels der Hermannschen gittertäuschung. Pflügers Arch Eur J Physiol 272(1):21–22
24. Fraser J (1908) A new visual illusion of direction. British J Psychol 1904–1920, 2(3):307–320
25. Cucker F (2013). Manifold mirrors: the crossing paths of the arts and mathematics. Cambridge University Press
26. Zollner F (1860) Ueber eine neue Art von Pseudoskopie und ihre Beziehungen zu den von Plateau und Oppel beschriebenen Bewegungsphanomenen. Ann. Phys. 186(7):500–523
27. Pierce AH (1898) The illusion of the kindergarten patterns. Psychol Rev 5(3):233
28. Gregory RL, Heard P (1979) Border locking and the Café Wall illusion. Perception 8(4):365–380
29. Pinna B (1990) Il dubbio sull'apparire. Upsel
30. Pinna B, Brelstaff GJ (2000) A new visual illusion of relative motion. Vision Res 40(16):2091–2096

31. Chi MT, Lee TY, Qu Y, Wong TT (2008) Self-animating images: illusory motion using repeated asymmetric patterns. ACM Trans Graph (TOG) 27(3):62
32. Kitaoka A (2003) Rotating snakes. a bitmap figure. http://www.ritsumei.ac.jp/akitaoka/index-e.html
33. Kitaoka A (2006) The effect of color on the optimized Fraser-Wilcox illusion. Gold prize at the 9th L'OR+ AL Art and Science of Color Prize
34. Conway BR, Kitaoka A, Yazdanbakhsh A, Pack CC, Livingstone MS (2005) Neural basis for a powerful static motion illusion. J Neurosci 25(23):5651–5656
35. Backus BT, Oruç I (2005) Illusory motion from change over time in the response to contrast and luminance. J Vision 5(11):10
36. Sohmiya S (2006) A wave-line colour illusion. Perception 36(9):1396–1398
37. Watanabe I, Anstis S (1997, March) Contour and shading range affect diamond illusion. In: Investigative ophthalmology & Visual science, vol 38, no 4. 227 East Washington SQ, Philadelphia, PA 19106, Lippincott-raven publ, p 4190
38. Shapiro A, Rose-Henig A (2013) Tusi or Not Tusi. http://www.shapirolab.net/IC2013/TusiorNotTusi_IC2013.pdf
39. http://www.ibiblio.org/expo/vatican.exhibit/exhibit/d-mathematics/Greek_astro.html
40. http://illusionoftheyear.com/?p=2955
41. Anstis S (2003) Moving objects appear to slow down at low contrasts. Neural Netw 16(5–6):933–938
42. Dürsteler MR (2006) The freezing rotation illusion. J Vis 6(6):547–547
43. Mach E (1865) On the effect of the spatial distribution of the light stimulus on the retina. Sitzungsber Math Nat Klasse Kaiserl Akad Wiss 52:303–332
44. Lotto RB, Williams SM, Purves D (1999) Mach bands as empirically derived associations. Proc Natl Acad Sci 96:5245–5250
45. Agaian SS (1999, March) Visual morphology. In: Electronic Imaging'99. International Society for Optics and Photonics, pp 139–150
46. Agaian SS, Panetta K, Grigoryan AM (2000) A new measure of image enhancement. In: IASTED International Conference on Signal Processing and Communication, pp 19–22

Path Generation for Software Testing: A Hybrid Approach Using Cuckoo Search and Bat Algorithm

Praveen Ranjan Srivastava

Abstract Software testing is the process of validating and verifying the computer program, application or product works according to its requirements. Computers have become an integral part of today's society in all the aspects; therefore, it is important that there exist no errors that could compromise safety, security or even financial investment. This chapter, focus on basis path testing as a part of white-box testing to provide code with a level of test coverage by generating all the independent paths for the given code by using its control flow graph. These paths are generated by applying a hybrid algorithm of existing Cuckoo Search Algorithm and Bat Algorithm. The main focus of this chapter is designing of this hybrid algorithm in which basic egg-laying property of cuckoo and echolocation and loudness property of bat is made use of.

Keywords Software testing · Basis path testing · Control flow graph · Cuckoo Search Algorithm · Bat Algorithm

1 Introduction

Software industry has been gaining importance since two decades until today. Software systems are almost used in all kinds of applications and various organizations everywhere around the world. In the present scenario, where customer is given the utmost importance in any business, clients are especially required to be provided with information about the quality of the product or service under test. Software testing [1] acts as a powerful tool in such situations. Software testing provides an independent view of the software to allow the organization to appreciate and understand the risks of software implementation. Software testing has become one of the

P.R. Srivastava (✉)
Information System and System Area, Indian Institute
of Management (IIM) Rohtak,
Rohtak, India
e-mail: praveenrsrivastava@gmail.com

© Springer International Publishing AG 2017
S. Patnaik et al. (eds.), *Nature-Inspired Computing and Optimization*,
Modeling and Optimization in Science and Technologies 10,
DOI 10.1007/978-3-319-50920-4_16

most important and integral parts of the software development life cycle. It is also a very important contributing factor in software quality assurance as it is in regard to identifying faults in software. Software testing methods are traditionally divided into white-box testing [2] and black-box testing [3, 4]. This chapter discusses basis path testing which is in concern with white-box testing.

Structural testing [2] (or white-box or path testing) is the testing of the internal structures of the code (code path identification) which in turn identifies the effective test cases (paths) [5]. Through this method of test design, many errors and problems could be uncovered. The approach for basis path testing [4], in this chapter, is based on metaheuristic principles [6] and has the advantage of both Cuckoo Search Algorithm [7–9] and Bat Algorithm [10–12]. Both algorithms, individually, have shown superiority over many other metaheuristic algorithms over a wide range of applications. The goal of this chapter is to generate all the independent paths giving control flow graph [2] and cyclomatic complexity [1] as an input. The number of linearly independent paths is equal to the cyclomatic complexity of the program. It determines the number of test cases that are necessary to achieve through test coverage of a particular module.

This chapter discusses the designing of the Hybrid of Cuckoo and Bat Algorithm and its implementation to various programs including open source. Result got by this chapter is very encouraging for illustration purpose and this chapter uses binary search code.

This chapter is divided into 6 sections. Section 1 discussed the introduction about software testing. Sect. 2 discusses the background work, and Sect. 3 discusses about the Cuckoo Algorithm and Bat Algorithm and gives a brief overview of these algorithms. Section 4 refers to the Hybrid Algorithm and its pseudocode and discusses in detail about this algorithm. Section 6 analyses the upper hand of the hybrid algorithm over Cuckoo and Bat algorithms individually. Section 7 concludes this chapter.

2   Related Work

The hybrid algorithm is inspired by the behaviour of Cuckoo shown in the reproduction and the echolocation behaviour of bats in finding their prey. These algorithms and other metaheuristic algorithms have been developed to carry out global search with three main advantages: solving problems faster, solving large problems and obtaining robust algorithms [7, 12]. For solving path problem, researcher used initially Genetic Algorithms (GA) [13] and Particle Swarm Optimization (PSO) [14] which are also some typical examples of the metaheuristic algorithms [15]. The efficiency of metaheuristic algorithms can be attributed to the fact that they imitate the best features in nature, especially the selection and survival of the fittest in biological systems which have evolved by natural selection over millions of years. The Cuckoo Search Algorithm has been proved efficient in solving several structural engineering optimization problems including stochastic test functions [16], and preliminary studies show that it is very promising and could outperform existing algorithms such as Genetic Algorithm and Particle Swarm Optimization. Firefly algorithm [17] has been applied to obtain the optimal solution for stochastic test functions, but Cuckoo Search Algorithm helps in obtaining it in simpler way.

Researches have already been done in optimal path generation [12] using ant principal, and in this paper, author tried to find out all the paths, but computational complexity and repetition are very high.

The methods such as Ant Colony Optimization (ACO) Algorithm [8–10], Genetic Algorithm (GA) [12, 13] and Search-based test data generation [14]. All these methods try to generate test data in an automated manner to facilitate the task of software testing. These approaches often use swarm intelligence [14, 15]. Recently for path generation optimization, a method based on Cuckoo Search (CS) has been used [16], and good results were achieved. But node redundancy was still an issue.

In 2011, an approach combining Genetic Algorithm along with tabu search [17] uses control-dependent graph. The objective function used in the approach fails to consider a number of factors such as distance from initial and final node and criticality. Another approach [18] uses Intelligent Water Drop Algorithm which attempted to minimize the repetitions and give the complete coverage. This algorithms promise complete traversal of the test domain, but lack in optimizing the test sequences generated by the respective approaches. For generating all paths through Firefly [19], Intelligent Water Drop [18], Bat Algorithm [12] and many more similar techniques [20] but computational complexity is still high.

In spite of a lot of work done for finding optimal or favourable test sequences, efforts have resulted in very little success due to the usage of various nature-inspired techniques and lack of proper objective function. The proposed approach tries to resolve the problem towards exhaustive state coverage and optimal sequence generation. In this regards, this chapter proposed more 'optimized algorithm' because it uses less number of variables in calculating the fitness function in an efficient way and it combines beauty of two nature-inspired techniques, and through this, repetition ratio is also minimized.

In the next section of the chapter, basic explanation of Cuckoo Search Algorithm and Bat Algorithm is given which are the motivational algorithms for this chapter.

3 Motivational Algorithm

3.1 Cuckoo Search Algorithm

Cuckoo Search is an optimization algorithm developed by Yang and Deb [7]. It is based on the parasitic behaviour of some Cuckoo species that lay their eggs in the nests of host birds of other species. Some of the eggs laid by Cuckoos are detected by host bird of that nest, and these alien eggs are either thrown away or the host bird abandons that nest to build a new nest elsewhere. Over the time, some Cuckoo species such as New World brood-parasitic Tapera have developed capabilities in which eggs laid by them are very close to the colours and patterns of the eggs of some selected host bird species [21]. The following representation scheme is chosen by Cuckoo Search Algorithm.

Each egg in a nest represents a solution, and a cuckoo egg represents a new solution. The aim is to use the new and possibly better egg to replace a not-so-good egg of cuckoo in the nests. However, this is the basic case, i.e. one cuckoo per nest, but the extent of the approach can be increased by incorporating the property that each nest can have more than one egg which represents a set of solutions.

Cuckoo Search is based on three idealized rules:

- Every Cuckoo lays one egg at a time in an arbitrarily selected nest,
- The best nests with superior eggs, i.e. the cuckoo eggs that are not detected by host bird as alien, will hatch and are carried over to the next generation.
- It is possible that the egg laid by cuckoo is discovered by host bird, and the discovered eggs are thrown away by the host bird. Thus, the discovered solutions are removed from future calculations.

Additionally, Yang [22] found out that the random-walk-style search used in CS should be performed via Levy flights as it is superior to simple random walk and is much closer to original behaviour of cuckoo.

Levy Flights

The purpose of Levy flight is to furnish a random walk, while Levy distribution is responsible for generating the random step length. The Levy distribution can be written as follows:

$$Levy \sim u = t^{-\lambda}, (1 < \lambda \leq 3)$$

It has an infinite variance with an infinite mean. The steps form a random-walk process with a power law step-length distribution with a heavy tail. Some of the new solutions should be generated by Levy flight around best solution obtained so far, and this will speed up the local search. To be safe from the problem of local optimum [23], the locations should be far enough from present best solution. This can be achieved by applying far-field randomization to calculate substantial fraction of new solutions.

3.2 Bat Algorithm [12]

This section throws light upon the behaviour of bats, the species from which the algorithm has been inspired from which is directly taken by following reference.

A. Behaviour of Bats

The idea of bat is directly taken by Prof. Yang's research paper. Bats are fascinating animals [11]. They are the only mammals with wings and also have advanced capability of echolocation. Most bats use echolocation to a certain degree among which microbats are a popular example as they use echolocation extensively, while megabats do not. This echolocation is a type of sonar to detect prey, avoid obstacles and locate their roosting crevices in the dark. These bats emit a very loud sound pulse and listen for the echo that bounces back from the surrounding objects. These pulses vary in properties and, depending on the species, change with their hunting strategies.

The loudness also varies from the loudest when searching for prey and decreases when homing towards the prey. The travelling range of such short pulses varies from a few metres, depending on the actual frequencies. Such echolocation behaviour of microbats can be defined in such a way that it can be associated with the objective function to be optimized, which makes it possible to formulate new optimization algorithms.

B. The Bat Algorithm

The Bat Algorithm was developed by Yang [11] and Srivastava et al. [12]. It is a nature-inspired metaheuristic algorithm ideal for optimizing mathematical problems of NP-hard complexity. He proposed that on idealizing the echolocation characteristics of microbars described above, we can develop Bat Algorithms with a few assumptions which are as follows [11]:

All bats use echolocation to sense distance, and they also 'know' the difference between food/prey and background barriers in some magical way.

Bats fly randomly with velocity vi at position xi with a fixed frequency fi and loudness A0 to search for prey. They can automatically adjust the frequency of their emitted pulses depending on the proximity of their target.

Even though the loudness can vary in many ways, we assume that the loudness varies from a large positive to a minimum constant value Amin.

The bats move about in a fashion as if each transition represents a local solution to close in the gap between them and their prey. A final state in a given state diagram represents a prey. Diagrams having multiple final states mean multiple prey for the bats, giving bats even the freedom to choose that prey which can be most easily captured [12].

As it is an iterative process, each bat moves about on its own and searches for prey. After each iteration, the paths followed by the bats are ranked according to a fitness function. The solutions generated are ordered, and the optimum of them is selected. The sequence stops when a bat captures its prey, or as in our case, a final state is encountered.

According to the algorithm, bat movement is performed through a random walk, given by a function dependent upon the loudness of emitted waves.

In the rest of the chapter, we will first outline the Hybrid Cuckoo–Bat Algorithm and then discuss it in detail to show the implementation of the algorithm.

4 Proposed Algorithm

As discussed in the previous section, the use of different random variables invoked
a need for a hybrid algorithm to be designed involving the benefits of the two best
metaheuristic algorithms, Cuckoo Search and Bat algorithms. Proposed schema is
depicted in Fig. 1, and the corresponding pseudocode is shown in Fig. 2.

The detailed explanation of the algorithm is given as follows:

Step 1.

Get the cuckoo and start it from nest 0 (node 0 in graph). It checks whether the
nest has optimal egg to be replaced. If the nest has only one egg, it moves ahead to
next nest as the probability of discovery of cuckoo egg is 1.

Step 2.

If cuckoo reaches a nest which has more than one egg (say, decision node), it
stops there to lay its egg. Meanwhile, it sends bat from the succeeding nests towards
the target nest (last nest, say, end node) to calculate the fitness of the path. Fitness is
needed so that cuckoo moves to the path where it can lay its maximum eggs.

Step 3.

As many numbers of succeeding nests are there, those bats fly from the nests
adjacent to the decision node. They return to the cuckoo with the minimum loudness

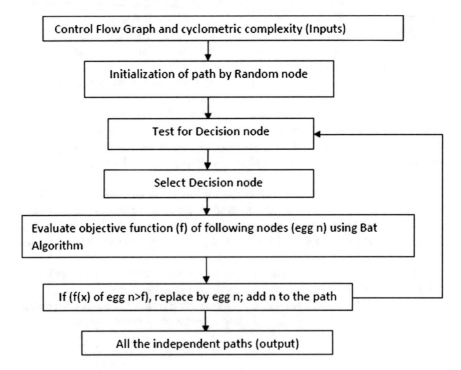

Fig. 1 Architecture for the hybrid algorithm

The pseudo code is given below:
BEGIN
Objective function F(x);
Initially a population of n host nests X(i) (i=0,1,2...n-1);
While(paths<cc)
Get a cuckoo randomly;
Let Cuckoo reach the host nest with eggs>1;
If (eggs in nest n>1)
      {
         Push n into stack;
Evaluate fitness of the paths of succeeding nests(say i,j) by sending bats from i and j towards the target node(say, end) ;
Bats return to cuckoo with their minimum loudness while going to the target node;
Cuckoo uses this measure of loudness to maximize F(x);
         If(F(i)>F(j)){
         Replace j by i;
         Pop n;
Cuckoo moves ahead to find next optimal nest;
         }
         end
}
End
Keep the nests with quality solutions;
Add them to the path traversed by cuckoo;
Prioritize the paths and find the best solution;
End while
Display all the independent paths in decreasing priority;
End

The F(x) in the above pseudo code is the fitness function pf Cuckoo which helps it to evaluate the fitness of a given path and choose the fittest path.

$F(x)= \{A(initial) - pulse\_loud\} * weight$

Where,

A(initial) is the initial loudness with which bat proceeds towards the end node.

Pulse_loud is the value returned by the pulse() of bat which returns the minimum loudness of bat while traversing the path towards the end node.

Weight is the summation of all the nodes encountered in the path taken by bat. This is also a measure returned by bat.

Fig. 2 Pseudocode of proposed approach

they emit while going to the end node. Here, echolocation property of bat is made use of.

Step 4.

Cuckoo makes use of the loudness returned by bats and calculates fitness of the paths. It proceeds towards the fitter path.

Step 5.

Similarly, cuckoo keeps the best solution. It does it by pushing the decision nodes into the stack, and when it reaches the end node, it pops it and starts once again from there to traverse other path.

Step 6.

Cuckoo keeps traversing till it covers all the paths with optimal nests and lays all its eggs.

Hence, all the independent paths are obtained (equal to cc) in the order of their software coverage. These paths are then displayed to the user.

Flow of the proposed approach is given above.

In the next section, implementation of hybrid algorithm on the binary search problem has been demonstrated.

5 Path Sequence Generation and Prioritization

The problems in software testing can be categorized into easy, average and difficult problems. The domain of the problems on which the proposed hybrid algorithm has been tested is given in Table 1.

In this chapter, the working of the algorithm is demonstrated by considering the example of the code for binary search (average problem) for better understanding of the algorithm. The code for binary search is given as follows:

```
Start
1. Int bottom = 0, top = Array.length − 1, mid;
2. r.found = false;
3. r.index = −1;
4. while(bottom <= top) {
5. mid = (top + bottom)/2;
6. if(Array[mid] == key) {
7. r.index = mid;
8. r.found = true; return;
9. }
10. else { if(Array[mid] < key)
11. bottom = mid + 1;
12. else top = mid − 1; }}
End
```

The input is control flow graph as given in Fig. 3.

The weights to be used in the fitness function are input by the user and have a constant value for every decision node and a different constant value for every non-decision node such that the value of weight for a decision node is higher than the value of weight for a non-decision node.

For this example, weights are shown in Table 2.

Table 1 Classification of problems

Easy	Factorial of n numbers
Average	Binary search code
Difficult	Class management system, enrolment system, telephone system

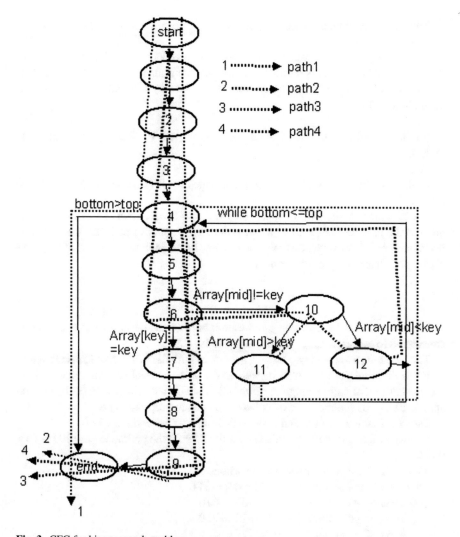

Fig. 3 CFG for binary search problem

Table 2 User-input weights for nodes

Node	Decision/non-decision	Weight
4, 6, 10	Decision node	10
1, 2, 3, 5, 7, 8, 9, 11, 12	Non-decision node	2
End node	–	0

The step-wise explanation of how the algorithm is implemented is as follows:

Step1.

Cuckoo chooses node 1 and checks whether that is decision node. The criteria for deciding a decision node is number of out-edges from the node > 1. Hence, at node 1, there is only 1 out-edge so the cuckoo moves forwards to node 2, and node 1 is added to path P1.

Step 2.

Similarly, node 2 and node 3 are also added to path P1. Now, cuckoo moves to node 4.

Step 3.

Node 4 is a decision node (number of out-edges > 1). Therefore, node 4 is pushed into stack.

The bats are sent on paths of succeeding nodes—1 to the path to end node and the other to the path to node 5. The bats traverse their respective paths till the end node is encountered. For every node they encounter, they calculate the factor of loudness at that node according to the formula:

$$A_i^{t+1} = \alpha \, A_i^t$$

Here, $\alpha = 0.9$, where A being the loudness. The loudness for the end node is considered to be zero. $A_{initial}$ we are taking to be 1000.

The bats return to the cuckoo which sent them with the value of the pulse function which is the minimum loudness as well as the sum of weights of the nodes which are present in that path. Using the information which bat brings with it, the cuckoo applies its fitness function and calculates the most appropriate path.

Step 3.1 Loudness of the end node $= 0$, weight of the end node $= 0$

Fitness of the path with the subsequent node being the end node $= (1000 - 0) * 0 = 0$

Step 3.2 For the path with subsequent node 5:

Loudness of bat at node $5 = 1000 * 0.9 = 900$

Loudness of bat at node $6 = 0.9 * 900 = 810$

Loudness of bat at node $7 = 0.9 * 810 = 729$

Loudness of bat at node $8 = 0.9 * 729 = 656.1$

Loudness of bat at node $9 = 0.9 * 656.1 = 590.49$

So the pulse function for this path $=$ value of minimum loudness $= 590.49$

Weight factor $= 2 + 10 + 2 + 2 + 2 = 18$

Hence, fitness $= (1000 - 590.49) * 18 = 7371.18$

Clearly this path has higher fitness.

Hence, the cuckoo–bat hybrid moves to the node 5.

Step 4.

Node 5 is not a decision node. So cuckoo–bat hybrid adds node 5 to the path and moves forwards to node 6.

Step 5.

Node 6 is a decision node. Hence, it is pushed to stack. Nodes 7 and 10 are not traversed subsequent nodes.

Calculating fitness in a similar way:

Loudness of bat at node $7 = 0.9*1000 = 900$

Node $8 = 0.9*900 = 810$

Node $9 = 0.9*810 = 729$

Sum of weights $= 2+2+2 = 8$

Fitness $= (1000 - 729)*8 = 2168$

Loudness of bat at node $10 = 0.9*1000 = 900$

Node $11 = 0.9*900 = 810$

Node $4 = 0.9*810 = 729$

Sum of weights $= 10 + 2 + 10 = 22$

Fitness $= (1000 - 729)*22 = 5962$

Hence, the path with subsequent node as node 10 is with higher fitness.

Node 6 is added to the path.

Step 6.

Cuckoo–bat hybrid moves to node 10 which is the decision node. Node 10 is pushed to stack. Node 11 and node 12 are its subsequent untraversed nodes.

Loudness of bat at node $11 = 0.9*1000 = 900$

Node $4 = 0.9*900 = 810$

Sum of weights $= 2 + 10 = 12$

Fitness $= (1000 - 810)*12 = 2280$

Loudness of bat at node $12 = 0.9*1000 = 900$

Node $4 = 0.9*900 = 810$

Sum of weights $= 2+10 = 12$

Fitness $= (1000 - 810)*12 = 2280$

Since the fitness of both the paths is equal, any path can be chosen.

Node 10 is added to the path.

Step 7.

Node 12 is not a decision node; hence, it is added to path and cuckoo–bat moves forwards to node 4. Node 4 again is a decision node. Since it has already been traversed, it is not pushed into stack. Fitness of its untraversed subsequent node (end in this case) is calculated.

Fitness of end node $= 0$

Now, node 4 is added to path and cuckoo–bat moves to next node which is end node.

Step 8.

As we have reached the target node (no. of out-edges $= 0$) so it completes path P1.

P1 $= \{1, 2, 3, 4, 5, 6, 10, 12, 4, \text{end}\}$

Step 9.

Pop node 10 from stack.

Copy in path P2 till last decision node (node 10 in this case).

Path P2 $= \{1, 2, 3, 4, 5, 6\}$

Now cuckoo–bat is in node 10, and it is a decision node so it calculates fitness of its subsequent untraversed nodes (node 11 in this case).

Loudness of bat at node $11 = 0.9*1000 = 900$

Node $4 = 0.9 * 900 = 810$

Sum of weights $= 2 + 10 = 12$

Fitness $= (1000 - 810) * 12 = 2280$

Now, Node 10 is added to path.

Step 10.

Cuckoo–bat moves to node 11 which is not a decision node. Hence, it is added to path and it moves to node 4.

Node 4 again is a decision node. Since it has already been traversed, it is not pushed into stack. Fitness of its untraversed subsequent node (end in this case) is calculated

Fitness of end node $= 0$

Now, node 4 is added to path and cuckoo–bat moves to next node which is an end node.

Step 11.

As we have reached the target node(no. of out-edges $= 0$), it completes path P2.

$P2 = \{1, 2, 3, 4, 5, 6, 10, 11, 4, end\}$

Step 12.

Pop node 6 from stack.

Copy in path P3 till last decision node (node 6 in this case).

Path P3= $\{1, 2, 3, 4, 5\}$

Now cuckoo–bat is in node 6, and it is a decision node so it calculates fitness of its subsequent untraversed nodes (node 7 in this case).

Loudness of bat at node $7 = 0.9 * 1000 = 900$

Node $8 = 0.9 * 900 = 810$

Node $9 = 0.9 * 810 = 729$

Sum of weights $= 2 + 2 + 2 = 8$

Fitness $= (1000 - 729) * 8 = 2168$

Now, Node 6 is added to path.

Step 13.

Nodes 7, 8, 9 are not the decision nodes, so they are added to path P3.

Cuckoo–bat moves to end node which is our target node.

Therefore,

$P3 = \{1, 2, 3, 4, 5, 6, 7, 8, 9, end\}$

Step 14.

Pop node 4 from stack.

Copy in path P4 till last decision node (node 4 in this case).

Path P4 $= \{1, 2, 3\}$

Now cuckoo–bat is in node 4, and it is a decision node, so it calculates fitness of its subsequent untraversed nodes (end node in this case).

Fitness of end node $= 0$

Now, node 4 is added to path P4; cuckoo–bat reaches end node which is the target node and it completes its traversal since stack is empty now.

Therefore,

$P4 = \{1, 2, 3, 4, end\}$

All the independent paths generated are given in Table 3 (in order of calculation).

Table 3 Finally generated independent paths

Path number	Nodes in path
Path 1	Start, 1, 2, 3, 4, 5, 6, 10, 12, 4, 5, 6, 7, 8, 9, end
Path 2	Start, 1, 2, 3, 4, 5, 6, 10, 11, 4, 5, 6, 7, 8, 9, end
Path 3	Start, 1, 2, 3, 4, 5, 6, 7, 8, 9, end
Path 4	Start, 1, 2, 3, 4, end

6 Analysis of Proposed Algorithm

This section fulfils its purpose of analysing the empirical results of the presented approach along with a few previous techniques. The techniques suggested by Genetic Algorithm [13], Ant Colony Optimization [20], Cuckoo Search Algorithm [8], the Intelligent Water Drop [18], and Bat Algorithm [13] have already provided feasible solutions to many software testing problems.

The results of the proposed approach have been compiled and statistically tested with those of the above-mentioned techniques, and they show promising results in continuing the research ahead.

Genetic Algorithm (GA)-based approach of test sequence generation is applied by creating objective function for each path. GAs also individually generates subpopulation for each objective function. The transitions, which are traversed and contained in the possible paths, are counted as a fitness value. The overall coverage transition is calculated by selecting the best solution from each subindividual and then executing the state machine diagram [13].

The Ant Colony Optimization Approach [20] can also be used for test sequence generation using state machine diagram where the selection of the transition depends upon the probability of the transition. The value of transition depends upon direct connection between the vertices, heuristic information of the transition and the visibility of a transition for an ant at the current vertex. The proposed algorithm uses only certain parameters of the Bat Algorithm which alone can generate adequate sequences. Moreover, the number of equations involved in calculating the fitness function only increases the efficacy of the algorithm.

Even though the GA algorithm focuses on critical nodes of the diagram, it fails to address the complete coverage problem. On the other hand, the ACO approaches complete coverage in all diagrams it was run on, yet its repetition ratio is fairly poor. The IWD, like ACO, also holds its guarantee to deliver 100% coverage, but it loses in deciding the critical sections which need be tested first. The Bat Algorithm-based approach flies higher than all these approaches by giving complete coverage, keeping in mind the critical sections of the diagram and also keeps a check on the repetition ratio, but repetition ratio was not very encouraging.

The proposed approach has a better result in all the condition as shown in Table 4.

Table 4 Comparison analysis

Algorithm	Case study					
	Class management system		Enrolment system		Telephone system	
	Complete coverage	Repetition ratio	Complete coverage	Repetition ratio	Complete coverage	Repetition ratio
Bat Algorithm	Yes	1.5	Yes	2.14	Yes	2.07
Cuckoo Search	Yes	1.67	Yes	3.0	Yes	2.11
Genetic Algorithm	No	2.33	No	4.43	No	2.89
Ant Colony	Yes	3.0	Yes	5.43	Yes	3.1
Intelligent Water Drop	Yes	2.6	Yes	2.9	Yes	2.13
Proposed approach	Yes	1.16	Yes	1.67	Yes	1.47

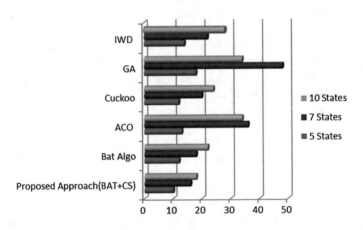

Figure 3 shows the comparison of the number of the nodes visited by the many approaches when applied to same state diagrams. The optimal transitions are obtained when all transitions can be covered with least number of states visited. The proposed algorithm-based approach shows much better results than all of the techniques. Only bat-based approach gives results comparable with the presented approach.

7 Conclusions and Future Scope

This chapter proposes a new algorithm for independent path generation by combining the benefits of Cuckoo Search and Bat algorithms. By applying proposed method, maximum software coverage is guaranteed, thus better software coverage. It takes into account the control flow graph corresponding to the program/code. So, a stronger soft computing technique for path generation is developed which may ultimately help software industry to a greater extent. At present, it is our firm hypothesis that the proposed approach would be more effective than existing heuristic alternatives.

As per the statistics given and tests conducted, the proposed approach yields results better than existing techniques of Genetic Algorithm, Ant Colony Optimization Techniques, Intelligent Water Drop and Cuckoo Search Algorithm in terms of coverage and optimality of generated sequences. This approach can readily be implemented and easily replace present techniques for sequencing industrial test cases.

The proposed algorithm is a hybrid algorithm combining the benefits of highly successful metaheuristic algorithms of Cuckoo Search and Bat. The proposed algorithm is significantly simpler than other metaheuristic algorithms and hence is superior. It is simple to implement in an industrial setting or to be reimplemented by a researcher. A meaningful metric to measure the complexity of an algorithm is the number of parameters used in the algorithm. The two parameters of this algorithm are the weight and the loudness factor per node. The loudness factor as calculated by the pulse function of the bat and the fitness value of a path computed by the cuckoo using the weight and the pulse function help us to priorities the paths with respect to maximum coverage such that the path with the maximum coverage of decision nodes has the highest priority.

References

1. Pressman RS (2007) Software engineering: a practitioners approach, 6th edn. McGraw Hill, Chapter 1[33–47], 13[387–406], 14[420–444]
2. Sommerville (2007) Software engineering, 8th edn. Pearson, Chapter 1[27–42], 11[265–288], 23[561–589]
3. Srikanth A, Nandakishore JK, Naveen KV, Singh P, Srivastava PR (2011) Test Case Optimization using artificial bee colony algorithm. Commun Comput Inf Sci 192. Adv Comput Commun 5:570–579
4. Briand LC (2002) Ways software engineering can benefit from knowledge engineering. In: Proceeding in 14th software engineering and knowledge engineering (SEKE), Italy, pp 3–6
5. Srivastava PR (2009) Test case prioritization. Int J Theoret Appl Inf Technol 4(2):178–181
6. Srivastava PR, Baby KM (2010) Automated software testing using metaheuristic technique based on an ant colony optimization. In: Electronic system design (ISED), 2010 international symposium, Bhubaneswar, pp 235–240
7. Yang XS, Deb S (2009) Cuckoo search via Lévy flights. In: Proceeding in world congress on nature and biologically inspired computing (NaBIC 2009), USA. IEEE, pp 210–214
8. Srivastava PR, Singh AT, Kumhar H, Jain M (2012) Optimal test sequence generation in state based testing using cuckoo search. Int J Appl Evol Comput 3(3):17–32. IGI global, USA
9. Gandomi AH, Yang XS, Alavi AH (2011) Cuckoo search algorithm: a metaheuristic approach to solve structural optimization problems. Eng Comput 27(1):17–35. Springer
10. Yang XS, Gandomi AH, Algorithm Bat (2012) A novel approach for global engineering optimization. Eng Comput 29(5):464–483
11. Yang XS (2010) Nature inspired cooperative strategies for optimization (NICSO). In: Studies in computational intelligence, vol 284. Springer

12. Srivastava PR, Pradyot K, Sharma D, Gouthami KP (2015) Favourable test sequence generation in state-based testing using bat algorithm. Int J Comput Appl Technol 51(4):334–343. Inderscience
13. Dahal DK, Hossain A, Suwannasart T (2008) GA-based automatic test data generation for UML state diagrams with parallel paths. In: Advanced design and manufacture to gain a competitive edge. Springer
14. Srikanth A, Kulkarni NJ, Naveen KV, Singh P, Srivastava PR (2011) Test case optimization using artificial bee colony algorithm. ACC (3):570–579. Springer
15. Srivastava PR (2010) Structured testing using an ant colony optimization. In: IITM 2010, Allahabad, ICPS. ACM, pp 205–209
16. Srivastava PR, Khandelwal R, Khandelwal S, Kumar S, Ranganatha SS (2012) Automated test data generation using cuckoo search and tabu search (CSTS) algorithm. J Intell Syst 21(2):195–224
17. Rathore A, Bohara A, Prashil RG, Prashanth TSL, Srivastava PR (2011) Application of genetic algorithm and tabu search in software testing. In: Proceedings of the 4th Bangalore annual compute conference, Compute 2011, Bangalore, India, March 25–26, 2011
18. Agarwal K, Goyal M, Srivastava PR (2012) Code coverage using intelligent water drop (IWD. Int J Bio-Inspired Comput 4(6):392–402. Inderscience
19. Srivastava PR, Mallikarjun B, Yang X-S (2013) Optimal test sequence generation using firefly algorithm, Swarm and evolutionary computation, vol 8, pp 44–53. Elsevier
20. Srivastava PR, Baby KM (2010) Automated software testing using metahurestic technique based on an ant colony optimization. Electronic system design (ISED), 2010 international symposium, Bhubaneswar, pp 235–240
21. Payne RB, Sorenson MD, Klitz K (2005) The cuckoos. Oxford University Press, USA
22. Yang XS (2010) Nature-inspired metaheuristic algorithms, 2nd edn. Luniver Press
23. Leccardi M (2005) Comparison of three algorithms for Lèvy noise generation. In: Fifth EUROMECH nonlinear dynamics conference (ENOC'05), Israel, pp 1–6

An Improved Spider Monkey Optimization for Solving a Convex Economic Dispatch Problem

Ahmed Fouad Ali

Abstract Spider monkey optimization (SMO) is a recent population-based swarm intelligence algorithm. It has powerful performance when it applied to solve global optimization problems. In this paper, we propose a new spider monkey optimization algorithm for solving a convex economic dispatch problem. Economic load dispatch (ELD) is a nonlinear global optimization problem for determining the power shared among the generating units to satisfy the generation limit constraints of each unit and minimizing the cost of power production. Although the efficiency of the spider monkey optimization algorithm, it suffers from slow convergence and stagnation when it applied to solve global optimization problems. We proposed a new hybrid algorithm in order to overcome this problem by invoking the multidirectional search method in the final stage of the standard spider monkey optimization algorithm. The proposed algorithm is called multidirectional spider monkey optimization algorithm (MDSMO). The proposed algorithm can accelerate the convergence of the proposed algorithm and avoid trapping in local minima. The general performance of the proposed MDSMO algorithm is tested on a six-generator test system for a total demand of 700 and 800 MW and compared against five Nature-Inspired algorithms. The experimental results show that the proposed algorithm is a promising algorithm for solving economic load dispatch problem.

1 Introduction

Spider monkey optimization (SMO) algorithm is a new swarm intelligence algorithm proposed by Bansal et al. [8]. SMO and other swarm intelligence algorithms are population-based algorithms such as ant colony optimization (ACO) [13], artificial bee colony optimization [22], bee colony optimization (BCO) [38], particle swarm

A.F. Ali (✉)
Faculty of Computers and Informatics, Department of Computer Science,
Suez Canal University, Ismailia, Egypt
e-mail: ahmed_fouad@ci.suez.edu.eg

© Springer International Publishing AG 2017
S. Patnaik et al. (eds.), *Nature-Inspired Computing and Optimization*,
Modeling and Optimization in Science and Technologies 10,
DOI 10.1007/978-3-319-50920-4_17

optimization (PSO) [23], and bat algorithm [41]. These algorithms are applied to solve global optimization problems and their applications [2, 5, 9, 17–19], due to their efficiency and powerful performance when they applied to solve these problems. In this paper, we propose a new hybrid algorithm to overcome the slow convergence of the standard SMO algorithm by invoking the multidirectional search in the final stage of the standard SMO algorithm in order to accelerate the search and avoid its slow conversion. The proposed algorithm is called multidirectional spider monkey optimization (MDSMO). Invoking the multidirectional search in the standard, SMO can accelerate the search by refining the best obtained solution from the standard SMO instead of running the algorithm for more iterations without any improvement in the results. The general performance of the proposed MDSMO algorithm is tested on a six-generator test system for a total demand of 700 and 800 MW, and it compared against Five Nature-Inspired algorithms. The rest of this paper is organized as follows. Some of the related works are summarized in Sect. 2. The economic load dispatch problem is presented in Sect. 3. In Sects. 4 and 5, we give an overview of the standard spider monkey optimization algorithm (SMO) and its main processes and the multidirectional search method. The proposed algorithm is presented in Sect. 6. The numerical results are described in Sect. 7. The conclusion of the paper and the future work make up Sect. 8.

2 Related Work

The mathematical form of the economic load dispatch problem (ELD) problem can be defined as a continuous optimization problem. Many researchers have applied the meta-heuristics and the nature-inspired algorithms to solve the continuous global optimization problems and their applications. The algorithms such as genetic algorithm [37], bat algorithm [1, 4, 6, 25, 27, 36], particle swarm optimization [3, 11, 12, 26, 36, 44], firefly algorithm [7, 42], cuckoo search [40, 41], and bacterial foraging algorithm [16, 24] are intensively used. Due to the efficiency of these algorithms when they applied to solve global optimization problems, they have been applied to solve economic load dispatch (ELD) problem such as many variant of genetic algorithm (GA) with good results to solve non-convex ELD problems [1]. The main advantage of GA over other algorithms that it can use a chromosome representation technique tailored to the specific problem. However, it suffers from the slow convergence and the long execution time. PSO with different variants has been applied to solve non-convex ELD problems in some works such as [10, 15, 34, 35]. The main advantage of the PSO algorithm is it is easy to perform and it used a few adjustable parameters and it is very efficient in exploring the search space (diversification process). However, it suffers from the slow convergence and its weak local search ability (intensifications process).

Also, variant bat algorithms (BA) have been used to solve the ELD problem [28, 31, 36]. The BA has a number of tunable parameters, which give a good control over the optimization process. BA gives a promising results when it applied for solving lower-dimensional optimization problem, but it becomes ineffective for solving high-dimensional problems because of its fast initial convergence [30].

In this paper, we propose a new algorithm to overcome the slow convergence of the meta-heuristics and the other nature-inspired algorithms by invoking the multidirectional search method as a local search method in the proposed algorithm in order to accelerate the convergence of it and to refine the best obtained solution in the final stage of the proposed algorithm.

3 Economic Dispatch Problem

The economic load dispatch (ELD) problem can be mathematically formulated as a continuous optimization problem. The goal of the economic dispatch problem is to find the optimal combination of power generation in such a way that the total production cost of the system is minimized. The cost function of a generating unit can be represented as a quadratic function with a sine component. The sin component denotes the effect of steam valve operation. The quadratic refers to a convex objective whereas the valve-point effect makes the problem non-convex. The convex and non-convex cost function of a generator can be expressed as follows.

$$F_i(P_i) = a_i P_i^2 + b_i P_i + c_i \tag{1}$$

$$F_i(P_i) = a_i P_i^2 + b_i P_i + c_i + |d_i sin[e_i * (P_i^{min} - P_i)]| \tag{2}$$

where P_i is the active power output, $F_i(P_i)$ is the generation cost, P_i^{min} is the minimum output limit of the generator, and the a_i, b_i, c_i, d_i, e_i are the cost coefficients of the generator. The fuel cost of all generators can be defined by the following equation:

$$Min F(P) = \sum_i^{Ng} a_i P_i^2 + b_i P_i + c_i + |d_i sin[e_i * (P_i^{min} - P_i)]| \tag{3}$$

where Ng is the number of generating units.

3.1 Problem Constraints

The total power generated should cover the power demand and the active power losses as follows.

$$\sum_{i=1}^{Ng} P_i = P_D + P_{loss} \tag{4}$$

where P_D is the total demand load and P_{loss} is the total transmission losses computed using the quadratic approximation as follows.

$$P_{loss} = \sum_{i=1}^{Ng} \sum_{j=1}^{Ng} P_i B_{ij} P_j \tag{5}$$

where B_{ij} is the loss coefficient matrix.

The real output power of each generator should be within a lower and an upper limit as follows.

$$P_i^{min} \leq P_i \leq P_i^{max} \tag{6}$$

where P_i^{min} and P_i^{max} are the lower and the upper limit of the ith generators.

3.2 Penalty Function

The penalty function technique is used to transform the constrained optimization problem to unconstrained optimization problem by penalizing the constraints and forming a new objective function. In this paper, the constraint of the ELD problem can be handled as follows.

$$\text{Min}(Q(P_i)) = F_c(P_i) + pen * G[h_k(P_i)] \tag{7}$$
$$\text{Subject to}: \ g_j(P_i) \leq 0, \ j = 1, \dots, J$$

where pen is the penalty factor and $G[h_k(P_i)]$ is the penalty function, which is calculated as follows.

$$G[h_k(P_i)] = \varepsilon^2 + \omega^2 \tag{8}$$

where ε and ω are equality and inequality constraint violation penalties, respectively, and can be calculated as follows.

$$\varepsilon = \left| P_D + P_{loss} - \sum_{j=1}^{Ng} P_j \right| \tag{9}$$

and

$$\omega = \begin{cases} |P_i^{min} - P_i| & P_i^{min} > P_i \\ 0 & P_i^{min} < P_i < P_i^{max} \\ |P_i - P_i^{max}| & P_i^{max} < P_i \end{cases} \tag{10}$$

4 Social Behavior and Foraging of Spider Monkeys

In the following sections, we highlight the spider monkeys fission–fusion and communication behaviors.

4.1 Fission–Fusion Social Behavior

Spider monkeys are living in a large community called unit-group or parent group. In order to minimize foraging competition among group individuals, spider monkeys divide themselves into subgroups. The subgroup members start to search for food and communicate together within and outside the subgroups in order to share information about food quantity and place. The parent group members search for food (forage) or hunt by dividing themselves in subgroups (fission) in different direction; then at night, they return to join the parent group (fusion) to share food and do other activities.

4.2 Social Organization and Behavior

The social behavior of spider monkeys is an example of fission–fusion system. The are living in group which is called parent group, each group contains of up to 50 members [33]. When they search for food, they divide themselves into small subgroups to forage within a core area of the parent group [33]. Spider monkeys use the different way to search for food. A female is a leader of the group, and she is responsible for finding the food. In case if she failed to find food resources, she divides the group into subgroups to forage separately [29]. At any time, 3 members can be found in each subgroup [29]. The males in each subgroup display aggressiveness behavior when two subgroups become closer. There are no physical contacts among subgroups members, and they respect distance between subgroups. Due to seasonal reasons, the group members can suffer from a shortage of food availability [29], so they fission the group into subgroups according to the food availability of each subgroups [32]. The members of the subgroups are still remained part of the parent group because of increasing mating chance and security from enemies.

4.3   Communication of Spider Monkeys

Spider monkeys are travailing in different direction to search for food. They interact
and communicate with each other using a particular call by emitting voice like a
horse's whinny. Each individual has its identified voice so that other members of the
group can distinguish who is calling. The long-distance communication helps spider
monkeys to stay away from predators, share food, and gossip. The group members
interact with each other by using visual and vocal communication [14].

4.4   Characteristic of Spider Monkeys

The characteristic of spider monkeys can be described in the following steps:

1. The spider monkeys as a fission–fusion social structure (FFSS)-based animals live
   in groups, where each group contains of 40–50 individuals. In FFSS, the group
   is divided into subgroups in order to reduce competition among group members
   when they search for foods.
2. The parent group leader is a female (global leader) who leads the group and
   responsible for searching food resources. If the group leader fails to get enough
   food, she divides the group into subgroups with 3–8 members to search for food
   independently.
3. Subgroups are also led by a female (local leader) who is responsible for selecting
   an efficient foraging route each day.
4. The members of each subgroup are communicated within and outside the subgroups depending on the availability of food and respect distinct territory boundaries.

4.5   The Standard Spider Monkey Optimization Algorithm

In the following sections, we highlight the main steps of the SMO algorithm.

4.5.1   The Main Steps of the Spider Monkey Optimization Algorithm

The SMO consists of 6 phases: local leader phase (LLP), global leader phase (GLP),
local leader learning (LLL) phase, global leader learning (GLP) phase, local leader
decision (LLD) phase, and global leader decision (GLD) phase. In global leader
phase, the position update process is inspired from the gbest-guided ABC [43] and
modified version of ABC [21]. The main steps of the SMO algorithm are described
as follows.

4.5.2 Population initialization

The initial population of the spider monkey optimization is generated randomly. The population contains n monkeys, each monkey SM_i represents a solution in the population and contains a D variables, SM_i, $(i = 1, 2, \ldots, n)$. The spider monkey SM_i is initialized as follows.

$$SM_{ij} = SM_j^{min} + U(0, 1) \times (SM_j^{max} + SM_j^{min}) \tag{11}$$

where SM_j^{min} and SM_j^{max} are bounds of SM_i and $U(0, 1)$ is a uniformly distributed random number, $U(0, 1) \in [0, 1]$.

4.5.3 Local Leader Phase (LLP)

In the local leader phase (LLP), each spider monkey SM_i updates its current position based on the information of the local leader experience. The fitness value of the new SM_i position is calculated, and if its value is better than the old position, the spider updates its position. The equation of the spider position update in a kth local group is shown as follows.

$$SM_{ij}^{new} = SM_{ij} + U(0, 1) \times (LL_{kj} - SM_{ij}) + U(-1, 1) \times (SM_{rj} - SM_{ij}) \tag{12}$$

where SM_{ij} is the jth dimension of the SM_i, LL_{kj} is the jth dimension of the kth local group leader. SM_{rj} represents the jth dimension of the rth SM which is selected randomly in kth group and $r \neq i$.

The main steps of the local leader phase are presented in Algorithm 1. In Algorithm 1, MG is the maximum groups number in the swarm and pr is the perturbation rate that controls the perturbation amount in the current position. The range of pr is $[0.1, 0.9]$.

Algorithm 1 Position Update Process in Local Leader Phase (LLP)

1: **for** $(k = 1; k \leq MG; k++)$ **do**
2:   **for** $(i = 1; i \leq N; i++)$ **do**
3:     **for** $(j = 1; j \leq D; j++)$ **do**
4:       **if** $(U(0, 1) \geq pr)$ **then**
5:         $SM_{ij}^{new} = SM_{ij} + U(0, 1) \times (LL_{kj} - SM_{ij}) + U(-1, 1) \times (SM_{rj} - SM_{ij})$
6:       **else**
7:         $SM_{ij}^{new} = SM_{ij}$
8:       **end if**
9:     **end for**
10:   **end for**
11: **end for**

4.5.4 Global Leader Phase (GLP)

In the global leader phase (GLP), each monkey SM updates its position using the experience of the global leader and local leader. The equation of the position update is shown as follows.

$$SM_{ij}^{new} = SM_{ij} + U(0, 1) \times (GL_j - SM_{ij}) + U(-1, 1) \times (SM_{rj} - SM_{ij}) \quad (13)$$

where GL_j is the jth dimension of the global leader position, j is chosen randomly and $j \in [1, 2, \ldots, D]$. In GLP phase, the spider monkey updates its position based on the probabilities $prob_i$ which is calculated as follows.

$$prob_i = 0.9 \times \frac{fitness_i}{max - fitness} + 0.1 \quad (14)$$

where $fitness_i$ is the fitness value of the SM_i and $max - fitness$ is the maximum fitness in the group.

The main steps of the GLP are presented in Algorithm 2.

Algorithm 2 Position Update Process in Global Leader Phase (GLP)

1: **for** $(k = 1; k \leq MG; k++)$ **do**
2:    $count = 1$
3:    $GS = k$th
4:   **while** $(count < GS)$ **do**
5:    **for** $(i = 1; i \leq GS; k++)$ **do**
6:     **if** $U(0, 1) < prob_i$ **then**
7:      $count = count + 1$
8:      Select random $j, j \in [1, D]$
9:      Select random SM_r from kth, $r \neq i$
10:      $SM_{ij}^{new} = SM_{ij} + U(0, 1) \times (GL_i - SM_{ij}) + U(-1, 1) \times (SM_{rj} - SM_{ij})$
11:     **end if**
12:    **end for**
13:    **if** $(i = GS)$ **then**
14:     $i = 1$
15:    **end if**
16:   **end while**
17: **end for**

4.5.5 Global Leader Learning (GLL) Phase

In the global leader learning phase (GLL), the global leader is updated by applying the greedy selection in the population (the position of the SM with the best position is selected. The GlobalLimitCount is incremented by 1 if the position of the global leader is not updated.

4.5.6 Local Leader Learning (LLL) Phase

The local leader updates its position in the group by applying the greedy selection. If the fitness value of the new local leader position is worse than the current position, then the LocalLimitCount is incremented by 1.

4.5.7 Local Leader Decision (LLD) Phase

If the local leader position is not updated for specific number of iterations which is called LocalLeaderLimit (LLL), then all the spider monkeys (solutions) update their positions randomly or by combining information from global leader and local leader as follows.

$$SM_{ij}^{new} = SM_{ij} + U(0, 1) \times (GL_j - SM_{ij}) + U(0, 1) \times (SM_{ij} - LLkj) \quad (15)$$

The main steps of the local leader decision (LLD) phase are shown in Algorithm 3.

Algorithm 3 Local Leader Decision Phase

1: **for** $(k = 1; k \leq MG; k++)$ **do**
2:   **if** $(LLC_k > LLL)$ **then**
3:     $LLC_k = 0$
4:     $GS = k$th
5:     **for** $(i = 1; i \leq GS; i++)$ **do**
6:       **for** $(j = 1; j \leq D; j++)$ **do**
7:         **if** $(U(0, 1) \geq pr)$ **then**
8:           $SM_{ij}^{new} = SM_j^{min} + U(0, 1) \times SM_j^{max} - SM_j^{min})$
9:         **else**
10:           $SM_{ij}^{new} = SM_{ij} + U(0, 1) \times (GL_j - SM_{ij}) + U(0, 1) \times (SM_{ij} - LL_{kj})$
11:         **end if**
12:       **end for**
13:     **end for**
14:   **end if**
15: **end for**

4.5.8 Global Leader Decision (GLD) Phase

In the global leader decision (GLD) phase, if the global leader is not updated for a specific number of iterations which is called GlobalLeaderLimit (GLL), then the global leader divides the (group) population into subpopulations (small groups). The population is divided into two and three subgroups and so on till the maximum number of groups MG. The operation of the group division is shown in Figs. 1, 2 and 3.

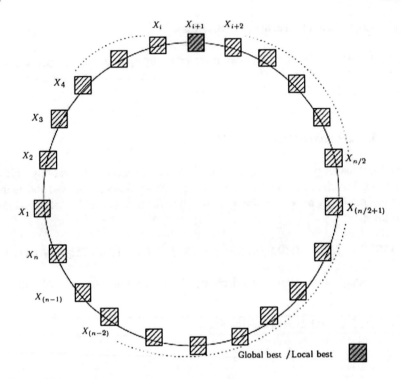

Fig. 1 SMO topology: single group

The main steps of the GLD phase are shown in Algorithm 4.

Algorithm 4 Global Leader Decision Phase

1: **if** $(GLC > GLL)$ **then**
2:    $GLC = 0$
3:    $GS = k$th
4:    **if** $(Number of groups < MG)$ **then**
5:       Divide the population into groups
6:    **else**
7:       Combine all the groups to make a single group
8:    **end if**
9:    Update Local Leaders position
10: **end if**

4.6 Spider Monkey Optimization Algorithm

The spider monkey optimization (SMO) algorithm is inspired from the foraging behavior of spider monkeys. The SMO algorithm applies four strategies to obtain the optimum solution during the search as follows.

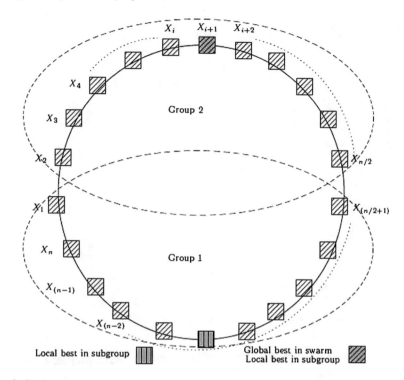

Fig. 2 SMO topology: population division into two groups

- The group members (solutions) start searching for food and evaluate their distance (fitness function) from the foods.
- The group members update their positions based on the distance from the foods, and the locale leader and global leader are assigned.
- The local leader updates its position within the group, and if its position is not updated for a specified number of iterations, then all group members start foraging in different directions.
- The global leader updates its position, and if its position is not updated for a specified number of iterations (stagnation), it divides the parent group into smaller subgroups.

The previous steps are repeated until termination criteria satisfied. The SMO algorithm uses two important control parameters in order to avoid stagnation, the first one is "GlobalLeaderLimit", while the second one is "LocalLeaderLimit".

The main steps of the standard SMO algorithm are presented in Algorithm 5.

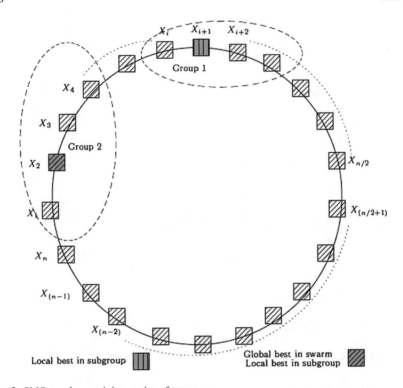

Fig. 3 SMO topology: minimum size of group

5 Multidirectional Search Algorithm

The multidirectional search algorithm is a direct search algorithm which is proposed by Dennis and Torczon in 1989 [39] as a general algorithm with a powerful properties for parallel computing. The Nelder–Mead method is a direct search method also, but it has a main drawback which is it can converge to non-minimizers when the dimension of the problem is large enough $n \geq 10$. On the other side, the multidirectional search has a promising behavior when it applied with the high-dimensional problems since it is backed by convergence theorems. The main steps of the multidirectional search algorithm are presented in Algorithm 6 and Fig. 4.

We can summarize the main steps of the multidirectional search as follows.

- **Step 1**. The algorithm starts by setting the initial values of the expansion factor μ, contraction factor θ and the maximum number of iterations parameter Max_{itr1} (line 1).
- **Step 2**. The multidirectional search algorithm begins with a simplex S with vertices x_i^0, where $i = 0, 1, n$ (line 2).
- **Step 3**. The vertices are sorted in ascending order where $f(x_0^0) \leq f(x_1^0) \leq \cdots \leq f(x_n^0)$ (line 3).

Algorithm 5 Spider Monkey Optimization (SMO) Algorithm

1: Set the initial values of the population size n, LocalLeaderLimit (LLL), GlobalLeaderLimit (GLL), perturbation rate(pr)

2: **for** ($i = 1 : i \leq n$) **do**

3:   Generate an initial population SM_i randomly.

4:   Evaluate the fitness function of each spider monkey (solution) SM_i.

5: **end for**

6: Assign the global leader and the local leader by applying greedy selection.

7: **repeat**

8:   **for** ($i = 1 : i \leq n$) **do**

9:     Generate the new position for each spider monkey SM_i by using self experience, local leader experience and global leader experience as shown in Algorithm 1.

10:     Evaluate the fitness function of the new spider monkeys positions.

11:     Apply greedy selection and select the best solution.

12:     Calculate the probability $prob_i$ for each SM_i as shown in Eq. (14).

13:     By using $prob_i$, self experience, global leader experience and group member experience as shown in Algorithm 2.

14:     If local group leader position is not updated for a specific number of iterations (LLL), then redirect all members for food searching as shown in Algorithm 3.

15:     If global leader position is not updated for a specific number of iterations (GLL), then she divides the group into subgroups as shown in Algorithm 4.

16:   **end for**

17: **until** Termination criteria satisfied.

18: Produce the best solution SM_i.

Fig. 4 Multidirectional search

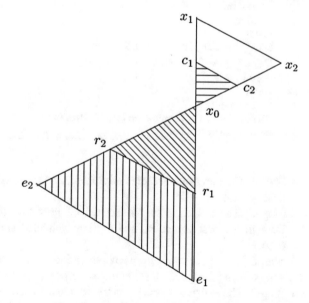

Algorithm 6 Multidirectional Search Algorithm

1: Set the values of expansion factor μ and contraction factor θ, where $\mu \in (1, +\infty)$, $\theta \in (0, 1) and Max_{itr1}$

2: Let x_i^0 denote the list of vertices in the current simplex, where $i = 0, 1, \ldots, n$ {**Simplex initialization**}

3: Order the n vertices from lowest function value $f(x_0^0)$ to highest function value $f(x_n^0)$ so that
$$f(x_0^0) \leq f(x_1^0) \leq f(x_2^0) \leq \ldots \leq f(x_n^0)$$

4: Set $k = 0$ {**Counter initialization**}

5: **repeat**

6:    **for** $(i = 1; i < n; i++)$ **do**

7:      Set $x_i^{k+1} = 2x_0^k - x_i^k$ {**Reflection step**}

8:      Evaluate the n vertices $f(x_i^{k+1})$

9:    **end for**

10:   **if** $(\min f(x_i^{k+1}, i = 1, \ldots, n < f(x_0^k)$ **then**

11:     **for** $(i = 1; i < n; i++)$ **do**

12:       Set $x_{ei}^{k+1} = (1 - \mu)x_0^k + \mu x_i^{k+1}$ {**Expansion step**}

13:       Evaluate the n vertices $f(x_{ei}^k)$

14:     **end for**

15:     **if** $(\min f(x_{ei}^k) < f(x_i^{k+1}), i = 1, \ldots, n)$ **then**

16:       Set $x_i^{k+1} = x_{ei}^k, i = 1, 2, \ldots, n$

17:     **else**

18:       **for** $(i = 1; i < n; i++)$ **do**

19:         Set $x_i^{k+1} = (1 - \theta)x_0^k - \theta x_i^{k+1}$ {**Contraction step**}

20:         Evaluate the n vertices $f(x_i^{k+1})$

21:       **end for**

22:     **end if**

23:   **end if**

24:   Set $x_{min}^{k+1} = \min f(x_i^{k+1}), i = 1, 2, \ldots, n$.

25:   **if** $(f(x_{min}^{k+1}) < f(x_0^k))$ **then**

26:     Set $x_0^k = x_{min}^{k+1}$.

27:   **end if**

28:   Set $k = k + 1$.

29: **until** $(k \geq Max_{itr1})$ {**Termination criteria stratified**}

30: Produce the best solution

- **Step 4**. The iteration counter is initialized, and the main algorithm loop is initialized (lines 4, 5).

- **Step 5**. The reflection step is started by reflecting the vertices $x_1, x_2, \ldots, x_n$ through the best vertex x_0, and the new reflected vertices are evaluated (lines 6–9).

- **Step 6**. If a reflect vertex is succussed and its value is better than the current best vertex, then the algorithm starts the expansion step (line 10).

- **Step 7**. The expansion process starts to expand each reflected edge by using the expansion factor μ, where $\mu = 2$ to create new expansion vertices. The new expansion vertices are evaluated in order to check the success of the expansion step (lines 10–14).

- **Step 8**. If the expansion vertex is better than the all reflection vertices, the new simplex will be the expansion simplex lines (15–16).

- **Step 9**. If the expansion and reflection steps were fail, then the contraction simplex starts by changing the size of the step we tack by using contraction factor θ, which reduce the reflection simplex to the half in the next iteration (lines 18–21).
- **Step 10**. The new vertices are evaluated and the vertices are sorted according to their evaluation function value and the new simplex is constructed (lines 24–27).
- **Step 11**. The iteration counter increases and the overall process are repeated till the termination criterion satisfied which is by default the maximum number of iterations Max_{itr1}. Finally, the best solution is produced (lines 28–30).

6 The Proposed MDSMO Algorithm

The main steps of the proposed MDSMO algorithm are the same steps of the standard spider monkey optimization; however, we invoke the multidirectional search in the last stage of the standard SMO algorithm in order to refine the best obtained solution from SMO algorithm. The objective of invoking the multidirectional search in the last stage of the standard SMO algorithm is to accelerate the convergence of the proposed algorithm instead of running the standard SMO algorithm for more iterations without any improvement in the results.

7 Numerical Experiments

In this section, we investigate the efficiency of the MDSMO algorithm by testing it on a convex load dispatch problem and comparing it against three nature-inspired algorithms. MDSMO was programmed in MATLAB. The parameter setting and the performance analysis of the proposed algorithm are presented in the following section as follows.

7.1 Parameter Setting

We summarize the parameter setting of the proposed MDSMO algorithm, and their values as shown in Table 1.

These values are based on the common setting in the literature of determined through our preliminary numerical experiments.

- **population size** n. The experimental tests show that the best number of the population size is $n = 40$, as increasing this number increases the function evaluations without notable improvement in the function value.

Table 1 Parameter setting

Parameters	Definitions	Values
n	Population size	40
pr	Perturbation rate	[0.1, 0.4]
MG	Maximum number of groups	5
LLL	Local leader limit	1500
GLL	Global leader limit	50
μ	Expansion factor	2
θ	Contraction factor	0.5
Max_{itr1}	Maximum iterations number for multidirectional search algorithm	100
N_{elite}	No. of best solution for multidirectional search	1

- pr, MG, LLL, GLL. The values of the perturbation rate pr, maximum number of groups MG, local leader limit (LLL), and global leader limit (GLL) are taken from the original paper of the standard SMO algorithm [8].
- μ, θ, Max_{itr1}. The experimental results of the direct search parameter show that the best values of the expansion factor is $\mu = 2$, the contraction factor $\theta = 0.5$, and the maximum number of the multidirectional search iterations is $Max_{itr1} = 100$.
- N_{elite} In the final intensification stage of the proposed algorithm, we applied a local search using the multidirectional method, starting from the elite solutions N_{elite} obtained in the previous search stage, and we set $N_{elite} = 1$. (Increasing the number of selected elite solutions increases the function evaluations.)

7.2 Six-Generator Test System with System Losses

The proposed MDSMO algorithm is tested on a six-generator test system. In order to simplify the problem, the values of parameters d, e in Eq. (2) have been set to zero. The value of B_{ij} is given by

$$B_{ij} = 10^{-4} \times \begin{bmatrix} 0.14 & 0.17 & 0.15 & 0.19 & 0.26 & 0.22 \\ 0.17 & 0.60 & 0.13 & 0.16 & 0.15 & 0.20 \\ 0.15 & 0.13 & 0.65 & 0.17 & 0.24 & 0.19 \\ 0.19 & 0.16 & 0.17 & 0.71 & 0.30 & 0.25 \\ 0.26 & 0.15 & 0.24 & 0.30 & 0.69 & 0.32 \\ 0.22 & 0.20 & 0.19 & 0.25 & 0.32 & 0.85 \end{bmatrix}$$

Table 2 Generator active power limits

Generator	1	2	3	4	5	6
$P_{min}(MW)$	10	10	35	35	130	125
$P_{max}(MW)$	125	150	225	210	325	315

Table 3 Fuel cost coefficients

No.	a	b	c
1	0.15240	38.53973	756.79886
2	0.10587	46.15916	451.32513
3	0.02803	40.39655	1049.9977
4	0.03546	38.30553	1243.5311
5	0.02111	36.32782	1658.5596
6	0.01799	38.27041	1356.6592

The proposed algorithm is carried out for a total demand of 700 and 800 MW. The generator active power limits and the fuel cost coefficients are given in Tables 2 and 3, respectively.

7.3 The General Performance of the Proposed MDSMO with Economic Dispatch Problem

The general performance of the proposed algorithm with economic dispatch problem is shown in Fig. 5 by plotting the number of iterations versus the cost ($/h$) for total system demand 700 and 800 MW, respectively. Figure 5 shows that the cost values are rapidly decreased with a few numbers of iterations. We can conclude from Fig. 5 that the proposed algorithm is a promising algorithm and can obtain the desired power with the minimum cost.

7.4 MDSMO and Other Algorithms

In this section, we highlight the three compared nature-inspired algorithms as follows.

- GA (Genetic algorithm). Genetic algorithm is a population-based meta-heuristics algorithm. It has been developed by Holland [20] to understand the adaptive processes of natural systems. GA usually applies a crossover operator by mating the parents (individuals) and a mutation operator that randomly modifies the individual contents to promote diversity to generate a new offspring. GA uses a probabilistic selection that is originally the proposed proportional selection. The

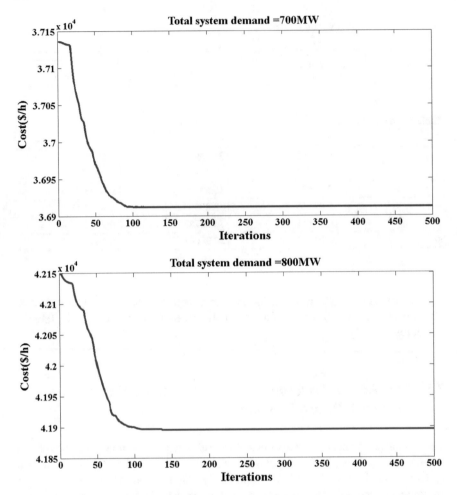

Fig. 5 The general performance of MDSMO with economic dispatch problem

replacement (survival selection) is generational, that is, the parents are replaced systematically by the offspring.

- PSO (Particle swarm optimization). Particle swarm optimization (PSO) is a population-based meta-heuristics algorithm that inspired from the behavior (information exchange) of the birds in a swarm, and it was proposed by Kennedy and Eberhart in 1995 [23]. PSO uses two techniques to find global optima, global best solution "gbest," and local best solution "lbest." The first technique is "gbest" technique; in this technique, all particles share information with each other and move to global best position. However, this technique has drawback, because it is easy to trap in local optima. The second technique is "lbest" technique, a specific number of particles are neighbors to one particle, but this technique also suffers from the slow of convergence.

- (Bat algorithm). Bat algorithm (BA) is a population-based meta-heuristics algorithm developed by Yang [41]. BA is based on the echolocation of microbats, which use a type of sonar (echolocation) to detect prey and avoid obstacles in the dark. The main advantage of the BA that it can provide a fast convergence at a very initial stage by switching from exploration to exploitation; however, switching from exploration to exploitation quickly may lead to stagnation after some initial stage.
- (Cuckoo search algorithm). Cuckoo search (CS) is a swarm intelligence algorithm developed by Yang and Deb [40] and mimic the reproduction strategy of the cuckoo birds.

7.4.1 Comparison Between GA, PSO, BA, CS, SMO, and MDSMO

The performance of the proposed MDSMO is tested on a six-generator test system and compared against five Nature-Inspired algorithms at total demand of 700 and 800 MW as shown in Tables 4 and 5, respectively. The parameters of other algorithms can be summarized as follows. The real coded GA was used with roulette wheel selection, arithmetic crossover, and uniform mutation. The probabilities of the crossover and mutation were set to 0.9 and 0.02, respectively. Also the inertia weight of the PSO decreasing linearly form 0.9 to 0.4, and the velocity constant was set to 2 for all the experiments. BA was implemented with initial loudness and pulse rate $A = 0.9$, $r = 0.1$. A fraction of worse nests of CS algorithm was set to 0.25. Finally, we applied the standard SMO parameter values.

The best value (best), average (Avg), and the standard divination (Std) of the results are reported over 30 runs in Tables 4 and 5. For all the experiments, all algorithms have applied the same termination criteria, which are the error tolerance is 0.01 MW or the maximum number of iterations is 500.

The results in Tables 4 and 5 show that the proposed algorithms outperform the other compared algorithms. Also it is a promising algorithm and can obtain the optimal or near-optimal results of the economic dispatch problem.

8 Conclusion and Future Work

In this paper, we presented an improved spider monkey optimization algorithm in order to solve economic dispatch problem. The proposed algorithm is called a multidirectional spider monkey optimization algorithm (MDSMO). Although the spider monkey optimization algorithm has a good ability to perform wide exploration and deep exploitation, it suffers from the slow convergence. The spider monkey optimization algorithm is running for number of iterations; then, the obtained solution passes to the multidirectional search algorithm to refine the best obtained solution and accelerate the search. We tested the MDSMO algorithm on a six-generator test system for a total demand of 700 and 800 MW and compared against Five Nature-

Table 4 Results for GA, PSO, BA, CS, SMO, and MDSMO at total system demand = 700 MW

Algorithm		P1 (MW)	P2 (MW)	P3 (MW)	P4 (MW)	P5 (MW)	P6 (MW)	Loss (MW)	Cost ($/h)
GA	Best	26.7997	15.8931	107.307	123.932	228.342	217.160	19.44	36924.15
	Avg	45.5736	48.619	105.805	106.478	211.450	200.676	18.61	37505.72
	Std	19.770	28.673	43.328	36.2062	45.62	45.043	1.325	382.88
PSO	Best	28.3023	9.9998	118.952	118.670	230.7563	212.737	19.431	36911.54
	Avg	28.3979	10.0233	119.0863	118.594	230.588	212.723	19.4262	36911.75
	Std	0.858	0.139	0.83555	0.6229	1.1889	0.494	0.02786	1.4869
BA	Best	28.0739	10.0569	119.985	117.772	231.133	212.391	19.423	36911.79
	Avg	28.3941	10.2677	119.159	119.0363	230.295	212.244	19.409	36912.54
	Std	0.6928	0.2676	2.2262	1.7091	2.9539	3.8	0.05999	1.0006
CS	Best	28.2907	10.005	118.959	118.623	230.315	212.524	19.432	36912.20
	Avg	28.2907	10.020	118.958	118.674	230.763	212.744	19.431	36912.33
	Std	0.148	0.021	0.154	0.245	0.184	0.015	0.024	1.25345
SMO	Best	28.2907	10.011	118.958	118.674	230.763	212.744	19.4318	36912.21
	Avg	28.2916	10.012	118.216	118.446	230.143	212.178	19.4516	36912.27
	Std	0.12	0.091	0.134	0.12	0.016	0.135	0.02	1.2543
MDSMO	Best	28.1484	10.0388	119.826	118.051	231.0218	212.418	19.4315	36910.82
	Avg	28.2646	10.0212	119.1860	118.695	231.2543	212.711	19.4237	36911.27
	Std	0.1491	0.0122	0.8521	0.42247	0.08144	0.3078	0.025	0.926

Table 5 Results for GA, PSO, BA, CS, SMO, and MDSMO at total system demand = 800 MW

Algorithm		P1 (MW)	P2 (MW)	P3 (MW)	P4 (MW)	P5 (MW)	P6 (MW)	Loss (MW)	Cost ($/h)
GA	Best	39.63015	13.23341	170.317	155.1286	232.4949	213.4204	24.2359	41976.08
	Avg	55.35765	54.95395	130.4268	1342949	230.3903	218.5409	23.9787	42614.68
	Std	25.9155	30.1187	45.3717	39.9879	49.7911	45.6905	1.3105	436.61
PSO	Best	32.59937	14.48227	141.5412	136.0392	257.6555	242.9997	25.3299	41895.98
	Avg	32.5959	14.51256	141.4859	135.9388	257.6442	243.1419	25.3322	41896.02
	Std	0.19817	0.2575	0.31681	0.66662	0.33471	0.86126	0.020216	0.23259
BA	Best	32.46774	14.34427	141.9097	135.7294	257.7276	243.1421	25.3359	41895.88
	Avg	32.58662	14.49149	141.7122	136.2057	257.3597	242.9548	25.3232	41896.17
	Std	0.38275	0.49502	0.97076	0.88628	1.2144	1.3829	0.037035	0.25826
CS	Best	32.5860	14.4839	141.5475	136.0435	257.6624	243.0073	25.3309	41896.70
	Avg	32.5915	14.5246	141.5488	136.1435	257.5613	243.0345	25.3319	41896.94
	Std	0.1241	0.24561	0.1452	0.8462	0.1258	0.7458	0.0624	0.08741
SMO	Best	32.5761	14.177	141.590	136.072	257.616	243.002	25.3356	41896.71
	Avg	32.5862	14.483	141.847	136.0436	257.663	243.117	25.3729	41896.95
	Std	0.0156	0.3567	0.0248	0.054	0.2587	0.8964	0.0489	0.0125
MDSMO	Best	32.4167	14.3442	141.9089	135.7588	257.7273	243.1418	25.3357	41891.82
	Avg	32.5898	14.4630	141.54390	135.7412	257.6587	243.0033	25.3399	41893.84
	Std	0.07545	0.0739	0.2145	0.05468	0.8348	0.9487	0.0811	0.29498

Inspired algorithms. The experimental results show that the proposed algorithm is a promising algorithm and is more successful than all of the compared algorithms. We should point out that the efficiency of the proposed algorithm does not ensure that it can always be better than other algorithms when it applies to solve other problems. In the future work, we will apply the proposed algorithm to solve other applications in optimization and we will compare it against other traditional ELD algorithms.

References

1. Abido MA (2003) A niched Pareto genetic algorithm for multi-objective environmental/economic dispatch. Int J Electr Power Energy Syst 25:97–105
2. Ali AF, Hassanien AE (2013) Minimizing molecular potential energy function using genetic Nelder-Mead algorithm. In: 8th international conference on computer engineering and systems (ICCES), pp 177–183
3. Ali AF (2014) A new hybrid particle swarm optimization with variable neighborhood search for solving unconstrained global optimization problems. In: The fifth international conference on innovations in bio-inspired computing and applications IBICA
4. Ali AF (2015) Accelerated bat algorithm for solving integer programming problems. Egypt Comput Sci J 39
5. Akhand MAH, Junaed ABM, Murase K (2012) Group search optimization to solve traveling salesman problem. In: 15th ICCIT 2012. University of Chittagong, pp 22–24
6. Amjady N, Nasiri-Rad H (2010) Solution of non-convex and non-smooth economic dispatch by a new adaptive real coded genetic algorithm. Expert Syst Appl 37:5239–5245
7. Apostolopoulos T, Vlachos A (2011) Application of the firefly algorithm for solving the economic emissions load dispatch problem. Int J Comb
8. Bansal JC, Sharma H, Jadon SS, Clerc M (2014) Spider monkey optimization algorithm for numerical optimization. Memetic Comput 6(1):31–47
9. Bansal JC, Deep SK, Katiyar VK (2010) Minimization of molecular potential energy function using particle swarm optimization. Int J Appl Math Mech 6(9):1–9
10. Cai J, Ma X, Li L, Haipeng P (2007) Chaotic particle swarm optimization for economic dispatch considering the generator constraints. Energy Convers Manage 48:645–653. doi:10.1016/j.enconman.2006.05.020
11. Chang WD (2009) PID control for Chaotic synchronization using particle swarm optimization. Chaos, Solutions Fractals 39(2):910–917
12. Coello CAC, Pulido GT, Lechuga MS (2004) Handling multiple objectives with particle swarm optimization. IEEE Trans Evol Comput 8(3):256–279
13. Dorigo M (1992) Optimization, learning and natural algorithms. PhD thesis, Politecnico di Milano, Italy
14. Fernandez R (2001) Patterns of association, feeding competition and vocal communication in spider monkeys, Ateles geoffroyi. Dissertations, University of Pennsylvania. http://repository.upenn.edu/dissertations/AAI3003685. Accessed 1 Jan 2001
15. Gaing ZL (2003) Particle swarm optimization to solving the economic dispatch considering the generator constraints. IEEE Trans Power Syst 18:1187–1195. doi:10.1109/tpwrs.2003.814889
16. Gazi V, Passino KM (2004) Stability analysis of social foraging swarms. IEEE Trans Syst Man Cybern Part B 34(1):539–557
17. Hedar A, Ali AF (2012) Tabu search with multi-level neighborhood structures for high dimensional problems. Appl Intell 37:189–206
18. Hedar A, Ali AF, Hassan T (2011) Genetic algorithm and tabu search based methods for molecular 3D-structure prediction. Int J Numer Algebra Control Optim (NACO)

19. Hedar A, Ali AF, Hassan T (2010) Finding the 3D-structure of a molecule using genetic algorithm and tabu search methods. In: Proceeding of the 10th international conference on intelligent systems design and applications (ISDA2010), Cairo, Egypt
20. Holland JH (1975) Adaptation in natural and artificial systems. University of Michigan Press, Ann Arbor
21. Karaboga D, Akayb B (2011) Amodified artificial bee colony (ABC) algorithm for constrained optimization problems. Appl Soft Comput 11(3):3021–3031
22. Karaboga D, Basturk B (2007) A powerful and efficient algorithm for numerical function optimization: artificial bee colony (ABC) algorithm. J Glob Optim 39(3):459–471. doi:10.1007/s10898-007-9149-x
23. Kennedy J, Eberhart RC (1995) Particle swarm optimization. Proc IEEE Int Conf Neural Netw 4:1942–1948
24. Kim DH, Abraham A, Cho JH (2007) A hybrid genetic algorithm and bacterial foraging approach for global optimization. Inf Sci 177:3918–3937
25. Komarasamy G, Wahi A (2012) An optimized K-means clustering technique using bat algorithm. Eur J Sci Res 84(2):263–273
26. Lei MD (2008) A Pareto archive particle swarm optimization for multi-objective job shop scheduling. Comput Ind Eng 54(4):960–971
27. Nakamura RYM, Pereira LAM, Costa KA, Rodrigues D, Papa JP, Yang XS (2012) BBA: a binary bat algorithm for feature selection. In: 25th SIBGRAPI conference on graphics, patterns and images (SIBGRAPI). IEEE Publication, pp 291–297
28. Niknam T, Azizipanah-Abarghooee R, Zare M, Bahmani-Firouzi B (2012) Reserve constrained dynamic environmental/economic dispatch: a new multi-objective self-adaptive learning bat algorithm. Syst J IEEE 7:763–776. doi:10.1109/jsyst.2012.2225732
29. Norconk MA, Kinzey WG (1994) Challenge of neotropical frugivory: travel patterns of spider monkeys and bearded sakis. Am J Primatol 34(2):171–183
30. Passino MK (2002) Biomimicry of bacterial foraging for distributed optimization and control. IEEE Control Syst 22(3):52–67
31. Ramesh B, Mohan VCJ, Reddy VCV (2013) Application of bat algorithm for combined economic load and emission dispatch. Int J Electr Eng Telecommun 2:1–9
32. Roosmalen VMGM (1985) Instituto Nacional de Pesquisas da Amaznia. Habitat preferences, diet, feeding strategy and social organization of the black spider monkey (ateles paniscus paniscus linnaeus 1758) in surinam. Wageningen, Roosmalen
33. Simmen B, Sabatier D (1996) Diets of some french guianan primates: food composition and food choices. Int J Primatol 17(5):661–693
34. Selvakumar AI, Thanushkodi KA (2007) A new particle swarm optimization solution to non-convex economic dispatch problems. IEEE Trans Power Syst 22:42–51
35. Selvakumar AI, Thanushkodi K (2008) Anti-predatory particle swarm optimization: solution to non-convex economic dispatch problems. Electr Power Syst Res 78:2–10
36. Sidi A (2014) Economic dispatch problem using bat algorithm. Leonardo J Sci 24:75–84
37. Subbaraj P, Rengaraj R, Salivahanan S (2011) Enhancement of self-adaptive real-coded genetic algorithm using Taguchi method for economic dispatch problem. Appl Soft Comput 11:83–92
38. Teodorovic D, DellOrco M (2005) Bee colony optimization a cooperative learning approach to complex transportation problems. In: Advanced OR and AI methods in transportation: Proceedings of 16th MiniEURO conference and 10th meeting of EWGT (13–16 September 2005). Publishing House of the Polish Operational and System Research, Poznan, pp 51–60
39. Torczon V (1989) Multi-directional search: a direct search algorithm for parallel machines. Rice University, Department of Mathematical Sciences, Houston
40. Yang XS, Deb S (2009) Cuckoo search via levy flights. In: World congress on nature and biologically inspired computing, 2009. NaBIC 2009. IEEE, pp 210–214
41. Yang XS (2010) A new metaheuristic bat-inspired algorithm. In: Nature inspired cooperative strategies for optimization (NICSO 2010), pp 65–74
42. Yang XS (2012) Swarm-based meta-heuristic algorithms and no-free-lunch theorems. In: Parpinelli R, Lopes HS (eds) Theory and new applications of swarm intelligence. Intech Open Science, pp 1–16

43. Zhu G, Kwong S (2010) gbest-guided artificial bee colony algorithm for numerical function optimization. Appl Math Comput 217(7):3166–3173
44. Zielinski K, Weitkemper P, Laur R (2009) Optimization of power allocation for interference cancellation with particle swarm optimization. IEEE Trans Evol Comput 13(1):128–150

Chance-Constrained Fuzzy Goal Programming with Penalty Functions for Academic Resource Planning in University Management Using Genetic Algorithm

Bijay Baran Pal, R. Sophia Porchelvi and Animesh Biswas

Abstract This chapter addresses grafting of penalty functions in the framework of fuzzy goal programming (FGP) for modeling and solving academic resource planning problem by employing genetic algorithm (GA). In model formulation, first incorporation of penalty functions to membership goals associated with percentage achievement of fuzzy objectives in different ranges is defined to obtain appropriate membership values of objectives in the decision horizon. Then, a set of chance constraints that are inherent to model is transformed into deterministic equivalents to solve the problem using FGP methodology. In solution search process, a GA scheme is employed iteratively to evaluate goal achievement function on the basis of assigned priorities to model goals of the problem. A sensitivity analysis with variations in priority structure of goals is also performed, and *Euclidean distance function* is used to identify appropriate priority structure to reach optimal decision. A demonstrative case example is considered to illustrate the approach.

Keywords Academic resource planning · Fuzzy goal programming · Genetic algorithm · Membership function · Penalty function

1 Introduction

Higher educational planning is actually the application of rational and systematic analyses to the process of educational development with the aim of making education more effective in response to the needs and goals of students and society. The emphasis on growth and expansion of higher education system has been taken into account across the countries to cope with growing concern in quality of human

B.B. Pal (✉) · A. Biswas
Department of Mathematics, University of Kalyani, Kalyani 741235, West Bengal, India
e-mail: bbpal18@hotmail.com

R.S. Porchelvi
Department of Mathematics, ADM College for Women, Nagapattinam 611001, India

© Springer International Publishing AG 2017
S. Patnaik et al. (eds.), *Nature-Inspired Computing and Optimization*,
Modeling and Optimization in Science and Technologies 10,
DOI 10.1007/978-3-319-50920-4_18

resources toward sustainable development of modern society. In recent years, a substantial change in the concept of educational planning has undergone extensively, and opening of a wide range of new study areas has been taken into account to explore the future development of higher education.

In context to the above, one of the more important characteristics of education system lies in its salient linkages with socioeconomic structure of a country. Since enhancement of educational planning and investment of more money are closely interconnected, both financial and human resource implications have to be carefully assessed and managerial decision needs to be systematically taken into account to cope with different problems of educational development in a rapidly changing academic world. Typically, since salary cost representing a high percentage of total recurring expenditure, recruitment of academic personnel is not always flexible enough owing to limitations on financial recourses to supply academic manpower for consistent academic activities in higher education institutions. As a matter of fact, establishments as well as development of new academic departments are always in deep crises for enhancement of higher education. However, in response to dynamic situation of growing interest in higher education, the shift of higher educational responsibility to regional/local non-government agencies are found in many countries, where self-financing system acts as major financial strength to run institutions under private agencies. But, since academic enhancement is not purely technical and have sensitive socioeconomic dimensions, national educational planning to make overall investment for education health in essence is needed, and tuition and fees need be properly justified, particularly in countries with poor economic health of general citizens regarding investment in education.

The history of education shows that the work of Greek philosopher Xenophon [34] on general education to fit military, social and economic objectives for public services during the fourth century BC, some 2,500 years ago, is an ancestry of today's higher educational planning across the globe. Also, the work of philosopher Plato on education plan [7] to render serves to various social needs is also a historic root to start education in modern society.

However, with the passing of time, higher education system was reformed and controlled under educational policy of countries in various social settings. It may be mentioned here that typical kind of educational planning went on in most places prior to the Second World War and for many generations before inauguration of decentralized system of education, and acceleration of it was observed globally more or less than the middle of last century. Then, from the socioeconomic point of view, Weber in 1947 [51] first suggested a constructive educational plan to capable of achieving the highest degree of efficiency for better services to society. The incorporation of organizational models of higher education was first suggested by Millett in 1962 [31]. The early works on educational planning with the use of quantitative methods in management science had also been documented [12, 40] in the literature. Thereafter, "International Institute for Educational Planning (IIEP)" was established by UNESCO in 1963 to contribute to the development of education throughout the world to expand both knowledge and supply of competent professionals to various service sectors.

The management science models for higher educational plan discussed previously was first surveyed by Rath [41], and further study was made by Hufner in 1968 [20]. A critical analysis on the practices and concepts of educational planning that have undergone substantial change to rationalize the process of educational development had been studied by Coombs in 1968 [10]. The conceptual framework of mathematical programming (MP) model on higher education was first proposed by McNamara [29]. A comprehensive bibliography on the state of the art of modeling academic planning problems studied from the 1960s to early 1970s for enrichment of higher education was first prepared by Schroeder [45] in 1973. Thereafter, formal academic planning models were studied [2] deeply in the literature. Then, during 1980s, extensive study was made [5, 13, 49] with due considerations of socioeconomic criteria of countries. It may be mentioned, however, that the root of modern educational planning can be traced in Douglas North's pioneering work [33] on institutional economics for social uplift in global system. Then, a constructive model for academic resource planning was well documented [48] widely in the literature. However, the last two decades have witnessed a paradigm shift in the balance of interest between educational planning and educational resource allocation policy making on the basis of needs in society.

Now, it is to be mentioned that the academic resource planning problems are with multiplicity of objectives, i.e., multiobjective decision-making (MODM) problems, the multicriteria optimization method for university administration towards meeting different socioeconomic goals was introduced by Geoffrion et al. [14]. The goal programming (GP) [21] method as target-oriented approach for multiobjective decision analysis in crisp environment and that based on satisficing philosophy [47] has been successfully implemented to university management system [23, 25, 26, 46, 51] in the past.

However, in the context of using convention GP methods to academic resource planning problems as well as to other MODM problems, it is worthy to mention that the primary objective of decision maker (DM) is to minimize overall deviations from target levels of goals in the decision situation. But, owing to limitation on utilization of resources, achievement of aspired levels of all model goals is a trivial case in actual practice. To cope with such a situation, marginal relaxations on the target levels of objectives and scaling of different penalties in terms of penalty weights regarding achievement of goal values in different ranges need to be essentially taken into account to make appropriate solution of a problem. The incorporation of such a conceptual frame in functional form, called penalty functions, to GP model as an extended version of it was introduced by Charnes and Collomb in 1972 [9]. The effective use of such an approach to MODM problems was also discussed [24, 32, 42] previously.

Now, in the context of modeling practical decision problems, it is to be observed that DMs are often faced with the difficulty of assigning exact values of different parameters owing to inherent inexactness of parameters themselves and imprecision in judgments of human. In such a case, fuzzy programming (FP) [56], which is based on theory of fuzzy sets [54], has appeared to solve imprecisely defined MODM problems. The FGP [37] has also been successfully implemented to MODM problems

[4, 35] and other decision problems in fuzzy environment. The GP formulation of university resource planning problem with interval parameter sets [22] has also been studied by Pal et al. [39] in the recent past. However, extensive study in the area is at an early stage.

Again, in university management system, it may be noted that some model data that are inherent to a problem are probabilistically uncertain [27] in nature. Therefore, stochastic programming (SP) methods were studied [44] extensively and implemented to practical problems. However, the use of such an approach to an academic planning problem is yet to be studied extensively in the literature. Further, fuzzy as well as probabilistic data are often involved in some practical problems. Eventually, both FP and SP modeling aspects need be taken into account in the framework of executable model of a problem in uncertain environment. The various fuzzy stochastic programming (FSP) approaches studied in the field have been surveyed [28] in the past. Although FSP methods to MODM problems have been studied previously, in-depth study in this area is at an early stage. Furthermore, the study on FGP approaches to probabilistically defined academic management problems is yet to circulate in the literature.

Now, to solve MODM problems in both crisp and uncertain environments, it may be mentioned that the uses of traditional MP approaches often lead to local optimal solutions owing to adaptation of single-point-based iterative techniques in solution search processes. Again, computational complexity [3] frequently arises there for the use of such a method to problems with nonlinear characteristics of functional expressions of objectives/constraints in actual practice. To overcome such difficulties, GAs as global random search approaches [15] in the field of natural computing have been successfully employed to MODM problems [16, 35]. Although GA solution search approach to educational planning model has been discussed [36] in a previous study, probabilistic aspects of model parameters has not been considered there and exploration of the field of study is inadequate in actual practice. Further, different soft-computing approaches to various academic management problems have also been studied [1] in the past. However, a paradigm shift of introducing bio-inspired computing to FGP with *penalty functions* approach to higher academic resource planning problems is yet to be widely discussed in the literature.

The objective of the study is to present how the notion of *penalty functions* can be used to FGP model of a university planning problem by adopting a GA-based solution search scheme in the process of executing the problem. In model formulation, first penalty functions are introduced to measure percentage achievement of fuzzy objectives in terms of membership values of associated goals in different ranges. The probabilistically defined certain academic performance measuring constraints in fractional form are also converted into their deterministic equivalents to execute the model of the problem in deterministic environment. Then, *priority-based* FGP [37] method is addressed to design model of the problem by grafting *penalty functions* to reach most satisfactory decision. In model execution, a GA scheme is employed in an iterative manner to achieve goal levels according to priorities assigned to them. Again, sensitivity analysis is performed on the priorities of model goals to derive different possible solutions of the problem. Then, *Euclidean distance function* [53] is used to obtain best solution for staff allocation decision under a given priority structure of model goals of the problem. The approach is illustrated through the academic resource allocation problem of University of Kalyani in West Bengal (WB) of India. A comparison of model solution is also made with the solution obtained by using the conventional *minsum* FGP approach [38] to highlight superiority of the proposed method.

Now, the general FGP problem formulation is presented in the Sect. 2.

2 FGP Problem Formulation

The generic form of a multiobjective FP problem can be presented as:

Find X so as to:

$$\text{satisfy } F_k(X) \begin{Bmatrix} \lesssim \\ \gtrsim \end{Bmatrix} b_k, k = 1, 2, ..., K$$

$$X \in S\{X \in R^n | Pr[f(X) \begin{pmatrix} \geq \\ \leq \end{pmatrix} c] \geq p, X \geq 0, c \in R^m\},$$

$$X^{min} \leq X \leq X^{max},$$

$$X \geq 0, \tag{1}$$

where X is the vector of decision variables, b_k is the imprecisely defined aspiration level of kth objective $F_k(X)$, $k = 1, 2, ..., K$. The signs $\lesssim$ and $\gtrsim$ indicate fuzziness [56] of $\leq$ and $\geq$ restrictions, and where *min* and *max* stand for maximum and minimum, respectively.

It is assumed that the feasible region $S(\neq \varphi)$ is bounded. "*Pr*" indicates probabilistic constraint, $f(X)$ is $(m \times 1)$ vector of constraints set, c is resource vector, and $p(0 < p < 1)$ is used to indicate vector of probability levels of satisfaction for randomness in model parameters.

Now, in FP approach, fuzzily described goals are characterized by their associated membership functions. The membership functions are defined in the Sect. 2.1.

2.1 *Membership Function Characterization*

The characterizations of membership functions are made by defining tolerance ranges for measuring degree of achievement of fuzzy aspiration levels of goals in the decision situation.

The membership function $\mu_k(X)$ (say) for fuzzy goal $F_k(X)$ can be characterized as follows.

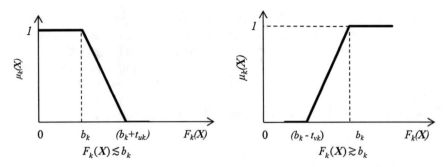

Fig. 1 Graphical representations of membership functions

For $\lesssim$ type restriction, if $(b_k + t_{uk})$ is taken as upper-tolerance limit for fuzzy goal achievement, then $\mu_k(X)$ can be defined as:

$$\mu_k(X) = \left\{ \begin{array}{ll} 1 & \text{, if } F_k(X) \leq b_k \\ \frac{(b_k+t_{uk})-F_k(X)}{t_{uk}} & \text{, if } b_k < F_k(X) < (b_k + t_{uk}) \\ 0 & \text{, if } F_k(X) \geq (b_k + t_{uk}) \end{array} \right\}, \; k \in K_1, \text{ (say)} \quad (2)$$

Again, for $\gtrsim$ type restriction, if $(b_k - t_{vk})$ is taken as lower-tolerance limit for fuzzy goal achievement, then $\mu_k(X)$ can be defined as:

$$\mu_k(X) = \left\{ \begin{array}{ll} 1 & \text{, if } F_k(X) \geq b_k \\ \frac{F_k(X)-(b_k-t_{vk})}{t_{vk}} & \text{, if } (b_k - t_{vk}) < F_k(X) < b_k \\ 0 & \text{, if } F_k(X) \leq (b_k - t_{vk}) \end{array} \right\}, \; k \in K_2, \text{ (say)} \quad (3)$$

where $t_{uk}(k \in K_1)$ and $t_{vk}(k \in K_2)$ represent upper- and lower-tolerance ranges, respectively, for achievement of aspired level b_k of kth fuzzy goal, and where $K_1 \cup K_2 = \{1, 2, ..., K\}$ with $K_1 \cap K_2 = \varphi$.

The graphical views of membership functions in (2) and (3) in their linear form are depicted in Fig. 1.

Now, in a general MP approach, the common practice is that of converting chance constraints into their deterministic equivalents to employ conventional solution method directly.

2.2 Deterministic Equivalents of Chance Constraints

The conversion of a chance constraint to its deterministic equivalent [6] depends on probability distribution of random parameters associated with the constraint. In the present decision situation, independent normally distributed random parameters are taken into consideration.

- The chance constraints with $\geq$ type in (1) in linear form can be explicitly presented as:

$$Pr[\sum_{j=1}^{n} a_{ij}x_j \geq c_i] \geq p_i, \ x_j \geq 0, i = 1, 2, ..., m \qquad (4)$$

where a_{ij} $(j = 1, 2, ..., n)$, and c_i are random coefficients and resource vector elements, respectively, associated with ith constraint, and p_i is the corresponding satisficing level of probability.

Then, using the standard probability rules, the deterministic equivalent of ith expression in (4) in quadratic form appears as:

$$E(y_i) + F_i^{-1}(1 - p_i)\sqrt{\{var\ (y_i)\}} \geq 0 \qquad (5)$$

where $y_i = (\sum_{j=1}^{n} a_{ij}x_j - c_i)$ and $F_i^{-1}(.)$ represents the inverse of probability distribution function $F(.)$, and where $E(y_i)$ and $var(y_i)$ designate mean and variance, respectively.

Here, it may be noted that if c_i is only random, then the expression in (5) would be linear in nature.

- When the chance constraints of above type with crisp coefficients are typically fractional in form, they can be represented as:

$$Pr\left[\frac{g_i(X)}{h_i(X)} \geq c_i\right] \geq p_i, \quad i = 1, 2, ..., m \qquad (6)$$

where $g_i(X)$ and $h_i(X)$ are linear in form, and where it is assumed that $h_i(X) > 0$ to preserve the feasibility of solution.

Then, the deterministic equivalent of ith expression in (6) in linear fractional form appears as:

$$\frac{g_i(X)}{h_i(X)} \geq E(c_i) + F_i^{-1}(p_i)\sqrt{\{var\ (c_i)\}}, \qquad (7)$$

where $E(c_i)$ and $var(c_i)$ represent mean and variance of c_i, respectively.

In an analogous way, the deterministic equivalents of chance constraints with "$\leq$" type restrictions can also be obtained.

Now, the formulation of an FGP model under a priority structure is presented in the Sect. 3.

3   Formulation of Priority Based FGP Model

In FGP method, the under-deviational variables associated with membership goals of the problem are minimized in goal achievement function (objective function) on the basis of importance of achieving aspired levels of fuzzy goal in the decision environment.

In the present decision situation, since a multiplicity of goals are involved and they conflict each other to achieve their aspired levels, the *priority-based* FGP [37] is considered to formulate the model of the problem. In *priority based* FGP, priorities are assigned to the goals according to their importance of achieving target levels, where a set of goals, which seems to be equally important concerning assignment of priorities, are included at the same priority level and numerical weights are given to them according to their relative importance of achieving aspired goal levels.
The generic form of a *priority based* FGP model can be presented as follows.

Find X so as to:

Minimize $Z = [P_1(\overline{\eta}), P_2(\overline{\eta}), ..., P_l(\overline{\eta}), ..., P_L(\overline{\eta})]$

and satisfy

$$\mu_k : \frac{(b_k + t_{uk}) - F_k(X)}{t_{uk}} + \eta_k - \rho_k = 1, \ k = 1, 2, ..., K_1,$$

$$\mu_k : \frac{F_k(X) - (b_k - t_{vk})}{t_{vk}} + \eta_k - \rho_k = 1, \ k = K_1 + 1, K_1 + 2, ..., K,$$

subject to the given system constraints in (1), (5) and (7),   (8)

where Z represents the vector of L priority achievement functions, η_k and $\rho_k (\geq 0)$ are under- and over-deviational variables, respectively, associated with kth membership goal. $P_l(\overline{\eta})$ is a linear function of weighted under-deviational variables, and where $P_l(\overline{\eta})$ is of the form:

$$P_l(\overline{\eta}) = \sum_{k \in K} w_{lk} \eta_{lk}, \ k = 1, 2, ..., K; l = 1, 2, ..., L; K \leq L \qquad (9)$$

where $\eta_{lk} (\geq 0)$ is renamed for η_k to represent it at lth priority level, $w_{lk} \ (>0)$ is the numerical weight associated with η_{lk} of kth goal relative to others which are grouped together at lth priority level.
The w_{lk} values are determined as [38]:

$$w_{lk} = \begin{cases} \frac{1}{(t_{uk})_l}, & \text{for the defined } \mu_k(X) \text{ in } (2) \\ \frac{1}{(t_{vk})_l}, & \text{for the defined } \mu_k(X) \text{ in } (3) \end{cases} \qquad (10)$$

where $(t_{uk})_l$ and $(t_{vk})_l$ are used to represent t_{uk} and t_{vk}, respectively, at lth priority level.

In the above formulation, the notion of pre-emptive priorities is that the goals which belong to lth priority level P_l are preferred most for the achievement of their aspired levels before considering achievement of goals included at the next priority P_{l+1}, regardless of any multiplier involved with P_{l+1}.

Also, the relationship among priorities can be represented as

$$P_1 >>> P_2 >>> \ldots >>> P_l >>> \ldots >>> P_L,$$

where ">>>" stands for "much greater than."

Now, in a decision situation, achievement of highest membership value of each of the fuzzy goals is a trivial one. Again, DM is often confused with that of assigning proper priorities to the goals for achievement of their individual aspired levels in an imprecise environment. To overcome the situation, the *Euclidean distance function for group decision analysis* [53] can be used to achieving an ideal-point-dependent solution and thereby selecting appropriate priority structure under which most satisfactory decision can be reached.

3.1 Euclidean Distance Function for Priority Structure Selection

Let $\{x_j^l; \ j = 1, 2, \ldots, n\}$ be the optimal solution obtained with the selection of lth priority structure, $l = 1, \ 2, \ldots, \ L$.

Consequently, the ideal solution can be recognized as $\{x_j^*; \ j = 1, 2, \ldots, n\}$, where $x_j^* = \underset{l=1}{\overset{L}{max}}\{x_j^l\}, \ j = 1, \ 2, \ldots, \ n.$

Using *Euclidean distance function*, the distance of lth solution from ideal point can be obtained as

$$E^{(l)} = [\sum_{j=1}^{n} (x_j^* - x_j^l)^2]^{1/2}, \tag{11}$$

where $E^{(l)}$ denotes distance function to measure distance between the solution achieved under lth priority structure and ideal solution.

Now, from the viewpoint of closeness of a solution to ideal point, the minimum distance always provides optimal decision.

Let $E^{(s)} = min\{E^{(l)}; l = 1, 2, \ldots, L\}$, where, "*min*" stands for minimum.

Then, sth priority structure is selected as the appropriate one to obtain optimal solution of the problem.

Now, from the viewpoint of practical implementation of the model in (8), it is to be observed that minimization of under-deviational variables of all goals to achieving highest membership value is not always possible owing to resource constraints, and membership values of goals are found to be marginal ones with respect to their tolerance ranges. Again, in a decision situation, deviations from goal levels in different ranges have different impacts on real system. Such a situation is not reflected as an output for the use of conventional FGP model to real-life problems. To overcome the difficulty, different scaling factors for measuring goal achievements at different levels of defined tolerance ranges would be an effective one to make appropriate decision. Here, the notion of using penalty functions [43] as employed in GP can be adapted to FGP model in (8) to create a balanced decision concerning achievements of fuzzy goals of the problem.

Now, the formulation of FGP model by incorporating penalty functions is discussed in the following section.

4   FGP Model with Penalty Functions

The notion of penalty functions is actually the scaling of penalties for goal achievements in different ranges instead of achieving goals uniquely with their deviations from aspired levels as defined in conventional fuzzy system.

The use of *penalty functions* is presented in the Sect. 4.1.

4.1   Penalty Function Description

In the field of GP, *penalty functions* are defined in terms of sum of deviational variables associated with the achievement of goal values in different ranges in the decision horizon. In goal achievement function, minimization of deviational variables with marginal penalties [42] defined for possible achievement of goal values in different ranges (called penalty scales) is considered.

Now, in general framework of FGP [38] model, only under-deviational variables associated with membership goals are involved in goal achievement function for minimizing them in making decision. As a matter of consequence, minimization of under-deviational variables of each membership goal arising out of grafting penalty functions need have to be taken into account in goal achievement function. In the sequel of model formulation, however, achievements of a fuzzy goal in terms of percentage [43] is considered to make commensurable decision. It is to be noted here that 100 % achievement of a fuzzy goal means its membership value is unity. In such a case, marginal penalty would be zero. Otherwise, marginal penalties as penalty weights need be introduced to find marginal values of membership functions in different ranges defined in imprecise environment. Actually, the defined penalty weights act as insurance against the violations of marginal achievements of goal values in different specified penalty scales. In actual practice, penalty scale prescription depends on the needs and desires of DM for searching solution in the domain of interest.

A system of penalty scales for marginal changes in membership values of a goal is summarized in Table 1.

Table 1 Description of penalty scales for membership goal achievement

Membership goal	Goal achievement range (in %)	Under-deviation (in %)	Marginal penalty
$\mu_k, k = 1, 2, ..., K$	Above 100	0	0
	100 to p_{k1}	$100 - p_{k1}$	α_{k1}
	p_{k1} to p_{k2}	$p_{k1} - p_{k2}$	α_{k2}
	....	....	...
	Below p_{kQ}	∞	∞

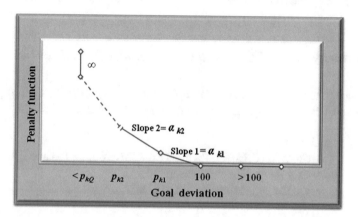

Fig. 2 Penalty function representation for fuzzy goal achievement

The schematic presentation of penalty scales for goal achievement is depicted in Fig. 2.

Now, grafting of *penalty functions* means redefining of $\eta_k (k = 1, 2, ..., K)$ by substituting a sum of different under-deviational variables associated with marginal achievements of membership values in different penalty ranges.

Let $\eta_{kq} (\geq 0)$, $(k = 1, 2, ..., K; q = 1, 2, ..., Q)$, designate under-deviational variables associated with marginal changes in goal achievement for penalty scales defined in Table 1.

Then, using the expression in (8), *penalty functions* grafted membership goals can be restated as:

$$\frac{(b_k + t_{uk}) - F_k(X)}{t_{uk}} + \sum_{q=1}^{Q} \eta_{kq} - \rho_k = 1, (k \in K_1)$$

$$\frac{F_k(X) - (b_k - t_{vk})}{t_{vk}} + \sum_{q=1}^{Q} \eta_{kq} - \rho_k = 1, (k \in K_2) \qquad (12)$$

Now, it is to be followed that the minimization of η_{kq} in (12) means further minimization of deviations from marginal goal values associated with defined goal achievement intervals. As such, η_{kq}, ($\forall k$ and q), in flexible form of goals appear as:

$$\eta_{k1} + \sigma_{k1}^{-} - \sigma_{k1}^{+} = \frac{p_{k1}}{100} - 1, \text{ and } \eta_{kq} + \sigma_{kq}^{-} - \sigma_{kq}^{+} = \frac{p_{kq-1} - p_{kq}}{100}, \forall k; \ q = 2, ..., Q$$

(13)

where σ_{kq}^{-}, $\sigma_{kq}^{+}(\geq 0)$ with $\sigma_{kq}^{-}.\sigma_{kq}^{+} = 0$ designate under- and over-deviational variables, respectively.

Now, priority-based FGP model with penalty functions is described in Sect. 4.2.

4.2 Priority Based FGP Model with Penalty Functions

In an analogous to conventional FGP formulation in (9), incorporation of *penalty function* to FGP model under a priority structure can be stated as follows:

> Find X so as to:
> Minimize $Z = [P_1(\bar{\eta}, \bar{\sigma}), P_2(\bar{\eta}, \bar{\sigma}), ..., P_l(\bar{\eta}, \bar{\sigma}), ..., P_L(\bar{\eta}, \bar{\sigma})]$,   (14)
> and satisfy the goal expressions in (13),
> subject to the constraints in (5) and (7),

where $P_l(\bar{\eta}, \bar{\sigma}) = \sum_{k \in K} w_{lk} \left\{ \sum_{q=1}^{Q} \alpha_{lkq}(\eta_{lkq} + \sigma_{lkq}^{-}) \right\}$, $k = 1, 2, ..., K; l = 1, 2, ..., L;$
$K \leq L$ and where $\alpha_{lkq}(> 0), q = 1, 2, ..., Q$, represent penalty weights associated with achievements of kth goal in different penalty ranges, which is included at lth priority level. The penalty weights can be numerically defined as in (10). The other notations can also be defined in an analogous to those defined in (8).

Now, the GA scheme to solve the proposed model is described in Sect. 4.3.

4.3 GA Scheme for FGP Model

In a GA solution search process [15, 19, 30], generation of a new population (i.e., new solution candidate) is made through execution of problem with the use of probabilistically defined operators: *selection, crossover,* and *mutation.* The real-value-coded chromosomes are considered here in evolutionary process of solving the problem. In genetic search process, simple roulette-wheel selection [11] and arithmetic crossover [18] for exploration of the promising regions of search space and uniform mutation [11] for exploitation of fittest chromosome are considered in the domain of feasible solution sets. The evaluation of fitness of a chromosome is made in the premises of optimizing objectives of the problem.

In the GA scheme, the function Z in (14) acts as an evaluation function to generate new population in course of searching decision.

The evaluation function can be represented as:

$$eval(S_g)_l = (Z_g)_l = (\sum_{k \in K} w_{lk} \left\{ \sum_{q=1}^{Q} \alpha_{lkq} (\eta_{lkq} + \sigma_{lkq}^{-}) \right\})_g, \qquad (15)$$

where S signifies survival function in the notion of nature-inspired computation, and $(Z_g)_l$ is renamed for Z_l to measure fitness value of gth chromosome, when evaluation for achievement of goals at lth priority level (P_l) is considered in the process of executing the problem.

The best value (Z_l^*) for fittest chromosome at a generation is determined as:

$$Z_l^* = min \{eval(S_g)_l | g = 1, 2, ..., G\}, \ l = 1, \ 2, \ ..., \ L \qquad (16)$$

The execution of the problem step by step on the basis of assigned priorities for goal achievement has been discussed [36] in a previous study.

Now, the formulation of FGP model of academic resource planning problem is described in the following section.

5   FGP Formulation of the Problem

The decision variables and different types of parameters involved with the proposed problem are introduced in Sect. 5.1.

5.1   Definitions of Decision Variables and Parameters

(i) Decision variables

$F_{drt} = $ number of full-time teaching staff (FTS) in department d, rank r, at time period t.

$P_{dt} = $ number of part-time teaching staff (PTS) employed in department d at time period t.

$N_{dt} = $ number of non-teaching staff (NTS) in department d at time period t.

(ii) Parameters

The following parameters are introduced to formulate the proposed model.

$[MF]_{drt} = $ minimum number of FTS required in department d, rank r, during time period t.

$[TF]_{dt} = $ total FTS required in department d at time period t.

$[MN]_{dt} =$ minimum number of non-teaching staff (NTS) required to running department d at time period t.

$[TS]_{dt} =$ total number of students (TS) enrolled in department d at time period t.

$[AST]_{drt} =$ annual (average) salary of a FTS in department d, rank r, at time period t.

$[ASN]_{dt} =$ annual (average) salary of a NTS in department d at time period t.

$[ARP]_{dt} =$ annual remuneration of a PTS in department d at time period t.

$[PRB]_{dt} =$ payroll budget allocated to department d at time period t.

$\alpha_d =$ ratio of PTS and FTS in department d.

$\beta_d =$ ratio of NTS and total teaching staff (TTS) [FTS and PTS] in department d.

$\gamma_d =$ ratio of total teaching staff (TTS) and TS in department d.

Now, fuzzy goals and constraints of the problem are described in the following section.

5.2 Descriptions of Fuzzy Goals and Constraints

(a) Fuzzy goal description

Two types of fuzzy goals that are involved with the problem are as follows:

- *FTS Goals*

  To measure academic performances of departments, an estimated number of total FTS should always be employed to each of the departments by management sector of university. But, due to limitation of budget, aspired level of FTS becomes fuzzy in nature during any plan period.

  The fuzzy goal expressions appear as:

$$\sum_{r=1}^{R} F_{drt} \gtrsim [TF]_{dt}, \quad d = 1, 2, ..., D \tag{17}$$

Now, as in conventional GP with *penalty functions* [42], the goals in (17) can be normalized to make them commensurable as well as to measure percentage achievements of goals.

Then, the goal expression in (17) takes the form:

$$\sum_{r=1}^{R} \left(\frac{100}{[TF]_{dt}} \right) F_{drt} \gtrsim 100, \quad d = 1, 2, ..., D \tag{18}$$

- *Budget goals*

  In the planning horizon, since budget is always limited to run a university, the payroll budget allocation to each department is fuzzily described.

  The budget goal expression appears as:

$$\sum_{r=1}^{R} [AST]_{drt} F_{drt} + [ASN]_{dt} N_{dt} + [ARP]_{dt} P_{dt} \lesssim [PRB]_{dt}, d = 1, 2, ..., D$$

(19)

In percentage scale, the expression in (19) takes the form:

$$\sum_{r=1}^{R} (\frac{100}{[PRB]_{dt}})[AST]_{drt} F_{drt} + (\frac{100}{[PRB]_{dt}})[ASN]_{dt} N_{dt} + (\frac{100}{[PRB]_{dt}})[ARP]_{dt} P_{dt} \lesssim 100, d = 1, 2, ..., D$$

(20)

Then, membership functions of the stated goals in (18) and (20) can be precisely stated as:

$$\mu_{F_d} : \frac{F_d(x_d) - (100 - t_{vd})}{t_{vd}} + \eta_d - \rho_d = 1, \quad d = 1, 2, ..., D$$

$$\mu_{B_d} : \frac{(100 + t_{ud}) - F_d(X_d)}{t_{ud}} + \eta_d - \rho_d = 1, \quad d = D+1, D+2, ..., 2D \quad (21)$$

where t_{vd} and t_{ud} represent lower- and upper-tolerance ranges, respectively, for achievement of aspired level 100 of the associated fuzzy goal, and where $F_d(x_d) = \sum_{r=1}^{R} \left(\frac{100}{[TF]_{dt}}\right) F_{drt}$, and

$$F_d(X_d) = \sum_{r=1}^{R} (\frac{100}{[PRB]_{dt}})[AST]_{drt} F_{drt} + (\frac{100}{[PRB]_{dt}})[ASN]_{dt} N_{dt} + (\frac{100}{[PRB]_{dt}})[ARP]_{dt} P_{dt},$$

where x_d is used to denote F_{drt}, and X_d is used to represent the vector of variables $(F_{drt}, N_{dt}, P_{dt})$ for concise representation of the expressions as appeared in (21).

Now, the system constraints are defined as follows:

(b) Crisp constraint description

The linear crisp constraints involved with model of the problem are as follows:

- *FTS constraints*

  To make effective running of academic curriculum, a minimum number of FTS at each rank need be provided to each of the departments.

  The FTS constraints appear as:

$$F_{drt} \geq [MF]_{drt}, r = 1, 2, ..., R; d = 1, 2, ..., D \quad (22)$$

- *NTS constraints*

  To execute official and teaching-related activities, a minimum number of NTS should be employed to each of the departments.

  The NTS constraints take the form:

$$N_{dt} \geq [MN]_{dt}, d = 1, 2, ..., D \quad (23)$$

(c) Probabilistic ratio constraint description
The different types of probabilistically defined constraints that are typically ratio in form are discussed as follows.

- *PTS-FTS constraints*
  In an uncertain situation, if it is not fit to employ required number of FTS at time period t, PTS at a certain ratio to total FTS should be provided in each of the departments.
  The ratio constraints appear as:

$$Pr[\frac{P_{dt}}{\sum\limits_{r=1}^{R} F_{drt}} \geq \alpha_d] \geq p_{\alpha_d}, \ d = 1, 2, ..., D \qquad (24)$$

- *NTS-TTS constraints*
  To assist the academic and official activities, a certain ratio of NTS to TTS should be maintained in each of the departments.
  The constraints take the form:

$$Pr[\frac{N_{dt}}{(\sum\limits_{r=1}^{R} F_{drt} + P_{dt})} \geq \beta_d] \geq p_{\beta_d}, \ d = 1, 2, ..., D \qquad (25)$$

- *TTS-TS constraints*
  To measure the performance against academic load, a certain ratio of TTS to TS should be considered for smooth running of departments.
  The constraints appear as:

$$Pr[\frac{(\sum\limits_{r=1}^{R} F_{drt} + P_{dt})}{[TS]_{dt}} \geq \gamma_d] \geq p_{\gamma_d}, \ d = 1, 2, ..., D \qquad (26)$$

Then, conversion of constraints set in (24)–(26) to their deterministic equivalents can be obtained by following the expression in (7).

Now, the executable FGP model of the problem by grafting *penalty functions* to model goals defined in (21) and thereby solving the problem by using GA is presented via a demonstrative case example in Sect. 6.

6 A Case Example

The academic resource allocation problem of the University of Kalyani (KU), WB in India is considered to demonstrate the effective use of the proposed approach.

Table 2 Data descriptions of FTS and payroll budget goal levels (2015–2016)

Department (d)	1	2	3	4	5	6
Goal level of FTS	12	10	8	8	12	12
Payroll Budget (in Rs.Lac)	96.68	88.5	62.87	62.37	121.74	109.79

Table 3 Data descriptions of providing minimum number of FTS and NTS

Department (d)	1	2	3	4	5	6
FTS: ranks [1, 2, 3]	(1, 2, 3)	(1, 2, 3)	(1, 3, 2)	(1, 2, 3)	(2, 4, 4)	(3, 3, 4)
NTS	6	5	6	5	4	5

The four departments, Computer Science & Engineering. (CSE), Master of Business Administration (MBA), Physiology (PHY), Microbiology (MB) established only few years back and two departments: Mathematics (MATH) and Statistics (STAT), established at the inception of KU are taken into account to demonstrate the model.

The data were collected from the Budget allocation programs published by KU [8]. The financial year (2015–2016) data along with four preceding years' (2011–2015) data concerned with fitting of probabilistic constraints were taken into account to formulate model of the problem. The departments are successively numbered 1, 2, 3, 4, 5, 6. The FTS ranks: professor, associate professor, and assistant professor are numbered 1, 2, and 3, respectively. Again, since staff allocation decision is a continuous updating process and decision making for the running period (2015–2016) is considered, specification of time (t) is omitted for model simplification in the decision-making situation. Again, for simplicity and without loss of generality, two categories of penalty scales individually associated with FTS and payroll budget for all the departments are taken into account to formulate the proposed model. The model data are presented in Tables 2, 3, 4, 5, and 6.

Now, following the expressions in (12), the model goals are constructed by using the data presented in Tables 2, 4 and 6.

Table 4 Data descriptions of annual average costs for FTS, NTS, and PTS

Rank (r)	1	2	3	PTS	NTS
Cost (Rs.Lac)	9.17	7.2	5.16	1.5	1.8

Table 5 Data descriptions of different ratios: From period (2011–2012) to (2015–2016)

Department → Staff ratio ↓	1	2	3	4	5	6
PTS-FTS	(0.31, 0.29, 0.30, 0.33, 0.27)	(0.26, 0.23, 0.26, 0.24, 0.26)	(0.3, 0.31, 0.33, 0.26, 0.30)	(0.33, 0.32, 0.27, 0.29, 0.29)	(0.25, 0.24, 0.23, 0.28, 0.25)	(0.27, 0.26, 0.22, 0.24, 0.26)
TTS-TS	(0.19, 0.17, 0.18, 0.24, 0.22)	(0.15, 0.16, 0.13, 0.15, 0.16)	(0.26, 0.22, 0.22, 0.28, 0.27)	(0.22, 0.22, 0.19, 0.17, 0.20)	(0.16, 0.13, 0.16, 0.15, 0.15)	(0.27, 0.26, 0.25, 0.21, 0.26)
NTS-TTS	(0.35, 0.35, 0.36, 0.33, 0.36)	(0.28, 0.32, 0.33, 0.30, 0.27)	(0.33, 0.37, 0.33, 0.34, 0.38)	(0.41, 0.40, 0.40, 0.40, 0.39)	(0.26, 0.27, 0.22, 0.22, 0.28)	(0.26, 0.26, 0.26, 0.23, 0.24)

Table 6 Data descriptions of penalty scales

Membership goal	Goal attainment range (in %)	Under-deviation (in %)	(Aspired Membership Value, Under-deviation)	Marginal penalty
$\mu_{F_d}, d = 1$ to 6	Above 100	0	$(>1, 0)$	0
	100–80	20	$(1, 0.20)$	0.05
	80–70	10	$(0.80, 0.10)$	0.10
	Below 70	∞	∞	∞
$\mu_{B_d}, d = 1$ to 6	Above 100	0	$(>1, 0)$	0
	100–85	15	$(1, 0.15)$	0.067
	85–75	10	$(0.75, 0.10)$	0.10
	Below 75	∞	∞	∞

Then, *penalty function* incorporated membership goals are obtained as follows.

$$
\begin{aligned}
0.05\left[8.33\left(F_{11} + F_{12} + F_{13}\right) - 80\right] + \eta_{11} + \eta_{12} - \rho_1 &= 1, \\
0.05\left[10\left(F_{21} + F_{22} + F_{23}\right) - 80\right] + \eta_{21} + \eta_{22} - \rho_2 &= 1, \\
0.05\left[12.5\left(F_{31} + F_{32} + F_{33}\right) - 80\right] + \eta_{31} + \eta_{32} - \rho_3 &= 1, \\
0.05\left[12.5\left(F_{41} + F_{42} + F_{43}\right) - 80\right] + \eta_{41} + \eta_{42} - \rho_4 &= 1, \\
0.05\left[8.33\left(F_{51} + F_{52} + F_{53}\right) - 80\right] + \eta_{51} + \eta_{52} - \rho_5 &= 1, \\
0.05\left[8.33\left(F_{61} + F_{62} + F_{63}\right) - 80\right] + \eta_{61} + \eta_{62} - \rho_6 &= 1 \qquad \text{(FTS goals)}
\end{aligned}
\tag{27}
$$

$$
\begin{aligned}
0.067\left[115 - \{1.034\left(9.17F_{11} + 7.2F_{12} + 5.16F_{13} + 1.8N_1 + 1.5P_1\right)\}\right] + \eta_{71} + \eta_{72} - \rho_7 &= 1, \\
0.067\left[115 - \{1.123\left(9.17F_{21} + 7.2F_{22} + 5.16F_{23} + 1.8N_2 + 1.5P_2\right)\}\right] + \eta_{81} + \eta_{82} - \rho_7 &= 1, \\
0.067\left[115 - \{1.591\left(9.17F_{31} + 7.2F_{32} + 5.16F_{33} + 1.8N_3 + 1.5P_3\right)\}\right] + \eta_{91} + \eta_{92} - \rho_9 &= 1, \\
0.067\left[115 - \{1.603\left(9.17F_{41} + 7.2F_{42} + 5.16F_{43} + 1.8N_4 + 1.5P_4\right)\}\right] + \eta_{10,1} + \eta_{10,2} - \rho_{10} &= 1, \\
0.067\left[115 - \{0.821\left(9.17F_{51} + 7.2F_{52} + 5.16F_{53} + 1.8N_5 + 1.5P_5\right)\}\right] + \eta_{11,1} + \eta_{11,2} - \rho_{11} &= 1, \\
0.067\left[115 - \{0.911\left(9.17F_{61} + 7.2F_{62} + 5.16F_{63} + 1.8N_6 + 1.5P_6\right)\}\right] + \eta_{12,1} + \eta_{12,2} - \rho_{12} &= 1 \quad \text{(Budget goals)}
\end{aligned}
\tag{28}
$$

The goals associated with incorporation of penalty scales appear as:

$$\eta_{d1} + \sigma_{d1}^- - \sigma_{d1}^+ = 0.20, \eta_{d2} + \sigma_{d2}^- - \sigma_{d2}^+ = 0.10, d = 1, 2, ..., 6;$$
$$\eta_{d1} + \sigma_{d1}^- - \sigma_{d1}^+ = 0.20, \eta_{d2} + \sigma_{d2}^- - \sigma_{d2}^+ = 0.10, d = 7, 8, ..., 12 \qquad (29)$$

Then, using the data presented in Table 3, the crisp constraints are obtained as:

$$F_{11} \geq 1, F_{12} \geq 2, F_{13} \geq 3, F_{21} \geq 1, F_{22} \geq 2, F_{23} \geq 3, F_{31} \geq 1, F_{32} \geq 3, F_{33} \geq 2, F_{41} \geq 1,$$
$$F_{42} \geq 2, F_{43} \geq 3, F_{51} \geq 2, F_{52} \geq 4, F_{53} \geq 4, F_{61} \geq 2, F_{62} \geq 2, F_{63} \geq 4 \quad \text{(FTS)} \qquad (30)$$

$$N_1 \geq 6, N_2 \geq 5, N_3 \geq 6, N_4 \geq 5, N_5 \geq 4, N_6 \geq 5 \qquad \qquad \text{(NTS)} \qquad (31)$$

Now, using the data in Table 5, the mean and variance pairs associated with the ratios of PTS-FTS, NTS-TTS, and TTS-TS for the six individual departments are successively obtained as:

$\{(0.30, 0.0005); (0.25, 0.0002); (0.30, 0.0006); (0.30, 0.0006); (0.25, 0.0003);$
$(0.25, 0.0004)\}, \{(0.35, 0.0002); (0.30, 0.0006); (0.35, 0.0006); (0.40, 0.0001);$
$(0.25, 0.0008); (0.25, 0.0002)\}$ and $\{(0.20, 0.0008); (0.15, 0.0002); (0.25, 0.0008);$
$(0.20, 0.0004); (0.15, 0.0002); (0.25, 0.0005)\}.$

Then, the deterministic equivalents of probabilistic constraints are obtained as:

$$P_1 \geq 0.30(F_{11} + F_{12} + F_{13}), P_2 \geq 0.25(F_{21} + F_{22} + F_{23}), P_3 \geq 0.30(F_{31} + F_{32} + F_{33}),$$
$$P_4 \geq 0.30(F_{41} + F_{42} + F_{43}), P_5 \geq 0.25(F_{51} + F_{52} + F_{53}), P_6 \geq 0.25(F_{61} + F_{62} + F_{63}) \quad (32)$$
$$N_1 \geq 0.35(F_{11} + F_{12} + F_{13} + P_1), N_2 \geq 0.30(F_{21} + F_{22} + F_{23} + P_2),$$

$$N_3 \geq 0.35(F_{31} + F_{32} + F_{33} + P_3), N_4 \geq 0.40(F_{41} + F_{42} + F_{43} + P_4),$$
$$N_5 \geq 0.25(F_{51} + F_{52} + F_{53} + P_5), N_6 \geq 0.25(F_{61} + F_{62} + F_{63} + P_6) \qquad (33)$$

$$(F_{11} + F_{12} + F_{13} + P_1) \geq 3.80, (F_{21} + F_{22} + F_{23} + P_2) \geq 4.05, (F_{31} + F_{32} + F_{33} + P_3) \geq 4.25,$$
$$(F_{41} + F_{42} + F_{43} + P_4) \geq 4, (F_{51} + F_{52} + F_{53} + P_5) \geq 13.20, (F_{61} + F_{62} + F_{63} + P_6) \geq 8.80$$
$$(34)$$

Now, the executable model can be constructed by using the expressions in (14).

In the execution process, four priority factors P_1, P_2, P_3, and P_4 are introduced to include the model goals in (27)–(29). Also, three priority structures are considered to execute the problem under three successive runs and then to perform sensitivity analysis on model solutions.

The executable model of the problem appears as:

Find $(F_{rd}, N_d, P_d | r = 1, 2, 3; d = 1, 2, 3, 4, 5, 6)$ so as to:

Minimize Z_l

and satisfy the goal expressions in (27)–(29), subject to the constraints set in (30)–(34), $\qquad \qquad (35)$

Table 7 Priority achievement functions under the three runs

Run	Priority achievement function (Z_l)
1	$Z_1 = \begin{bmatrix} P_1 \left(\sum_{d=1}^{4} w_{d1}\{0.05(\eta_{d1} + \sigma_{d1}^{-}) + 0.10(\eta_{d2} + \sigma_{d2}^{-})\} \right), \\ P_2 \left(\sum_{d=5}^{6} w_{d2}\{0.05(\eta_{d1} + \sigma_{d1}^{-}) + 0.10(\eta_{d2} + \sigma_{d2}^{-})\} \right), \\ P_3 \left(\sum_{d=7}^{10} w_{d3}\{0.067(\eta_{d1} + \sigma_{d1}^{-}) + 0.10(\eta_{d2} + \sigma_{d2}^{-})\} \right), \\ P_4 \left(\sum_{d=11}^{12} w_{d4}\{0.067(\eta_{d1} + \sigma_{d1}^{-}) + 0.10(\eta_{d2} + \sigma_{d2}^{-})\} \right) \end{bmatrix}$
2	$Z_2 = \begin{bmatrix} P_1 \left(\sum_{d=1}^{4} w_{d1}\{0.05(\eta_{d1} + \sigma_{d1}^{-}) + 0.10(\eta_{d2} + \sigma_{d2}^{-})\} \right), \\ P_2 \left(\sum_{d=7}^{10} w_{d2}\{0.067(\eta_{d1} + \sigma_{d1}^{-}) + 0.10(\eta_{d2} + \sigma_{d2}^{-})\} \right), \\ P_3 \left(\sum_{d=5}^{6} w_{d3}\{0.05(\eta_{d1} + \sigma_{d1}^{-}) + 0.10(\eta_{d,2} + \sigma_{d2}^{-})\} \right), \\ P_4 \left(\sum_{d=11}^{12} w_{d4}\{0.067(\eta_{d1} + \sigma_{d1}^{-}) + 0.10(\eta_{d2} + \sigma_{d2}^{-})\} \right) \end{bmatrix}$
3	$Z_3 = \begin{bmatrix} P_1 \left(\sum_{d=5}^{6} w_{d1}\{0.05(\eta_{d1} + \sigma_{d1}^{-}) + 0.10(\eta_{d2} + \sigma_{d2}^{-})\} \right), \\ P_2 \left(\sum_{d=1}^{4} w_{d2}\{0.05(\eta_{d1} + \sigma_{d1}^{-}) + 0.10(\eta_{d2} + \sigma_{d2}^{-})\} \right), \\ P_3 \left(\sum_{d=11}^{12} w_{d3}\{0.067(\eta_{d1} + \sigma_{d1}^{-}) + 0.10(\eta_{d2} + \sigma_{d2}^{-})\} \right), \\ P_4 \left(\sum_{d=7}^{10} w_{d4}\{0.067(\eta_{d1} + \sigma_{d1}^{-}) + 0.10(\eta_{d2} + \sigma_{d2}^{-})\} \right) \end{bmatrix}$

where F_{rd}, N_d, P_d (≥ 0) are all integers.

The three priority achievement functions Z_l, $(l = 1, 2, 3)$, under three successive runs are presented in Table 7.

Then, in course of solving the problem in (35), the function Z_l is considered fitness function in evolutionary process of using GA. The execution is made step by step according to priorities assigned to model goals of the problem. The GA-based program is designed in Language C^{++}.

The chromosome length 30 with population size 100 as in standard GA scheme is considered to searching of feasible solution in the domain of interest. The number of generations = 300 is initially taken into account to conduct experiments with different values of p_c and p_m. The execution is performed in Intel Pentium IV with 2.66 GHz clock pulse and 1GB RAM. It was found that $p_c = 0.8$ and $p_m = 0.08$ are successful to reach the optimal solution with limited number of generations 200.

The staff allocation decisions under the three runs are displayed in Table 8.

Now following the results in Table 8, the ideal solution point is found as (Professor, Associate professor, Assistant professor, PTS, NTS) = (5, 6, 9, 4, 7).

Table 8 Staff allocation decisions under the three runs

Run	Department → Staff ↓	CSE	MBA	PHY	MB	MATH	STAT
1	Professor	1	1	1	1	2	3
	Associate Professor	6	4	3	2	6	3
	Assistant Professor	5	5	4	5	4	6
	PTS	4	3	3	3	3	3
	NTS	6	5	6	5	4	5
2	Professor	1	1	1	1	2	3
	Associate Professor	6	5	3	3	6	3
	Assistant Professor	5	4	4	4	4	6
	PTS	4	3	3	3	4	3
	NTS	6	5	6	5	4	5
3	Professor	1	5	1	1	2	3
	Associate Professor	2	2	3	2	5	3
	Assistant Professor	9	3	4	5	5	6
	PTS	4	3	3	3	3	4
	NTS	7	5	6	5	4	5

Then, the *Euclidean distances* obtained under three successive runs are obtained as $E^{(1)} = 15.81$, $E^{(2)} = 15.74$, $E^{(3)} = 16$.

The result shows that the minimum distance corresponds to $E^{(2)} = 15.74$.

Therefore, the priority structure under Run 2 would be the appropriate one to reach optimal staff allocation decision in the planning horizon.

6.1 An Illustration for Performance Comparison

If priority structure of the model goals is not taken into account, i.e., *minsum* FGP [38] model (where minimization of the sum of weighted deviational variables in the achievement function is considered) with *penalty functions* is considered, then the evaluation function appears as:

Table 9 Staff allocation decisions under the two FGP approaches

Approach	Department → Staff ↓	CSE	MBA	PHY	MB	MATH	STAT
Proposed FGP	Professor	1	1	1	1	2	3
	Associate Professor	6	5	3	3	6	3
	Assistant Professor	5	4	4	4	4	6
	PTS	4	3	3	3	4	3
	NTS	6	5	6	5	4	5
Minsum FGP	Professor	1	1	1	1	1	3
	Associate Professor	2	2	3	2	2	3
	Assistant Professor	9	7	4	5	9	6
	PTS	4	3	3	3	4	3
	NTS	6	5	6	5	6	5

$$eval(S_g) = (Z_g) = \left(\begin{array}{l} w_1\{0.05(\eta_{11} + \sigma_{11}^-) + 0.10(\eta_{12} + \sigma_{12}^-)\} + w_2\{0.05(\eta_{21} + \sigma_{21}^-) + 0.10(\eta_{22} + \sigma_{22}^-)\} + \\ w_3\{0.05(\eta_{31} + \sigma_{31}^-) + 0.10(\eta_{32} + \sigma_{32}^-)\} + w_4\{0.05(\eta_{41} + \sigma_{41}^-) + 0.10(\eta_{42} + \sigma_{42}^-)\} + \\ w_5\{0.05(\eta_{51} + \sigma_{51}^-) + 0.10(\eta_{52} + \sigma_{52}^-)\} + w_6\{0.05(\eta_{61} + \sigma_{61}^-) + 0.10(\eta_{62} + \sigma_{62}^-)\} + \\ w_7\{0.067(\eta_{71} + \sigma_{71}^-) + 0.10(\eta_{72} + \sigma_{72}^-)\} + w_8\{0.067(\eta_{81} + \sigma_{81}^-) + 0.10(\eta_{82} + \sigma_{82}^-)\} + \\ w_9\{0.067(\eta_{91} + \sigma_{91}^-) + 0.10(\eta_{92} + \sigma_{92}^-)\} + w_{10}\{0.067(\eta_{10,1} + \sigma_{10,1}^-) + \\ 0.10(\eta_{10,2} + \sigma_{10,2}^-)\} + w_{11}\{0.067(\eta_{11,1} + \sigma_{11,1}^-) + 0.10(\eta_{11,2} + \sigma_{11,2}^-)\} + \\ w_{12}\{0.007(\eta_{12,1} + \sigma_{12,1}^-) + 0.10(\eta_{12,2} + \sigma_{12,2}^-)\} \end{array} \right)_g ,$$

$$(36)$$

where the best value (Z^*) for fittest chromosome is determined as:

$$Z^* = min \{eval(S_g)|g = 1, 2, ..., G\}.$$

The solution obtained here in the same decision environment along with the resultant decision reached under the proposed approach is presented in Table 9.
The decisions obtained under the two approaches are displayed in Fig. 3.
A comparison shows that a better staff allocation plan is achieved under the proposed approach in the decision-making environment.

NOTE 3: In the context using simple GA to the problem, it may be mentioned that since feasible solution space $S(\neq \varphi)$ is bounded and objectives are explicitly linear in nature, the searching of solution converges finitely. Again, since the model execution works step by step on the basis of priorities of goals, the execution process always terminates after a finite number of generations. Here, termination condition occurs when either $Z^* = 0$ is achieved or a value of $Z^* > 0$ cannot be improved further at a next generation number.

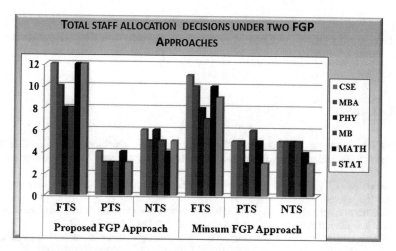

Fig. 3 Staff allocation decisions under the two approaches

7 Conclusions

The main merit of using the proposed approach to academic resource planning problem is that the possible instances of fuzzy data can be accommodated in the framework of model without involving any computational difficulty. Again, the model is a flexible enough with regard to the incorporation of various fuzzy/probabilistic data to arrive at a satisfactory decision on the basis of needs and desires of university administration regarding enhancement of higher education in the premises of uncertainty. The major advantage of using GA search method to the proposed model is that computational load with linearization of model goals/constraints along with computational complexity [17] with arriving at suboptimal (local) solution does not arise here in the process of searching decision in uncertain environment.

It may be pointed out here that although the proposed model is an effective one to make proper staff allocation decision for academic enrichment, measuring of output /benefit from an educational programme would have to be a part of such problem from the perspective of growing concern about both qualitative and quantitative education in society. But, measuring of output of such problem is one of the most difficult problems, because the problem of expanding quality of education and human resource generation is typically inexact in nature in a dynamic situation of evolving academic awareness in society. Although study on input/output analysis in education have been made [52, 55] in the past, an effective approach for evaluation of such criterion is an emerging research problem.

Finally, it may be concluded that the model discussed in this chapter may focus on a more general understanding of modeling education planning and implementing natural computing tools to expand education culture to fit various objectives in society in the current academic world.

Acknowledgements The authors would like to thank the editors of the NICO2016 Book and anonymous reviewers for their valuable comments and suggestions to improve the quality of presentation of this chapter.

References

1. Al-Yakoob SM, Sherali HD (2006) Mathematical programming models and algorithms for a class-faculty assignment problem. Eur J Oper Res 173(2):488–507. doi:10.1016/j.ejor.2005. 01.052
2. Archer M (1979) Social origins of educational systems. Sage Publications, Beverley Hills (Calif), London
3. Awerbach S, Ecker JG, Wallace WA (1976) A note: hidden nonlinearities in the application of goal programming. Manag Sci 22(8):918–920. doi:10.1287/mnsc.22.8.918
4. Biswas A, Pal BB (2005) Application of fuzzy goal programming technique to land use planning in agricultural system. Omega 33(5):391–398. doi:10.1016/j.omega.2004.07.003
5. Bleau BL (1981) Planning model in higher education: historical review and survey of currently available models. High Educ 10(2):153–168. doi:10.1007/BF00129129
6. Blumenfeld D (2010) Operations research calculations handbook. CRC Press, New York
7. Bobonich C (2011) Plato's laws: a critical guide. In: Bobonich C (ed) Plato's Laws: a critical guide (Cambridge critical guides), Cambridge University Press, Cambridge
8. Budget Estimates (both revenue and development), the financial year (2014-2015) and revised estimates (2015-2016), University of Kalyani, West Bengal (India), Published by Finance officer, Kalyani University Press, Kalyani (2015)
9. Charnes A, Collomb B (1972) Optimal economic stabilization policy: linear goal-interval programming models. Socio-Econ Plann Sci 6(4):431–435. doi:10.1016/0038-0121(72)90016-X
10. Coombs PH (1968) The world educational crisis: a systems analysis. Oxford University Press, New York, London and Toronto
11. Deb K (2002) Multi-objective optimization using evolutionary algorithm. Wiley, United States
12. Gani J (1963) Formulae of projecting enrollments and degrees awarded in universities. J R Stat Soc (Series A) 126(3):400–409. doi:10.2307/2982224
13. Garvin DA (1980) The economics of university behavior. Academic Press, New York
14. Geoffrion AM, Dyer JS, Feinberg A (1971) Academic departmental management: an application of an interactive multi-criterion optimization approach. In: Ford research program in university administration (Grant No. 680-G267A) P-25, University of California, Berkeley (1971) ERIC No.: ED081402
15. Goldberg DE (1989) Genetic algorithms in search, optimization, and machine learning. Addison-Wesley Publishing Company, Boston
16. Gong D, Zhang Y, Qi C (2010) Environmental/economic power dispatch using a hybrid multi-objective optimization algorithm. Electr Power Eng Syst 32(6):607–614. doi:10.1016/j.ijepes. 2009.11.017
17. Hannan EL (1981) On an interpretation of fractional objectives in goal programming as related to papers by Awerbuchet et al., and Hannan. Manag Sci 27(7):847–848. doi:10.1287/mnsc.27. 7.847
18. Hasan BHF, Saleh MSM (2011) Evaluating the effectiveness of mutation operators on the behaviour of genetic algorithms applied to non-deterministic polynomial problems. Informatica 35(4):513–518
19. Holland JH (1973) Genetic algorithmsand optimal allocationof trials. SIAM J Comput 2(2):88–105. doi:10.1137/0202009
20. Hufner K (1968) Economics of higher education and educational planning- a bibliography. Socio-Econ Plann Sci 2(1):25–101. doi:10.1016/0038-0121(68)90031-1

21. Ignizio JP (1976) Goal programming and extensions. Lexington Books, Lexington, Massachusetts
22. Inuiguchi M, Kume Y (1991) Goal programming problems with interval coefficients and target intervals. Euro J Oper Res 52(3):345–360. doi:10.1016/0377-2217(91)90169-V
23. Keown AJ, Taylor BW III, Pinkerton JM (1981) Multiple objective capital budgeting within the university. Comput Oper Res 8(2):59–70. doi:10.1016/0305-0548(81)90034-4
24. Kvanli AH, Buckley JJ (1986) On the use of U- shaped penalty functions for deriving a satisfactory financial plan utilizing goal programming. J Bus Res 14(1):1–18. doi:10.1016/0148-2963(86)90052-4
25. Kwak NK, Lee C (1998) A multicriteria decision making approach to university resource allocations and information infrastructure planning. Eur J Oper Res 110(2):234–242. doi:10.1016/S0377-2217(97)00262-2
26. Lee SM, Clayton ER (1972) A goal programming model academic resource planning. Manag Sci 18(8):B395–B408. doi:10.1287/mnsc.18.8.B395
27. Liu B (2009) Theory and practice of uncertain programming (2nd edn). Springer, Berlin. doi:10.1007/978-3-540-89484-1
28. Luhandjula MK (2006) Fuzzy stochastic linear programming: survey and future research directions. Eur J Oper Res 174(3):1353–1367 (2006). doi:10.1016/j.ejor.2005.07.019
29. McNamara JF (1971) Mathematical programming models in educational planning. Rev Educ Res 41(5):419–446. doi:10.3102/00346543041005419
30. Michalewicz Z (1996) Genetic algorithms + Data structures = Evolution programs, 3rd edn. Springer, New York
31. Millett JD (1962) The academic community: an essay on organizaiton. McGraw Hill, New York
32. Minguez MI, Romero C, Domingo J (1988) Determining optimum fertilizer combinations through goal programming with penalty functions: an application to sugar beet production in Spain. J Oper Res Soc 39(1):61–70. doi:10.2307/2581999
33. North DC (1990) Institutions, institutional change and economic performance. Cambridge University Press, New York
34. O'Flannery J (2003) Xenophon's (the education of Cyrus) and ideal leadership lessons for modern public administration. Public Adm Q 27(1/2):41–64 (Spring, 2003)
35. Pal BB, Biswas P (2014) GA based FGP to solve BLP model of EEPGD problem. In: Wang J (ed) Encyclopedia of business analytics and optimization. IGI Global, United States, pp 494–510. doi:10.4018/978-1-4666-5202-6
36. Pal BB, Chakraborti D, Biswas P (2010) A genetic algorithm method to fuzzy goal programming formulation based on penalty function for academic personnel management in university system. IEEE Explore Digit Libr 1–10:2010. doi:10.1109/ICCCNT.5591805
37. Pal BB, Moitra BN (2003) A goal programming procedure for solving problems with multiple fuzzy goals using dynamic programming. Eur J Oper Res 144(3):480–491. doi:10.1016/S0377-2217(01)00384-8
38. Pal BB, Moitra BN, Maulik U (2003) A goal programming procedure for fuzzy multiobjectivel fractional programming problem. Fuzzy Sets & Syst 139(2):395–405. doi:10.1016/S0165-0114(02)00374-3
39. Pal BB, Kumar M, Sen S (2012) A priority based goal programming method for solving academic personnel planning problems with interval-valued resource goals in university management system. Int J Appl Manag Sci 4(3):284–312. doi:10.1504/IJAMS.2012.047678
40. Platt WJ (1962) Education-rich problems and poor markets. Manag Sci 8(4):408–418. doi:10.1287/mnsc.8.4.408
41. Rath GJ (1968) Management science in university operation. Manag Sci 14(6):B-373–B-384. doi:10.1287/mnsc.14.6.B373
42. Rehman T, Romero C (1987) Goal programming with penalty functions and livestock ration formulation. Agric Syst 23(2):117–132. doi:10.1016/0308-521X(87)90090-4
43. Romero C (1991) Handbook of critical issues in goal programming. Pergamon Press, Oxford

44. Sahinidis NV (2004) Optimization under uncertainty: state-of- the- art and opportunities. Comput Chem Eng 28(6–7):971–983. doi:10.1016/j.compchemeng.2003.09.017
45. Schroeder RG (1973) A survey of management science in university operations. Manag Sci 19(8):895–906. ERIC No.: EJ075626
46. Schroeder RG (1974) Resource planning in university management by goal programming. Oper Res 22(4):700–710. doi:10.1287/opre.22.4.700
47. Simon HA (1945) Administrative behavior. Free Press, New York
48. Thomas HG (1996) Resource allocation in higher education: a cultural perspective. Res Post-Compuls Educ 1(1):35–51. doi:10.1080/1359674960010104
49. Verspoor AM (1989) Pathways to change: improving the quality of education in developing countries. World Bank, Washington, D.C
50. Walters AJ, Mangold J, Haran EGP (1976) A comprehensive planning model for long-range academic strategies. Manag Sci 22(7):727–738. ERIC No.: EJ138043
51. Weber M (1964) Max Weber economy and society. In: Henderson AR, Parsons T (eds) The theory of social and economic organisation. The Free Press, New York
52. Woodhall M (1987) Economics of education: a review. In: Psacharopoulos G (ed) Economics of education- research and studies. Pergamon Press, New York
53. Yu PL (1973) A class of solutions for group decision problems. Manag Sci 19(8):936–946. doi:10.1287/mnsc.19.8.936
54. Zadeh LA (1965) Fuzzy Sets Inf Control 8(3):338–353. doi:10.1016/S0019-9958(65)90241-X
55. Zhu K, Yu S, Dio F (2007) Soft computing applications to estimate the quantitative contribution of education on economic growth. Appl Math Comput 187(2):1038–1055. doi:10.1016/j.amc.2006.09.088
56. Zimmermann H-J (1987) Fuzzy sets, decision making and expert systems. Kluwer Academic Publisher, Boston. doi:10.1007/978-94-009-3249-4

Swarm Intelligence: A Review of Algorithms

Amrita Chakraborty and Arpan Kumar Kar

Abstract Swarm intelligence (SI), an integral part in the field of artificial intelligence, is gradually gaining prominence, as more and more high complexity problems require solutions which may be sub-optimal but yet achievable within a reasonable period of time. Mostly inspired by biological systems, swarm intelligence adopts the collective behaviour of an organized group of animals, as they strive to survive. This study aims to discuss the governing idea, identify the potential application areas and present a detailed survey of eight SI algorithms. The newly developed algorithms discussed in the study are the insect-based algorithms and animal-based algorithms in minute detail. More specifically, we focus on the algorithms inspired by ants, bees, fireflies, glow-worms, bats, monkeys, lions and wolves. The inspiration analyses on these algorithms highlight the way these algorithms operate. Variants of these algorithms have been introduced after the inspiration analysis. Specific areas for the application of such algorithms have also been highlighted for researchers interested in the domain. The study attempts to provide an initial understanding for the exploration of the technical aspects of the algorithms and their future scope by the academia and practice.

Keywords Swarm intelligence · Machine learning · Bio-inspired algorithms · Intelligent algorithms · Literature review · Nature-inspired computing

A. Chakraborty (✉)
Department of Electronics and Telecommunication Engineering,
Jadavpur University,
Kolkata 700032, West Bengal, India
e-mail: amrita.chakraborty2@gmail.com

A.K. Kar
Information Systems, DMS, Indian Institute of Technology,
Delhi 110016, New Delhi, India

© Springer International Publishing AG 2017
S. Patnaik et al. (eds.), *Nature-Inspired Computing and Optimization*,
Modeling and Optimization in Science and Technologies 10,
DOI 10.1007/978-3-319-50920-4_19

475

1 Introduction

Swarm intelligence (SI) (or bio-inspired computation in general) refers to a subset of artificial intelligence (AI). It has been identified as an emerging field which was coined for the first time by Gerardo Beni and Jing Wang in 1989 in the context of developing cellular robotic systems. There are multiple reasons responsible for the growing popularity of such SI-based algorithms, most importantly being the flexibility and versatility offered by these algorithms. The self-learning capability and adaptability to external variations are the key features exhibited by the algorithms which has attracted immense interest and identified several application areas. In recent times, swarm intelligence has grown in popularity with the increasing prominence of NP-hard problems where finding a global optima becomes almost impossible in real-time scenario. The number of potential solutions which may exist in such problems often tends to be infinite. In such situations, finding a workable solution within time limitations becomes important. SI finds its utility in solving nonlinear design problems with real-world applications considering almost all areas of sciences, engineering and industries, from data mining to optimization, computational intelligence, business planning, in bioinformatics and in industrial applications. Some high-end application areas include navigation control, interferometry, planetary motion sensing, micro-robot control, malignant tumour detection and control and image processing technologies. Being an emerging topic of research, not many publications are available which relate to swarm intelligence, except for few of the dominant approaches, which again has been over applied. Hence, the authors aim to present a review which discusses certain handpicked swarm intelligence algorithms and their future scope.

This study emphasizes on different SI-based algorithms entirely governed by the behavioural aspect of biological organisms which, in the present case, are restricted to insects and animals (mammals and amphibians) only. Both well-known and emerging swarm algorithms are discussed to impart an idea regarding the operation strategy and identify the potential application areas of each. Eight different swarm algorithms are discussed, and the scope of research for each of them are identified. However, certain swarm intelligence-based algorithms such as cuckoo search, flower pollination algorithm and particle swarm optimization are not included in the present scope of study since these algorithms have been already included in another publication authored by the same group [1]. This work is limited to the inspiration analysis, literature survey and identification of suitable application domains only, and in-depth study with the implementation (using pseudocodes) of each algorithm is currently excluded from the scope of this study.

The study is structured into different sections starting with a basic introductory background followed by the research methodology adopted in preparing the review. This is followed by sections separately dealing with insect-based and animal-based algorithms in detail. The algorithms discussed among the insect-based category are bee-inspired, ant colony optimization, firefly-based and glow-worm-based studies. Animal-based algorithms however include lion-, monkey-, bat- and wolf-based studies. After an inspiration analysis, followed by the literature review, the potential application areas for each algorithm are identified. A future course of action discusses the probability of certain algorithms gaining prominence in research and practice for future research.

2 Research Methodology

The research was conducted in multiple stages. Initially, the particular algorithms to be studied in depth were identified. Authors have made an effort to throw light upon lesser known algorithms which are identified to be in their rudimentary development stage. Such a study would be of interest to the readers as it was indeed challenging to identify such algorithms since they lack enough supporting publications and resources as compared to the well-known classical algorithms such as genetic algorithms or neural network which have been widely studied since the 1960s, resulting in several publications available in support of the same. Development in the fields of such lesser known algorithms has been identified majorly from the Internet resources, such as the Google search engine using keywords such as bio-inspired algorithms, swarm intelligence, heuristics, machine learning, intelligent algorithms, meta-heuristics and nature-inspired algorithms. Few publications in the conference proceedings and in the chapters of this book have been identified to collect necessary information regarding the recently developed algorithms. The next stage comprises of the detailed literature survey conducted for each of the identified algorithms to highlight its associated pros and cons and provide a better understanding of the subject matter. It also gives an overview of the potential scope of applications of these algorithms. Adopting the methodology discussed as above, a handful of algorithms (twelve in number) were identified which can be broadly categorized under insect-based and animal-based algorithm categories, as an evident from Fig. 1. Collective insect-based algorithms such as bee-inspired and ant colony, individual insect-based algorithms such as fruit fly, wasp and glow-worm [2–5] and individual animal-based algorithms such as monkey, wolf, lions and bats [6–11] have been explored for their variety of applications in different domains, and a detailed literature review has been conducted. In the subsequent sections, the various algorithms of swarm intelligence would be highlighted, followed by a brief description, before indulging in their application modes in varied problem domains. Figure 1 shows a basic hierarchy and categorization of swarm intelligence-based algorithms which are discussed in the present study.

It is important to note that there are other swarm intelligence algorithms, which may be classified as under bird-based swarm algorithms (e.g. Levy flights, Cuckoo search), which have been intentionally excluded from the current study due to too much increase in scope for a detailed discussion. Figure 2 shows the number of publications specific to each of the algorithms under insect-based and animal-based categories. This gives the readers an idea regarding the popularity and the extent to which each algorithm has been explored. The newly emerging algorithms can also be identified to pursue active research in the unexplored areas.

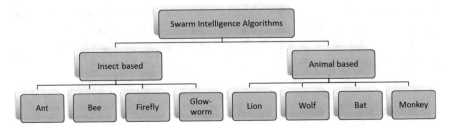

Fig. 1 Hierarchy of swarm intelligence-based algorithms as adopted in the present study

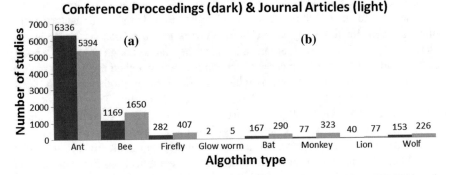

Fig. 2 No. of publications for different SI-based algorithms: **a** Insect based, **b** Animal based

This analysis indicates that for most of the algorithms on swarm intelligence, conferences have become as important as journals, in terms of popularity among researchers for knowledge dissemination. Probably, the faster turnaround time for conference proceedings as compared to the longer review period for most journals is the reason for such a preference.

3   Insect-Based Algorithms

3.1   Ant Colony Optimization Algorithm

3.1.1   Inspiration Analysis

Artificial ant colony systems are inspired from the real-life behaviour of ant colonies and are gaining gradual prominence in the domain of artificial intelligence or, more commonly, the swarm intelligence. A single ant is considered to be the most common biological entity whose abilities are well known and limited. However, if an ant colony is taken into account, then, their behavioral aspect can be considered to be much better organized. Such artificially intelligent algorithms possess an inherent ability to deal with complicated situations effectively. Common applications include development of intelligent solutions for simplified transportation of heavy goods and finding the shortest route between the source and destination points. Such communication between groups of individuals is self-organized and indirect, meaning that individual ants do not need to directly communicate with every other member, and the process is commonly known as stigmergy. Ant colonies are known to communicate via a chemical substance, known as pheromone, in which each agent (ant, in the present scenario) discards during its traversal path. Other agents simply follow the pheromone trail to further reinforce it. Similar problem-solving approach is adopted by artificially intelligent systems, where each unorganized individual (or agent) indirectly communicates and tends to follow the path of the pheromone trail. A group of ants propagate by the application of a stochastic local decision policy depend on two primary factors: the pheromone and heuristic values. The intensity of pheromones detects the probable number of ants following the particular trail, and the heuristic value is problem dependent. Sequences of iterative steps are conducted from a sample of solutions, and the heuristic value is updated to obtain the optimal solution. Among multiple paths, the agent chooses to follow the reinforced pheromone trail (higher quality solutions) and discards the path having diminishing pheromone intensity. Similar multiple iterations are computed to finally obtain the best-fit solution in a reasonable amount of search time. The ant colony optimization is one of the most popular search algorithms, for solving combinational optimization problems.

However, in addition to the standard ant colony optimization algorithm, its variations have also gained prominence in addressing specific problems. The commonly known variants are Elitist Ant System (EAS), Max-Min Ant System (MMAS), Rank-based Ant System (RAS), Continuous Orthogonal Ant Colony (COAC) and the Recursive Ant Colony (RAC) optimization technique. EAS is employed for solving route allocation problem, MMAS is deployed in software project scheduling problem [12], RAS is suitable in dealing with thermal generator maintenance scheduling problem, COAC addresses continuous optimization problems [13], and RAC optimization technique is implemented for parameter estimation of a function [14].

3.1.2 Applications on Ant-Based Algorithms

Based on a variety of applications of the ant colony-based algorithm, an extensive literature review has been conducted to identify the potential application areas and visualize the gradual development of the topic through several years, dating since 1995 till date. Artificial ant colony has applications in data mining [15–17] where ant colony clustering is employed and [2, 18–20] where classification rule is applied in artificial ant colonies. Artificial ant colonies have been utilized to solve the travelling salesman problem [21–27] employing local search using Max-Min ant systems. Vehicle routing problem is solved using similar ant swarm-based approach [28–30]. Quadratic assignment problem [31–34] and scheduling problems (job shop scheduling [35], bus driver scheduling [36], university timetabling [37]) also form crucial application areas in the field of artificial ant swarm-based methods. Some studies have also been conducted to apply such algorithms in the area of telecommunication networking systems [38, 39]. The ant-based algorithm [40–42] is currently being employed for solving newer problem approaches such as membership functions [43], predictive control for nonlinear processes [44], congestion control for MIMO detection [45] and automated recognition of plant species [46]. Ant colony algorithm also has applications in manufacturing industry which also needs seamless cooperation with the logistics company to form a perfect supply chain [47]. It also finds application in market segmentation [48] and feature selection [49]. It has also been used in cloud computing [50] and resource allocation [51]. Reviews of the domain highlight that ant colony algorithms have been applied in the domain of data compression in the domain of image processing, parameter estimation in dynamic systems, economic dispatch problems, probabilistic gaming theory, option selection and prioritization, medical diagnostics, congestion control, telecommunications, aerospace control, social networks mining and target tracking problems [1].

3.2 Bee-Inspired Algorithms

3.2.1 Inspiration Analysis

The artificial bee colony algorithm can be defined as a meta-heuristic method which adopts the technique employed by an intelligent swarm of bees to identify their food source. The nature of honeybees are studied based on their communication, selection of nest location, mating, task allocation, reproduction, dance, placement of pheromone and movement to subsequently modify the algorithm according to the requirements of the problem. The artificial bee colony algorithm performs optimization by iteratively searches for the best-fit (sometimes number based) solution among a large number of data while attempting to solve critical problems. The members of the bee swarm are sectored into three different categories, viz. the employed, the onlooker and the scouts. Scout bees are entrusted the job of random search for fresh food resources. Upon the identification of a food source (which could be a candidate solution), it is marked with a fitness quotient. In subsequent steps, if a fresh food source is located by employed bees with a higher degree of fitness, the fresh source is selected for further processing else it is neglected. The employed bees constantly update their database with information of newer and better food sources and discard the previous data and transfer the degree of fitness to the onlooker bees in their hives. Finally, it is the task of the onlooker bees to identify the best food source (fittest solution) by the frequency of presence of food (obtained as a ratio of the fitness value of a source of food to the sum of fitness values of all the other sources of food). In case the bees are unable to improve the fitness quotient of the food source, in that case, the solutions are nullified.

A popular variant of artificial bee colony algorithm is the Fitness-Scaled Chaotic Artificial Bee Colony (FSCABC) algorithm which was proposed in 2011 [52]. It has applications in system identification of a small-scale unmanned helicopter [53].

3.2.2 Applications of Artificial Bee Colony (ABC) Algorithm

The previous publication reporting the utilization of various problem-solving approaches by the utilization of artificial bee colony is available such as [3, 54–57]. The ABC algorithm finds application in [55] as a single-objective numerical value optimizer. Other application areas include searching, assignment, task allocation, multi-level thresholding, routing problem and maximization and/or minimization problems. The algorithm is also applied in collective decision-making for dealing with multi-criteria selection problems. The application domains would be characterized by evolutionary computation requirement with high level of scalability and differential change within the potential solution space. It has not only been used for unconstrained optimization but also for constrained optimization domains. Further, it has been used in domains such as multidimensional numeric problems, discrete and continuous optimization problems, evolutionary, differential evolution and multiple-objective optimization problems.

Artificial bee colony algorithm also finds innovative applications in customer segmentation in mobile e-commerce environment [58], in marketing and fraudulent activity detection [11] and in agriculture-focused expert systems [59].

3.3 Firefly-Based Algorithms

3.3.1 Inspiration Analysis

Firefly algorithms were introduced to deal with complicated problems having either equality- or inequality-based criteria. Firefly algorithms treat multi-modal functions with better efficiency as compared to other swarm algorithms. Similar to the ant-based and bee-based algorithms, firefly algorithm also adopts a basic random population-based search, thus promoting intelligent learning from a group of varied solutions and resulting in maximum convergence and error-free outcome. The algorithm utilizes the natural behaviour of fireflies whereby bioluminescence or flashing signals to other fireflies for the purpose of finding prey, finding mates or only mutual communication. These fireflies exhibit characteristics similar to that of swarm intelligence because of their self-organization and decentralized decision-taking capability. The intensity of flashing is considered to be an indicator of fitness for the male firefly. However, in the conventional algorithm, all fireflies are considered to be unisex, and hence, all fireflies are known to be mutually attracted in a similar manner. The attractiveness of the firefly is directly proportional to the light intensity (or flashing), which in turn further acts as an indication of fitness for a potential "candidate solution".

The attribute of attractiveness is proportional to brightness, and both are known to be inversely proportional to the distance traversed by the fireflies. According to a basic problem-solving approach, a random population of fireflies is selected. After this step, one of the fitness function parameters is updated and the consequent fitness value evaluated, for each firefly in the problem domain. This iteration continues, and subsequent fireflies are selected to execute similar steps such that best or the fittest individuals of a solution pool may be forwarded for the following round of evaluation. The iteration is controlled by the number of computations which are pre-determined. Yet another important capability of this algorithm is its ability of usage in conjunction with other algorithms to obtain enhanced outcome.

There are several variants of firefly-based algorithms which have multiple application areas in almost all domains of science and communication. Popular among the list of variants are Adaptive Firefly Algorithm (AFA), Discrete Firefly Algorithm (DFA), Multi-Objective FA (MOFA), Lagrangian FA (LFA), Chaotic FA (CFA), Hybrid FA (HFA) and FA-based Memetic Algorithm (FAMA), to name a few. DFA is employed to solve NP-hard scheduling problems, and in image segmentation, MOFA is used for solving multi-objective load dispatch problems and LFA finds utility in power system optimization unit commitment problems, whereas FAMA is applied for electrical load forecasting.

3.3.2 Applications of Firefly-Based Algorithms

Firefly-based algorithms are known to have a host of applications, some of which have been specified underneath. Constrained [60] and multi-modal function optimization [61] in stochastic algorithms and stochastic optimization [62] are efficiently dealt with firefly based algorithms. Such algorithms also find usage in eagle-based Levy flight systems [63], in mixed variable structural optimization [4] and for solving non-convex economic dispatch problems [64]. The literature review by [65] also aids in providing better insight regarding the algorithm, [66] in 2014 for accurate short-term load forecasting, [67] for greyscale image watermarking, [68] for capacitated facility location problem, and [69] and [70] for heart disease prediction and the introduction of oppositional and dimensional based firefly algorithms.

Firefly algorithms may also be employed for solving NP-hard problems, with equality- and/or inequality-driven constraints, highly complex classification problems, continuous and discrete search space domains, combinatorial optimization domains, parallel computational domains and multi-objective search challenges. The firefly algorithm can be used in conjunction with other methods such as multi-valued logic (e.g. rough set theory), cellular learning automata and artificial neural networks to develop hybrid approaches. Firefly-based algorithms are also employed to solve load dispatch problems [71], stock market price forecasting [72], image compression [73], manufacturing cell formation [74], job shop scheduling [75] and energy conservation [76].

3.4 Glow-Worm-Based Algorithms

3.4.1 Inspiration Analysis

Glow-worm-based algorithm is a swarm intelligence-based algorithm for optimizing multi-modal functions, developed by imbibing the behaviour of glow-worms in artificial intelligent systems. Glow-worms are known to possess the capability to modify the intensity of a chemical, called luciferin emission (similar to pheromones in the case of ant-based algorithms) which helps them glow at various intensities [5, 77, 78]. The glow-worms communicate with other members by glowing. The luciferin-induced glow of a glow-worm attracts mates during reproduction or preys for feeding. If a glow-worm emits more luciferin, it glows more brightly and it attracts more number of other glow-worms or preys. The brighter the glow intensity, the better the attraction is. The artificial glow-worm-based algorithm relies on the behaviour of glow-worms. In the artificial glow-worm swarm, initially a number of glow-worms are randomly selected to organize an artificial swarm. Each agent represents potential solution to the optimization problem, where each member in the swarm utilizes the search domain to select its neighbours and decides its direction of movement by the strength of the luciferin intensity picked up from them. A glow-worm is attracted to another neighbour glow-worm if the luciferin level of the latter is higher than that of the former. Each glow-worm in a swarm is assigned an objective function and a certain value of luciferin intensity based on its current location. During traversal, each glow-worm compares its intensity with the those glow-worms in the neighbourhood and changes course of travel if a glow-worm with higher intensity of luciferin is encountered. This process is followed iteratively till the maximum convergence, and the fittest solution candidate is reached.

3.4.2 Applications of Glow-Worm-Based Algorithm

Glow-worm-based algorithms can be utilized for a number of applications, similar to the other insect-based algorithms reported as above. Common applications include detection of multiple source locations [5], parameter optimization for multi-modal parameter search [77] and searching of multiple local optima for multi-modal functions [78]. Glow-worm-based algorithms find application in robotics [79], signal source localization [80], image processing [81], knapsack problems [82], travelling salesman problem [83], distributed energy resource management [84] and in Map Reduce problem in big data applications [85]. The review of existing literature highlights that the glow-worm-based algorithm needs significant exploration in terms of both theory development and application in diverse problem domains.

4   Animal-Based Algorithms

4.1   Bat-Based Algorithm

4.1.1   Inspiration Analysis

The bat algorithm [86] is one of the recently developed animal group/herd-based learning algorithms which utilizes the echo-based location prediction mechanism employed by bats or other nocturnal animals to obtain solutions for single- and multi-objective domains within continuous solution space. This process, known as echolocation, is used to refer to the process by which bats use echo signals emitted by them and other obstacle in the neighbourhood to navigate in the surroundings. This technique enables bats to precisely judge the exact location of any object or prey, even in the absence of light. Similar to the RADAR principle, echolocation enables bats to estimate the range or the probable distance at which the target object/prey is located. Bats can adjust their flight speed, the frequency and intensity of their echo while hunting for prey. Vector algebra is the tool employed to solve problem domains iteratively and by breaking it into smaller sub-problems. With single iteration, the data consisting of the intensity of scream and frequency needs to be modified so that the frequency increases and cry intensity reduces once a bat gets into the proximity of a potential food source. Although majority of the application area of bat algorithm is restricted to the continuous problem domain, a binary bat type was introduced to address discrete decision-making [6]. Studies are being conducted to combine this algorithm with the older and more established intelligent algorithms such as artificial neural networks [87].

Popular variants of bar algorithm include Multi-objective Bat Algorithm (MOBA), Directed Artificial Bat Algorithm (DABA) and Binary Bat Algorithms (BBA). MOBA is employed in solving multi-objective engineering tasks [10] and DABA finds application in obstacle tracking robots [88], whereas BBA has widespread application in feature selection for many optimization techniques [89].

4.1.2   Applications of Bat-Based Algorithm

The bat algorithm has been extensively employed in published articles for multi-objective optimization [90], inverse parameter estimation, constrained optimization search [91], combinatorial optimization, multi-job scheduling, classification, vector matching, clustering [92] and multi-valued system-based optimization [93]. The bat algorithm is beneficial for complicated multidimensional domains where convergence is a major issue like in structural design optimization problems [94], chaotic multi-objective problems, Doppler effect applications [95] and position controlling for piezoelectric actuator [96].

Bat-based algorithms are also used in problem domains such as power system stabilizers [97], wireless multi-hop network management [98], scheduling problems

[99], global search problems [100], feature selection and prioritization problems [101], structural engineering [102] and office workplace management [103].

4.2 Monkey-Based Algorithm

4.2.1 Inspiration Analysis

Monkey algorithm is efficient in solving the optimization of multi-variate systems. This method is derived from simulations of mountain-climbing techniques used by monkeys. Considering a number of mountains in a sample space (within the feasible problem space of the domain) in order to locate a mountain peak of the highest elevation (i.e. the maximal solution of the function), monkeys will shift to an elevated location as compared to their respective positions (this function is called climbing). Each monkey, after reaching a mountaintop, would further look out to search for mountaintops with yet higher elevation. If available, the monkey would relocate from its current position to reach a higher elevation (this step is known as watch–jump process). This iterative process continues till the monkey reaches the mountain peak. After subsequent repetitions of the watch–jump and climb functions being repeated, each monkey will find a locally maximum mountaintop which may be a local maximum. In order to locate a yet higher mountaintop, it is important for each monkey to somersault to a new search domain (somersault function) to move towards global optima. After several iterations of the climb, watch, jump and somersault processes, the highest mountain peak found by the monkeys is reported as an optimal solution. Asynchronous climb monkey algorithm is one of the major variants utilized for sensor placement in canton towers for health monitoring hubs [104].

4.2.2 Applications of Monkey-Based Algorithm

This algorithm allows for significant reduction in the cost of high complex optimization, especially in multidimensional problems [7]. Monkey-based algorithms were originally utilized for optimization problems with continuous variables. It is also used for solving transmission network expansion problem, which is categorized as a discrete optimization problem [105]. Monkey algorithm has powerful computational capabilities and is capable of solving multidimensional problems' expansion and planning effectively with a smaller population size [106, 107].

While exploring the application of the monkey-based algorithms, it was seen that it has been used across some innovative problem domains such as image processing [108], graph mining [109] and health monitoring systems [110]. It is important to note that the monkey-based algorithms have not been tested across too many application domains, such as in industrial engineering domain, where typically a lot of these algorithms get applied.

4.3 Lion-Based Algorithm

4.3.1 Inspiration Analysis

In the lion-based algorithm, an initial population is formed by a set of randomly generated solutions called lions. Some of the lions in the initial population are selected as nomad lions, and rest population (resident lions) is randomly partitioned into subsets called prides. A percentage of the pride's members are considered as female and the rest are considered as male, while this sex rate in nomad lions is vice versa. For each lion, the best obtained solution in passed iterations is called best visited position, and during the optimization process, it is updated progressively. In this algorithm, a pride territory refers to an area which consists of each member's best visited position. In each pride, some randomly selected females go for hunting. Hunters move towards the prey to encircle and catch it. The rest of the females move towards different positions of territory, whereas the male lions in pride, roam within the territory. In each pride, young males are excluded from their maternal pride and become nomad, and on reaching maturity, their power is less than resident males. Also, a nomad lion (both male and female) moves randomly in the search space to obtain a better place (solution). If the strong nomad male invades the resident male, the resident male is driven out of the pride by the nomad lion. The nomad male becomes the resident lion. In this evolution, some resident females immigrate from one pride to another or switch their lifestyles and become nomad and vice versa, and some nomad female lions join prides. Due to numerous factors such as lack of food and competition, weakest lion generally die out or get killed. The above process continues until the terminating condition is reached.

4.3.2 Applications of Lion-Based Algorithm

Lion algorithms are employed for nonlinear system identification process and bilinear system identification processes. Lion-based algorithms [8] also prove to be better than genetic algorithms in cases where a large-scale bilinear model is employed [111]. These algorithms have also seen some innovative applications in domains such as contour modelling [112]. However a lot of scope is presented in exploring the developments in the algorithm itself and also in the application of this algorithm in different domains and problems.

4.4 Wolf-Based Algorithm

4.4.1 Inspiration Analysis

Wolf-based algorithm is one of the very newly introduced meta-heuristic algorithms by Simon Fong [113]. It is based on the hunting behaviour of a pack of wolves where each individual searching wolf (an intelligent agent) hunts for a prey individually, silently (no physical communication with the rest of the pack), and they merge by shifting their present positions to the positions of the other wolves in the pack in case the fresh locations are more suitable than the older ones. Wolves move in levy flights during the food-searching mode. A random hunter is elected from a pack where a wolf will move out of its current line of sight to a newly generated random position upon locating a potential food prey. This random jump enables the wolves to remain out of a local sub-optimal solution and gain global optimality. Wolf-based algorithm has sometimes shown to be more superior to some of existing bio-inspired algorithms when sub-optimality is very high in a continuous problem domain [9].

4.4.2 Applications of Wolf-Based Algorithm

The wolf-based algorithm is used in fault system estimation problem in power systems [114], in optimal operation of hydropower station [115]. A variant of the wolf-based algorithm, i.e. the grey wolf optimization technique, has several applications in multi-layer perceptron [116], in training of q-Gaussian radial basis functional-link nets [117], in solving economic dispatch problems [118], solving combined economic emission dispatch problems [119], optimum allocation of STATCOM devices on power system grid to minimized load buses voltage deviations and system power losses [120] and for solving evolutionary population dynamics [121]. It is also used in problem domains such as control of DC motors [122], feature prioritization and selection problems [123], multi-input multi-output contingency management problems [124] and for detection of faulty sections in power systems [125], to name a few.

5 Future Research Directions

Insect-based algorithms such as ant colony and bee colony, firefly-based algorithm and bat-based algorithm from the animal category form the older group of optimization techniques, where immense research has been conducted and potential application areas explored, supported with a adequate number of references in the present study. Better and fresh application areas need to be sought out for the other algorithms such as lion-, shark-, wolf- and glow-worm-based algorithms, since these are comparatively newer and still unexplored. The research focus is shifted to these lesser explored algorithms to identify advanced and simplified, time-effective techniques for complicated computation problems. Further, future research could also focus on the development of individual bird-based algorithms and review them both in terms of application and scope. Within these individual algorithms also, there is immense scope of fine-tuning to enhance the outcome based on context specific requirements. Further synergies may be explored among such swarm intelligence algorithms with other bio-inspired algorithms (such as genetic algorithm, neural networks and multi-valued logic). This will enhance the scope of exploring the improvements in performance through application of meta-heuristics and hybrid algorithms.

6 Conclusions

The study presents a detailed review on selective popularly known swarm intelligence-based algorithms, from which only insect-based and animal-based algorithms are chosen. It is evident from the study that few algorithms discussed like ant colony, bee colony, firefly- and bat-based algorithms are very popular and well explored having several publications relating to the subject. These algorithms have been applied in numerous diverse problem domains from supplier selection in industrial engineering to process automation in electrical engineering. While the introduction of these algorithms have been made in pure science (e.g. mathematics) and engineering (e.g. computer science and electrical engineering), it is exciting to see the applications highlighted in social science and business studies. In the business application domain, a particular area which has highlighted many applications is in the domain of supplier network management, supplier selection problem, facility layout problem, allocation of resources and scheduling of jobs.

Authors have also reported details of the lesser known algorithms among which glow-worm-, monkey-, lion- and wolf-based algorithms form an important category. Basic insights to all such algorithms have been provided followed by an up-to-date literature review and identified potential applications for each of them. It is however important to note that these algorithms are yet to be introduced extensively in many of the more application-oriented domains, as has been highlighted earlier in the review. This throws up future scope of application-specific studies in domains such as social science, business studies and medical domains.

References

1. Kar AK (2016) Bio-inspired computing—a review of algorithms and scope of applications. Expert Syst Appl 59:20–32
2. Parpinelli RS, Lopes HS, Freitas AA (2001) An ant colony based system for data mining:Applications to medical data. In: Lee S, Goodman E, Wu A, Langdon WB, Voigt H, Gen M, Sen S, Dorigo M, Pezeshk S, Garzon M, Burke E (eds) Proceedings of the genetic and evolutionary computation conference (GECCO-2001), San Francisco, California, USA, 7–11. Morgan Kaufmann, pp 791–797
3. Karaboga D (2005) An idea based on honey bee swarm for numerical optimization, vol 200. Technical report-tr06, Erciyes University, Engineering Faculty, Computer Engineering Department
4. Gandomi AH, Yang XS, Alavi AH (2011) Mixed variable structural optimization using firefly algorithm. Comput Struct 89(23):2325–2336

5. Krishnanand KN, Ghose D (2005) Detection of multiple source locations using a glow-worm metaphor with applications to collective robotics. In: IEEE swarm intelligence symposium, Pasadena, CA, pp 84–91
6. Mirjalili S, Mirjalili SM, Yang XS (2014) Binary bat algorithm. Neural Comput Appl 25 (3–4):663–681
7. Mucherino A, Seref O (2007) Monkey search: a novel metaheuristic search for global optimization. In: AIP conference proceedings, vol 953, pp 162–173
8. Yazdani M, Jolai F (2015) Lion optimization algorithm (LOA): a nature-inspired metaheuristic algorithm. J Comput Des Eng (in press)
9. Raton FL, USA, pp 351–392. Liu C, Yan X, Liu C, Wu H (2011) The wolf colony algorithm and its application. Chin J Electron 20:212–216
10. Yang XS (2011) Bat algorithm for multi-objective optimisation. Int J Bio-Inspired Comput 3:267–274
11. Prabha MS, Vijayarani S (2011) Association rule hiding using artificial bee colony algorithm. Int J Comput Appl 33(2):41–47
12. Crawford B, Soto R, Johnson F, Monfroy E, Paredes F (2014) A max-min ant system algorithm to solve the software project scheduling problem. Expert Syst Appl 41(15):6634–6645
13. Hu XM, Zhang J, Yun Li Y (2008) Orthogonal methods based ant colony search for solving continuous optimization problems. J Comput Sci Technol 23(1):2–18
14. Gupta DK, Arora Y, Singh UK, Gupta JP (2012) Recursive ant colony optimization for estimation of parameters of a function. In: 1st international conference on recent advances in information technology (RAIT), pp 448–454
15. Abraham A, Ramos V (2003) Web usage mining using artificial ant colony clustering. In: Proceedings of congress on evolutionary computation (CEC2003), Australia, IEEE Press, pp 1384–1391. ISBN 0780378040
16. Handl J, Knowles J, Dorigo M (2003) Ant-based clustering: a comparative study of itsrelative performance with respect to k-means, average link and 1d-som. Technical ReportTR/IRIDIA/2003-24, Universite Libre de Bruxelles
17. Schockaert S, De Cock M, Cornelis C, Kerre EE (2004) Efficient clustering with fuzzy ants. Appl Comput Intell
18. Parpinelli RS, Lopes HS, Freitas AA (2002) Data mining with an ant colony optimizationalgorithm. IEEE Trans Evol Comput 6(4):321–332
19. Ramos V, Abraham A (2003) Swarms on continuous data. In: Proceedings of the congress on evolutionary computation. IEEE Press, pp 1370–1375
20. Liu B, Abbass HA, McKay B (2004) Classification rule discovery with ant colonyoptimization. IEEE Comput Intell Bull 3(1):31–35
21. Gambardella LM, Dorigo M (1995) Ant-q: A reinforcement learning approach to the traveling salesman problem. In: Proceedings of the eleventh international conference on machine learning, pp 252–260
22. Dorigo M, Maniezzo V, Colorni A (1996) The ant system: optimization by a colony of cooperating agents. IEEE Trans Syst Man Cybern Part B: Cybern 26(1):29–41
23. Gambardella LM, Dorigo M (1996) Solving symmetric and asymmetric tsps by ant colonies. In: Proceedings of the IEEE international conference on evolutionary computation (ICEC'96), pp 622–627
24. Stutzle T, Hoos HH (1997) The MAX-MIN ant system and local search for the traveling salesman problem. In: Proceedings of the IEEE international conference on evolutionary computation (ICEC'97), pp 309–314
25. Stutzle T, Hoos HH (1998) Improvements on the ant system: introducing the MAX-MIN ant system. In: Steele NC, Albrecht RF, Smith GD (eds) Neural Artificial networks and genetic, algorithms, pp 245–249
26. Stutzle T, Hoos HH (1999) MAX-MIN ant system and local search for combinatorial optimization problems. In: Osman IH, Voss S, Martello S, Roucairol C (eds) Meta-heuristics: advances and trends in local search paradigms for optimization, pp 313–329

27. Eyckelhof CJ, Snoek M (2002) Ant systems for a dynamic tsp. In: ANTS '02: Proceedings of the third international workshop on ant algorithms, London, UK. Springer, pp 88–99
28. Bullnheimer B, Hartl RF, Strauss C (1999) Applying the ant system to the vehicle routing problem. In: Roucairol C, Voss S, Martello S, Osman IH (eds) Meta-heuristics, advances and trends in local search paradigms for optimization
29. Cicirello VA, Smith SF (2001) Ant colony control for autonomous decentralized shop floor routing. In: The fifth international symposium on autonomous decentralized systems, pp 383–390
30. Wade A, Salhi S (2004) An ant system algorithm for the mixed vehicle routing problem with backhauls. In: Metaheuristics: computer decision-making, Norwell, MA, USA, 2004. Kluwer Academic Publishers, pp 699–719
31. Maniezzo V (1998) Exact and approximate nondeterministic tree-search procedures for the quadratic assignment problem. Research CSR 98-1, Scienze dell'Informazione, Università di Bologna, Sede di Cesena, Italy
32. Maniezzo V, Colorni A (1999) The ant system applied to the quadratic assignment problem. IEEE Trans Knowl Data Eng
33. Gambardella LM, Taillard E, Dorigo M (1999) Ant colonies for the quadratic assignment problem. J Oper Res Soc 50:167–176
34. Stutzle T, Dorigo M (1999) ACO algorithms for the quadratic assignment problem. In: Dorigo M, Corne D, Glover F (eds) New ideas in optimization
35. Colorni A, Dorigo M, Maniezzo V, Trubian M (1994) Ant system for job shop scheduling. J Oper Res Stat Comput Sci 34(1):39–53
36. Forsyth P, Wren A (1997) An ant system for bus driver scheduling. Research Report 97.25, University of Leeds School of Computer Studies
37. Socha K, Knowles J, Sampels M (2002) A MAX-MIN ant system for the university timetabling problem. In: Dorigo M, Di Caro G, Sampels M (eds) Proceedings of ANTS2002—third international workshop on ant algorithms. Lecture notes in computer science, vol 2463. Springer, Berlin, Germany, pp 1–13
38. Schoonderwoerd R, Holland OE, Bruten JL, Rothkrantz LJM (1996) Ant-based loadbalancing in telecommunications networks. Adapt Behav 2:169–207
39. Di Caro G, Dorigo M (1998) Antnet: distributed stigmergetic control forcommunications networks. J Artif Intell Res 9:317–365
40. Dorigo M, Blum C (2005) Ant colony optimization theory: a survey. Theoret Comput Sci 344(2):243–278
41. Dorigo M, Birattari M, Stützle T (2006) Ant colony optimization. IEEE Comput Intell Mag 1(4):28–39
42. Dorigo M, Birattari M (2010) Ant colony optimization. In: Encyclopedia of machine learning. Springer US, pp 36–39
43. Hong TP, Tung YF, Wang SL, Wu YL, Wu MT (2012) A multi-level ant-colony mining algorithm for membership functions. Inf Sci 182(1):3–14
44. Bououden S, Chadli M, Karimi HR (2015) An ant colony optimization-based fuzzy predictive control approach for nonlinear processes. Inf Sci 299:143–158
45. Mandloi M, Bhatia V (2015) Congestion control based ant colony optimization algorithm for large MIMO detection. Expert Syst Appl 42(7):3662–3669
46. Ghasab MAJ, Khamis S, Mohammad F, Fariman HJ (2015) Feature decision-making ant colony optimization system for an automated recognition of plant species. Expert Syst Appl 42(5):2361–2370
47. Kuo RJ, Chiu CY, Lin YJ (2004) Integration of fuzzy theory and ant algorithm for vehicle routing problem with time window. In: IEEE annual meeting of the fuzzy information, 2004. Processing NAFIPS'04, vol 2, pp 925–930. IEEE
48. Chiu CY, Kuo IT, Lin CH (2009) Applying artificial immune system and ant algorithm in air-conditioner market segmentation. Expert Syst Appl 36(3):4437–4442
49. Hua XY, Zheng J, Hu WX (2010) Ant colony optimization algorithm for computing resource allocation based on cloud computing environment [J]. J East China Normal Univ (Nat Sci) 1(1):127–134

50. Chiu CY, Lin CH (2007) Cluster analysis based on artificial immune system and ant algorithm. In: Third international conference on natural computation (ICNC 2007), vol 3, pp 647–650. IEEE
51. Abraham A, Ramos V (2003) Web usage mining using artificial ant colony clustering and linear genetic programming. In: The 2003 congress on evolutionary computation, 2003. CEC'03, vol 2, pp 1384–1391. IEEE
52. Wu L (2011) UCAV path planning based on FSCABC. Inf–Int Interdiscip J 14(3):687–692
53. Ding L, Hongtao W, Yu Y (2015) Chaotic artificial bee colony algorithm for system identification of a small-scale unmanned helicopter. Int J Aerosp Eng 2015, Article ID 801874:1–12
54. Karaboga D, Basturk B (2007) A powerful and efficient algorithm for numerical function optimization: artificial bee colony (ABC) algorithm. J Glob Optim 39:459–471
55. Karaboga D, Akay B (2009) A survey: algorithms simulating bee swarm intelligence. Artif Intell Rev 31(1–4):61–85
56. Gao WF, Liu SY (2012) A modified artificial bee colony algorithm. Comput Oper Res 39(3):687–697
57. Karaboga D, Gorkemli B, Ozturk C, Karaboga N (2014) A comprehensive survey: artificial bee colony (ABC) algorithm and applications. Artif Intell Rev 42(1):21–57
58. Deng X (2013) An enhanced artificial bee colony approach for customer segmentation in mobile e-commerce environment. Int J Adv Comput Technol 5(1)
59. Babu MSP, Rao NT (2010) Implementation of artificial bee colony (ABC) algorithm on garlic expert advisory system. Int J Comput Sci Res 1(1):69–74
60. Lukasik S, Zak S (2009) Firefly algorithm for continuous constrained optimization tasks. In: Computational collective intelligence. Semantic web, social networks and multiagent systems. Springer, Berlin, Heidelberg, pp 97–106
61. Yang XS (2009) Firefly algorithms for multimodal optimization. In: Stochastic algorithms: foundations and applications. Springer, Berlin, Heidelberg, pp 169–178
62. Yang XS (2010) Firefly algorithm, stochastic test functions and design optimisation. Int J Bio-Inspired Comput 2(2):78–84
63. Yang X-S, Deb S (2010) Eagle strategy using Lévy walk and firefly algorithms for stochastic optimization. In: Gonzalez JR (ed) Nature inspired cooperative strategies for optimization (NISCO 2010), SCI 284. Springer, Berlin, pp 101–111
64. Yang XS, Hosseini SSS, Gandomi AH (2012) Firefly algorithm for solving non-convex economic dispatch problems with valve loading effect. Appl Soft Comput 12(3):1180–1186
65. Fister I, Yang XS, Brest J (2013) A comprehensive review of firefly algorithms. Swarm Evol Comput 13:34–46
66. Kavousi-Fard A, Samet H, Marzbani F (2014) A new hybrid modified firefly algorithm and support vector regression model for accurate short term load forecasting. Expert Syst Appl 41(13):6047–6056
67. Mishra A, Agarwal C, Sharma A, Bedi P (2014) Optimized gray-scale image watermarking using DWT-SVD and firefly algorithm. expert syst appl 41(17):7858–7867
68. Rahmani A, MirHassani SA (2014) A hybrid firefly-genetic algorithm for the capacitated facility location problem. Inf Sci 283:70–78
69. Long NC, Meesad P, Unger H (2015) A highly accurate firefly based algorithm for heart disease prediction. Expert Syst Appl 42(21):8221–8231
70. Verma OP, Aggarwal D, Patodi T (2015) Opposition and dimensional based modified firefly algorithm. Expert Syst Appl
71. Apostolopoulos T, Vlachos A (2010) Application of the firefly algorithm for solving the economic emissions load dispatch problem. Int J Comb 2011
72. Kazem A, Sharifi E, Hussain FK, Saberi M, Hussain OK (2013) Support vector regression with chaos-based firefly algorithm for stock market price forecasting. Appl Soft Comput 13(2):947–958
73. Horng MH (2012) Vector quantization using the firefly algorithm for image compression. Expert Syst Appl 39(1):1078–1091

74. Sayadi MK, Hafezalkotob A, Naini SGJ (2013) Firefly-inspired algorithm for discrete optimization problems: an application to manufacturing cell formation. J Manuf Syst 32(1):78–84
75. Karthikeyan S, Asokan P, Nickolas S, Page T (2015) A hybrid discrete firefly algorithm for solving multi-objective flexible job shop scheduling problems. Int J Bio-Inspired Comput 7(6):386–401
76. dos Santos Coelho L, Mariani VC (2013) Improved firefly algorithm approach applied to chiller loading for energy conservation. Energy Build 59:273–278
77. Krishnanand KN, Ghose D (2009a) Glowworm swarm optimization: a new method foroptimizing multi-modal functions. Int J Comput Intell Stud 1(1):84–91
78. Krishnanand KN, Ghose D (2009b) Glowworm swarm optimization for simultaneous capture of multiple local optima of multimodal functions. Swarm Intell 3(2):87–124
79. Krishnanand KN, Ghose D (2005) Detection of multiple source locations using a glowworm metaphor with applications to collective robotics. In: Proceedings 2005 IEEE swarm intelligence symposium, 2005, pp 84–91
80. Krishnanand KN, Ghose D (2009) A glowworm swarm optimization based multi-robot system for signal source localization. In: Design and control of intelligent robotic systems. Springer, Berlin, Heidelberg, pp 49–68
81. Senthilnath J, Omkar SN, Mani V, Tejovanth N, Diwakar PG, Shenoy AB (2012) Hierarchical clustering algorithm for land cover mapping using satellite images. IEEE J Sel Top Appl Earth Obs Remote Sens 5(3):762–768
82. Gong Q, Zhou Y, Luo Q (2011) Hybrid artificial glowworm swarm optimization algorithm for solving multi-dimensional knapsack problem. Procedia Eng 15:2880–2884
83. Zhou YQ, Huang ZX, Liu HX (2012) Discrete glowworm swarm optimization algorithm for TSP problem. Dianzi Xuebao (Acta Electronica Sinica) 40(6):1164–1170
84. Di Silvestre ML, Graditi G, Sanseverino ER (2014) A generalized framework for optimal sizing of distributed energy resources in micro-grids using an indicator-based swarm approach. IEEE Trans Ind Inform 10(1):152–162
85. Al-Madi N, Aljarah I, Ludwig SA (2014) Parallel glowworm swarm optimization clustering algorithm based on MapReduce. In: 2014 IEEE symposium on swarm intelligence (SIS). IEEE, pp 1–8
86. Yang XS (2010). A new metaheuristic bat-inspired algorithm. In: Nature inspired cooperative strategies for optimization (NICSO 2010). Springer, Berlin, Heidelberg, pp 65–74
87. Jaddi NS, Abdullah S, Hamdan AR (2015) Multi-population cooperative bat algorithm-based optimization of artificial neural network model. Inf Sci 294:628–644
88. Rekaby A (2013) Directed artificial bat algorithm (DABA): a new bio-inspired algorithm. In: International conference on advances in computing, communications and informatics (ICACCI), Mysore
89. Mirjalili S, Mirjalili SM, Yang X (2014) Binary bat algorithm, neural computing and applications (in press) (2014). Springer. doi:10.1007/s00521-013-1525-5
90. Yang XS (2011) Bat algorithm for multi-objective optimization. Int J Bio-Inspired Comput 3(5):267–274
91. Gandomi AH, Yang XS, Alavi AH, Talatahari S (2012) Bat algorithm for constrained optimization tasks. Neural Comput Appl doi:10.1007/s00521-012-1028-9
92. Yang XS, He X (2013) Bat algorithm: literature review and applications. Int J Bio-Inspired Comput 5(3):141–149
93. Gandomi AH, Yang XS (2014) Chaotic bat algorithm. J Comput Sci 5(2):224–232
94. Rodrigues D, Pereira LA, Nakamura RY, Costa KA, Yang XS, Souza AN, Papa JP (2014) A wrapper approach for feature selection based on bat algorithm and optimum-path forest. Expert Syst Appl 41(5):2250–2258
95. Meng XB, Gao XZ, Liu Y, Zhang H (2015) A novel bat algorithm with habitat selection and Doppler effect in echoes for optimization. Expert Syst Appl 42(17):6350–6364
96. Svečko R, Kusić D (2015) Feed-forward neural network position control of a piezoelectric actuator based on a BAT search algorithm. Expert Syst Appl 42(13):5416–5423

97. Ali ES (2014) Optimization of power system stabilizers using BAT search algorithm. Int J Electr Power Energy Syst 61:683–690
98. Li L, Halpern JY, Bahl P, Wang YM, Wattenhofer R (2005) A cone-based distributed topology-control algorithm for wireless multi-hop networks. IEEE/ACM Trans Netw 13(1):147–159
99. Musikapun P, Pongcharoen P (2012) Solving multi-stage multi-machine multi-product scheduling problem using bat algorithm. In: 2nd international conference on management and artificial intelligence, vol 35. IACSIT Press Singapore, pp 98–102
100. Wang G, Guo L (2013) A novel hybrid bat algorithm with harmony search for global numerical optimization. J Appl Math (2013)
101. Nakamura RY, Pereira LA, Costa KA, Rodrigues D, Papa JP, Yang XS (2012) BBA: a binary bat algorithm for feature selection. In 2012 25th SIBGRAPI conference on graphics, patterns and images. IEEE, pp 291–297
102. Hasançebi O, Teke T, Pekcan O (2013) A bat-inspired algorithm for structural optimization. Comput Struct 128:77–90
103. Khan K, Nikov A, Sahai A (2011) A fuzzy bat clustering method for ergonomic screening of office workplaces. In: Third international conference on software, services and semantic technologies S3T 2011. Springer, Berlin, Heidelberg, pp 59–66
104. Yi T-H, Li H-N, Zhang X-D (2012) Sensor placement on Canton Tower for health monitoring using asynchronous-climb monkey algorithm. Smart Mater Struct 21. doi:10.1088/0964-1726/21/12/125023
105. Ramos-Frenańdez G, Mateos JL, Miramontes O, Cocho G, Larralde H, Ayala-Orozco B (2004) Levy walk patterns in the foraging movements of spider monkeys (Atelesgeoffroyi). Behav Ecol Sociobiol 55(223):230
106. Zhao R, Tang W (2008) Monkey algorithm for global numerical optimization. J Uncertain Syst 2(3):165–176
107. Wang J, Yu Y, Zeng Y, Luan W (2010). Discrete monkey algorithm and its application in transmission network expansion planning. In: IEEE conference on power and energy society general meeting, July 2010, pp 1–5
108. Vu PV, Chandler DM (2012) A fast wavelet-based algorithm for global and local image sharpness estimation. IEEE Signal Process Lett 19(7):423–426
109. Zhang S, Yang J, Cheedella V (2007) Monkey: approximate graph mining based on spanning trees. In: 2007 IEEE 23rd international conference on data engineering. IEEE, pp 1247–1249
110. Yi TH, Li HN, Zhang XD (2012) Sensor placement on Canton Tower for health monitoring using asynchronous-climb monkey algorithm. Smart Mater Struct 21(12):125023
111. Rajkumar BR (2014) Lion algorithm for standard and large scale bilinear system identification: A global optimization based on Lion's social behaviour. In: IEEE congress on evolutionary computation, July 2014, pp 2116–2123
112. Shah-Hosseini H, Safabakhsh R (2003) A TASOM-based algorithm for active contour modeling. Pattern Recogn Lett 24(9):1361–1373
113. Tang R, Fong S, Yang X.-S, Deb S (2012) Wolf search algorithm with ephemeral memory. In: IEEE seventh international conference on digital information management (ICDIM 2012), Aug 2012, pp 165–172
114. Wang J, Jia Y, Xiao Q (2015). Application of wolf pack search algorithm to optimal operation of hydropower station. Adv Sci Technol Water Resour 35(3):1–4 & 65
115. Mirjalili S, Mirjalili SM, Lewis A (2014) Grey wolf optimizer. Adv Eng Softw 69:46–61
116. Mirjalili S (2015) How effective is the Grey Wolf optimizer in training multi-layer perceptrons. Appl Intell 1–12
117. Nipotepat M, Sunat K, Chiewchanwattana S (2014) An improved grey wolf optimizer for training q-Gaussian radial basis functional-link nets. In: IEEE international conference in computer science and engineering
118. Wong LI et al (2014) Grey wolf optimizer for solving economic dispatch problems. In: IEEE international conference on power and energy
119. Mee SH, Sulaiman MH, Mohamed MR (2014) An application of grey wolf optimizer for solving combined economic emission dispatch problems. Int Rev Modell Simul (IREMOS) 7(5):838–844

120. El-Gaafary Ahmed AM et al (2015) Grey wolf optimization for multi input multi output system. Generations 10:11
121. Saremi S, Mirjalili SZ, Mirjalili SM (2015) Evolutionary population dynamics and grey wolf optimizer. Neural Comput Appl 1–7
122. Madadi A, Motlagh MM (2014) Optimal control of DC motor using grey wolf optimizer algorithm. TJEAS J-2014-4-04/373-379, 4(4):373–379
123. Emary E, Zawbaa HM, Grosan C, Hassenian AE (2015) Feature subset selection approach by gray-wolf optimization. In: Afro-European conference for industrial advancement. Springer International Publishing, pp 1–13
124. El-Gaafary AA, Mohamed YS, Hemeida AM, Mohamed AAA (2015) Grey wolf optimization for multi input multi output system. Univ J Commun Netw 3(1):1–6
125. Huang SJ, Liu XZ, Su WF, Tsai SC, Liao CM (2014) Application of wolf group hierarchy optimization algorithm to fault section estimation in power systems. In: IEEE international symposium on circuits and systems (ISCAS), June 2014, pp 1163–1166

Printed in the United States
By Bookmasters